AF539698

ADVANCED DIFFERENTIAL EQUATIONS

DPH MATHEMATICS SERIES

ADVANCED DIFFERENTIAL EQUATIONS

(For M.A., M.Sc., I.A.S., P.C.S.)

By

A.K. Sharma

DISCOVERY PUBLISHING HOUSE
NEW DELHI-110002

Published by:
Namit Wasan
DISCOVERY PUBLISHING HOUSE PVT. LTD.
4383/4B, Ansari Road, Darya Ganj
New Delhi-110 002 (India)
Phone : +91-11-23279245; 23253475; 43596065
E-mail : discoverybooksindia@gmail.com
discoverypublishinghouse@gmail.com
namitwasan9@gmail.com
web : www.discoverypublishinggroup.com

***Edition:* 2020**

ISBN: 978-81-7141-826-8

Advanced Differential Equations

Printed at:
Infinity Imaging Systems
Delhi

Preface

Differential Equation (For post graduate students), The field which has ever explaining-frontiers and seek even to prepare a sequential compendium of all the developments being reported from day to day in this field and to include these all in a small volume like, this is truly a monumental task, we therefore had no choice but to make a selection of a subject matter to suit some specific requirements. Our choice of selection of the subject matter for the present volume has been chiefly government by our adherence to the aim of covering the syllabi prescribed by the various Indian Universities for the students offering mathematics at the post graduate level. All along, it has been our endeavour to present the subject matter in a simple and lucid style and to arrange. It in a sequential way to that the reader may not feel any discreteness in going through the book from beginning to the end.

Suggestions for the improvement of the book shall be grateful acknowledged and incorporated in the next edition.

The author are highly thankful to Mr. Tilak Wasan (Managing Director Discovery Publishing House, and Mr. Prempal Singh (Prop. Unique Computers) for bringing out this volume expeditiously in an elegant form.

A.K. Sharma

Contents

Page

Preface

1. Solution in Series 1

Introduction, General Method of Solving a Differential Equation by the Method of 'Solution of Series', Case I: Roots of Indicial Equation Equal, Case II: Roots of Indicial Equation Unequal and Differing by a Quantity not an Integer, Case III: Roots of Indicial Equation Differing by an Integer Making Coefficient of y Infinity, Method of Differentiation, Some Cases where the Method Fails, The Particular Integral, Series Solution about a Particular Point.

2. Picard's Iteration Methods (Uniqueness and Existence Theorem) 44

Picard Iteration Method, Existence and Uniqueness of Solution, The Lipschitz Condition, Existence Theorem, Uniqueness Theorem, Existence and Uniqueness Theorem (The General Case).

3. Partial Differential Equations First Order 73

Introduction, Derivation of Partial Differential Equation, Definitions, Linear Partial Differential Equations of Order One, Lagrange's Linear Equation, Lagrange's Solution of the Linear Equation, Geometrical Interpretation of Lagrange's Linear Equations, The Linear Equations with a Independent Variables, Special Types of Equations, Standard I. Equations Involving only p and q and no x, y, z, Standard II. Equations Involving only p, q and z i.e., Equations of the Form f(z, p, q) = 0, Standard III. i.e., Equation of the Form f(x, p) = F (y, q), Standard IV, General Method of Solution, Two Independent Variables, Three or more Independent Variables.

4. **Partial Differential Equations with Constant Coefficients** 144

Introduction, Homogeneous Linear Equations with Constant Coefficients, Solution of Linear Partial Differential Equation, To Find the Complementary Function, When the Auxiliary Equation has Equal (Repeated) Roots, The Particular Integral, Short Method, Exceptional Case. When f(a, b) = 0, General Method, Non Homogeneous Linear Equations With Constant Coefficients, Particular Integrals, Equations Reducible to Homogeneous Linear Form.

5. **Partial Differential Equations of Order Two** 189

Introduction, Canonical Forms.

6. **Monge's Methods of Integration** 217

Monge's Method of Integrating Rr + Ss + Tt = V, Monge's Method of Integrating.

7. **Linear Partial Differential Equations** 253

Linear Partial Differential Equation, Homogeneous Linear Equations, Method of Solution, Equations Reducible to Homogeneous Form.

8. **Singular Solution** 275

Singular Solution, Discriminant, Extraneous Loci, Important.

9. **Exact Differential Equations (Particular Forms)** 299

Exact Differential Equations, Condition of Exactness for a Linear Differential Equation of Order n, Integrating Factors, Non-Linear Equation, An Equation Which does not Contain x Directly, An Equation Which does not Contain y Directly, An Equation of the Form $d^ny/dx^n = f(x)$, An Equation of the Form $d^2y/dx^2 = f(y)$, An Equation of the Form, Equations in Which Order of the Differential Coefficients Differ by Unity.

10. **Numerical Integration** 319

Introduction, Simpson's Rule.

11. **Wave, Heat, Laplace and Diffusion Equations** 324

One-Dimensional Wave Equation, Two-Dimensional Wave Equation, Heat Equation, Laplace's Equation, Laplace's Equation in Terms of Spherical Coordinates, Laplace's Equation in Terms of Cylindrical Coordinates, Diffusion Equation.

12. Boundary Value Problems **336**

A Boundary Value Problem, Solution by Separation of Variables, Solution of One Dimensional Wave Equation, Solution of Two Dimensional Wave Equation, Vibration of a Circular Membrane, Solution of One-dimensional Heat Equation, Solution of Two Dimensional Laplace's Equation, Solution of Two Dimensional Heat Equation, Solution of Two-Dimensional Laplace's Equation Given n the Cylindrical Coordinates, Solution of the Laplace's Equation Given in Cylindrical Coordinates, Solution of Laplace's Equation Given in Spherical Coordinates, Solution of Three Dimensional Wave Equation in the Spherical Polar Coordinates, Solution of the Diffusion Equation. (By the Method of the Separation of Variables), Solution of the Diffusion Equation, Solution of the Diffusion Equation.

13. Linear Differential Equations (Matrix Method) **418**

Differential Equation, To Write a Differential Equation as a Relation Among the Value of y, Linear Differential Equation (Definition), Linear Differential Equation with Constant Coefficients, Order of a Linear Differential Equation, Solution of a Differential Equation, Solution of the Equation, Finite Linear Combination of Solution, Fundamental Set of Solution (or Linearly Independent Solution), General Solution of the Homogeneous Equation of Order 2, General Solution of the Homogeneous Differential Equation of Order n, Particular Solution of the Complete Diff. Equation, Method of Undetermined Coefficients to Find the Particular Solution, Special Operator Methods to Find the Particular Solution, Solution of Simultaneous Differential Equation, Matrix Method for Solving a System of Linear Differential Equation, Working Method for Solving a Second Order Homogeneous Differential Equation with Constant Coefficients.

14. Cobweb Phenomenon and Generating Functions **467**

Cobweb Phenomenon (or Cobweb Cycles), Generating Functions, Some Special Generating Functions, The Linearly Property of the Generating Function Transformation, Generating Function Method of Solving a Linear Differential Equation,

1

Solution in Series

1.1 Introduction

For solving such equations we start with the assumption that it is possible to find a convergent series arranged according to the powers of the independent variable, which approximately satisfies the equation i.e., express the value of the variable.

Let us consider linear equation

$$P_0 \frac{d^n y}{dx^n} + P_1 \frac{d^{n-1} y}{dx^{n-1}} + \ldots P_n y = 0$$

have a solution of the form

$$y = A_0 x^{m_0} + A_1 x^{m_1} + A_2 x^{m_2} + \ldots + A_r x^{m_r}$$

where the series is either finite or an infinite convergent series for some value or values of x. Now we have to find to initial term, the relation between the powers of x and the relation between the co-efficients.

Example 1:

Integrate in series

$$2x^2 \frac{d^2 y}{dx^2} - x \frac{dy}{dx} + (1 - x^2)\, y = x^2. \qquad \ldots(1)$$

(Rohilkhand 97; Meerut 94(P), 91(P), 96(P), 96(BP), 97(BP))

Solution:

First we shall find the C.F. of the given equation (1) i.e., the solution of the equation

$$2x^2 \frac{d^2y}{dx^2} - x \frac{dy}{dx} + \left(1 - x^2\right) y = 0 . \qquad ...(2)$$

Putting x^m for y in the L.H.S. of the equation, we have

$$2x^2.m\ (m - 1)\ x^{m-2} - x.mx^{m-1} + (1 - x^2)\ x^m$$

$$\therefore \quad (2m^2 - 3m + 1)\ x^m - x^{m+2}$$

Clearly the common difference of powers is two

$\therefore$ Let the solution of (2) be

$$y = \sum_{r=0}^{\infty} A_r\, x^{m+2r}$$

so that $\dfrac{dy}{dx} = \sum_{r=0}^{\infty} A_r\ (m + 2r)\ x^{m+2r-1}$

and $\dfrac{d^2y}{dx^2} = \sum_{r=0}^{\infty} A_r\ (m + 2r)\ (m + 2r - 1)\ x^{m+2r-2}$.

Substituting in (2), we have

and $\dfrac{d^2y}{dx^2} = \sum_{r=0}^{\infty} A_r\ (m + 2r).(m + 2r - 1)\ x^{m+2r-2}$

Substituting these values in the given equation (1), we have

$$\sum_{r=0}^{\infty} A_r\ [x^2\ (m + 2r)\ (m + 2r - 1)\ x^{m+2r-2}]$$

$$+ x\ (m + 2r)\ x^{-m+2r-1} + (x^2 - 4)\ x^{m+2r}] = 0$$

$$\therefore \sum_{r=0}^{\infty} A_r\ [\{(m + 2r)^2 - 4\}\ x^{m+2r} + x^{m+2r+2}] = 0$$

$$\therefore \quad \sum_{r=0}^{\infty} A_r\ [(m + 2r - 2)\ (m + 2r + 2)\ x^{m+2r} + x^{m+2r+2}] = 0 \qquad ...(3)$$

which being an identity, we can equate to zero the coefficients of various powers of x.

$\therefore$ Equating to zero the coefficients of the lowest power of x, i.e., of x^m, we have

$$A_0\ (m - 2)\ (m + 2) = 0.$$

Now, $A_0 \neq 0$ as it is the coefficients of the first term with which we start to write the series

$$\therefore \quad m = -2 \Rightarrow 2$$

the difference of the roots of indicial equation = 4 (an integer).

Again equating to zero the coefficient of the general term i.e., of x^{m+2p}, we have

$$A_p (m + 2p - 2)(m + 2p + 2) + A_{p-1} = 0$$

$$A_p = \frac{-1}{(m+2p-2)(m+2p+2)} A_{p-1} \quad ...(4)$$

Putting p = 1, 2, ..., in (3), we have

$$A_1 = \frac{1}{m.(m+4)} A_0$$

$$A_2 = \frac{1}{(m+2)(m+6)} A_1$$

$$= (-1)^2 \frac{1}{m(m+2)(m+4)(m+6)} A_0 \text{ etc.}$$

$$y = \sum_{r=0}^{\infty} A_r x^{m+2r} = A_0 x^m + A_1 x^{m+2} + A_2 x^{m+4} + ...$$

$$= A_0 x^m \left[1 - \frac{x^2}{m(m+4)} + \frac{x^4}{m(m+2)(m+4)(m+6)} - \frac{x^6}{m(m+2)(m+4)^2(m+6)(m+8)} ....\right]$$

∴ When m = –2, the coefficient becomes infinite. To overcome this difficulty replace

A_0 by (m + 2) k where k ≠ 0.

The solution becomes

$$y = k x^m \left[(m+2) - \frac{(m+2)x^2}{m(m+4)} + \frac{x^4}{m(m+4)(m+6)} \frac{x^6}{m(m+4)^2(m+6)(m+8)} ...\right] \quad ...(4)$$

Now, putting m = – 2, one solution is

$$y = kx^{-2} \left[-\frac{x^4}{2^2.4} + \frac{x^6}{2^3.4.6} - \frac{x^8}{2^3.4^2.6.8}\right] = ku \text{ (say)}$$

Now, substituting (4) in the given differential equation (1) and on simplification, we have

$$x^2 \frac{d^2y}{dx^2} + x\frac{dy}{dx} + (x^2 - 4)\, y = kx^m\,(m+2)^2\,(m-2).$$

There is a factor $(m + 2)^2$. Proceeding as in we see that y as well as $\frac{\partial y}{\partial m}$, satisfies the differential equation when $m = -2$. Also putting $m = 2$ in (4), we get a solution. Thus, we have found three solutions of the given differential equation which is of order two only.

Now, $$\frac{\partial y}{\partial m} = kx^m \log x \left[(m+2) - \frac{(m+2)}{m\,(m+4)}x^2 \ldots\right]$$

$$+\, kx^m \left[1 - \left(\frac{1}{m\,(m+4)} - \frac{(m+2)}{m^2\,(m+4)} - \frac{(m+2)}{m\,(m+4)^2}\right)x^2 + \ldots\right]$$

$$\therefore \left(\frac{\partial y}{\partial m}\right)_{m=-2} = ku \log x + kx^{-2}\left[1 - \left\{\frac{1}{-2.2} - 0 - 0\right\}x^2 \ldots\right]$$

$$= ku \log x + kx^{-2}\left[1 + \frac{x^2}{2^2} + \frac{x^4}{2^2.4^2}\ldots\right] = kv \text{ (say)}$$

and $$(y)_{m=2} = kx^2\left[4 - \frac{4}{2.6}x^2 + \frac{x^4}{2.6.8}\ldots\right]$$

$$= -\,2^2.4^2.kx^{-2}\left[-\frac{x^4}{2^2.4} + \frac{x^6}{2^3.4.6} - \frac{x^8}{2^3.4^2.6.8}\ldots\right] = k\omega \text{ (say)}$$

$$\sum_{r=0}^{\infty} A_r\,[2\,(m+2r)(m+2r-1)x^{m+2r} - (m+2r)\,x^{m+2r} + (1-x^2)\,x^{m+2r}] = 0$$

$$\Rightarrow \sum_{r=0}^{\infty} A_r\,[-x^{m+2r+2} + (2m+4r-1)\,(m+2r-1)\,x^{m+2r}] = 0 \qquad \ldots(3)$$

which being in identity, we can equate to zero the coefficients of various powers of x.

$\therefore$ Equating to zero the coefficient of the power of x i.e. of x^m we have

$$A_0\,(2m-1)\,(m-1) = 0.$$

Now, $A_0 \neq 0$ as it is the coefficient of the first term with which we start to write the series

$$\therefore m = 1, \frac{1}{2}.$$

Now, equating to zero the coefficient of the general term i.e. of x^{m+2p+2}, we have

$$-A_p + (2m + 4p + 3)(m + 2p + 1)A_{p+1} = 0$$

$$\therefore A_{p+1} = \frac{1}{(m+2p+1)(2m+4p+3)} A_p. \qquad ...(4)$$

Putting p = 0, 1, 2, 3, ... in (4) we have $A_1 = \frac{1}{(m+3)(2m+3)} A_0$

$$A_2 = \frac{1}{(m+3)(2m+7)} A_1 = \frac{1}{(m+1)(m+3)(2m+3)(3m+7)} A_0$$

$$A_3 = \frac{1}{(m+5)(2m+11)} A_2$$

$$= \frac{1}{(m+1)(m+3)(m+5)(2m+3)(2m+7)(2m+11)} A_0, \text{ etc.}$$

$$\therefore \ y = \sum_{r=0}^{\infty} A_r \ x^{m+2r} = A_0 x^m + A_1 x^{m+2} + A_2 x^{m+4} + A_3 x^{m+6} = ...$$

$$= A_0 x^m \left[1 + \frac{1}{(m+1)(2m+3)} x^2 + \frac{1}{(m+1)(m+3)(2m+3)(2m+7)} x^4 \right.$$

$$\left. + \frac{1}{(m+1)(m+3)(m+5)(2m+3)(2m+7)(2m+11)} x^6 ... \right] \qquad ...(5)$$

Putting m = 1 and taking A_0 = a, we have

$$y = ax\left[1 + \frac{x^2}{2.5} + \frac{x^4}{2.4.5.9} + \frac{x^6}{2.4.6.5.9.13} + ...\right]$$

which is one solution of (2).

Again putting $m = \frac{1}{2}$ and taking A_0 = b, we have

$$y = bx^{1/2}\left[1 + \frac{x^2}{2.3} + \frac{x^4}{2.4.3.7} + \frac{x^6}{2.4.6.7.11} + ...\right]$$

which is another solution of the equation (2).

∴ Complete solution of equation (2) is

$$y = ax\left[1+\frac{x^2}{2.5}+\frac{x^4}{2.4.5.9}+\frac{x^6}{2.4.6.5.9.13}+...\right]$$

$$+ bx^{1/2}\left[1+\frac{x^2}{2.3}+\frac{x^4}{2.4.3.7}+\frac{x^6}{2.4.6.3.7.11}+...\right]$$

which is the C.F. of the given equation (1).

Now, to find the P.I. substitute $y = c_0\, x^m$, then from (1)

$$2c_0 m\,(m-1)\,x^m - c_0 m x^m + c_0\,(1-x^2)\,x^m = x^2$$

i.e. we should have $c_0[2m^2 - 3m + 1]\,x^m = x^2$

i.e., $m = 2$ and $c_0.3 = 1$ $\quad \therefore\ c_0 = \dfrac{1}{3}$.

$\therefore$ Putting $m = 2$, $A_0 = c_0 = \dfrac{1}{3}$ in (5). We have

$$\text{P.I.} = \frac{1}{3}x^2\left[1+\frac{x^2}{3.7}+\frac{x^4}{3.5.7.11}+...\right] = \frac{x^2}{1.3}+\frac{x^4}{1.3.3.7}+\frac{x^6}{1.3.5.3.7.11}+...$$

Hence the complete primitive of the given equation (1) is

$$y = ax\left(1+\frac{x^2}{2.5}+\frac{x^4}{2.4.5.9}+...\right) + bx^{1/2}\left(1+\frac{x^2}{2.3}+\frac{x^4}{2.4.3.7}+...\right)$$

$$+\frac{x^2}{1.3}+\frac{x^4}{1.3.3.7}+\frac{x^6}{1.3.5.3.7.11}+...$$

1.2 General Method of Solving a Differential Equation by the Method of 'Solution in Series'

Substituting $y = x^m$, **(Kanpur, 91)**

so that $\dfrac{dy}{dx} = mx^{m-1}$, $\dfrac{d^2y}{dx^2} = m\,(m-1)\,x^{m-2}$

etc., in the given equation, let us have

$$f_1(m)\ x^{m'} + f_2(x)\ x^{m''} + ...$$

Now, the successive differences of the powers of x in the series in general should be $m'' - m' = s$ (say). Then the series can be written as

$$y = A_0x^m + A_1x^{m+s} + A_2x^{m+2s} + ... + A_rx^{m+rs} + ...$$

$$\Rightarrow \qquad y = \sum_{r=0}^{\infty} A_r x^{m+rs}$$

so that $$\frac{dy}{dx} = \sum_{r=0}^{\infty} A_r (m+rs)\, x^{m+rs-1}$$

$$\frac{d^2y}{dx^2} = \sum_{r=0}^{\infty} A_r (m+rs)(m+rs-1)\, x^{m+rs-2} \text{ etc.}$$

Substituting these values in the given differential equation, we obtain an identity with its right hand member equal to zero. Equating to zero the coefficients of various powers of x we get a number of equations which determine the value or values of m and the relation among the various A's which result in giving a law by means of which A_{r+1} can be determined in terms of A_r whatever may be. Thus, all the A's are determined in terms of A_0 which remains undetermined.

This is the integration is series in ascending powers of x. Had we taken the difference m' – m" = s the resulting series would have been in descending powers of x.

For determining the initial power of x for an equation of nth order, put that coefficient equal to zero which is of nth degree in m. Thus we shall get a quadratic equation as **Indicial Equation.** From this we shall get value of m.

Now there arise following cases depending upon the nature of two roots of **Indicial equation:**

(i) The roots of indicial equation *equal.*

(ii) The roots of indicial equation *unequal* and differing by a quantity not an integer.

(iii) The roots of indicial equation *unequal* and differing by an integer making the coeff. of y infinity.

(iv) The roots of indicial equal *unequal* and differing by an integer, making a coeff. of y indeterminate.

We shall discuss these cases one by one taking examples of each case.

In case two coefficients are of nth degree, two sets of series be derived one in ascending powers of x and the other in descending powers.

In case a third coefficient is also of nth degree the terms of the series may be successively deduced : but the process will be difficult and tedious. Also with difficulty we may find very few terms if we apply this method to non linear equations.

1.3 Case I: Roots of Indicial Equation Equal

Example 1:

Solve $\left(x-x^2\right)\frac{d^2y}{dx^2}+\left(1-5x\right)\frac{dy}{dx}-4y=0.$ ...(1)

(Meerut, 95 (BP), 96; Rohilkhand, 92, 94; Agra, 91, 95, 98; Kanpur, 92, 94, 96)

Solution:

Substituting $y = x^m$ is the left hand side of the given equation, we have $(x - x^2)\, m\,(m - 1)\, x^{m-2} + (1 - 5x)\, m\, x^{m-1} - 4x^m$

$\Rightarrow \quad (m^2 + 4m + 4)\, x^m - m^2x^{m-1}$

i.e. Common difference of the powers is $m - (m - 1) = 1$.

$\therefore$ Let us assume that the solution of the given equation (1) is

$$y=\sum_{r=0}^{\infty}A_r\,x^{m+r}$$

so that $$\frac{dy}{dx}=\sum_{r=0}^{\infty}A_r\left(m+r\right)x^{m+r-1}$$

and $$\frac{d^2y}{dx^2}=\sum_{r=0}^{\infty}A_r\left(m+r\right)\left(m+r-1\right)x^{m+r-2}$$

Substituting these in the given equation (1), we have

$$\sum_{r=0}^{\infty}A_r\,[(x - x^2)\,(m + r)\,(m + r - 1)\,x^{m+r-2} + (1 - 5x)\,(m + r)\,x^{m+r-1} - 4x^{m+r}] = 0$$

$$\Rightarrow \quad \sum_{r=0}^{\infty}A_r\,[-(m + r + 2)^2\,x^{m+r} + (m + r)^2\,x^{m+r-1}] = 0. \qquad ...(2)$$

Now, equation (2), being an identity, we can equate to zero the coefficients of various powers of x.

$\therefore$ Equating to zero the coefficient of lowest power of x i.e., of x^{m-1}, we have $\quad m^2 A_0 = 0$.

Now $A_0 \neq 0$, as it is the coefficient of the first term with which we start to write the series.

$\therefore$ m = 0 (two identical values).

Equating to zero the coefficient of the next higher power of x i.e., of x^m, we have

$$-A_0\,(m + 2)^2 + A_1\,(m + 1)^2 = 0$$

$$\therefore \quad A_1 = \frac{(m+2)^2}{(m+1)} A_0.$$

Again equating to zero the coefficient of the general term i.e., of x^{m+p}, we have

$$-A_p(m+p+2)^2 + A_{p+1}(m+p+1)^2 = 0$$

$$\therefore A_{p+1} = \frac{(m+p+1)^2}{(m+p+1)} A_p. \qquad \text{...(3)}$$

Now putting p = 1, 2, ..., in (2) we have

$$A_2 = -\frac{(m+3)^2}{(m+2)^2} A_1 = \frac{(m+3)^2}{(m+1)^2} A_0.$$

$$A_3 = -\frac{(m+4)^2}{(m+3)^2} A_2 = \frac{(m+4)^2}{(m+1)^2} A_0, \text{ etc.}$$

$$\therefore y = \sum_{r=0}^{\infty} A_r \ x^{m+r} = A_0 x^m + A_1 x^{m+1} + A_2 \ x^{m+2} + \ldots$$

$$= A_0 x^m \left[1 + \left(\frac{m+2}{m+1}\right)^2 x + \left(\frac{m+3}{m+1}\right)^2 x^2 + \left(\frac{m+4}{m+1}\right)^2 x^3 + \ldots\right] \qquad \text{...(4)}$$

Putting m = 0,

$$y = A_0 \ [1 + 2^2 x + 3^2 x^2 + 4^2 x^3 + \ldots] \qquad \text{...(5)}$$

which is a solution of the given series instead of two.

Now, proceeding as in we see that $\frac{\partial y}{\partial m}$ is a second solution of the differential equation if m is put equal to zero after differentiation.

Differentiating (4), partially w.r.t. m we have

$$\frac{\partial y}{\partial m} = A_0 x^m \log x \left[1 + \left(\frac{m+2}{m+1}\right)^2 x + \left(\frac{m+3}{m+1}\right)^2 x^2 + \left(\frac{m+4}{m+1}\right)^2 x^3 + \ldots\right]$$

$$+ A_0 x^m \left[2\left(\frac{m+2}{m+2}\right)\left\{\frac{1}{m+1} - \frac{m+2}{(m+1)^2}\right\} x\right.$$

$$\left. + 2\left(\frac{m+3}{m+1}\right)\left\{\frac{1}{m+1} - \frac{m+3}{(m+1)^2}\right\} x^2 + \ldots\right]$$

$$= y \log x + A_0 x^m \left[2\left(\frac{m+2}{m+1}\right)\left\{\frac{1}{m+1} - \frac{m+2}{(m+1)^2}\right\} x + 2\left(\frac{m+3}{m+1}\right)\left\{\frac{1}{m+1} - \frac{m+3}{(m+1)^2}\right\} x^2 + \ldots \right]$$

Putting m = 0 and A_0 = a and b respectively in the two series, we have

$$y = a\,[1 + 2^2x + 3^2x^2 + 4^2x^3 + \ldots] = au \text{ (say)}$$

and $\left(\frac{\partial y}{\partial m}\right)_{m=0}$ = bu log x 2b [1.2. (1 – 2) x + 1.3. (1 – 3) x^2 + ...]

$$= bu \log x - 2b\,[1.2x + 2.3x^2 + 3.4\,x^3 + \ldots] = bv \text{ (say)}.$$

∴ The complete solution is y = au + bv.

Example 2:

Solve $x\frac{d^2y}{dx^2} + \frac{dy}{dx} + xy = 0$. *(Bassel's equation when n = 0).*

(Meerut, 98, 92, 96(P); Kanpur, 97; Raj., 91; Jiwaji, 92; Rohilkhand, 90)

Solution:

Substituting $y = x^m$ in the left hand side of the given equation, we have

$$xm(m-1)\,x^{m-2} + mx^{m-1} + x.x^m \quad \Rightarrow \quad x^{m+1} + m^2\,x^{m-1}$$

i.e., common difference of the powers is (m + 1) – (m – 1) = 2.

Let us assume that the solution of the given equation (1) is

$$y = \sum_{r=0}^{\infty} A_r\, x^{m+2r}$$

so that $\frac{dy}{dx} = \sum_{r=0}^{\infty} A_r\,(m + 2r)\, x^{m+2r-1.}$

and $\frac{dy}{dx} = \sum_{r=0}^{\infty} A_r\,(m + 2r)\,(m + 2r - 1)\, x^{-m+2r-2.}$

Substituting these values in the given equation (1), we have

$$\sum_{r=0}^{\infty} A_r\,[(m + 2r)\,(m + 2r - 1)\,x^{-m+2r-1}\,(m + 2r)\,x^{m+2r-1} + x^{m+2r+1}] = 0$$

$$\Rightarrow \quad \sum_{r=0}^{\infty} A_r\,[x^{m+2r+1} + (m + 2r)\,(m + 2r - 1 + 1)\,x^{m+2r-1}] = 0$$

$$\Rightarrow \quad \sum_{r=0}^{\infty} A_r \left[x^{m+2r+1} + (m+2r)^2 x^{m+2r-1}\right] = 0. \qquad ...(2)$$

Now, equation (2) being an identity, we can equate to zero the coefficients of various powers of x.

$\therefore$ Equating to zero the coefficient of lowest power of x i.e. of x^{m-1}, we have $A_0 m^2 = 0$.

Now, $A_0 \neq 0$, as it is the coefficient of the first term with which we start to write the series.

$\therefore \quad m = 0$ (two coincident values).

Equating to zero the co-efficients of the next higher power of x i.e. of x^{m+1}, we have

$$A_0 + A_1 (m+2)^2 = 0 \qquad \therefore \quad A_1 = -\frac{1}{(m+2)^2} A_0.$$

Again equating to zero the coefficient of the general term i.e., of x^{m+2p+1}, we have

$$A_p + (m + 2p + 2)^2 A_{p+1} = 0$$

$$\therefore \quad A_{p+1} = \frac{1}{(m+2p+2)^2} A_p. \qquad ...(3)$$

Now, putting, p = 1, 2, ... in (3), we have

$$A_2 = \frac{1}{(m+4)^2} A_1 = (-1)^2 \frac{1}{(m+2)^2 (m+4)^2} A_0$$

$$A_3 = -\frac{1}{(m+6)^2} A_2 = (-1)^2 \frac{1}{(m+2)^2 (m+4)^2 (m+6)^2} A_0, \text{ etc.}$$

$$\therefore \sum_{r=0}^{\infty} A_r^{m+2r} = A_0 x^m + A_1 x^{m+2} + A_2 x^{m+4} + ...$$

$$= A_0 x^m \left[1 - \frac{x^2}{(m+2)^2} + \frac{x^4}{(m+2)^2 (m+4)^2} ...\right] \qquad ...(4)$$

Putting m = 0.

$$y = A_0 \left[1 - \frac{x^2}{2^2} + \frac{x^4}{2^2.4^2} - \frac{x^6}{2^2.4^2.6^2} + ...\right] \qquad ...(5)$$

which is a solution of the given series.

This gives only one series instead of two.

Now, from equation (4), we have

$$\frac{dy}{dx} = A_0\left[mx^{m-1} - \frac{(m+2)\,x^{m+1}}{(m+2)^2} + \frac{(m+4)\,x^{m+3}}{(m+2)^2\,(m+4)^2}\cdots\right]$$

and $$\frac{d^2y}{dx^2} = A_0\left[m\,(m-1)\,x^{m-2} - \frac{(m+2)\,(m+1)\,x^m}{(m+2)^2} + \frac{(m+4)\,(m+3)\,x^{m+2}}{(m+2)^2\,(m+4)^2}\cdots\right]$$

Substituting these in the left hand side of the given equation, we get

$$x.A_0\left[m\,(m-1)\,x^{m-2} - \frac{(m+2)\,(m+1)\,x^m}{(m+2)^2} + \frac{(m+4)\,(m+3)\,x^{m+2}}{(m+2)^2\,(m+4)^2}\cdots\right]$$

$$+\,A_0\left[mx^{m-1} - \frac{(m+2)\,x^{m+1}}{(m+2)^2} + \frac{(m+4)\,x^{m+3}}{(m+2)^2\,(m+4)^2}\cdots\right]$$

$$+\,xA_0\left[x^m - \frac{x^{m+2}}{(m+2)^2} + \frac{x^{m+4}}{(m+2)^2\,(m+4)^2}\cdots\right]$$

i.e., $A_0\, m^2x^{-1}$ (All other terms being cancelled).

As the involves the square of m, its partial differential coefficient w.r.t. m, i.e. $2A_0\, mx^{m-1} + A_0m^2x^{m-1}\log x$, will also vanish when m = 0.

i.e., $$\frac{\partial}{dm}\left[x\frac{d^2}{dx^2} + \frac{d}{dx} + x\right]y = 2A_0mx^{m-1} + A_0m^2x^{m-1}\log x.$$

Differential operators being commutative, may be written a

$$\left[x\frac{d^2}{dx^2} + \frac{d}{dx} + x\right]\frac{\partial y}{\partial m} = 2A_0mx^{m-1} + A_0m^2x^{m-1}\log x.$$

Hence, $\frac{\partial y}{\partial m}$ is a second solution of the given differential equation if m equal to zero after differential. Differentiating (4), partially w.r.t. m, we have

$$\frac{\partial y}{\partial m} = A_0x^m\log x.\left[1 - \frac{x^2}{(m+2)^2} + \frac{x^4}{(m+2)^2\,(m+4)^2}\cdots\right]$$

$$+A_0x^m\left[\frac{2x^2}{(m+2)^3}=\left\{\frac{2}{(m+2)^2(m+4)^2}+\frac{2}{(m+2)^2(m+4)^2}-\right\}x^4+\ldots\right]$$

$$= y\log x + A_0x^m\left[\frac{2x^2}{(m+2)^2}-\frac{2x^4}{(m+2)^2(m+4)^2}\left\{\frac{1}{m+2}+\frac{1}{m+4}\right\}+\ldots\right]$$

Putting m = 0 and A_0 = a and b respectively in the two series, we have

$$y = a\left\{1-\frac{x^2}{2^2}+\frac{x^4}{2^2.4^2}\ldots\right\}=au \quad \text{(say)}$$

and $$\left(\frac{\partial y}{\partial m}\right)_{m=0} = bu\log x + b\left\{\frac{x^2}{2^2}-\frac{1}{2^2.4^2}\left(1+\frac{1}{2}\right)x^4+\ldots\right\} = bv \text{ (say)}$$

∴ The complete solution is $y = au + bv$.

Note: The equation $x\frac{d^2y}{dx^2}+\frac{dy}{dx}+xy=0$ is called Bessels equation of order zero.

Example 3:

Solve $x\frac{d^2y}{dx^2}=(1+x)\frac{dy}{dx}+2y=0$. ...(1)

(Kanpur, 97, 99)

Solution:

Substituting $y = x^m$ in the left hand side of the given equation, we have

$$xm(m-1)x^{m-2}+(1+x)mx^{m-1}+2.x^m$$

$$\Rightarrow \quad (m+2)x^m+m^2x^{m-1}$$

i.e., common difference of the powers is m – (m – 1) = 1.

∴ Let us assume that the solution of the given equation (1) is

$$y = \sum_{r=0}^{\infty} A_r x^{m+r}$$

so that $$\frac{dy}{dx}=\sum_{r=0}^{\infty}A_r(m+r)x^{m+r-1}$$

and $$\frac{d^2y}{dx^2} = \sum_{r=0}^{\infty} A_r\ (m + r)\ (m + r - 1\ 1)\ x^{m+r-2}.$$

Substituting these values in the given equation (1), we have

$$\sum_{r=0}^{\infty} A_r\ [(m + r)\ (m + r - 1)\ x^{m+r-1} + (1 + x)\ (m + r)\ x^{m+r-1} + 2x^{m+r}] = 0$$

$$\sum_{r=0}^{\infty} A_r\ [(m + r + 2)\ x^{m+r} + (m + r)^2\ x^{m+r-1}] = 0.$$

Now, equation (2) being an identity, we can equate to zero the coefficients of various powers of x.

$\therefore$ Equating to zero the coefficient of lowest power of x i.e. of x^{m-1}, we have $$A_0\ m^2 = 0$$

Now, $A_0 \neq 0$, as it is the coefficient of the first term with which we start to write the series,

$\therefore$ $m = 0$ (two coincident values).

Equating to zero the coefficient of the next higher power of x, i.e., of x^m, we have

$$A_0\ (m + 2) + A_1\ (m + 1)^2 = 0$$

$$\therefore \quad A_1 = -\frac{m+2}{(m+1)^2} A_0.$$

Again equating to zero the coefficient of the general term i.e., of x^{m+p}, we have

$$A_p\ (m + p + 2) + (m + p + 1)^2\ A_{p+1} = 0$$

$$\therefore \quad A_{p+1} = -\frac{(m+p+2)}{(m+p+1)^2} A_p. \qquad ...(3)$$

Again equating to zero the coefficient of the general term i.e. of x^{m+p}, we have

$$A_p\ (m + p + 2) + (m + p + 1)^2\ A_{p+1} = 0$$

Now, putting p = 1, 2, ... in (3), we have

$$A_2 = -\frac{(m+3)}{(m+2)^2} A_1 = (-1)^2 . \frac{(m+2)(m+3)}{(m+1)^2 (m+2)^2} A_0$$

$$A_3 = -\frac{(m+4)}{(m+3)^3} A_2 = (-1)^3 . \frac{(m+2)(m+3)(m+4)}{(m+1)^2 (m+2)^2 (m+3)} A_0 \text{ etc.}$$

$$\therefore \sum_{r=0}^{\infty} A_r\, x^{m+r} = A_0x^m + A_1x^{m+1} + A_2x^{m+2} + \ldots$$

$$= A_0x^m \left[1 - \frac{(m+2)}{(m+1)^2}x + \frac{(m+2)(m+3)}{(m+1)^2(m+2)^2}x^2 \ldots\right] \quad \ldots(4)$$

Putting m = 0

$$y = A_0 \left[1 - 2x + \frac{3}{2!}x^2 + \frac{4}{3!}x^3 + \ldots\right] \quad \ldots(5)$$

which is a solution of the given series.

This gives only one series instead of two.

Now, if we substitute the series (4) in the lift hand side of the given differential equation (without putting m = 0), we shall get the single term $A_0m^2 x^{m-1}$.

As this involves the square of m, its partial differential coefficient w.r.t m, i.e., $2A_0mx^{-1} + A_0m^2x^{m-1} \log x$ will also vanish when m = 0.

i.e.,
$$\frac{\partial}{\partial m}\left[x\frac{d^2}{dx^2} + (1+x)\frac{d}{dx} + 2\right]y = 2A_0mx^{m-1} + A_0m^2x^{m-1}\log x.$$

Differential operators being commutative, this may be written as

$$\left[x\frac{d^2}{dx^2} + (1+x)\frac{d}{dx} + 2\right]\frac{\partial y}{\partial m} = 2A_0\, mx^{m-1} + A_0m^2x^{m-1} \log x$$

Hence, $\frac{\partial y}{\partial m}$ is a second solution of the given differential equation if m is put equal to zero after differentiation.

Differentiating (4) partially w.r.t. m, we have

$$\frac{\partial y}{\partial m} = A_0x^m \log x \left[1 - \frac{(m+2)}{(m+1)^2}x + \frac{(m+2)(m+3)}{(m+1)^2(m+2)^2}x^2 \ldots\right]$$

$$+ A_0x^m\left[\left\{2.\frac{(m+2)}{(m+1)^3} - \frac{1}{(m+1)^2}\right\}x\right.$$

$$\left. + x^2\left\{\frac{1}{(m+1)^2(m+2)} - \frac{2(m+3)}{(m+1)^3(m+2)} - \frac{(m+3)}{(m+1)^2(m+2)^2}\right\} + \ldots\right]$$

$$= y \log x + A_0 x^m \left[2\left\{ \frac{(m+2)}{(m+1)} - \frac{1}{2} \right\} \frac{x}{(m+1)^2} \right.$$

$$\left. + \frac{x^2}{(m+1)^2 (m+2)} \left\{ 1 - \frac{2(m+3)}{(m+1)} - \frac{(m+3)}{m+2} \right\} \ldots \right]$$

Putting m = 0 and A_0 = a and b respectively in the two series, we have

$$y = a\left[1 - 2x + \frac{3}{2!}x^2 - \frac{4}{3!}x^3 + \ldots\right] = au \quad \text{(say)}$$

and $$\left(\frac{\partial y}{\partial m}\right)_{m=0} = bu \log x + b\left[2\left(2 - \frac{1}{2}\right)x - \frac{3}{2!}\left(-\frac{1}{3} + 2 + \frac{1}{2}\right)x^2 + \ldots\right]$$

$$= bu \log x + b\left(3x - \frac{13}{2.2!}x^2 + \ldots\right) = bv \text{ (say)}$$

∴ The complete solution is y = au + bv.

1.4 Case II: Roots of Indicial Equation Unequal and Differing by a Quantity not an Integer

Example 1:

Solve $9x(1 - x)\dfrac{d^2y}{dx^2} - 12\dfrac{dy}{dx} + 4y = 0$ ...(1)

(Meerut, 97; Agra 90; Jiwaji, 92)

Solution:

Putting x^m for y in left hand side of the equation, we have

$$9x(1 - x)\, m\,(m - 1)\, m^{m-2} - 12mx^{m-1} + 4x^m$$

$$\Rightarrow \quad \{-9m(m - 1) + 4\}\, x^m + \{9m(m - 1) - 12m\}\, x^{m-1}$$

Clearly the common difference of powers is one.

∴ Let the solution be $y = \sum_{r=0}^{\infty} A_r\, x^{m+r}$

so that $\dfrac{dy}{dx} = \sum_{r=0}^{\infty} A_r\,(m + r)\, x^{m+r-1}$

and $\dfrac{d^2y}{d^2x} = \sum_{r=0}^{\infty} A_r\,(m + r)(m + r - 1)\, x^{m+r-2}$.

Substituting these values in the given equation, we have

$$\sum_{r=0}^{\infty} A_r\,[9x(1-x)(m+r)(m+r-1)x^{m+r-2} - 12(m+r)\,x^{m+r-1} + 4x^{m+r}]=0$$

$$\Rightarrow \quad \sum_{r=0}^{\infty} [\{-9(m+r)(m+r-1)+4\}\,x^{m+r}$$

$$+ \{9(m+r)(m+r-1) - 12(m+r)\,x^{m+r-1}\}] = 0$$

$$\Rightarrow \quad \sum_{r=0}^{\infty} [\{-9(m+r)(m+r-1)+4\}\,x^{m+r-1}$$

$$-3(m+r)(3m+3r-7)\,x^{m+r-1}] \qquad ...(2)$$

Now, (2) being an identity, we can equate to zero the coefficients of various power of x.

∴ Equating to zero the coefficient of lower power of x i.e. of x^{m-1}, we have

$$-3A_0\,(m)\,(3m-7) = 0.$$

Now, A_0 ¹ 0, as it is the coefficient of the first term with which we start to write the series,

$$\therefore m = 0 \qquad \Rightarrow \qquad m = \frac{7}{3}$$

Again equating to zero the coefficient of the general term i.e. of x^{m+p}, we have

$$A_p(3m+3p-4)(3m+3p+1) - 3(3m+p+1)$$

$$(3m+3p+3-7)\,A_{p+1} = 0$$

$$\therefore A_{p+1} = A_{p+1} = \frac{(3m+3p-4)}{3(m+p+1)(3m+3p+4)}\,A_p$$

$$\Rightarrow \quad A_{p+1} = \frac{(3m+3p+1)}{3(m+p+1)}\,A_p.$$

Putting p = 0, 1, 2, 3, ... we have $A_1 = \dfrac{(3m+1)}{3(m+1)}\,A_0$

$$A_2 = \frac{(3m+4)}{3(m+2)}\,A_1 = \frac{(3m+4)(3m+1)}{3^2(m+2)(m+1)}\,A_0$$

$$A_3 = \frac{(3m+7)}{3(m+3)}\,A_2 = \frac{(3m+7)(3m+4)(3m+1)}{3(m+3)(m+2)(m+1)}\,A_0, \text{ etc.}$$

$\therefore \quad y = \sum_{r=0}^{\infty} A_r\, x^{m+r} = A_0x^m + A_1x^{m+1} + A_2x^{m+2} + A_3x^{m+3} + \ldots$

$$= A_0 x^m \left[1 + \frac{3m+1}{3(m+1)} x + \frac{(3m+1)(3m+4)}{3^2 (m+1)(m+2)} x^2 + \frac{(3m+1)(3m+4)(3m+7)}{3^3 (m+1)(m+2)(m+3)} x^3 + \ldots\right]$$

where $m = 0$, taking $A_0 = a$, we have

$$y = a\left[1 + \frac{1}{3}x + \frac{1.4}{3.6}x^2 + \frac{1.4.7}{3.6.9}x^3 + \ldots\right] = au \text{ (say)}$$

which is one solution of the given equation.

Again when $m = \frac{7}{3}$, taking $A_0 = b$, we have

$$y = b\, x^{7/3}\left[1 + \frac{8}{10}x + \frac{8.11}{10.13}x^2 + \frac{8.11.14}{10.13.16}x^3 \ldots\right] = bv \text{ (say)}$$

which is the other solution of the given equation.

Hence, the complete primitive is $y = au + bv$.

Example 2:

Solve completely in series the equation

$$x^2 \frac{d^2y}{dx^2} + x\frac{dy}{dx} + (x^2 - n^2)\, y = 0 \qquad \ldots(1)$$

(Bassel's equation) taking 2n as non-integral.

(Kanpur, 98; Meerut, 97(P))

Solution:

Substituting x^m for y in the left hand side of the given equation, we have

$$x^2 m(m-1)\, x^{m-2} + x.mx^{m-1} + (x^2 - n^2).x^m$$

$$\Rightarrow \quad \{m(m-1) + m - n^2\}\, x^m + x^{m+2}.$$

Clearly difference of powers is 2.

$\therefore$ Let the solution be $y = \sum_{r=0}^{\infty} A_r\, x^{m+2r}$

so that $\frac{dy}{dx} = \sum_{r=0}^{\infty} A_r\ (m + 2r)\ x^{m+2r-1}$

and $\frac{d^2y}{d^2x} = \sum_{r=0}^{\infty} A_r\ (m + 2r)\ (m + 2r - 1)\ x^{m+2r-2}.$

Substituting these values in the given equation, we have

$$\sum_{r=0}^{\infty} A_r\ [(m + 2r)(m + 2r - 1)\ x^{m+2r} + (m + 2r)x^{m+2r} + (x^2 - n^2)\ x^{m+2r}] = 0$$

$$\Rightarrow \quad \sum_{r=0}^{\infty} A_r\ [\{m + 2r)^2 - n^2\}\ x^{m+2r} + x^{m+2r+2}] = 0$$

$$\Rightarrow \quad \sum_{r=0}^{\infty} A_r\ [(m + 2r + n)\ (m + 2r - n)\ x^{m+2r} + x^{m+2r+2}] = 0 \qquad ...(2)$$

which being an identity, we can equate to zero the coefficients of various powers of x.

$\therefore$ Equating to zero the coefficient of lowest power of x i.e. of x^m, we have $A_0\ (m + n)\ (m - n) = 0$

Now, $A_0 \neq 0$, as it is the coefficient of the first term with which we start the series

$\therefore \quad m = n \Rightarrow - n$

difference of roots of indicial equation is 2n (not an integer).

Again equating to zero the coefficient of general term i.e. of x^{m+2p}, we have

$$A_p\ (m + 2p + n)\ (m + 2p - n) + A_{p-1} = 0$$

$$\therefore \quad A_p = -\frac{1}{(m+2p+n)\,(m+2p-n)} A_{p-1} \qquad ...(2)$$

Putting $p = 1, 2, ...,$ we have $A_1 = -\frac{1}{(m+2+n)\,(m+4-n)} A_0$

$$A_2 = -\frac{1}{(m+4+n)\,(m+4-n)} A_1$$

$$= (-1)^2 \frac{1}{(m+4+n)\,(m+4-n)\,(m+2+n)\,(m+2-n)} A_0, \text{ etc.}$$

$$\therefore\ y = \sum_{r=0}^{\infty} A_r\ x^{m+2r} = A_0x^m + A_1x^{m+2} + A_2x^{m+4} + ...$$

$$= A_0 x^m \left[1 - \frac{1}{(m+2+n)(m+2-n)} x^2 + \frac{1}{(m+4+n)(m+4-n)(m+2+n)(m+2-n)} x^4 \ldots \right]$$

$\therefore$ When m = n, taking A_0 = a, we have

$$y = ax^n \left[1 - \frac{1}{2(2n+2)} x^2 + \frac{1}{2.4(2n+2)(2n+4)} x^4 \ldots \right]$$

$$= ax^n \left[1 - \frac{x^2}{2^4 .1!(n+1)} + \frac{x^4}{2^4 2!(n+1)(n+2)} \ldots \right] = au \text{ (say)}$$

which is one solution of the given equation.

Again when m = – n, taking A_0 = b, we have

$$y = bx^{-n} \left[1 - \frac{1}{2(2n+2)} x^2 + \frac{2}{2.4(-2n+2)(-2n+4)} x^4 \ldots \right]$$

$$= bx^{-n} \left[\frac{x^2}{2^2 .1!(-n+1)} + \frac{x^4}{2^4 .2!(-n+1)(-n+2)} \ldots \right] = bv \text{ (say)}$$

which is the other solution of the given equation.

Hence, the complete primitive is y = au + bv.

Note: If $a = \dfrac{1}{2^n \Gamma(n+1)}$ then au

i.e. $$\frac{1}{2^n \Gamma(n+1)} x^n \left[1 + \frac{x^2}{2^2 .1!(n+1)} + \frac{x^4}{2^4 .2!(n+1)(n+2)} \ldots \right]$$

is called the Bassel's function of order n and is denoted by J_n (x).

Example 3:

Solve $(x - x^2) \dfrac{d^2 y}{dx^2} + (1-x)\dfrac{dy}{dx} = y = 0$. ...(1)

(Meerut, 96)

Solution:

Substituting $y = x^m$ in the left hand side of the given equation, we have

$(x - x^2)\, m(m-1)\, x^{m-2} + (1-x)\, mx^{m-1-x^n}$

$\Rightarrow \quad (m^2 + 1)x^m - m^2x^{n-1}$

i.e., common difference of the powers is $m - (m - 1) = 1$.

$\therefore$ Let us assume that the solution of the given equation (1) is

$$y = \sum_{r=0}^{\infty} A_r x^{m+r}$$

so that $$\frac{dy}{dx} = \sum_{r=0}^{\infty} A_r (m+r) x^{m+r-1}$$

and $$\frac{d^2y}{d^2x} = \sum_{r=0}^{\infty} A_r (m+r)(m+r-1) x^{m+r-2}$$

Substituting these values in the given equation (1), we have

$$\sum_{r=0}^{\infty} A_r [(x - x^2)(m + r)(m + r - 1) x^{m+r-2} + (1 - x)(m + r)x^{m+r-1} - x^{m+r}] = 0$$

$$\Rightarrow \quad \sum_{r=0}^{\infty} A_r [-\{(m + r)^2 + 1\} x^{m+r} + (m + r)^2 x^{m+r-1}] = 0 \qquad ...(2)$$

Now, equation (2) being an identity, we can equate to zero the coefficient of various powers of x.

$\therefore$ Equating to zero the coefficient of lowest power of x i.e. of x^{m-1}, we have $\quad A_0 m^2 = 0$.

Now, $A_0 \neq 0$, as it is the coefficient of the first term with which we start to write the series.

$\therefore \quad m = 0$ (two identical values).

Equating to zero the coefficient of the next higher power of x i.e. of x^m, we have

$$-A_0 (m^2 + 1) + A_1 (m + 1)^2 = 0$$

$$\therefore A_1 = \frac{(m^2+1)}{(m+1)^2} A_0.$$

Again equating to zero the coefficient of the general term i.e. of x^{m+p}, we have

$$-A_p \{(m + p)^2 + 1\} + A_{p+1} (m + p + 1)^2 = 0$$

$$\therefore \quad A_{p+1} = \frac{(m+p)^2+1}{(m+p+1)^2} A_p. \qquad ...(3)$$

Now, putting p = 1, 2... in (3) we have

$$A_2 = \frac{(m+1)^2}{(m+2)^2} A_1 = \frac{\{(m+1)^2+1\}(m^2+1)}{(m+2)^2(m+1)^2} A_0$$

$$A_3 = \frac{(m+1)^2}{(m+3)^2} A_1 = \frac{(m^2+1)\{(m+1)^2+1\}\{(m+2)^2+1\}}{(m+1)^2.(m+2)^2(m+3)^2} A_0 \text{ etc.}$$

$$\therefore\ y = \sum_{r=0}^{\infty} A_r x^{m+r} = A_0x^m + A_1x^{m+1} + A_2x^{m+2} + \ldots$$

$$= A_0x^m \left[1 + \frac{(m^2+1)}{(m+1)^2}x + \frac{(m^2+1)\{(m+1)^2+1\}}{(m+1)^2(m+2)^2}x^2 + \ldots\right] \qquad \ldots(4)$$

Putting m = 0

$$y = A_0\left[1 + x + \frac{2}{4}x^2 + \frac{2.5}{4.9}x^3 + \ldots\right] \qquad \ldots(5)$$

which is a solution of the given series.

This gives only one series instead of two.

Now, proceeding as in we see that $\frac{\partial y}{\partial m}$ is a second solution of the given differential equation if m is put equal to zero after differentiation.

Differentiation (4) partially w.r.t. m, we have

$$\frac{\partial y}{\partial m} = A_0x^m \log x\left[1 + \frac{(m^2+1)}{(m+1)^2}x + \frac{(m^2+1)\{(m+1)^2+1\}}{(m+1)^2(m+2)^2}x^2 \ldots\right]$$

$$+ A_0x^m\left[\left\{\frac{2m}{(m+1)^2} - \frac{2(m^2+1)}{(m+1)^3}\right\}x\right.$$

$$+\left\{\frac{2m\{(m+1)^2+1\} - (m^2+1)2(m+1)}{(m+1)^2(m+2)^2} - \frac{2(m^2+1)\{(m+1)^2+1\}}{(m+1)^3(m+2)^2}\right.$$

$$\left.\left. - \frac{2(m^2+1)\{(m+1)^2+1\}}{(m+1)^2(m+2)^3}\right\}x^2 + \ldots\right]$$

$$= y \log x + A_0 x^n \left[\frac{2}{(m+1)^2} \left\{ m - \frac{(m^2+1)}{(m+1)} \right\} x \right.$$

$$+ \frac{2x^2}{(m+1)^2 (m+2)^2} \left\{ m \left\{ (m+1)^2 + 1 \right\} + (m^2+1)(m+1) \right.$$

$$\left. \left. - \frac{(m^2+1)\left\{(m+1)^2+1\right\}}{(m+1)} - \frac{(m^2+1)\left\{(m+1)^2+1\right\}}{(m+2)} \right\} + \ldots \right]$$

Putting m = 0 and A_0 = a and b respectively in the two series we have

$$y = a\left[1 + x + \frac{2}{4}x^2 + \frac{2.5}{4.9}x^3 + \ldots\right] = au \text{ (say)}$$

and $$\left(\frac{\partial y}{\partial m}\right)_{m=0} = bu \log x + b\left[2(0-1)x + \frac{2x^2}{4}\left(0+1-2-\frac{2}{2}\right) + \ldots\right]$$

$$= bu \log x + b\left[-2x - x^2 - \frac{14}{27}x^3 \ldots\right] = bv \text{ (say)}$$

∴ The complete primitive is y = au + bv.

Note: In general, if the indicial equation has two equal roots m = α, we get the two independent solutions by substituting m = α, in y and $\frac{\partial y}{\partial m}$. The second solution will always consist of the product of the first solution (or its numerical) value and log x, added to another series.

Example 4:

Solve $$(2x + x^3)\frac{d^2y}{dx^2} - \frac{dy}{dx} - 6xy = 0 \quad \ldots(1)$$

(Rohilkhand, 93, 95, 98; Meerut, 91, 92(P); Agra, 92)

Solution:

Substituting x^m for y in left hand side of the given equation, we have

$(2x \quad x^3)\, m(m-1)\, x^{m-2} - mx^{m-1} - 6x^{m+1}$

⇒ $(2m^2 - 2m - m)\, x^{m+1} + (m^2 - m - 6)\, x^{m+1}$.

Clearly difference of powers is 2.

∴ Let the solution be $\sum_{r=0}^{\infty} A_r\, x^{m+2r}$

so that $\frac{dy}{dx} = \sum_{r=0}^{\infty} A_r (m + 2r) x^{m+2r-1)}$

and $\frac{d^2y}{d^2x} = \sum_{r=0}^{\infty} A_r (m + 2r) (m + 2r - 1) x^{m+2r-1}$.

Substituting these values in the given equation (1), we have

$$\sum_{r=0}^{\infty} A_r [(2x + x^3) (m + 2r) (m + 2r - 1) x^{m+2r-2} - (m + 2r) x^{m+2r-1} - 6x^{m+2r+1}] = 0$$

$$\Rightarrow \sum_{r=0}^{\infty} A_r [\{(m + 2r)^2 - (m + 2r) - 6\} x^{m+2r+1} + \{2 (m + 2r)^2 - 3 (m + 2r)\} x^{m+2r-1}] = 0$$

$$\Rightarrow \sum_{r=0}^{\infty} A_r [(m + 2r - 3) (m + 2r + 2) x^{m+2r+1} + (m + 2r) (2m + 4r - 3) x^{m+2r-1}] = 0 \quad ...(2)$$

which being an identity, we can equate to zero the coefficients of various powers of x.

∴ Equating to zero the coefficient of lower power of x i.e. of x^{m-1}, we have

$$A_0(m) (2m - 3) = 0.$$

Now, $A_0 \neq 0$, as it is the coefficient of the general term i.e. x^{m+2p+1}, we have

$$A_p(m + 3p - 3)(m + 2p + 2) + (m + 2p + 2)(2m + 4p + 4 - 3)A_{p+1}=0$$

$$A_{p+1} = -\frac{(m+2p-3)}{(m+2p+2)} \frac{(m+2p+2)}{(2m+4p+1)} A_p$$

$$\Rightarrow A_{p+1} = -\frac{m+2p-3}{2m+4p+1} A_p \quad ...(3)$$

Putting p = 0, 1, 2, ..., we have $A_1 = -\frac{m-3}{2m+1} A_0$

$$A_2 = -\frac{m-1}{2m+5} A_1 = (-1)^2 \frac{(m-3)(m-1)}{(2m+1)(2m+5)} A_0$$

$$A_3 = -\frac{m-1}{2m+9} A_2 = (-1)^3 \frac{(m-3)(m-1)(m+1)}{(2m+1)(2m+5)(2m+9)} A_0, \text{ etc.}$$

$$\therefore \sum_{r=0}^{\infty} A_r\, x^{m+2r} = A_0x^m + A_1x^{m+2} + A_2x^{m+4} + A_3x^{m+6} + \ldots$$

$$= A_0x^m\left[1-\frac{m-3}{2m+1}x^2+\frac{(m-3)(m+1)}{(2m+1)(2m+5)}x^4\right.$$

$$\left.-\frac{(m-3)(m-1)(m+1)}{(2m+1)(2m+5)(2m+9)}x^4\ldots\right]$$

$\therefore$ When $m = 0$, taking $A_0 = a$ we have

$$y = a\left[1+3x^2+\frac{3}{5}x^4-\frac{1}{15}x^6-\ldots\right] = au \quad \text{(say)}$$

which is one solution of the given equation

Again when $m = \frac{3}{2}$, taking $A_0 = b$, we have

$$y = bx^{3/2}\left[1+\frac{3}{8}x^2-\frac{1.3}{8.16}x^4+\frac{1.3.5}{8.16.24}x^6\ldots\right] = bv \text{ (say)}$$

which is other solution of the given equation.

Hence, the complete primitive is $y = au + bv$.

Example 5:

Solve in descending powers of x

$$\left(1-x^2\right)\frac{d^2y}{dx^2}-2x\frac{dy}{dx}+n(n+1)y=0 \qquad \ldots(1)$$

(Legendre's equation).

(Rohilkhand, 99; Meerut, 90, 94, 97 (R); Osmania, 98; Kanpur, 94; Raj., 90)

Solution:

Substituting x^m for y in the left hand side of the equation, we have

$$(1 - x^2)\, m\, (m - 1)\, x^{m-2} - 2x.mx^{m-1} + n\,(n + 1)\, x^m$$

$$\Rightarrow \quad (- m^2 - m + n^2 + n)\, x^m + m\,(m - 1)\, x^{m-2}.$$

Clearly the common difference of powers is 2.

$\therefore$ Let the solution be $y = \sum_{r=0}^{\infty} A_r\, x^{m-2r}$

so that $\dfrac{dy}{dx} = \sum_{r=0}^{\infty} A_r\,(m - 2r)\, x^{m-2r-1}$

and $$\frac{d^2y}{dx^2} = \sum_{r=0}^{\infty} A_r\,(m-2r)(m-2r-1)\,x^{m-2r-2}.$$

Substituting these values in the given equation, we have

$$\sum_{r=0}^{\infty} A_r\,[(1-x^2)(m-2r)(m-2r-1)\,x^{m-2r-2} - 2x\,(m-2r)\,x^{m-2r-1} + n(n+1)\,x^{m-2r}] = 0$$

$$\Rightarrow \sum_{r=0}^{\infty} A_r\,[\{-(m-2r)(m-2r-1) - 2(m-2r) + n(n+1)\}x^{m-2r} + (m-2r)(m-2r-1)\,x^{m-2r-2}] = 0$$

$$\Rightarrow \sum_{r=0}^{\infty} A_r\,[\{n^2-(m-2r)^2 + (n-m+2r)\}x^{m-2r} + (m-2r)(m-2r-1)\,x^{m-2r-2}] = 0$$

$$\Rightarrow \sum_{r=0}^{\infty} A_r\,[(n-m+2r)(n+m-2r+1)\,x^{m-2r} + (m-2r)(m-2r-1)\,x^{m-2r-2}] = 0 \qquad ...(2)$$

which being an identity, we can equate to zero the coefficients of various powers of x.

∴ Equating to zero the coefficient of highest power of x i.e. of x^m, we have

$$A_0\,(n-m)(n+m+1) = 0.$$

Now $A_0 \neq 0$, as it is the coefficient of the first term with which we start to write the series

$$\therefore \quad m = n$$

$$\Rightarrow \quad -(n+1).$$

Again equating to zero the coefficient of the general term i.e. of x^{m-2p}, we have

$$A_p(n-m+2p)(n+m-2p+1) + (m-2p+2)(m-2p+1)A_{p-1} = 0.$$

$$\therefore\ A_p = \frac{(m-2p+2)(m-2p+1)}{(n-m+2p)(n+m-2p+1)}\,A_{p-1} \qquad ...(3)$$

Putting, p = 1, 2, ..., we have

$$A_1 = \frac{m.(m-1)}{(n-m+2)(n+m-1)}\,A_0$$

$$A_2 = -\frac{(m-2)(m-3)}{(n-m+4)(n+m-3)} A_0$$

$$= (-1)^2 \frac{m(m-1)(m-2)(m-3)}{(n-m+2)(n-m+4)(n+m-1)(n+m-3)} A_0 \text{ etc.}$$

$$\therefore\ y = \sum_{r=0}^{\infty} A_r x^{m-2r} = A_0 x^m + A_1 x^{m-2} + A_2 x^{m-4} + \ldots$$

$$= A_0 \left[x^m - \frac{m.(m-1)}{(n-m+2)(n+m-1)} x^{m-2} + \frac{m(m-1)(m-2)(m-3)}{(n-m+2)(n-m+4)(n+m-1)(n+m-3)} x^{m-4} \ldots \right]$$

when $m = n$, taking $A_0 = a$, we have

$$y = a\left[x^n - \frac{n(n-1)}{2(2n-1)} x^{n-2} + \frac{n(n-1)(n-2)(n-3)}{2.4.(2n-1)(2n-3)} x^{n-4} + \ldots \right] = au \quad \text{(say)}$$

which is one solution of the given equation

Again when $m = -n-1$, taking $A_0 = b$, we have

$$y = b\left[x^{-n-1} + \frac{(n+1)(n+2)}{2.(2n+3)} x^{-n-3} + \frac{(n+1)(n+2)(n+3)(n+4)}{2.4.(2n+3)(2n+5)} x^{-n-5} + \ldots \right] = bv \text{ (say)}$$

which is the other solution of the given equation.

Hence the complete primitive is $y = au + bv$.

Note: If we replace a by $\frac{1.3\ldots(2n-1)}{n!}$ and b by $\frac{n!}{1.3\ldots(2n+1)}$ then au and bv are defined by $P_n(x)$ and $Q_n(x)$ respectively and they are called Lengendre's Polynomials.

Example 6:

Solve $\quad (2+x^2)\dfrac{d^2y}{dx^2} + x\dfrac{dy}{dx} + (1+x)\,y = 0.$ ...(1)

Solution :

Let us assume the solution of (1) as $y = \sum_{r=0}^{\infty} A_r x^{m+r}$,

so that $\frac{dy}{dx} = \sum_{r=0}^{\infty} A_r (m + r) x^{m+r-1}$

and $\frac{d^2y}{d^2x} = \sum_{r=0}^{\infty} A_r (m + r)(m + r - 1) x^{m+r-2}$

Substituting these values in (1), we have

$$\sum_{r=0}^{\infty} A_r [(2 + x^2)(m + r)(m + r - 1) x^{m+r-2}$$
$$+ x.(m + r) x^{m+r+1} + (1 + x) x^{m+r}] = 0$$

$$\Rightarrow \sum_{r=0}^{\infty} A_r [(m + r)(m + r - 1) + (m + r) + 1\} x^{m+r} + x^{m+r+1}$$
$$+ 2(m + 2)(m + r - 1) x^{m+r-2}] = 0$$

$$\Rightarrow \sum_{r=0}^{\infty} A_r [x^{m+r+1} + \{(m + r)^2 + 1\} x^{m+r}$$
$$+ 2(m + r)(m + r - 1) x^{m+r-2}] = 0 \qquad ...(2)$$

Now (2) being an identity, we can equate to zero the coefficients of various powers of x.

∴ Equating to zero the coefficient of the lowest power of x, i.e. of x^{m-2}, we have

$$2A_0 m (m - 1) = 0.$$

Since A_0 ' 0, as it is the coefficient of the first term with which we start to write the series.

∴ $m = 0, m = 1$.

Equating to zero the coefficient of next higher power of x, i.e. of x^{m-1}, we have

$$2A_1 (m + 1) m = 0 \qquad ...(3)$$

the coefficient of A_1 vanishes when $m = 0$.

∴ A_1 is indeterminant

Equating to zero the coefficient of x^m, we have

$A_0 (m^2 + 1) + 2A_2 (m + 2)(m + 1) = 0$

$$\therefore A_2 = -\frac{(m^2 + 2)}{2(m+1)(m+2)} A_0.$$

Again equating to zero the coefficient of the general term, i.e., of x^{m+p}, we have

$A_{p-1} + \{(m + p)^2 + 1\} A_p + 2(m + p + 2)(m + p + 1) A_{p+2} = 0$

$2(m + p + 1)(m + p + 2) A_{p+2} = -A_{p-1} - \{(m + p)^2 + 1\} A_p$ ...(4)

Putting p = 1, 2, ... in (4), we have

$2(m + 2)(m + 3) A_3 = -A_0 - \{(m + 1)^2 + 1\} A_1$

$$\therefore A_3 = \frac{-1}{2(m+2)(m+3)} A_0 - \frac{\left\{(m+1)^2+1\right\}}{2(m+2)(m+3)} A_1$$

and $2(m + 3)(m + 4) A_4 = -A_1 - \{(m + 2)^2 + 1\} A_2$

$$= -A_1 + \frac{(m^2+4m+5)(m^2+1)}{2(m+1)(m+2)} A_0$$

$$\therefore A_4 = -\frac{1}{2(m+3)(m+4)} A_1 + \frac{(m^2+1)(m^2+4m+5) A_0}{2^2(m+1)(m+2)(m+3)(m+4)}, \text{ etc.}$$

$$\therefore y = \sum_{r=0}^{\infty} A_r x^{m+r} = A_0x^m + A_1x^{m+1} + A_2x^{m+2} + A_3x^{m+3} + A_4x^{m+4} + \ldots$$

$$= A_0x^m + A_1x^{m+1} - \frac{(m^2+1)}{2(m+1)(m+2)} A_0 x^{m+2}$$

$$-\left[\frac{A_0}{2(m+2)(m+3)} + \frac{(m+1)^2+1}{2(m+2)(m+3)} A_1\right] x^{m+3}$$

$$+\left[-\frac{A_1}{2(m+3)(m+4)} + \frac{(m^2+1)(m^2+4m+5)}{2^2(m+1)(m+2)(m+3)(m+4)} A_0\right] x^{m+4} + \ldots$$

$$= A_0 x^m \left[1 - \frac{(m^2+1)}{2(m+1)(m+2)} x^2 - \frac{1}{2(m+2)(m+3)} x^3 + \frac{(m^2+1)(m^2+4m+5)}{2^2(m+1)(m+2)(m+3)(m+4)} x^4 \ldots\right]$$

$$+ A_1 x^m \left[x - \frac{(m+1)^2+1}{2(m+2)(m+3)} x^3 - \frac{1}{2(m+3)(m+4)} x^4 \ldots\right]$$

Putting m = 0 and taking A_0 = a and A_1 = b, we have

$$y = a\left[1-\frac{1}{4}x^2-\frac{1}{12}x^3+\frac{5}{96}x^4...\right]+b\left[x-\frac{1}{6}x^3-\frac{1}{24}x^4...\right] \quad ...(5)$$

which contains two arbitrary constants, so it may be taken as the complete primitive.

If we take m = 1, then from (3) we have $A_1 = 0$

$$\therefore \quad y = A_0 x\left[1-\frac{1}{6}x^2-\frac{1}{24}x^3...\right]$$

which is a constant multiple of the second series in the solution given by equation (5).

Hence, equation (5) is the complete primitive.

Note: In general if the Indicial Equation has two roots m_1 and m_2 (say) $m_1 > m_2$ differing by an integer, and if one of the coefficients of y becomes indeterminate when $m = m_2$, the complete primitive is given by putting $m = m_1$, in y, which contains two arbitrary constants. By putting $m = m_1$, in y, we get series which is merely a constant multiple of one of the series contained in the first solution.

Example 7:

Solve $x^2\dfrac{d^2y}{dx^2}+x\dfrac{dy}{dx}+(x^2-1)y=0$. ...(1)

(Bassel's equation of order unity)

(Gorakhpur, 94; Rohilkhand, 90; Meerut, 95, 97; Kanpur, 92)

Solution:

Substituting x^m for y, in L.H.S. of the given equation

we have $x^2.m\,(m-1)\,x^{m-2} + x.m\,x^{m-1} + (x^2+1)\,x^m$

$\Rightarrow \quad (m^2-1)\,x^m + x^{m+2}$.

Clearly the difference of powers is 2

$\therefore$ Let the solution be $y = \sum_{r=0}^{\infty} A_r\, x^{m+2r}$

so that $\dfrac{dy}{dx} = \sum_{r=0}^{\infty} A_r\,(m+2r)\,x^{m+2r-1}$

so that $\dfrac{d^2y}{dx^2} = \sum_{r=0}^{\infty} A_r\ (m + 2r)(m + 2r - 1)\ x^{m+2r-2}$.

Substituting these in the given equation (1), we have

$$\sum_{r=0}^{\infty} A_r\ [x^2.(m + 2r)(m + 2r - 1).x^{m+2r-2} + x.(m + 2r)\ x^{m+2r-1}$$
$$+ (x^2 - 1)x^{m+2r}] = 0$$

$$\Rightarrow \quad \sum_{r=0}^{\infty} A_r\ [\{(m + 2r)^2 -\}\ x^{m+2r} + x^{m+2r+2}] = 0$$

$$\Rightarrow \quad \sum_{r=0}^{\infty} A_r\ [(m + 2r - 1)(m + 2r + 1)\ x^{m+2r} + x^{m+2r+2}] = 0 \qquad ...(2)$$

Now, equation (2) being an identity, we can equate to zero the coefficients of various powers of x.

∴ Equating to zero the coefficient of the lowest power of x.

i.e., of x^m. we have $\quad A_0(m - 1)(m + 1) = 0$

Now, A_0 ¹ 0 as it is the coefficient of the first with which we start to write the series

∴ $\quad m = -1 \Rightarrow 1$,

the difference of the roots of indicial equation is 2 (an integer).

Again equating to zero the coefficient of the general term, i.e. of x^{m+2p}, we have $\quad A_p(m + 2p - 1)(m + 2p + 1) + A_{p-1} = 0$

$$\therefore A_p = -\frac{1}{(m+2p-1)(m+2p+1)} A_{p-1}$$

Putting p = 1, 2, ..., in (3), we have $A_1 = -\dfrac{1}{(m+1)(m+2)} A_0$

$$A_2 = \frac{1}{(m+3)(m+5)} A_1 = (-1)^2 \frac{1}{(m+1)(m+3)^2(m+5)} A_0$$

$$A_3 = -\frac{1}{(m+5)(m+7)} A_2$$

$$= (-1)^3 \frac{1}{(m+1)(m+3)^2(m+5)^2.(m+7)} A_0, \text{ etc.}$$

$$\therefore y = \sum_{r=0}^{\infty} A_r\ x^{m+2r} = A_0x^m + A_1x^{m+2} + A_2x^{m+4} + A_3x^{m+6} + \ldots$$

$$= A_0 x^m \left[\frac{x^2}{(m+1)(m+3)} + \frac{x^4}{(m+1)(m+3)^2(m+5)} - \frac{x^6}{(m+1)(m+3)^2(m+5)^2(m+7)} \dots \right]$$

When m = – 1, the coefficient become infinite; to overcome this difficulty replace

A_0 by (m + 1) k where k ≠ 0.

The solution becomes

$$y = kx^m \left[(m+1) - \frac{1}{(m+3)} x^2 + \frac{1}{(m+3)^2(m+5)} x^4 - \frac{1}{(m+3)^2(m+5)^2(m+7)} x^6 + \dots \right] \quad \dots(4)$$

Now putting m = – 1, one solution is

$$y = kx^{-1} \left[-\frac{1}{2} x^2 + \frac{1}{2^2.4} x^4 - \frac{1}{2^2.4^2.6} x^6 \dots \right] = ku \text{ (say)}$$

Now, substituting equation (4) in the given differential equation (1) on simplification, we have

$$x^2 \frac{d^2y}{dx^2} + x \frac{dy}{dx} + (x^2 - 1) y = kx^m (m+1)^2 (m-1).$$

There is a factor $(m + 1)^2$, proceeding as in we see that was well $\frac{\partial y}{\partial m}$ satisfies the differential equation when m = – 1.

Also putting m = 1 in (4), we get a solution. Thus we have found three solution of the given differential equation which is of order two only.

Now

$$\left(\frac{\partial y}{\partial m}\right) = kx^m \log x \left[(m+1) - \frac{1}{(m+3)} x^2 + \frac{1}{(m+3)^2(m+5)} x^4 \dots \right]$$

$$+ kx^m \left[1 + \frac{1}{(m+3)^2} x^2 - \left\{ \frac{2}{(m+3)^3(m+5)} + \frac{1}{(m+3)^2(m+5)^2} \right\} x^4 \dots \right]$$

$$\left(\frac{\partial y}{\partial m}\right)_{m=-1} = ku \log x + kx^{-1}\left[1+\frac{1}{2^2}.x^2 - \left(\frac{2}{2^2 4}+\frac{1}{2^2.4^2}\right)x^4 \ldots\right]$$

$$= ku \log x + kx^{-1}\left[1+\frac{1}{2^2}x^2 - \frac{1}{2^2.4}\left(\frac{2}{2}+\frac{1}{4}\right)x^4 \ldots\right] = kv \text{ (say)}$$

and $(y)_{m=1} = kx\left[2-\frac{1}{4}x^2 - \frac{2}{4^2.6}x^4 - \frac{1}{4^2.6^2.7}x^6 \ldots\right] = k\omega$ (say)

Clearly $w = -4u$; we have found only two linearly independent solutions.

Hence putting $k = a$ and b in the two series the complete primitive is $y = au + bv$.

1.5 Case III: Roots of Indicial Equation Differing by an Integer Making Coefficient of y Infinity

Example 1:

Solve $(1 + x^2)\, y_2 + 2xy_1 = 0$ *where* $x = 0,\ y_1 = 1,\ y = 0.$ ...(1)

Solution:

Differentiating the given equation (1), n times by Leibnitz's theorem, we have

$$(1 + x^2)\, y_{n+2} + 2\,(n + 1)\, xy_{n+1} + n\,(n + 1)\, y_n = 0.$$

When $x = 0$, we have $(y_{n+2})_0 = -\,n\,(n + 1)\,(y_n)_0$ ...(2)

From the given equation, when $x = 0$, we have $(y_2)_0 = 0$.

Putting $n = 2, 4, \ldots$ in (2), we have $(y_2)_0 = (y_4)_0 = \ldots = 0$ (each).

Again putting $n = 1, 3, \ldots$ in (2) we have

$(y_3)_0 = 1.2\,(y_1)_0 = -\,2!$

$(y_5)_0 = -\,3.4.\,(y_3)_0 = -\,4!$

$(y_7)_0 = -\,5.6\,(y_5)_0 = -\,6!$ etc.

$$\therefore y = (y)_0 + (y_1)_0.x + (y_2)_0.\frac{x^2}{2!} + \ldots$$

$$= x - 2!.\,x - 2!.\frac{x^3}{3!} + 4!\frac{x^5}{5!} - 6!\frac{x^7}{7!} \ldots = x - \frac{x^3}{3} + \frac{x^5}{5} - \frac{x^7}{7} + \ldots$$

Example 2:

Solve $x^2\frac{d^2y}{dx^2} + x\frac{dy}{dx} + (x^2 - 4)\,y = 0$. ...(1)

(Bassel's equation of order 2) **(Meerut, 95; Kanpur, 90)**

Solution:

Substituting x^m for y, in L.H.S. of the given equation, we have

$x^2 m(m-1)x^{m-2} + x.mx^{m-1} + (x^2-4)x^m$ Þ $(m^2-4)x^m + x^{m+2})$.

Clearly the difference of powers is 2.

$\therefore$ Let the solution by $y = \sum_{r=0}^{\infty} A_r x^{m+2r}$

so that $\dfrac{dy}{dx} = \sum_{r=0}^{\infty} A_r (m+2r)x^{m+2r-1}$

Clearly $w = -64v$, so we have only two linearly independent solutions. Hence putting $k = a$ and b in the two series, complete primitive is

$y = au + bv.$

Note: The two series obtained are convergent for all values of x.

Example 3:

Transform Besseus Equation

$$x^2 \frac{d^2y}{dx^2} + x\frac{dy}{dx} + \left(x^2 - n^2\right) = y = 0 \text{ by substitution } x + \frac{1}{z}.$$

Hence show that it has no integrals that are regular in descending powers of x.

Solution:

When $x + \dfrac{1}{z}$ as, in the last example, we have

$$\frac{dy}{dx} = -z^2\frac{dy}{dz} \text{ and } \frac{d^2y}{dx^2} = z^4\frac{d^2y}{dz^2} + 2z^3\frac{dy}{dz}.$$

Substituting in (1), we have

$$\frac{1}{z^2}\left(z^4\frac{d^2y}{dz^2} + 2z^3\frac{dy}{dz}\right) + \frac{1}{z}.\left(-z^2\frac{dy}{dz}\right) + \left(\frac{1}{z^2} - n^2\right)y = 0$$

$$\Rightarrow \quad z^4\frac{d^2y}{dz^2} + z^3\frac{dy}{dz} + \left(1 - n^2z^2\right)y = 0.$$

Now, let the solution of (2) be $y = \sum_{r=0}^{\infty} A_r z^{m+r}$

so that $\dfrac{dy}{dz} = \sum_{r=0}^{\infty} A_r (m+r) z^{m+r-1}$.

and $\frac{d^2y}{dz^2} = \sum_{r=0}^{\infty} A_r (m + r)(m + r - 1) z^{m+r-2}$.

Substituting these values in equation (2), we have

$$\sum_{r=0}^{\infty} A_r [(m + r)(m + r - 1) z^{m+r+2} + (m + r) z^{m+r+2} + (1 - n^2z^2) z^{m+r}]$$

$$\Rightarrow \sum_{r=0}^{\infty} A_r [\{(m + r)^2 - n^2 z^{m+r+2} + z^{m+r}] = 0. \qquad ...(3)$$

Equating to zero the coefficient of lowest power of z, i.e. of z^m the indicial equation is

$A_0 = 0$ which has no roots as $A_0 \neq 0$.

∴ The differential equation (2) has no regular integrals in ascending powers of z.

Hence, the given equation (1) has no regular integrals in descending powers of x.

1.6 Method of Differentiation

Example 1:

$$\text{Solve}\left(1 - x^2\right)\frac{d^2y}{dx^2} - x\frac{dy}{dx} + m^2y = 0 \qquad ...(1)$$

where $x = 0,\ y = 0,\ \frac{dy}{dx} = m$. **(Meerut, 1995(BP))**

Solution:

The given equation can be written as

$(1 - x^2) y_2 - xy_1 + m^2y = 0$.

Differentiating in n times by Leibnitz's theorem, we have

$$(1 - x^2) y_{n+2} - (2n + 1) xy_{n+1} + (m^2 - n^2) y_n = 0 \qquad ...(2)$$

when x = 0, we have $(y_{n+2})_0 = (n^2 - m^2)(y_n)_0$ ...(3)

Putting n = 0, we have $(y_2)_0 = - m^2 (y)_0 = 0$.

∴ Putting n = 2, 4, ... in (3) we have

$(y_2)_0 = (y_4)_0 = (y_6)_0 = ... = 0$ (each).

Putting n = 1, 3, 5, ... etc. in (3) we have

$(y_3)_0 = (1^2 - m^2)(y_1)_0 = (1^2 - m^2) m$

$(y_5)_0 = (3^2 - m^2)(y_3)_0 = (3^2 - m^2)(1^2 - m^2) m$, etc.

$$\therefore y = (y)_0 + x\,(y_1)_0 + \frac{x^2}{2!}(y_2)_0 + \ldots \qquad \text{(Maclaurin's Theorem)}$$

$$\therefore y = mx + (1^2 - m^2)\, m\frac{x^3}{3!} + \left(3^2 - m^2\right)\left(1 - m^2\right)m\frac{x^5}{5!} + \ldots$$

1.7 Some Cases where the Method Fails

Example 1:

Transform the equation $\frac{d^2y}{dx^2} - y = 0,$ ...(1)

by the substitution $x = \frac{1}{z}$ *and show that it has no integrals that are regular in descending powers of x.*

Solution:

$$x + \frac{1}{z} \quad \therefore \quad \frac{dx}{dz} = -\frac{1}{z^2}$$

$$\frac{dy}{dx} = \frac{dy}{dz}\cdot\frac{dz}{dx} = -z^2\frac{dy}{dz}$$

and $$\frac{d^2y}{dx^2} = \frac{d}{dx}\left(-z^2\frac{dy}{dz}\right) = \frac{d}{dz}\left(-z^2\frac{dy}{dz}\right)\cdot\frac{dz}{dx}$$

$$= \left(-z^2\frac{d^2y}{dz^2} - 2z\frac{dy}{dz}\right)\left(-z^2\right) = z^4\frac{d^2y}{dz^2} + 2z\frac{dy}{dz}.$$

Substituting these values in the given equation (1), the transformed equation is

$$z^4\frac{d^2y}{dz^2} + 2z^3\frac{dy}{dz} - y = 0. \qquad \text{...(2)}$$

Let $y = \sum_{r=0}^{\infty} A_r\, z^{m+r}$ be the solution of (2)

so that $$\frac{dy}{dz} = \sum_{r=0}^{\infty} A_r\,(m + r)\, z^{m+r-1}$$

and $$\frac{d^2y}{dz^2} = \sum_{r=0}^{\infty} A_r\,(m + r)(m + r - 1)\, z^{m+r-2}.$$

Substituting these values in (2) we have

$$\sum_{r=0}^{\infty} A_r \left[(m + r)(m + r - 1) z^{m+r+2} + 2(m + r) z^{m+r+2} - z^{m+r}\right] = 0$$

$$\Rightarrow \sum_{r=0}^{\infty} A_r \left[\{(m + r)(m + r + 1)\}\right] z^{m+r+2} - z^{m+r}] = 0. \qquad ...(3)$$

Equating to zero the coefficients of the lowest power of z i.e. of z^m, the indicial equation is, $-A_0 = 0$, which has no roots as $A_0 \neq 0$.

$\therefore$ The differential equation (2) has no regular integrals in ascending powers of z.

Hence the given equation (1) has no regular integrals in descending powers of x.

1.8 The Particular Integral

If in a linear equation, right hand side is of the form x^m, then the usual method to determine the particular integral corresponding to x^m, is followed. For the clear understanding of the method see the following example.

Example 1:

Solve $x^4 \dfrac{d^2y}{dx^2} + x\dfrac{dy}{dx} + y = x^{-1}$. **(Meerut, 92; Osmania, 99)**

Solution:

Substituting $y = x^m$ in the left hand side of the given equation, we have

$$x^4.m(m - 1)x^{m-2} + x.mx^{m-1} + x^m \Rightarrow m(m-1)x^{m+2} + (m+1)x^m$$

i.e. the common difference of the powers is $(m + 2) - m = 2$.

$\therefore$ Let us assume $y = \sum_{r=0}^{\infty} A_r x^{m-2r}$ ...(1)

the C.F. of the given equation i.e. the solution of the equation

$$x^4 \frac{d^2y}{dx^2} + x\frac{dy}{dx} + y = 0 \qquad ...(2)$$

Thus, substituting the value of y from (1)(in (2), we get

$$\sum_{r=0}^{\infty} [(m-2r)(m-2r-1) A_r x^{m-2r+2} + (m-2r+1) A_r x^{m-2r}] = 0 \qquad ...(3)$$

which is an identity.

$\therefore$ Equating to zero the coefficient of highest power of x i.e. of x^{m+2}, we get $m(m - 1) A_0 = 0$

$\therefore m = 0, 1, \quad \because \quad A_0 \neq 0.$

Now, equating to zero the coefficient of the general term i.e. of x^{m-2r}, we get $(m - 2r - 2)(m - 2r - 3) A_{r+1} + (m - 2r + 1) A_r = 0$

$$\Rightarrow \quad A_{r+1} = \frac{(m-2r+1)}{(m-2r-3)(m-2r-2)} A_r . \qquad ...(3)$$

Case I. when m = 0, $A_{r+1} = \frac{2r-1}{(2r+2)(2r+3)} A_r$ ...(4)

Now putting r = 0, 1, 2, 3, ... in (4), we get

$$A_1 = -\frac{1}{2.3} A_0 = -\frac{1}{3!} A_0, \; A_2 = \frac{1}{4.5} A_1 = -\frac{1}{5!} A_0,$$

$$A_3 = \frac{3}{6.7} A_2 = -\frac{1.3}{7!} A_0, \text{ etc.}$$

$$\therefore \quad y = \sum_{r=0}^{\infty} A_r \, x^{m-2r} = \sum_{r=0}^{\infty} A_r \, x^{-2r} \qquad (\because m = 0)$$

$$= A_0 + A_1 x^{-2} + A_2 x^{-4} + A_3 x^{-6} + ...$$

$$= A\left[1 - \frac{1}{3!} x^{-2} - \frac{1}{5!} x^{-4} - \frac{1.3}{7!} x^{-6} ...\right] \text{ Taking } A_0 = A_r \qquad ...(4)$$

Case II. When m = 1, from (3), we have

$$A_{r+1} = \frac{r-1}{(r+1)(2r+1)} A_r$$

Now putting r = 0, $A_1 = - A_0$

putting r = 1, $A_2 = 0$

putting r = 2, 3, ..., $A_3 = A_4 = ... = 0$ (each)

$$\therefore y = \sum_{r=0}^{\infty} A_r \, x^{m-2r} = \sum_{r=0}^{\infty} A_r \, x^{1-2r} \qquad (\because m = 1)$$

$$= A_0 x + A_1 x^{-1} + A_2 x^{-3} + ...$$

$$= A_0 \left(1 - \frac{1}{x}\right) = B\left(x - \frac{1}{x}\right) \text{ Taking } A_0 = B.$$

$\therefore$ the C.F. of the equation is

$$A\left(1 - \frac{1}{3!} x^{-2} - \frac{1}{5!} x^{-4} - \frac{1.3}{7!} x^{-6} ...\right) + B\left(x - \frac{1}{x}\right).$$

To find P.I. To find the P.I., put $y = C_0 x^m$. Thus, we must have

$m(m-1)C_0 x^{m+2} = x^{-1}$

$\Rightarrow \quad m(m-1)C_0 = 1$ and $m + 2 = -1$

$\Rightarrow \quad m = -3$ and $C_0 = \dfrac{1}{12}$

$\therefore \quad$ Let $y = \sum_{r=0}^{\infty} C_r x^{m-2r} = \sum_{r=0}^{\infty} C_r x^{-3-2r}$

be the P.I. of the given equation.

Substituting $y = \sum_{r=0}^{\infty} C_r x^{-3-2r}$ in (2) and proceeding as above, we get

$$C_{r+1} = \frac{2(r+1)}{(2r+5)(2r+6)} C_r$$

Putting r = 0, 1, 2, ...

$$C_1 = \frac{2}{5.6} C_0,\ C_2 = \frac{4}{7.8} C_1 = \frac{2.4}{5.6.7.8} C_0, \text{ etc.}$$

$$\therefore \quad \text{P.I.} = \sum_{r=0}^{\infty} C_r x^{-3-2r} = C_0 x^{-3} + C_1 x^{-5} + C_2 x^{-7} + \ldots$$

$$= C_0 x^{-3}\left[1 + \frac{2}{5.6}x^{-2} + \frac{2.4}{5.6.7.8}x^{-4} + \ldots\right]$$

$$= \frac{1}{12}x^{-3}\left[1 + \frac{2}{5.6}x^{-2} + \frac{2.4}{5.6.7.8}x^{-4} + \ldots\right]$$

$$= 2x^{-3}\left[\frac{1}{4!} + \frac{2}{6!}x^{-2} + \frac{2.4}{8!}x^{-4} + \ldots\right]$$

Hence, the solution of the given equation is

$$y = A\left(1 - \frac{1}{3!}x^{-2} - \frac{1}{5!}x^{-4} - \frac{1.3}{7!}x^{-6} \ldots\right) + B\left(x - \frac{1}{x}\right)$$

$$+ 2x^{-3}\left(\frac{1}{4!} + \frac{2}{6!}x^{-2} + \frac{2.4}{8!}x^{-4} + \ldots\right).$$

Example 2:

Solve $\left(1 - x^2\right)\dfrac{d^2y}{dx^2} + 2x\dfrac{dy}{dx} + y = 0.$ ...(1)

(Meerut, 91(P), 97, 97(P), 97(BP))

Solution:

Let the solution of (1) be

$$y = \sum_{r=0}^{\infty} A_r x^{m+r}$$

so that $\dfrac{dy}{dx} = \sum_{r=0}^{\infty} A_r (m + r) x^{m+r-1}$

and $\dfrac{d^2y}{dx^2} = \sum_{r=0}^{\infty} A_r (m + r)(m + r - 1) x^{m+r-2}$.

Substituting in the given equation (1), we have

$$\sum_{r=0}^{\infty} A_r [(1 - x^2)(m + r - 1) x^{m+r-2} + 2x (m + r) x^{m+r-1} + x^{m+r}] = 0$$

$$\Rightarrow \quad \sum_{r=0}^{\infty} A_r [\{-(m + r)(m + r - 1) + 2(m + r) + 1\} x^{m+r} + (m + r)(m + r - 1) x^{m+r-2}] = 0$$

$$\Rightarrow \quad \sum_{r=0}^{\infty} A_r [-\{(m + r)(m + r - 3) - 1\} x^{m+r} + (m + r)(m + r - 1) x^{m+r-2}] = 0 \qquad ...(2)$$

which being an identity, we can equate to zero the coefficients of various powers of x.

Equating to zero the coefficient of lowest power of x i.e. of x^{m-2}, we have

$A_0 m (m - 1) = 0$,

Since $A_0 \neq 0$. $\therefore m = 0, 1$.

Again equating to zero the coefficient of next higher power of x, i.e. x^{m+p}, we have

$$-A_p \{(m + p)(m + p - 3) - 1\} + A_{p+2} (m + p + 2)(m + p + 1) = 0.$$

$$A_{p+2} = \frac{(m+p)(m+p-3)-1}{(m+p+1)(m+p+2)} A_p. \qquad ...(3)$$

Putting p = 0, 1, 2, ..., we have

$$A_2 = \frac{m(m-3)-1}{(m+1)(m+2)} A_0; \quad A_3 = \frac{(m+1)(m-2)-1}{(m+2)(m+3)} A_1$$

$$A_4 = \frac{(m+2)(m-1)-1}{(m+3)(m+4)} A_2 = \frac{\{(m+2)(m-1)\}\{(m+1)(m-2)-1\}}{(m+1)(m+2)(m+3)(m+4)} A_1$$

$$A_5 = \frac{(m+3)\,m-1}{(m+4)\,(m+5)}\,A_3 = \frac{\{m(m+3)-1\}\,\{(m+1)\,(m-2)-1\}}{(m+2)\,(m+3)\,(m+4)\,(m+5)}\,A_1, \text{ etc.}$$

$$\therefore\ y = \sum_{r=0}^{\infty} A_r\, x^{m+r}$$

$$= A_0x^m + A_1x^{m+1} + A_2x^{m+2} + A_3x^{m+3} + A_4x^{m+4} + A_5x^{m+5} + \ldots$$

$$= A_0x^m\left[1+\frac{m\,(m-3)-1}{(m+1)\,(m+2)}\,x^2+\frac{\{(m+2)\,(m-1)-1\}\,\{m\,(m-3)-1\}}{(m+1)\,(m+2)\,(m+3)\,(m+4)}\,x^4\ldots\right]$$

$$+ A_1x^m\left[x+\frac{(m+1)\,(m-2)-1}{(m+2)(m+3)}\,x^3\right.$$

$$\left.+\frac{\{m\,(m+3)-1\}\,\{(m+1)(m-2)-1\}}{(m+2)\,(m+3)\,(m+4)\,(m+5)}\,x^5\ldots\right]$$

Putting $m = 0$ and taking $A_0 = a$ and $A_1 = b$ we have

$$y = a\left[1-\frac{1}{2}x^2+\frac{1}{8}x^4+\ldots\right]+b\left[x-\frac{1}{2}x^3+\frac{1}{40}\,x^5\ldots\right] \qquad \ldots(4)$$

which contains two arbitrary constants and hence is the complete primitive of the given equations.

If we take $m = 1$, then, we shall get a series which is constant multiple of the second series in (4).

Hence, equations (4) is the complete primitive of (1).

1.9 Series Solution about a Particular Point

We can also find the series solution of a diff. equation about a point other than $x = 0$. To find the series solution about a point $x = a$ i.e. in powers of $x - 1$, put $x = a + z$ in the given equation and find its series solution (expansion). Finally replace $z = x - a$ to obtain the required solution.

Example 1:

Solve $\dfrac{d^2y}{dx^2} - y = x$. ...(1)

(Meerut, 1992)

Solution:

Let $y = \sum_{m=0}^{\infty} A_m x^m$ ...(2)

be the solution of (1)

$\therefore$ Substituting in equation (1), we get

$$\sum_{m=0}^{\infty} A_m [m(m-1)x^{m-2} - x^m] - x = 0 \qquad ...(3)$$

Equating coefficient of x^0, we get

$$A_2.2(2-1) - A_0 = 0$$

$$\Rightarrow \qquad A_2 = \frac{1}{2} A_0.$$

Equating coefficient of x, we get

$$A_3.3(3-1) - A_1 - 1 = 0$$

$$\Rightarrow \qquad A_3 = \frac{1}{6} + \frac{1}{6} A_1$$

Now, equating to zero, the coefficient of the general term i.e., of x^{m-2}, we get, $A_m.m(m-1) - A_{m-2} = 0$

$$\Rightarrow \qquad A_m = \frac{A_{m-2}}{m(m-1)}$$

Putting m = 4, 5, ..., we get

$$A_4 = \frac{1}{12} A_2 = \frac{1}{24} A_0$$

$$A_5 = \frac{1}{20} A_3 = \frac{1}{20}\left(\frac{1}{6} + \frac{1}{6} A_1\right), \text{ etc.}$$

$$\therefore \qquad y = \sum_{m=0}^{\infty} A_n x^n = A_2 + A_1 x^n A_2 x^2 + A_3 x^3 + ...$$

$$= A_0 + A_1 x + \frac{1}{2} A_0 x^2 + \left(\frac{1}{6} + \frac{1}{6} A_1\right) x^3$$

$$+ \frac{1}{24} A_0 x^4 + \frac{1}{20}\left(\frac{1}{6} + \frac{1}{6} A_1\right) x^5 + ...$$

$$= a\left(1 + \frac{1}{2}x^2 + \frac{1}{24}x^4 + ...\right) + b\left(x + \frac{1}{6}x^3 + \frac{1}{120}x^5 + ...\right)$$

$$+ \left(\frac{1}{6}x^3 + \frac{1}{120}x^5 + ...\right)$$

Taking A_0 = a and A_1 = b. Hence the required solution **Ans.**

EXERCISES

Solve completely in series the following equations:

1. $2x(1-x)\dfrac{d^2y}{du^2}+(1-x)\dfrac{dy}{du}+3y=0.$

[**Ans.** : $y = a\left(1 - 3x + \dfrac{3x^2}{1.3} + \dfrac{3x^3}{3.5} + \dfrac{3x^4}{5.7} + \ldots\right) + bx^{1/2}(1 - x)$]

2. $(1-x^2)\dfrac{d^2y}{du^2}-2x\dfrac{dy}{du}+2y=0.$

[**Ans.** : $y = a(1 - x^2 \dfrac{1}{2}x^4 - \dfrac{1}{5}x^6 \ldots) + bx$]

3. $\dfrac{d^2y}{du^2}-3x\dfrac{dy}{du}-3y=0.$ [**Ans.** : $y = 1+\dfrac{3}{2!}x^2+\dfrac{2}{2!}x^4+\dfrac{5.3^4}{6!}x^6 e\ldots$

Given $y = 1$, $\dfrac{dy}{du} = 1$ when $= 2$.

4. $4x\dfrac{d^2y}{du^2}+2\dfrac{dy}{du}+y=0.$ [**Ans.** : $a \log\sqrt{x} + b \sin\sqrt{x}$].

2

Picards Iteration Methods
(Uniqueness and Existence Theorem)

2.1 Picard Iteration Method (Meerut, 97, 93)

In many of the engineering problems, we are often confronted with a differential equation whose solution cannot be found by standard methods.

In such problems, it is sufficient to obtain an approximate solution only. We shall mention here the **Picards Iteration Method** for giving an approximate solution of thc initial value problem of the form

$$\frac{dy}{dx} = f(x, y),\ y(x_0) = y_0. \qquad ...(A)$$

By an iteration method we mean a method which consists of a repeated application of exactly the same type of steps where in each step we use the result of proceeding step (or steps). We now explain various steps in Picard's Method. By integrating, we may write (A) in the from

$$y(x) = y_0 + \int_{x_0}^{x} f[t,\ y(t)]\, dt \qquad ...(B)$$

where t is the variable of integration. It is easy to check that the integral is zero when $x = x_0$, so that $y = y_0$. Thus, (B) satisfies the initial condition in (A). Also if we differentiate (B), we obtain the given differential equation.

In order to obtain a solution y (x) of (B), we proceed stepwise as follows:

Put, $y = y_0$ = cons. on the right. This gives

$$y_1(x) = y_0 + \int_{x_0}^{x} f[t,\ y_0]\, dt.$$

We now substitute y_1 (x) in the same manner and get

$$y_2(x) = y_0 + \int_{x_0}^{x} f\left[t,\ y_1(t)\right] dt$$

Continuing in this way, at the nth step of integration process, we shall get

$$y_n(x) = y_0 + \int_{x_0}^{x} f\left[t,\ y_{n-1}(t)\right] dt. \qquad \text{...(C)}$$

Thus, we obtain a sequence of approximate solutions

$y_1(x), y_2(x), \ldots, y_n(x)$.

Example 1:

Apply Picard method of the initial value problem

$$\frac{dy}{dx} = xy + 1, \quad y(0) = 0$$

and find the successive approximations.

Solution:

Here $x_0 = 0$, $y_0 = 0$, $f(x.y) = xy + 1$, so that equation (C) of proceeding article becomes

$$y_n(x) = y_0 + \int_{x_0}^{x} \left[y_{n-1}(t) + 1\right] dt$$

Starting from $x_0 = 0$ and $y_0 = 0$, we get

$$y_1(x) = y_0 + \int_{x_0}^{x} \left[t\ y_0(t) + 1\right] dt = \int_0^x dt = x$$

$$y_2(x) = y_0 + \int_{x_0}^{x} \left[t\ y_1(t) + 1\right] dt = \int_0^x \left(t^2 + 1\right) dt = \frac{x^3}{3} + x$$

$$y_3(x) = y_0 + \int_{x_0}^{x} \left[t\ y_2(t) + 1\right] dt$$

$$= \int_{x_0}^{x} \left[t\left(t + \frac{1}{3}t^3\right) + 1\right] dt = x + \frac{1}{3}x^3 + \frac{1}{3.5}x^5$$

Hence, we have, $y_0 = 0$, $y_1 = x$, $y_2 = x + \frac{x_3}{3}$, $y_3 = x + \frac{1}{3}x^3 + \frac{1}{3.5}x^5$, etc.

Ans.

Example 2:

Find the third approximation of the solution of the equation

$$\frac{d^2y}{dx^2} = x^2\ d\frac{dy}{dx} + x^4 y,$$

where $y = 5$, *and* $\frac{dy}{dx} = 1$ *when* $x = 0$. **(Rohilkhand, 98)**

Solution:

Let $\frac{dy}{dx} = z$

$\therefore$ $\frac{dz}{dx} = \frac{d^2y}{dx^2} = x^2z + x^4y$

and $x_0 = 0$, $y_0 = 5$, and $z_0 = 1$

which is the same problem as.

Third approximation

$$y_3(x) = 5 + x + \frac{1}{12}x^4 + \frac{1}{6}x^6 + \frac{2}{63}x^7 + \frac{1}{72}x^9.$$

Example 3:

Find the third approximation of the solution of the equations

$$\frac{dy}{dx} = z,\ \frac{dz}{dx} = x^3(y+z)$$

by Picard's Method where $y = 1$, $z = \frac{1}{2}$, *when* $x = 0$.

(Meerut, 97, 97(P); Rohilkhand, 93, 99)

Solution:

Here the given simultaneous equations are

$$\frac{dy}{dx} = z = f(x, y, z)$$

$$\frac{dz}{dx} = x^3(y + z) = g(x, y, z) \text{ and } x_0 = 0,\ y_0 = 1,\ z_0 = \frac{1}{2}$$

$$y_n = y_0 + \int_{x_0}^{x} f\left[t, y_{n-1}(t), z_{n-1}(t)\right] dt$$

$$= y_0 + \int_{x_0}^{x} z_{n-1}(t)\, dt$$

and $$z_n = z_0 + \int_{x_0}^{x} g\left[t, y_{n-1}(t), z_{n-1}(t)\right] dt$$

$$z_0 + \int_{x_0}^{x} t^3\left[y_{n-1}(t) + z_{n-1}(t)\right] dt$$

First approximation: Taking $x_0 = 0$, $y_0 = 1$, $z_0 = \frac{1}{2}$,

$$y_1 = y_0 + \int_{x_0}^{x} z_0 \, dt = 1 + \int_0^x \frac{1}{2} \, dt = 1 + \frac{1}{2}x$$

$$z_1 = z_0 + \int_{x_0}^{x} t^3 \left(y_0 + z_0\right) dt = \frac{1}{2} + \int_0^x t^3 \left(1 + \frac{1}{2}\right) dt = \frac{1}{2} + \frac{3}{8}x^4 .$$

Second approximation :

$$y_2 = y_0 + \int_{x_0}^{x} z_1(t) \, dt = 1 + \int_0^x \left(\frac{1}{2} + \frac{3}{8}t^4\right) dt = 1 + \frac{1}{2}x + \frac{3}{40}x^5$$

$$z_2 = z_0 + \int_{x_0}^{x} \left[t^3 \, y_1(t) + z_1(t)\right] dt$$

$$= \frac{1}{2} + \int_0^x t^3 \left(\frac{3}{2} + \frac{t}{2} + \frac{3}{8}t^4\right) dt = \frac{1}{2} + \frac{3}{8}x^4 + \frac{1}{10}x^5 + \frac{3}{64}x^8 .$$

Third approximation :

$$y_3 = y_0 + \int_{x_0}^{x} z_2(t) \, dt$$

$$= 1 + \int_0^x \left(\frac{1}{2} + \frac{3}{8}t^4 + \frac{1}{10}t^5 + \frac{3}{64}t^8\right) dt$$

$$= 1 + \frac{1}{2}x + \frac{3}{40}x^5 + \frac{1}{60}x^6 + \frac{1}{192}x^9$$

$$z_3 = z_0 + \int_{x_0}^{x} \left[t^3 \, y_2(t) + z_2(t)\right] dt$$

$$= \frac{1}{2} + \int_0^x t^3 \left(\frac{3}{2} + \frac{t}{2} + \frac{3}{8}t^4 + \frac{7}{40}t^5 + \frac{3}{64}t^8\right) dt$$

$$= \frac{1}{2} + \frac{3}{8}x^4 + \frac{1}{10}x^5 + \frac{3}{64}x^8 + \frac{7}{360}x^9 + \frac{1}{256}x^{12} .$$

Ans.

Example 4:

Apply Picard's Iteration Method to the initial value problem

$$\frac{dy}{dx} = y, \; y(0) = 1$$

and show that the successive approximations tend to the limit $y = e^x$*, the exact solution.*

Solution:

Here $h_0 = 0$, $y_0 = 1$, $f(x, y) = y$

so the equation (C) of proceeding article becomes

$$y_n(x) = y_0 + \int_{x_0}^{x} y_{n-1}(t)\, dt$$

Taking $y_0 = 1$, we obtain,

$$y_1(x) = y_0 + \int_{x_0}^{x} y_0(t)\, dt = 1 + \int_0^x 1.dt = 1 + x$$

$$y_2(x) = y_0 + \int_{x_0}^{x} y_1(t)\, dt = 1 + \int_0^x (1+t)\, dt = 1 + x + \frac{x^2}{2}$$

$$y_3(x) = y_0 + \int_{x_0}^{x} y_2(t)\, dt = 1 + \int_0^x \left(1 + t + \frac{t^2}{2}\right) dt = 1 + x + \frac{x^2}{2!} + \frac{x^3}{3!}$$

In general, we shall get

$$y_n(x) = 1 + x + \frac{x^2}{2!} + \dots + \frac{x^n}{n!}.$$

Thus, the successive approximations tend to the limit $y = e^x$ as $n \to \infty$, which is the exact solution.

Example 5:

Using Picard's Method find the third approximation of the solution of the equation

$$\frac{dy}{dx} = 2y - 2x^2 - 3 \text{ where } y = 2 \text{ when } x = 0.$$

(Meerut, 91(P), 93, 98; Agra 92(P) 99; Rohilkhand, 95)

Solution:

Here $x_0 = 0,\ y_0 = 2$

and $f(x, y) = 2y - 2x^2 - 3$

we have $y_n(x) = y_0 + \int_{x_0}^{x} \left[2y_{n-1}(t) - 2t^2 - 3\right] dt$

First approximation : Taking $x_0 = 0$ and $y_0 = 2$,

$$y_1(x) = y_0 + \int_0^x (2y_0 - 2t^2 - 3)\, dt$$

$$= 2 + \int_0^x (4 - 2t^2 - 3)\, dt = 2 + x - \frac{2}{3}x^3.$$

Second approximation :

$$y_2(x) = 2 + \int_0^x (2y_1 - 2t^2 - 3)\, dt$$

$$= 2 + \int_0^x \left(4 + 2t - \frac{4}{3}t^3 - 2t^2 - 3\right) dt = 2 + x + x^2 - \frac{1}{3}x^4 - \frac{2}{3}x^3$$

Third approximation :

$$y_3(x) + 2 \mid \int_0^x \left(2y_2 - 2t^2 - 3\right) dt$$

$$= 2 + \int_0^x \left(4 + 2t + 2t^2 - \frac{2}{3}t^4 - \frac{4}{3}t^3 - 2t^2 - 3\right) dt$$

$$= 2 + x + x^2 - \frac{1}{3}x^4 - \frac{2}{15}x^5.$$ **Ans.**

Example 6:

Apply Picard's Method to find the third approximation of the solution of the equation

$$\frac{dy}{dx} = x + y^2 \quad \textit{where } y = 0 \textit{ when } x = 0.$$

(Rohilkhand, 92, 94; Meerut 99, 94; Agra 93, 95)

Solution:

Here $x_0 = 0$, $y_0 = 0$ and $f(x, y) = x + y^2$

We have $y_n(x) = y_0 + \int_{x_0}^x f\left[t, y_{n-1}(t)\right] dt$

$$\therefore y_n(x) = \int_0^x \left[t + y^2{}_{n-1}(t)\right] dt \qquad ...(1)$$

First approximation : From (1), we have

$$y_1(x) = \int_0^x \left[t + y^2{}_0(t)\right] = \int_0^x t\, dt = \frac{1}{2}x^2.$$

Second approximation : From (1), we have

$$y_2(x) = \int_0^x \left[t + y_1^2(t)\right] dt = \int_0^x \left(t + \frac{1}{4}t^4\right) dt = \frac{1}{2}x^2 + \frac{1}{20}x^5$$

Third approximation : From (1), we have

$$y_3(x) = \int_0^x \left[t + y_2^2(t)\right] dt = \int_0^x \left[t + \left(\frac{1}{2}t^2 + \frac{1}{20}t^5\right)^2\right] dt$$

$$= \int_0^x \left[t + \frac{1}{4}t^4 + \frac{1}{20}t^7 + \frac{1}{400}t^{10}\right] dt$$

$$= \frac{1}{2}x^2 + \frac{1}{20}x^5 + \frac{1}{160}x^8 + \frac{1}{4400}x^{11}.$$ **Ans.**

Example 7:

Find the third approximation of the solution of the equation

$$\frac{dy}{dx} = 2 - \frac{y}{x}.$$

by Picard's Method, where $y = 2$ when $x = 1$.

(Meerut, 92, 94, 95, 97 (R); Agra, 92, 98; Rohilkhand, 90, 96)

Solution:

Here $x_0 = 1$, $y_0 = 2$ and $f(x, y) = 2 - \frac{y}{x}$.

we have $y_n(x) = y_0 + \int_{x_0}^{x} \left(2 - \frac{y_{n-1}(t)}{t}\right) dt$,

First approximation : Taking $x_0 = 1$ and $y_0 = 2$

$$y_1(x) = y_0 + \int_{x_0}^{x} \left(2 - \frac{y_0}{t}\right) dt = 2 + \int_1^x \left(2 - \frac{2}{t}\right) dt$$

$$= 2 + [2x - 2 \log t]_1^x = 2x - 2 \log x.$$

Second approximation,

$$y_2(x) = 2 + \int_1^x \left[2 - \frac{1}{t} y_1(t)\right] dt$$

$$= 2 + \int_1^x \left[2 - \frac{1}{t}(2t - 2 \log t)\right] dt = 2 + \int_1^x \frac{2}{t} \log t \, dt = 2 + (\log x)^2$$

Third approximation :

$$y_3(x) = 2 + \int_1^x \left(2 - \frac{1}{t} y_2(t)\right) dt = 2 + \int_1^x \left[2 - \frac{1}{t}.\left\{2 + (\log t)^2\right\}\right] dt$$

$$= 2 + \int_1^x \left[2 - \frac{2}{t} - \frac{1}{t}\left\{(\log t)^2\right\}\right] dt = 2x - 2 \log x - \frac{1}{3}(\log x)^3.$$

Example 8:

Find the third approximation of the solution of the equation

$$\frac{dy}{dx} = zm \; \frac{dz}{dx} = x^2 z + x^4 y$$

by Picard's Method, $y = 5$ and $z = 1$ when $x = 0$.

(Meerut, 96(P), 97(BP))

Solution:

Here the given simultaneous equations are

$$\frac{dy}{dx} = z = f(x, y, z); \quad \frac{dz}{dx} = x^2 z + x^4 y = g(x, y, z)$$

and $\quad x_0 = 0, y_0 = 5, z_0 = 1$

$$y_n = y_0 + \int_{x_0}^{x} f[t, y_{n-1}(t), z_{n-1}(t)]\, dt = y_0 + \int_{x_0}^{x} z_{n-1}(t)\, dt$$

and $\quad z_n = z_0 + \int_{x_0}^{x} g[t, y_{n-1}(t), z_{n-1}(t)]\, dt$

$$z_0 + \int_{x_0}^{x} \left[t^2 z_{n-1}(t) + t^4 y_{n-1}(t)\right] dt$$

First approximation : Taking $y_0 = 5$, $z_{=1 \text{ and } x}{}^0{}_{=0,}$

$$y_1 = y_0 + \int_{x_0}^{x} z_0(t)\, dt = 5 + \int_0^x 1.\, dt = 5 + x$$

$$z_1 = z_0 + \int_{x_0}^{x} \left[t^2 z_0(t) + t^4 . y_0(t)\right] dt$$

$$= 1 + \int_0^x \left(t^2 .1 + t^4 .5\right) dt = 1 + \frac{1}{3}x^3 + x^5 .$$

Second approximation :

$$y_2 = y_0 + \int_0^x z_1(t)\, dt = 5 + \int_0^x \left(1 + \frac{t^3}{3} + t^5\right) dt = 5 + x + \frac{1}{12}x^4 + \frac{1}{6}x^6$$

$$z_2 = z_0 + \int_0^x \left[t^2 z_1(t) + t^4 y_1(t)\right] dt$$

$$= 1 + \int_0^x \left[t^2 \left(1 + \frac{1}{3}t^3 + t^5\right) + t^4 (5 + t)\right] dt$$

$$= 1 + \frac{1}{3}x^3 + x^5 + \frac{2}{9}x^6 + \frac{1}{8}x^8 .$$

Third approximation :

$$y_3 = y_0 + \int_0^x z_2(t)\, dt = 5 + x + \frac{1}{12}x^4 + \frac{1}{6}x^6 + \frac{2}{63}x^7 + \frac{1}{72}x^9$$

$$z_3 = z_0 + \int_0^x \left[t^2 z_2(t) + t^4 y_2(t)\right] dt$$

$$= 1 + \int_0^x \left[t^2 \left(1 + \frac{1}{3}t^3 + t^5 + \frac{2}{9}t^6 + \frac{1}{8}t^8\right)\right.$$

$$\left. + t^4 \left(5 + t + \frac{1}{12}t^4 + \frac{1}{6}t^6\right)\right] dt$$

$$= 1+\frac{1}{3}x^3+x^5+\frac{2}{9}x^6+\frac{1}{8}x^8+\frac{11}{224}x^9+\frac{7}{264}x^{11}.$$ **Ans.**

2.2 Existence and Uniqueness of Solution

It may happen that an initial value problem has no solution or it may have exactly one solution of it may have more than one solution. Our aim in this section is to find under what conditions an initial value problem has at least one solution and under what conditions does that problem have one and only one solution, that is, a Unique Solution. This leads us to the **Existence Theorem** and **Uniqueness Theorems** respectively. These existence and uniqueness theorems will play an important role when a differential equation cannot be solved by elementary standard methods. In a more advanced course on differential equations, considerations concerning the existence, uniqueness and general behaviour of solutions are of paramount importance.

We now prove these theorems.

2.3 The Lipschitz Condition

If f (x, y) be a function defined for (x, y) is a domain D in x–y plane, then the function f (x, y) is said to satisfy the **Lipschitz Condition** in D if there exists a positive constant K such that

$$|f(x, y_2) - f(x, y_1)| \le K\,|y_2 - y_1|,$$

for every pair of points (x, y_1), $(x, y_2) \in D$. The constant K being independent of x, y_1 and y_2 and is called the **Lipschitz Constant.**

2.4 Existence Theorem

The initial value problem

$$\frac{dy}{dx} = y\ (x, y),\ y(x_0) = y_0 \qquad \text{...(1)}$$

has at least one solution y (x) provided the function f (x, y) is continuous and bounded for all values of x, in a domain D and there exists positive constants M and K such that

$$|f(x, y) \le M \qquad \text{...(2)}$$

and satisfies the Lipschitz condition

$$|f(x, y_2) - f(x, y_1)| \le K\,|y_2 - y_1|, \qquad \text{...(3)}$$

for all points in domain D.

Proof by Picard's Method: Consider the iterative sequence

$$y_n(x) = y_0 + \int_{x_0}^{x} f[t, y(t)]\,dt \qquad ...(5)$$

$n = 1, 2, 3, ...$

with $y_0(t) = y_0$ for the initial value problem (1). In order that the initial value problem (1) may have a solution, it is necessary that the sequence $\{y_n(x)\}$ of functions converges to a limiting function y (x) which is a solution of (1) or of the equivalent internal equation

$$y(x) = y_0 + \int_{x_0}^{x} f[t, y(t)]\,dt \qquad ...(5)$$

To ensure the existence of the limiting function

$$y = (x) = \lim_{n\to\infty} y_n(x). \qquad ...(6)$$

we use the fact that y_n may be written as a sum of successive differences:

$$y_n = y_0 + \sum_{i=0}^{n-1}(y_{i+1} - y_i) \qquad ...(7)$$

This follows that the sequence $\{y_n\}$ will converge if the infinite series $\sum(y_{i+1} - y_i)$ converges.

From (4) we have

$$y_i(x) = y_0 + \int_{x_0}^{x} f[t, y_{i-1}(t)]\,dt$$

and $\quad y_{i+1}(x) = y_0 + \int_{x_0}^{x} f[t, y_i(t)]\,dt$

$$\therefore \quad y_{i+1}(x) - y_i(x) = \int_{x_0}^{x} \{f[t, y_i(t)] - f[t, y_{i-1}(t)]\}\,dt \qquad ...(8)$$

The equation (8) is true for all integers $i = 1, 2, 3, ...$

Also from (4) $\quad y_1(x) = y_0 + \int_{x_0}^{x} f[t, y_0]dt$

$$\therefore \quad y_1(x) - y_0 = \int_{x_0}^{x} f[t, y_0]dt \qquad ...(9)$$

The condition (2) ensures the existence of integrals (8) and (9),

Considering (9), we have

$$|y_1(x) - y_0| \le \int_{x_0}^{x} |f[t, y_0]|\,|dt|$$

$$\le \int_{x_0}^{x} M\,|dt| \text{ by (2)}$$

$$= M\,|x - x_0| \qquad ...(10)$$

Again making use of Lipschitz condition (3), we get from (8),

$$|y_2(x) - y_1(x)| £ \int_{x_0}^{x} \left| f\{t, y_1(t)\} - f(t, y_0) \right| dt |$$

$$\leq \int_{x_0}^{x} K \left| y_1(t) - y_0 \right| |dt| \qquad \text{by (3)}$$

$$\leq \int_{x_0}^{x} K.M \left| t - x_0 \right| |dt| \qquad \text{by (10)}$$

$$= KM \frac{|x - x_0|^2}{2!} \qquad ...(11)$$

Similarly, $|y_3(x) - y_2(x)| \leq \dfrac{K^2 M |x - x_0|^3}{3!}$

In general, we shall have

$$|y_i(x) - y_{i-1}(x)| \leq M K^{i-1} \frac{|x - x_0|^i}{i!} \qquad ...(12)$$

The truthness of result (12) for all values of i can be established by the mathematical induction method.

$\therefore$ We must show that the inequality (12) holds when i is replaced by i + 1. For this purpose, we again make use of (8) and (3). We have

$$| y_{i+1}(x) - y_i(x) | \leq \int_{x_0}^{x} |f[t, y_i(t)] - f[t, y_{i-1}(t)]| \, | dt |$$

$$\leq \int_{x_0}^{x} K |. y_i(t) - y_{i-1}(t) | \, | dt |$$

$$\leq \int_{x_0}^{x} K.MK^{i-1} \frac{|x - x_0|}{i!} |dt| = MK^i \frac{|x - x_0|^{i+1}}{(i+1)!} \qquad ...(13)$$

The relation (13) establishes the validity of (12) for all values of i.

From (13), we see that the absolute values of terms in the series (7) are term by term smaller that corresponding terms in the series,

$$y_0 + M|x - x_0| + M.K \frac{|x - x_0|^2}{2!} + MK^2 \frac{|x - x_0|^3}{3!} + ...$$

whose sum is $y_0 + \dfrac{M}{K}$ [exp. $\{K|x - x_0|\} - 1$]

Now, the above Taylor's series converges for all values of $(x - x_0)$, and so the function $y_n(x)$ converges uniformly to a function y(x) for all values of x in any finite interval, i.e., $\lim_{n\to\infty} y_n(x) \to y(x)$

Now, Proceeding to the limits $n \to \infty$, we get from (4)

$$\lim_{n\to\infty} y_n(x) = y_0 + \lim_{n\to\infty} \int_{x_0}^{x} f[t, y_{n-1}(t)] \, dt$$

$$\Rightarrow \quad y(x) = y_0 + \lim_{n\to\infty} \int_{x_0}^{x} f\left[t, y_{n-1}(t)\right] dt. \qquad ...(14)$$

Since f (x, y) is continuous function of both x and y in the range of values considered and hence y_n (x) converges to y (x) uniformly over the interval in question, the following interchanges of limiting operations are justified

$$\lim_{n\to\infty} \int_{x_0}^{x} f\left[t, y_{n-1}(t)\right] dt = \int_{x_0}^{x} \lim_{n\to\infty} f\left[t, y_{n-1}(t)\right] dt.$$

$$= \int_{x_0}^{x} f\left[t, \lim_{n\to\infty} y_{n-1}(t)\right] dt = \int_{x_0}^{x} f\left[t, y(t)\right] dt \qquad ...(15)$$

Thus, from (14) and (15), we get

$$y(x) = y_0 + \int_{x_0}^{x} f\left[t, y(t)\right] dt$$

This shows that iterative sequence (4) converges to a solution of the differential equation, problem (1) for all values of x ∈ D under the given conditions.

Thus, the theorem is completely established.

2.5 Uniqueness Theorem

The initial value problem

$$\frac{dy}{dx} = f(x, y), \; y(x_0) = y_0 \qquad ...(1)$$

has a unique solution provided the function f (x, y) is continuous and bounded for all values of x in a domain D and there exists positive constants M and K such that $| f(x, y) | \le M$

and satisfies the Lipschitz condition

$$|f(x, y_2) - f(x, y_1)| \le K\, |y_2 - y_1|$$

for all points in domain D. **(Agra, 94; Meerut, 94, 95, 91, 95)**

Proof by Picard's Method: Suppose if possible the initial value problem (1) has two distinct solutions y(x) and u(x). Then

$$u(x) = y_0 + \int_{x_0}^{x} f\left[t, u(t)\right] dt \text{ and } y(t) = y_0 + \int_{x_0}^{x} f\left[t, y(t)\right] dt$$

$$\therefore y(x) - u(x) = \int_{x_0}^{x} \left[f\{t, y(t)\} - f\{t, u(t)\}\right] dt. \qquad ...(2)$$

Since f (x, y) is bounded and satisfies Lipschitz condition, we have

$$|f(x, y)| \le M \qquad ...(3)$$

and $\quad |f(x, y_2) - f(x, y_1)| \le K\, |y_2 - y_1|. \qquad ...(4)$

Using (2) and (3), we have

$$|y(x) - u(x)| \leq \int_{x_0}^{x} |f[t, y(t)] - f[t, u(t)]||dt|$$

$$\leq \int_{x_0}^{x} [|f[t, y(t)]| + |-f[t, u(t)]]|dt|$$

$$\leq \int_{x_0}^{x} (M + M)|dt| = 2M|x - x_0|$$

i.e., $|y(x) - u(x)| \leq 2M|(x - x_0|.$ ...(5)

Again using (2) and (4), we have

$$|y(x) - u(x)| \leq \int_{x_0}^{x} K|y(t) - u(t)||dt| \quad ...(6)$$

Combining (5) and (6), we obtain

$$|y(x) - u(x)| \leq \int_{x_0}^{x} K.2M|t - x_0||dt|$$

$$= 2MK\frac{|x - x_0|^2}{2!} \quad ...(7)$$

Employing inequality equation (7) on the right hand side of (6), we have

$$|y(x) - u(x)| \leq \int_{x_0}^{x} K.2MK\frac{|t - x_0|^2}{2!}|dt|$$

$$= 2MK^2\frac{|t - x_0|^3}{3!} \quad ...(8)$$

Continuing in this way, we shall obtain

$$|y(x) - u(x)| \leq \frac{MK^{n-1}|x - x_0|^n}{n!}, \; n = 1, 2, 3, \quad ...(9)$$

Now, the right hand side of (9) tends to zero as n tends to infinity for all finite values of x.

Thus, $|y(x) - u(x)| = 0$

$\Rightarrow$ $y(x) = u(x)$

for all finite values of x. This shows that the solution is unique.

2.6 Existence and Uniqueness Theorem *(The General Case)*

The initial value problem

$$\frac{dy}{dx} = f(x, y), \; y(x_0) = y_0$$

has a unique solution for all values of x in the range

$$|x - x_0| \leq a,$$

provided the function f (x, y) is continuous and satisfy the conditions (i) $|f(x, y)| \leq M$

and (ii) $|f(x, y_2) - f(x, y_1)| \leq K|y_2 - y_1|$

(Lipschitz Condition) for all values x and y, M, K being positive constants, in the rectangle R defined by

$$|x - x_0| \leq a \text{ and } |y - y_0| \leq Ma.$$

Proof by Picard's Method: The point of difference in two theorems is that here we are considering the limited range

$$|x - x_0| \leq a \qquad ...(1)$$

instead of considering all values of x so that theorem may be applicable to a wider class of functions f (x, y). We have seen that the proof of the existence theorem depended upon obtaining the inequality

$$|y_{i+1}(x) - y_i(x) \leq MK^i \frac{|x - x_0|^{i+1}}{(i+1)!} \qquad ...(2)$$

for all values of x. Here our aim to establish the inequality (2) for limited range (1).

For this we again consider the relation

$$y_{i+1}(x) - y_i(x) = \int_{x0}^{x} [f\{t, y_i(t)\} - f\{t, y_{i-1}(t)\}]\, dt. \qquad ...(3)$$

$$i = 1, 3, \ldots$$

Since we do not require the conditions (i) and (ii) to be applicable for all values of y but merely in a suitable neighbourhood of y_0, we shall consider the possibility of obtaining bounds not for $y_{i+1} - y_i$ but for $y_i - y_0$.

$$\text{Now, } y_i - y_0 = \int_{x0}^{x} f[t, y_{i-1}(t)]\, dt, \qquad ...(4)$$

so long as $|f(t, y)| \leq M$ for $|t - x_0| \leq a$ and $y = y_0$.

We get from (4), $|y_1(x) - y_0| \leq \int_{x0}^{x} |f(t, y)|\, |dt|$

$$\Rightarrow \quad |y_1(x) - y_0| \leq M|x - x_0| \leq Ma \qquad ...(5)$$

Using the condition $|f(x, y)| \leq M$

for $\quad |x - x_0| \leq a$ and $|y - y_0| \leq Ma,$

we conclude from $\quad y_2(x) - y_0 = \int_{x0}^{x} f[t, y_1, (t)]\, dt$

$$|y_2(x) - y_0| \leq \int_{x0}^{x} |f[t, y_1(t)]|\, |dt|$$

$$\leq \int_{x0}^{x} M\, |dt| = M|x - x_0|$$

$$\leq M.a. \qquad ...(6)$$

Then by induction, we get $|y_i(x) - y_0| \leq Ma$ ...(7)

for $|x - x_0| \leq a$ and all t.

After finding the bound (7) for y_i we formulate our Lipschitz Condition for (3) as follows :

$$|f(x, y_2) - f(x, y_1)| \leq M |y_2 - y_1| \quad ...(8)$$

in the range $|x - x_0| \leq a$ and $|y - y_0| \leq Ma$. We then have

$$|y_{i+1}(x) - y_i(x)| \leq \int_{x_0}^{x} K |y_i(t) - y_{i-1}(t)| |dt| \quad ...(9)$$

Using (5) in (9) with i = 1, we obtain

$$|y_2(x) - y_1(x)| \leq \int_{x_0}^{x} MK |t - x_0| |dt| = MK \frac{|x - x_0|^2}{2!} \quad ...(10)$$

Continuing in this way, we shall arrive at

$$|y_{i+1}(x) - y_i(x)| \leq MK^i \frac{|x - x_0|^{i+1}}{(i+1)!}$$

valid in the interval $|x - x_0| \leq a$.

The rest of the existence proof and also the uniqueness proof is the same as in the proceeding two theorems given in.

Remark:

If f (x, y) satisfies the condition $\left|\frac{\partial f}{\partial y}\right| \leq M$ ...(A)

for all values of x, y in the given range, then the Lipschitz condition is also satisfied with the same constant M. For, we have by the mean value theorem of differential calculus

$$f(x, y_2) - f(x, y_1) = (y_2 - y_1) \left|\frac{\partial f}{\partial y}\right| y = \bar{y}, y_1 < \bar{y} < y_2 \quad ...(B)$$

where (x, y_1) and (x, y_2) are assumed in the given range. From (A) and (B), we have

$$|f(x, y_2) - f(x, y_1) \leq M |y_2 - y_1|$$

Which is Lipschitz Condition. Thus, the Lipschitz Condition can be replaced 6y the strange condition (A).

Theorem:

If S is either a rectangle $|x - x_0| \leq a$, $|y - y_0| \leq b$, *(a, b > 0) or as strip* $|x - x_0| \leq a$, $|y| < \infty$, *(a > 0), and if is a real of valued function defined on S such that* $\frac{\partial f}{\partial y}$ *exists, is continuous on S, and*

$$\left|\frac{\partial}{\partial y} f(x, y)\right| \le K, \ (x, y) \in S,$$

for a positive constant K, then f satisfies a Lipschitz Condition on S with Lipschitz Constant K. **(Meerut, 1994)**

Solution:

We have

$$f(x, y_1) - f(x, y_2) = \int_{y_1}^{y_2} \frac{\partial}{\partial y} f(x, y).dy$$

$$\Rightarrow \quad |f(x, y_1) - f(x, y_2)| = \left| \int_{y_1}^{y_2} \frac{\partial}{\partial y} f(x, y)\, dy \right|$$

$$\le \int_{y_1}^{y_2} \left|\frac{\partial}{\partial y} f(x, y)\right| |dy|$$

$$\le K \int_{y_1}^{y_2} |dy|$$

$$\Rightarrow \quad |f(x, y_1) - f(x, y_2)| \le K \ |y_1 - y_2|$$

for all $(x, y_1), (x, y_2) \in S$

$\Rightarrow$ f (x, y) satisfy Lipschitz Condition on S and Lipschitz Constant is K.

Example 1:

Show that for the problem $\frac{dy}{dx} = y, \ y(0) = 1,$

the constant a in Picard's Theorem must be smaller than unity.

(Meerut, 1993)

Solution:

Here the condition for boundness on f i.e. $|f(x, y)| \le M$ for $|y - y_0| \le Ma$ takes the form

$|y| \le M$ for $|y - 1| \le Ma$.

And if we choose $M \ge 1$, Lipschitz Condition is also satisfied since in this case the Lipschitz Condition assumes the form

$|f(x, y_2) - (x, y_1)| = |y_2 - y_1| \le M |y_2 - y_1| \qquad \because m \ge 1.$

Now $|y - 1| \le Ma$ implies that $|y| - 1 \le Ma$.

Hence the inequality $|y| \le M$ will be satisfied for all values of $|y - 1| \le Ma$ provided it is satisfied for $|y| = 1 + Ma$.

Accordingly, we must have

$$1 + Ma \le M \Rightarrow a \le \frac{M-1}{M} = 1 - \frac{1}{M} \Rightarrow a < 1.$$

[Since M is positive finite constant].

Example 2:

Illustrate by an example that a continuous function may not satisfy a Lipschitz Condition on a rectangle. **(Meerut, 95)**

Solution:

Let us consider the function

$$f(x, y) = y^{2/3}, \text{ on the rectangle } |x| \le 1, |y| \le 1.$$

Obviously f (x, y) is continuous in the rectangle as it is a polynomial in y.

But $$\left|\frac{\partial}{\partial y} f(x, y)\right| = \left|\frac{2}{3y^{1/3}}\right|$$

which does not exist at y = 0, which is a point of the rectangle.

Hence the Lipschitz Condition is not satisfied on the rectangle.

Example 3:

For the initial value problem

$$\frac{dy}{dx} = e^y,\ y(0) = 0$$

find the largest interval $|x| \le a$ in which the Picard's theorem guarantees existence of a unique solution. **(Meerut, 97)**

Solution:

The condition for boundedness of f (x, y) i.e.

$|f(x, y)| \le M$ for $|y - y_0| \le Ma$ takes the form

$e^y \le M$, for $|y - 0| \le Ma$.

If y_1, y_2 with $y_1 < y_2$ lies in the range $|y| \le Ma$,

we have by the mean value theorem,

$$e^{y_2} - e^{y_1} = (y_2 - y_1)\left|\frac{\partial}{\partial y} e^y\right|_{y=\bar{y}} \quad \text{where } y_1 < \bar{y} < y_2$$

$$\Rightarrow \quad e^{y_2} - e^{y_1} \le (y_2 - y_1) M \qquad [\because e^y \le M].$$

So that Lipschitz Condition is also satisfied. Now, the inequality $e^y \le M$ will be satisfied for all values of

$| y | \le Ma$

provided it is satisfied if y = Ma. Accordingly, we must have

$$e^{Ma} \le M \Rightarrow a \le \frac{\log M}{M}$$

Since a is positive, M lies in the range $1 \le M < \infty$. It is easy to see that $\frac{\log M}{M}$ is maximum when M = e = 2.718.

The Picard's Theorem then assures existence of a unique solution in the interval $|x| \le a$

where $a = \frac{1}{e} = 0.308$.

Example 4:

If S is defined by the rectangle $|x| \le a$, $|y| \le b$, show that the function $f(x, y) = x^2 + y^2$, satisfies the Lipschitz Condition. Find the Lipschitz Constant.

(Meerut, 91 (P), 92(P), 93(P))

Solution:

Let (x, y_1) and (x, y_2) be two points in the rectangle S, then

$$|f(x, y_2) - f(x, y_1)| = |(x^2 + y_2^2) - (x^2 + y_1^2)|$$

$$= |y_2^2 - y_1^2| = |y_2 + y_1|\,|y_2 - y_1|$$

$\Rightarrow$ $|f(x, y_2) - f(x, y_1)| \le 2b\,|y_2 - y_1|$

$\Rightarrow$ f (x, y) satisfy the Lipschitz Condition, and the Lipschitz Constant K = 2b.

Aliter. Here $f(x, y) = x^2 + y^2$

$$\left|\frac{\partial}{\partial y} f(x, y)\right| = |2y| \le 2b, \text{ for } (x, y) \in S.$$

Thus, $\frac{\partial}{\partial y}$ exists is continuous and bounded for all $(x, y) \in S$. Hence by theorem f (x, y) satisfies Lipschitz condition and the Lipschitz constant is 2b.

Example 5:

Show by considering the initial value problem

$$\frac{dy}{dx} = \sqrt{|y|},\ y(0) = 0$$

that the continuity of f (x, y) is not enough the uniqueness of the solution of the initial value problem

$$\frac{dy}{dx} = f(x, y),\ y(x_0) = y_0.$$

Solution:

Although the function f (x, y) = $\sqrt{|y|}$ is continuous for all values of y, the initial value problem (1) has two solutions.

$$y = 0 \text{ and } y = \begin{cases} \frac{x^2}{4} & \text{for } x \geq 0 \\ -\frac{x^2}{4} & \text{for } x \leq 0 \end{cases}$$

The reason is that Lipschitz Condition is violated in any region which includes y = 0, since for $y_1 = 0$ and positive y_2, we have

$$\frac{|f(x, y_2) - f(x, y_1)|}{|y_2 - y_1|} = \frac{\sqrt{y_2}}{y_2} = \frac{1}{\sqrt{y_2}} \quad (y_2 > 0)$$

and this can be made as large as we please by choosing y_2 sufficiently small whereas Lipschitz Condition requires that $1/\sqrt{y_2}$ should exceed a fixed constant M.

Example 6:

Examine existence and uniqueness of the solution of the initial value problem

$$\frac{dy}{dx} = y^2,\ y(1) = -1.$$ **(Meerut, 92(P), 93(P), 94, 96(BP)]**

Solution:

Here f (x, y) = $\frac{dy}{dx} = y^2$ and $\frac{\partial f}{\partial y} = 2y$.

Obviously f and $\frac{\partial f}{\partial y}$ are both continuous for all (x, y). We consider the rectangle R

$|x - 1| \leq a$, $(y + 1) \leq b$, about the point (1, – 1).

Obviously in this rectangle

$$|f(x, y_2) - f(x, y_1)| = |y_2^2 - y_1^2|$$

$$= |y_2 + y_1|\,|y_2 - y_1|$$

$\leq (2 + 2b) |y_2 - y_1|$

Thus, the Lipschitz Condition is satisfied in the rectangle R. Now let $M = \max. |f(x, y)|$ for $x, y \in R$

and $h = \min. (a, b/m)$,

Then the given problem possesses a unique solution for $|x - 1| \leq h$.

In this case

$M = \max. |f(x, y)| = \max. |y^2| = |(-1 - b)^2| = (1 + b)^2$

$\therefore \quad h = \min. \{a, b/M\} = \min. \{a, b/(1 + b)^2\}$

Now let $b/(1 + b)^2 = \phi(b)$.

$$\therefore \quad \phi'(b) = \frac{1-b}{(1+b)^3} \text{ and } \phi''(b) = \frac{2b-4}{(1+b)^4}$$

For max or min. of $\phi(b)$, $\phi'(b) = 0 \Rightarrow b = 1$.

when $b = 1$, $\phi''(b) = -$ ve, $\therefore \phi(b)$ is max. when $b = 1$

and max. $\phi(b) = \phi(1) = \dfrac{1}{4}$.

$$\therefore \quad \text{if } a > \frac{1}{4}, \ \phi(b) = \frac{b}{(1+b)^2} \leq a, \text{ for all } b > 0.$$

$$\Rightarrow \quad h = \frac{b}{(1+b)^2} \leq \frac{1}{4}.$$

and if $a < \dfrac{1}{4}$, then $h < \dfrac{1}{4}$, $\therefore$ max $\dfrac{b}{(1+b)^2}$ is $\dfrac{1}{4}$.

Thus, we have $h \leq \dfrac{1}{4}$.

Now, for $b = 1$, $a \geq \dfrac{1}{4}$, $h = \min. \left\{a, \dfrac{b}{(1+b)^2}\right\} = \min. \left\{a, \dfrac{1}{4}\right\} = \dfrac{1}{4}$

Hence, the given problem possesses a unique solution when $|x - 1| \leq \dfrac{1}{4}$

i.e., in the interval $\dfrac{3}{4} \leq x \leq \dfrac{5}{4}$.

Example 7:

Show that $f(x, y) = xy^2$ satisfies the Lipschitz Condition on the rectangle $|x| \leq 1$, $|y| \leq 1$ but does not satisfy a Lipschitz Condition on the strip $|x| \leq 1$, $|y| < \infty$. **(Meerut, 1992, 94(P), 95(BP))**

Solution:

Here $f(x, y) = xy^2$.

$|f(x, y_2) - f(x, y_1)| = \left|xy_2^2 - xy_1^2\right| = |x|\,|y_2 + y_1|\,|y_2 - y_1|$

$\therefore$ In the rectangle $|x| \le |, (y) \le 1$

$|f(x, y_2) - f(x, y_1)| \le (1.2)|\, y_2 - y_1|$

Hence Lipschitz Condition is satisfied.

Also $$\left|\frac{d(x, y_2) - f(x, 0)}{y_2 - 0}\right| = |x|\,|y_2| \to \infty$$

when $|y_2| \to \infty$ if $|x| \ne 0$.

Hence, the Lipschitz Condition is not satisfied on the strip $|x| \le 1, |y| < \infty$.

Aliter. Here $f(x, y) = xy^2$.

$$\left|\frac{\partial}{\partial y} f(x, y)\right| = |2xy| \le 2, \text{ for } (x, y) \text{ on rectangle } |x| \le 1, |y| \le 1.$$

Thus, $\frac{\partial f}{\partial y}$ exists, is continuous bounded in the rectangle $|x| \le 1$, $|y| \le 1$, hence by theorem, Lipschitz Condition is satisfied on $|x| \le 1$, $|y| \le 1$.

But in the strip $|x| \le 1$, $|y| \le \infty$, there does not exist any positive constant K, such that

$$\left|\frac{\partial}{\partial y} f(x, y)\right| = |2xy| \le K.$$

Hence, Lipschitz Condition is not satisfied in the strip $|x| \le 1$, $|y| < \infty$.

EXERCISES

1. $(x - x^2)\dfrac{d^2y}{dx^2} + 4\dfrac{dy}{dx} + 2y = 0.$ [**Ans.** $y = A\left(1 - \frac{x}{2} + \frac{1}{10}x^2\right) + Bx^{-3}$

$(1 - 5x + 10x^2 - 10x^3 - 5x^4 - x^5)$

$y = Ax^2 (1 - 5x^{-1} + 10x^{-2} - 10x^{-3} + 5x^{-4} - x^{-5})$

$+ Bx^{-1}\left(1 - \frac{1}{2}x^{-1} + \frac{1}{10}x^{-2}\right)$

2. $(2x^2 + 1)\dfrac{d^2y}{dx^2} + x\dfrac{dy}{dx} + 2y = 0.$

$$\textbf{[Ans. } y = A\left[1-\frac{2}{2!}x^2+\frac{2(2.1.2+4)}{4\,!}x^4-\frac{2\,(2.1.2+4)\,(2.3.4+6)}{6!}x^6+\ldots\right]$$

$$+B.\left[x-\frac{3}{3!}x^3+\frac{3.(2.2.3+5)}{5!}x^5-\frac{3\,(2.2.3+5)\,(2.4.5+7)}{7\,!}x^7+\ldots\right]$$

3. $\dfrac{d^2y}{dx^2}-3x\dfrac{dy}{dx}-3y=0.$

Given y = 1, $\dfrac{dy}{dx}=0$ when x = 0. **(Meerut 93; Rohilkhand, 90)**

$$\textbf{[Ans. } y=\sum_{n=0}^{\infty}\frac{3^n}{n!\,2^n}x^{2n}\,]$$

4. $(1-x^2)\dfrac{d^2y}{dx^2}-x\dfrac{dy}{dx}-a^2y=0.$ When x = 0, y = 1 and $\dfrac{dy}{dx}=a.$

$$\textbf{[Ans. } y = 1 + ax + \frac{a^2}{2!}x^2+\frac{a\left(a^2+1^2\right)}{3\,!}x^3+\frac{a^2\left(a^2+2^2\right)}{4\,!}x^4+\ldots]$$

5. $x\dfrac{d^2y}{dx^2}-xy=1$. When $(y)_0=1$, $\left(\dfrac{dy}{dx}\right)_0=0$. **(Meerut, 94(P))**

$$\textbf{[Ans. } y=1+\frac{1}{3}x^3+\frac{4}{6\,!}x^6+\frac{4.7}{9!}x^9+\ldots.$$

$$+\frac{1}{2!}x^2+\frac{3}{5!}x^4+\frac{3.6}{8!}x^8+\frac{3.68}{11!}x^{11}+\ldots]$$

6. $x\dfrac{d^2y}{dx^2}+\dfrac{dy}{dx}+my=0.$ Given y = 1, when x = 0. **(Meerut, 94, 95)**

[Ans. $y = e^{-mx}$]

7. Solve $\dfrac{d^2y}{dx^2}+(x-1)^2\dfrac{dy}{dx}-4\,(x-1)\,y=0,$

in series about the point x = 1.

$$\textbf{[Ans. } y = A\left[1+\frac{2}{3}(x-1)^3+\frac{1}{45}(x-1)^6-\frac{1}{1620}(x-1)^9\ldots\right]$$

$$+B\left[(x-1)+\frac{1}{4}(x-1)^4\right]$$

8. $\dfrac{d^2y}{dx^2} - x\dfrac{dy}{dx} - py = 0.$ **(Meerut, 90)**

$$\text{[Ans. } y = A\left[1 + p\frac{x^2}{2!} + p(p+2)\frac{x^4}{4!} + p(p+2)(p+4)\frac{x^6}{6!} + \ldots\right]$$

$$+ B\left[x + (p+1)\frac{x^3}{3!} + (p+1)(p+3)\frac{x^5}{5!} + \ldots\right]$$

9. $x\dfrac{d^2y}{dx^2} + (x+n)\dfrac{dy}{dx} + (n+1)y = 0.$ **(Meerut, 99, 93; Agra 91)**

$$\text{[Ans. } y = A\left[n - (n+1)x + (n+2)\frac{x^2}{2!} - (n+3)\frac{x^3}{3!} + \ldots\right]$$

$$+ Bx^{1-n}\left[1 + \frac{2}{n-2}x + \frac{3}{(n-2)(n-3)}x^2 + \frac{4}{(n-2)(n-3)(n-4)}x^3 + \ldots\right]$$

10. Integrate in series. Gauss's Hypergeometric Equation,

$x(1-x)\dfrac{d^2y}{dx^2} + \{c - (a+b+1)x\}\dfrac{dy}{dx} - aby = 0.$ [Raj. 80, 81]

[**Ans.** $y = A.F(a, b, c, x) + B.F(a+1-c, b+1-c, 2-c, x)$]

11. $4(x^4 - x^2)\dfrac{d^2y}{dx^2} + 8x^3\dfrac{dy}{dx} - y = 0.$ [**Ans.** $y = au + bv$

where $u = x^{1/2}\left\{1 + \dfrac{1.3}{4^2}x^2 + \dfrac{1.3.5.7}{4^2.8^2}x^4 + \dfrac{1.3.5.7.9.11}{4^2.8^2.12^2}x^6 + \ldots\right\}$]

12. $(x + x^2 + x^3)\dfrac{d^2y}{dx^2} + 3x^3\dfrac{dy}{dx} - 2y = 0.$

[**Ans.** $y = au + bv$ where $u + 2x + 2x^2 - x^3 - x^4 + \dfrac{4}{5}x^5 \ldots$]

13. $x\dfrac{d^2y}{dx^2} + \dfrac{dy}{dx} + y = 0.$ **(Meerut, 92)**

$$\text{[Ans. } y = (A + B\log x)\left[1 - \frac{x}{1^2} + \frac{x^2}{1^2.2^2} - \frac{x^3}{1^2.2^2.3^2} + \ldots\right]$$

$$+ 2B\left[\frac{x}{1^2} - \frac{x^2}{1^2.2^2}\left(1 + \frac{1}{2}\right) + \frac{x^2}{1^2.2^2.3^2}\left(1 + \frac{1}{2} + \frac{1}{3}\right)\ldots\right]$$

14. $\dfrac{d^2y}{dx^2}+xy=0.$ **(Meerut, 92(P), 93(P))**

[**Ans.** $y=A\left[1-\frac{1}{3!}x^3+\frac{1.4}{6!}x^6-\frac{1.4.7}{9!}x^9+\ldots\right]$

$+Bx\left[1-\frac{2}{4!}x^3+\frac{2.5}{7!}x^6-\frac{2.5.8}{10!}x^9+\ldots\right]$

15. $\dfrac{d^2y}{dx^2}-x^2\dfrac{dy}{dx}-y=0.$ **(Agra, 96, 97)**

[**Ans.** $y=A\left(1+\frac{1}{2}x^2+\frac{1}{24}x^4+\frac{1}{20}x^5\ldots\right)$

$+B\left(x+\frac{1}{6}x^3+\frac{1}{12}x^4+\frac{1}{12}x^5\ldots\right)$

16. $\dfrac{d^2y}{dx^2}+ax^2y=0$ when $x=0$, $y=1$, $\dfrac{dy}{dx}=2.$ **(Kanpur, 95)**

[**Ans.** $y=\left\{1-\frac{ax^4}{3.4}+\frac{a^2x^8}{3.4.7.8}-\frac{a^3x^{12}}{3.4.7.8.11.12}+\ldots\right\}$

$+2x\left\{1-\frac{ax^4}{4.5}+\frac{a^2x^8}{4.5.8.9}-\frac{a^3x^{12}}{4.5.8.9.12.13}+\ldots\right\}$

17. $(1-x^2)\dfrac{d^2y}{dx^2}-2x\dfrac{dy}{dx}+2y=0.$ *(Legendre's Equation of order unity)*

(Kanpur, 93)

[**Ans.** $y=a\left(1-x^2-\frac{1}{3}x^4-\frac{1}{5}x^6\ldots\right)+bx$]

18. $\dfrac{d^2y}{dx^2}+x^2y=0.$ **(Kanpur, 95)**

[**Ans.** $y=a\left[1-\frac{x^4}{3.4}+\frac{x^8}{3.4.7.8}+\ldots\right]+b\left[x-\frac{x^5}{4.5}+\frac{x^9}{4.5.8.9}+\ldots\right]$

19. $x(1-x^2) \dfrac{d^2y}{dx^2} + (1-3x^2)\dfrac{dy}{dx} - xy = 0.$

[**Ans.** y = au + bv. where $u = 1 + \dfrac{1^2}{2^2}x^2 + \dfrac{1^2.3^2}{2^2.4^2}x^4 + \ldots$

and $v = u \log x + \left\{\dfrac{1}{42}x^2 + \dfrac{21}{128}x^4 + \ldots\right\}$

20. $x(1-x)\dfrac{d^2y}{dx^2} - 3x\dfrac{dy}{dx} - y = 0.$

[**Ans.** y = au + bv. where $u = x + 2x^2 + 3x^3 + \ldots = x(1-x)^{-2}$
and $v = u \log x + 1 + x^2 + \ldots = u \log x + (1-x)^{-1}$]

21. $x(1-x)\dfrac{d^2y}{dx^2} - (1+3x)\dfrac{dy}{dx} - y = 0.$

[**Ans.** y = au + bv. where $u = 1.2x^2 + 2.3x^2 + 3.4x^4 + \ldots$
and $v = u + u \log x + (-1 + x + 3x^2 + 4x^3 + \ldots)$]

22. $2x(1-x)\dfrac{d^2y}{dx^2} + (1-x)\dfrac{dy}{dx} + 3y = 0.$

(Rohilkhand, 96; Meerut, 96; Kanpur, 90)

[**Ans.** $y = a\left[1 - 3x + \dfrac{3x^2}{1.3} + \dfrac{3x^3}{3.5} + \dfrac{4x^4}{5.7} + \ldots\right] + bx^{1/2}(1-x)$]

23. $4x\dfrac{d^2y}{dx^2} + 2\dfrac{dy}{dx} + y = 0.$ [**Ans.** $y = a\cos\sqrt{x} + b\sin\sqrt{x}$]

24. $\dfrac{d^2y}{dx^2} + y = 0.$ [**Ans.** $y = a\cos x + b\sin x$]

25. Apply Picard's Method upto third approximation to solve the equation:

$$\frac{dy}{dx} = x + z,\ \frac{dz}{dx} = x - y^2,$$

given that y = 2, z = 1 when x = 0. **(Meerut, 96(BP); Agra, 90)**

[**Ans.** $y_0 = 2,\ y_1 = 2 + x + \dfrac{1}{2}x^2,\ y_2 = 2 + x - \dfrac{3}{2}x^2 + \dfrac{1}{6}x^3,$

$y_3 = 2 + x - \dfrac{3}{2}x^2 - \dfrac{1}{2}x^3 - \dfrac{1}{4}x^4 - \dfrac{1}{20}x^5 - \dfrac{1}{120}x^6,$

$$z_0 = 1,\ z_1 = 1 - 4x + \frac{1}{2}x^2,\ z_2 = 1 - 4x - \frac{3}{2}x^2 - x^3\ \frac{1}{4}x^4 - \frac{1}{20}x^5$$

$$z_3 = 1 - 4x - \frac{3}{2}x^2 + \frac{5}{3}x^3 + \frac{7}{12}x^4 - \frac{31}{60}x^5 + \frac{1}{12}x^6 - \frac{1}{252}x^7\ ]$$

26. Apply Picard's Method to find the solution of the problem

$\frac{dy}{dx} = y - x,\ y(0) = 2.$ **(Meerut, 90, 95; Agra, 96)**

$$[\textbf{Ans.}\ y_0 = 2,\ y_1 = 2 + 2x - \frac{1}{2}x^2;\ y_2 = 2 + 2x + \frac{x^2}{2!} - \frac{x^3}{3!}$$

$$y_3 = 2 + 2x + \frac{x^2}{2!} + \frac{x^3}{3!} - \frac{x^4}{4!};\ y_4 = 2 + 2x + \frac{x^2}{2!} + \frac{x^3}{3!} + \frac{x^4}{4!} - \frac{x^5}{5!}\ ]$$

tending to $y = 2 + 2x + \frac{x^2}{2!} + \ldots = 1 + x + e^2$

27. Find the first three approximations in the solution of the following equation $\frac{dy}{dx} = 1 + xy,\ y(0) = 2.$ **(Meerut, 90, 91; Agra, 90)**

$$[\textbf{Ans.}\ y_0 = 2,\ y_1 = 2 + x + x^2,\ y_2 = 2 + x + x^2 + \frac{x^3}{3} + \frac{x^4}{4}.$$

$$y_3 = 2 + x + x^2 + \frac{x^3}{3} + \frac{x^4}{4} + \frac{x^5}{15} + \frac{x^6}{24}\ ]$$

28. Find the third approximation of the solution of the following equation

$\frac{dy}{dx} = 2x + z,\ \frac{dz}{dx} = 3xy + x^2z$

where $y = 2$ and $z = 0$ when $x = 0$.

$$[\textbf{Ans.}\ y_3 = 2 + x^2 + x^3 + \frac{3}{20}x^5 + \frac{1}{10}x^6.$$

$$z_3 = 3x^2 + \frac{3}{4}x^4 + \frac{6}{5}x^5 + \frac{3}{28}x^7 + \frac{3}{40}x^8\ ]$$

29. Find the third approximation of the solution of the equation

$\frac{d^2y}{dx^2} = x^3\left(\frac{dy}{dx} + y\right),$

where y = 1, and $\frac{dy}{dx} = \frac{1}{2}$ when x = 0. **(Agra, 91; Rohilkhand, 97)**

[**Ans.** $y_3 = 1 + \frac{1}{2}z + \frac{3}{40}x^5 + \frac{1}{60}x^6 + \frac{1}{92}x^9$]

30. Use Picard's Method to approximate the solution of the equation $\frac{dy}{dx} + 2xy^2 = 0$ with y = 1 when x = 0 and hence show that $y = 1/(1 + x^2)$. **(Meerut, 95(BP)**

[**Ans.** $y_1 = 1 - x^2$, $y_2 = 1 - x^2 + x^4 - \frac{1}{3}x^6$

$y_3 = 1 - x^2 + x^4 - x^6 + \frac{2}{3}x^8 - \frac{1}{3}x^{10} + \frac{1}{9}x^{12} - \frac{1}{63}x^{14}$]

31. Obtain the solution of the equation $\frac{dy}{dx} = x^2 + y^2$; y(0) = 1 by Picard's Method, as far as the term involving x^4. **(Meerut, 96)**

[**Ans.** $y_1 = 1 + x + \frac{1}{2}x^3$, y_2

$= 1 + x + x^2 + \frac{2}{3}x^3 + \frac{1}{6}x^4 + \frac{1}{15}x^5 + \frac{1}{63}x^7$]

32. Use Picard's Method to obtain a solution of the differential equation: $\frac{dy}{dx} = x^2 - y$, y(0) = 0. Find at least the fourth approximation to each solution. **(Meerut, 96)**

[**Ans.** $y_1 = \frac{1}{3}x^3$, $y_2 = \frac{1}{3}x^3 - \frac{1}{12}x^4$, $y_3 = \frac{1}{3}x^3 - \frac{1}{12}x^4 + \frac{1}{60}x^5$

$y_4 = \frac{1}{3}x^3 - \frac{1}{12}x^4 + \frac{1}{60}x^5 - \frac{1}{360}x^6$]

33. $\frac{dy}{dx} = 2x - y^2$ where y = 0 at x = 0. **(Agra, 97)**

[**Ans.** $y_1 = x^2$, $y_2 = x^2 - \frac{1}{5}x^5$, $y_3 = x^2 - \frac{1}{5}x^5 + \frac{1}{20}x^8 - \frac{1}{275}x^{11}$]

34. Solve the differential equation $\frac{dy}{dx} = x - y$ with the condition y = 1 when x = 0, and show that the sequence of approximations given

by Picard's Method tends to the exact solution as a limit.

[**Ans.** $y_0 = 1$, $y_1 = 1 - x + \frac{x^2}{2!}$, $y_2 = 1 - x + \frac{2x^2}{2!} - \frac{x^3}{3!}$

$y_3 = 1 - x + \frac{2x^2}{2!} - \frac{2x^3}{3!} + \frac{x^4}{4!}$; $y_4 = 1 - x + \frac{2x^2}{2!} - \frac{2x^3}{3!} + \frac{x^4}{4!} - \frac{x^5}{5!}$]

35. $\frac{dy}{dx} = 3e^x + 2y$, $y(0) = 0$. **(Meerut, 93(P), 94)**

[**Ans.** $y_0 = 0$, $y_1 = 3(e^x - 1)$, $y_2 = 9e^x - 6x - 9$,

$y_3 = 21e^x - 6x^2 - 18x - 21$]

36. $\frac{dy}{dx} = y^2$, $y(0) = 1$.

[**Ans.** $y_0 = 1$, $y_1 = 1 + x$, $y_2 = 1 + x + x^2 + \frac{1}{3}x^3$

$y_3 = 1 + x + x^2 + x^3 + \frac{2}{3}x^4 + \frac{1}{3}x^5 + \frac{1}{9}x^6 + \frac{1}{63}x^7$]

37. $\frac{dy}{dx} = 2xy - 1$, $y(0) = 0$.

[**Ans.** $y_0 = 0$, $y_1 = -x$, $y_2 = -x - \frac{2}{3}x^3$, $y_3 = -x - \frac{2}{3}x^3 - \frac{4}{15}x^5$]

38. $\frac{dy}{dx} = x + y$, $y(0) = 1$. **(Meerut, 93, 95)**

[**Ans.** $y_1 = 1 + x + \frac{1}{2}x^2$, $y_2 = 1 + x + x^2 + \frac{1}{6}x^3$

$y_3 = 1 + x + x^2 + \frac{1}{3}x^3 + \frac{1}{24}x^4$]

39. $\frac{dy}{dx} = x + y$, $y(0) = -1$. **(Meerut, 94(P)]**

[**Ans.** $y_0 = -1$, $y_1 = -1 - x + \frac{1}{2}x^2$,

$y_2 = -1 - x + \frac{1}{6}x^3$, $y_3 = -1 - x + \frac{1}{24}x^4$]

40. $\frac{dy}{dx} = xy$, $y(0) = 2$.

[**Ans.** $y_0 = 2$, $y_n = 2\left[1 + \frac{1}{2}x^2 + \frac{1}{2!}\left(\frac{1}{2}x^2\right)^2 + \ldots + \frac{1}{n!}\left(\frac{1}{2}x^2\right)^n\right]$

41. $\dfrac{dy}{dx} = 1 + y^2, \; y(0) = 0.$

[**Ans.** $y_0 = 0, \; y_1 = x, \; y_2 = x + \dfrac{1}{3}x^3, \; y_3 = x + \dfrac{1}{3}x^3 + \dfrac{2}{15}x^5 + \dfrac{1}{63}x^7$ etc.]

42. Show that the Picard's Theorem, ensures existence of a Unique Solution in the interval $|x| \le \dfrac{1}{2}$ for the initial value problem

$\dfrac{dy}{dx} = x + y^2, \; y(0) = 0.$

43. Examine the existence and uniqueness of solutions of the initial value problem.

$\dfrac{dy}{dx} = y^{1/3}, \; y(0) = 0.$ **(Meerut, 92, 97(P), 97(BP))**

44. If S is defined by the rectangle $|x| \le a, \; |y| \le b$, show that the function $f(x, y) = x \sin y + y \cos x.$ satisfy the Lipschitz condition. Find Lipschitz Constant.

[**Ans.** $a + 1$]

45. If $(x, y) = y^{2/3}$, show that Lipschitz Condition is not satisfied in any region containing the origin and that the solution of the differential equation

$\dfrac{dy}{dx} = f(x, y)$

satisfying the initial condition $y = 0$ when $x = 0$ is not unique.

3

Partial Differential Equations (First Order)

3.1 Introduction

A partial differential equation is a relation between the dependent variable and its partial derivatives with regard to them as $r + ka^2t + 2as$ where

$$r = \frac{\partial^2 z}{\partial x^2},\ t = \frac{\partial^2 z}{\partial y^2},\ s = \frac{\partial^2 z}{\partial x \partial y}$$ are partial differetial equations.

The order of a partial differential equation is the order of the partial differential coefficient of the highest order appearing in it. Here in this chapter we shall discuss the Partial Differential Equations (First Order).

3.2 Derivation of Partial Differential Equation

(a) By the Elimination of Arbitrary Constants

Let us have a function

$$f(x, y, z, a, b) = 0 \qquad ...(1)$$

where a and b are arbitrary constants.

Differentiating (1) partially. w.r.t. x and y, we have

$$\frac{\partial f}{\partial x} + \frac{\partial f}{\partial z} \cdot \frac{\partial z}{\partial x} = 0 \Rightarrow \frac{\partial f}{\partial x} + \frac{\partial f}{\partial z} p = 0 \qquad ...(2)$$

and $$\frac{\partial f}{\partial y} + \frac{\partial f}{\partial z} \cdot \frac{\partial z}{\partial y} = 0 \Rightarrow \frac{\partial f}{\partial y} + \frac{\partial f}{\partial z} . q = 0 \qquad ...(3)$$

Eliminating a and b from (1), (2) and (3), we have differential equation of the form

$$F(x, y, z, p, q) = 0$$

which is the partial differential equation of the first order.

Note: If the number of arbitrary constants more than the number of independent variables then the partial differential equation thus obtained will be of higher order then the first.

(b) By the Elimination of Arbitrary Functions

(Kanpur, 99; Meerut, 95(BP))

Let u and v be the two functions of (x, y, which are connected by the relation

$$f(u, v) = 0. \qquad ...(1)$$

Differentiating equation (1) w.r.t. x and y, we have

$$\frac{\partial f}{\partial u}\left(\frac{\partial u}{\partial x}+\frac{\partial u}{\partial z}p\right)+\frac{\partial f}{\partial v}\left(\frac{\partial v}{\partial x}+\frac{\partial v}{\partial z}p\right)=0 \qquad ...(2)$$

and $$\frac{\partial f}{\partial u}\left(\frac{\partial u}{\partial y}+\frac{\partial u}{\partial z}.p\right)+\frac{\partial f}{\partial v}\left(\frac{\partial v}{\partial y}+\frac{\partial v}{\partial z}q\right)=0 \qquad ...(3)$$

Eliminating $\frac{\partial f}{\partial u}$ and $\frac{\partial f}{\partial v}$ from equation (2) and (3), we have

$$\left(\frac{\partial u}{\partial x}+\frac{\partial u}{\partial z}p\right)\left(\frac{\partial v}{\partial y}+\frac{\partial v}{\partial z}q\right)=\left(\frac{\partial u}{\partial y}+\frac{\partial u}{\partial z}q\right)\left(\frac{\partial v}{\partial x}+\frac{\partial v}{\partial z}p\right)$$

which can be written as

$$\left(\frac{\partial u}{\partial y}\frac{\partial v}{\partial z}-\frac{\partial v}{\partial y}\frac{\partial u}{\partial z}\right)p+\left(\frac{\partial v}{\partial x}.\frac{\partial u}{\partial z}-\frac{\partial u}{\partial x}\frac{\partial v}{\partial z}\right)q=\frac{\partial u}{\partial x}.\frac{\partial v}{\partial y}-\frac{\partial u}{\partial y}.\frac{\partial v}{\partial x}$$

$$\Rightarrow \quad Pp + Qq = R. \qquad ...(4)$$

where $P = \frac{\partial u}{\partial y}\frac{\partial v}{\partial z}-\frac{\partial v}{\partial y}\frac{\partial u}{\partial z}=\frac{\partial(u, v)}{\partial(y, z)}$

$$Q = \frac{\partial v}{\partial x}\frac{\partial u}{\partial z}-\frac{\partial u}{\partial x}\frac{\partial v}{\partial z}=\frac{\partial(u, v)}{\partial(z, x)}$$

and $$R = \frac{\partial v}{\partial x}\frac{\partial v}{\partial y}-\frac{\partial u}{\partial y}\frac{\partial v}{\partial x}=\frac{\partial(u, v)}{\partial(x, y)}$$

Equation (4) is the required partial differential equation.

Example 1:

Form a partial differential equation by eliminating the function f from

$$z = y^2 + 2f\left(\frac{1}{f}+\log y\right). \qquad ...(1)$$

Solution:

We have $z = y^2 + 2f\left(\frac{1}{f}+\log y\right)$

Differentiating (1) partially w.r.t. x and y, we have

$$\frac{dz}{dx} = p = 2f'\left(\frac{1}{x}+\log y\right).\left(-\frac{1}{x^2}\right)$$

$$\Rightarrow \qquad - px^2 = 2f'\left(\frac{1}{x}+\log y\right)$$

and $$\frac{dz}{dx} = q = 2y + 2f'\left(\frac{1}{x}+\log y\right).\frac{1}{y}$$

$$\Rightarrow \qquad qy - 2y^2 = 2f'\left(\frac{1}{x}+\log y\right). \qquad ...(2)$$

From (2) and (3), we have

$$- px^2 = qy - 2y^2 \qquad \Rightarrow x^2p + yq + 2y^2.$$ **Ans.**

Example 2:

Form a partial differential equation by eliminating function f and from

$$z = f\cdot(x + iy) + F\,(x - iy). \qquad ...(1)$$

(Kanpur, 90, 91)

Solution:

We have $z = f\,(x + iy) + F\,(x - iy)$.

Differentiating (1), we have

$$\frac{dz}{dx} = f'\,(x + iy) + F'\,(x - iy)$$

$$\frac{dz}{dy} = if'\,(x + iy) - F'\,(x - iy)$$

$$\frac{d^2z}{dx^2} = f''\,(x + iy) + F''\,(x - iy)$$

and $$\frac{d^2z}{dy^2} = - f''\,(x + iy) - F''\,(x - iy)$$

$$\therefore \frac{d^2z}{dx^2}+\frac{d^2z}{dy^2}=0.$$ **Ans.**

Example 3:

Form a partial differential equation by eliminating a, b, c from

$$\frac{x^2}{a^2}+\frac{y^2}{b^2}+\frac{z^2}{c^2}=1. \qquad ...(1)$$

Solution:

Differentiating equation (1) partially w.r.t. x and y, we have

$$\frac{2x}{a^2}+\frac{2z}{c^2}.\frac{\partial z}{\partial x}=0$$

$$\Rightarrow \quad c^2x + a^2z\,\frac{\partial z}{\partial x}=0 \qquad ...(2)$$

and $$\frac{2y}{b^2}+\frac{2z}{c^2}.\frac{\partial z}{\partial y}=0$$

$$\Rightarrow \quad c^2y + b^2z\,\frac{\partial z}{\partial y}=0 \qquad ...(3)$$

Again differentiating equation (2) partially w.r.t. x, we have

$$c^2 + a^2\left(\frac{\partial z}{\partial x}\right)^2+a^2z\frac{\partial^2 z}{\partial x^2}=0$$

$$\Rightarrow \quad \frac{c^2}{a^2}+\left(\frac{\partial z}{\partial x}\right)^2+z\frac{\partial^2 z}{\partial x^2}=0$$

Substituting $\frac{c^2}{a^2}-z\frac{z\,\partial z}{x\,\partial x}$ from (2), we have

$$-\frac{z}{x}\frac{\partial z}{\partial x}+\left(\frac{\partial z}{\partial x}\right)^2+z\frac{\partial^2 z}{\partial x^2}=0$$

$$\Rightarrow \quad -z\frac{\partial z}{\partial x}+x\left(\frac{\partial z}{\partial x}\right)^2+xz\frac{\partial^2 z}{\partial x^2}=0$$

Similarly differentiating (3) partially w..r.t y and substituting the value of $\frac{c^2}{b^2}$ from (3), we shall get

$$-z\frac{\partial z}{\partial y}+\left(\frac{\partial z}{\partial y}\right)^2+yz\frac{\partial^2 z}{\partial y^2}=0.$$ **Ans.**

Example 4:

Form a partial differential equation by eliminating a, b from

z = (x + a) (y + b).

Solution:

We have $z = (x + a)(y + b)$ ...(1)

$$\therefore \quad \frac{\partial z}{\partial x} = p = (y + b) \quad \text{and} \quad \frac{\partial z}{\partial y} = q = (x + a).$$

Substituting in (1), we have $z = pq$

which is the required differential equation. **Ans.**

3.3 Definitions

Complete Integrals **(Kanpur, 97, 91; IAS, 95)**

If from the partial differential equation. $f(x, y, z, p, q) = 0$.

We can find a relation $F(x, y, z, a, b) = 0$ which contains as many arbitrary constants as there are independent variables, the relation $F(x, y, z, a, b) = 0$ is known as **Complete Integral** of the given equation.

Particular Integral **(Kanpur, 97, 91; IAS, 95)**

Giving the particular values to the constants a and b occurring in the complete integral, we obtain a solution of the given partial differential equation which we shall call as **Particular Integral.**

Singular Integral **(Kanpur, 97, 91)**

The equation of the envelope of the surfaces represented by the complete integral, of the given partial differential equation is called its **Singular Integral.** Thus, if $F(x, y, z, a, b) = 0$ is the complete integral then singular integral is obtained by eliminating a and b from

$$f = 0, \ \frac{\partial f}{\partial a} = 0 \ \text{and} \ \frac{\partial f}{\partial b} = 0.$$

General Integral **(Kanpur, 97)**

We have seen in that if two functions u, v, of x, y, z are connected by an arbitrary function $f(u, v) = 0$ then eliminating f we get a partial differential equation of the form $Pp + Qq + R$

The solution of the equation is $f(u, v) =$ which is called the **General Integral** of the differential equation.

3.4 Linear Partial Differential Equations of Order One

A differential equation involving partial derivatives p and q only and no higher is called of order one. If the degree of p and q are unity then it is called a **Linear Partial Differential Equation of Order One.**

3.5 Lagrange's Linear Equation

The partial differential equation of the form Pp + Qq = R where P, Q, R are functions of x, y, z is the standard form of the linear partial differential equation of the order one and is called **Lagrange's Linear Equation.**

3.6 Lagrange's Solution of the Linear Equation

(Kanpur, 96, 91; Delhi, Hons 99, 95; IAS, 99)

In by eliminating an arbitrary function f from

$$f(u, v) = 0 \qquad ...(1)$$

connecting two functions u and v we get the partial differential equation

$$Pp + Qq = R \qquad ...(2)$$

where $$P = \frac{\partial u}{\partial y}.\frac{\partial v}{\partial z} - \frac{\partial v}{\partial y}.\frac{\partial u}{\partial z}$$

$$Q = \frac{\partial v}{\partial x}.\frac{\partial u}{\partial z} - \frac{\partial u}{\partial x}.\frac{\partial v}{\partial z}$$

and $$R = \frac{\partial u}{\partial x}.\frac{\partial v}{\partial y} - \frac{\partial u}{\partial y}.\frac{\partial v}{\partial x}$$

Thus, equation (1) is the general integral of (2) and so we have to find the values of u and v.

Let u = a and v = b be two equations where a and be are arbitrary constants.

Differentiating them, we have

$$\frac{\partial u}{\partial x}.dx + \frac{\partial u}{\partial y}.dy + \frac{\partial u}{\partial z}.dz = 0.$$

and $$\frac{\partial v}{\partial x}.dx + \frac{\partial v}{\partial y}.dy + \frac{\partial y}{\partial z}.dz = 0.$$

Solving these, we have

$$\frac{dx}{\frac{\partial u}{\partial y}.\frac{\partial v}{\partial z} - \frac{\partial v}{\partial y}.\frac{\partial u}{\partial z}} = \frac{dy}{\frac{\partial u}{\partial z}.\frac{\partial v}{\partial x} - \frac{\partial u}{\partial x}.\frac{\partial v}{\partial z}} = \frac{dz}{\frac{\partial u}{\partial x}.\frac{\partial v}{\partial y} - \frac{\partial u}{\partial y}.\frac{\partial v}{\partial x}}$$

i.e., $$\frac{dx}{P} = \frac{dy}{Q} = \frac{dz}{R} \qquad ...(3)$$

Solution of the above differential equations are u = a and v = b.

Thus, the solution of the given equation (1) is found is f (u, v) = 0.

Note: Equations given by (3) are called **Lagrange's Auxiliary Equations or Subsidiary Equations.**

Working Method: For the solution of the partial differential equation

$$Pp + Qq = R$$

form the auxiliary equations $\frac{dx}{P} = \frac{dy}{Q} = \frac{dz}{R}$

Find two independent integrals of auxiliary equations say u = a and v = b. Then the general integral of the equation is given by f (u, v) = 0 when f is an arbitrary function.

3.7 Geometrical Interpretation of Lagrange's Linear Equations

(Kanpur, 96; G.N.U.A., 90)

Lagrange's linear equation is

$$Pp + Qq = R \qquad ...(1)$$

which can be written as

$$Pp + Qq + R(-1) = 0.$$

We know that the d.c's of the normal at a point on the surface f (x, y, z) = 0 are proportional to

$$\frac{\partial f}{\partial x} : \frac{\partial f}{\partial y} : \frac{\partial f}{\partial z}$$

$$\Rightarrow \quad -\frac{\partial f}{\partial x}\bigg|\frac{\partial f}{\partial z} : -\frac{\partial f}{\partial y}\bigg|\frac{\partial f}{\partial z} : -1$$

$$\Rightarrow \quad \frac{\partial z}{\partial x} : \frac{\partial z}{\partial y} : -1 \qquad \Rightarrow p : q : -1.$$

Hence, the geometrical interpretation of (1) is that they normal to a certain surface is perpendicular to a line whose direction cosines are in the ratio P : Q : R.

We saw that the simultaneous equations

$$\frac{dx}{P} = \frac{dy}{Q} = \frac{dz}{R} \qquad ...(2)$$

represented a family of curves that the tangent at any point had direction cosines in the ratio P : Q : R and that f (u, v) = 0 represented a surface through such curves, where u = cons. and v = cons. are two particular integrals of (2). Through every point of such a surface passes a curve of the family, lying wholly on the surface. Hence, the normal to this surface at any point must be perpendicular to the tangent to this curve, i.e., perpendicular to a line whose dc's are proportional to P : Q : R. This is just what is required by the partial differential equation.

Thus, equations (1) and (2) define the same of surfaces and are thus equivalent.

3.8 The Linear Equations with a Independent Variables

Let the linear equation with n independent variables be

$$P_1p_1 + P_2p_2 + P_3p_3 + \dots + P_np_n = R \quad \text{...(1)}$$

where $p_i = \dfrac{\partial z}{\partial x_i}$, i = 1, 2, ..., n

and $P_1, P_2, \dots, P_n$ and R are functions of $x_1, x_2, \dots x_n$ and z.

The general integral of (1) is

$$f(u_1, u_2, \dots, u_n) = 0$$

where u_1 = const., u_2 = const., ..., u_n = const.

are n independent integrals of subsidiary equations.

$$\frac{dx_1}{P_1} = \frac{dx_2}{P_2} = \dots = \frac{dx_n}{P_n} = \frac{dz}{R}.$$

Example 1:

Solve $(z^2 - 2yz - y^2)\, p + (xy + zx)\, q = xy - zx.$

(Meerut, 94, 98, 92 (P), 93, 94)

Solution:

The subsidiary equations are

$$\frac{dx}{z^2 - 2yz - y^2} = \frac{dy}{xy + zx} = \frac{dz}{xy - zx}$$

Taking x, y, z as multiples, we have

$$\text{Each fraction} = \frac{x\,dx + y\,dy + z\,dz}{0}$$

$\therefore \qquad x\,dx + y\,dy + z\,dz = 0.$

Integrating, $x^2 + y^2 + z^2 = c_1$.

Again, taking the last two members, we have

$$\frac{dy}{y+z} = \frac{dz}{y-z}$$

$\Rightarrow \quad (y - z)\, dy = (y + z)\, dz$

$\Rightarrow \quad y\, dy - (z\, dy + y\, dz) - z\, dz = 0.$

Integrating, we have $y^2 - 2yz - z^2 = c_2$

$\therefore$ The general solution is

$$f\,(x^2 + y^2 + z^2,\ y^2 - 2yz - z^2) = 0.$$ **Ans.**

Example 2:

Solve $(y^3x - 2x^4)\, p + (2y^4 - x^3y)\, q = 9z\,(x^3 - y^3)$. **(Kanpur, 98, 90)**

Solution:

We have $(y^3x - 2x^4)\, p + (2y^4 - x^3y)\, q = 9z\,(x^3 - y^3)$.

The subsidiary equations are

$$\frac{dx}{y^3x - 2x^4} = \frac{dy}{2y^4 - x^3y} = \frac{dz}{9z\left(x^3 - y^3\right)}$$

Proceeding as in the two integrals are

$$\frac{y}{x^2} + \frac{x}{y^2} = c_1 \text{ and } xyz/^{1/3} = c_2.$$

$\therefore$ The general solution is $f\left(\dfrac{y}{x^2} + \dfrac{x}{y^2},\ xyz^{1/3}\right) = 0$. **Ans.**

Example 3:

Solve $(x^2 - yz)\, p + (y^2 - zx)\, q = z^2 - xy$.

Solution:

We have $(x^2 - yz)\, p + (y^2 - zx)\, q = z^2 - xy$.

The subsidiary equations are $\dfrac{dx}{x^2 - yz} = \dfrac{dy}{y^2 - zx} = \dfrac{dz}{z^2 - xy}$.

$$\therefore \frac{dx - dy}{(x-y)\,(x+y+z)} = \frac{dy - dz}{(y-z)\,(x+y+z)} = \frac{dz - dx}{(z-x)\,(x+y+z)}$$

Taking the first two members, we have

$$\frac{dx - dy}{x - y} = \frac{dy - dz}{y - z}.$$

$\therefore \qquad \log(x - y) = \log(y - z) + \log c_1$

$\Rightarrow \qquad \dfrac{x-y}{y-z} = c_1.$

Similarly taking the last two members, we have $\dfrac{z-x}{y-z} = c_2$.

The general solution is $f\left(\dfrac{x-y}{y-z}, \dfrac{z-x}{y-z}\right) = 0$. **Ans.**

Example 4:

Solve $(y^2 + z^2 - x^2)p - 2xyq + 2zx = 0$.

(Kanpur, 95, 92; Meerut, 91, 92(P), 95; PCS (UP) 91)

Solution:

The given equation can be written as

$$(y^2 + z^2 - x^2)p - 2xyq = -2zx.$$

The subsidiary equations are

$$\frac{dx}{y^2 + z^2 - x^2} = \frac{dy}{-2xy} = \frac{dz}{-2zx}$$

Taking the last two members, we have $\dfrac{dy}{y} = \dfrac{dz}{z}$

$\therefore \qquad \log\dfrac{y}{z} = \log c_1 \qquad \Rightarrow \dfrac{y}{z} = c_1.$

Again using x, y, z as multiples

Each fraction $= \dfrac{x\,dx + y\,dy + z\,dz}{-x(x^2 + y^2 + z^2)} = \dfrac{dz}{-2zx}$

$$\Rightarrow \qquad 2\frac{x\,dx + y\,dy + z\,dz}{x^2 + y^2 + z^2} = \frac{dz}{z}$$

Integrating, we have $\log(x^2 + y^2 + z^2) = \log z + \log c_2$

$\Rightarrow \qquad x^2 + y^2 + z^2 = zf\left(\dfrac{y}{z}\right).$ **Ans.**

Example 5:

Solve $(y + z)p + (z + x)q = x + y$.

Solution:

The subsidiary equations are

$$\frac{dx}{y+z} = \frac{dy}{z+x} = \frac{dz}{x+y}.$$

$$\therefore \frac{dx+dy+dz}{2(x+y+z)} = \frac{dx-dy}{(-x-y)} = \frac{dy-dz}{-(y-z)}$$

Taking the first two members, we have

$$\frac{dx+dy+dz}{x+y+z} + 2\frac{dx-dy}{x-y} = 0.$$

$\therefore \log(x + y + z) + 2 \log(x - y) = \log c_1$

$(x + y + z)(x - y)^2 = c_1.$

Again taking the last two members, we have

$$\frac{dx-dy}{x-y} = \frac{dy-dz}{y-z} \qquad \therefore \log(x - y) = \log(y - z) + \log c_2.$$

$$\Rightarrow \quad \frac{x-y}{y-z} = c_2.$$

The general integral is $f\left[\frac{x-y}{y-z}, (x-y)^2(x+y+z)\right] = 0$. **Ans.**

Example 6:

Solve $p + q = \frac{z}{a}$.

Solution:

Here P = 1, Q = 1 and $R = \frac{z}{a}$.

$\therefore$ Subsidiary equations are $\frac{dx}{1} = \frac{dy}{1} = \frac{dz}{z/a}$

Taking the first two terms we have

$dx = dy \qquad \therefore z = c_2 e^{y/a}$

Again taking the last two terms we have

$$\frac{dy}{a} = \frac{dz}{z}. \qquad \therefore z = c_2 e^{y/a}.$$

$\therefore$ The general integral is $\quad z = e^{y/a} f(x - y)$. **Ans.**

Example 7:

Solve $pz - qz = z^2 + (x + y)^2$.

Solution:

The subsidiary equations are $\frac{dx}{z} = \frac{dy}{-z} = \frac{dz}{z^2 + (x+y)^2}$.

Taking the first and the second members, we have

$dx + dy = 0$ $\therefore\ x + y = c_1$.

Taking the first and second members, we have

$$\frac{zdz}{z^2+(x+y)^2} = dx \Rightarrow \frac{2zdz}{z^2+c_1^2} = 2dx$$

Integrating, we have

$$\log(z^2 + c_1^2) = 2x + c_2;\ \log(z^2 + x^2 + y^2 + 2xy) - 2x = c_2$$

$\therefore$ The general integral is

$$f[x + y, \log(z^2 + x^2 + y^2 + 2xy) - 2x] = 0.$$ **Ans.**

Example 8:

Solve $yzp + zxq = xy$.

Solution:

We have $yzp + zxq = xy$.

The subsidiary equations are $\dfrac{dx}{yz} = \dfrac{dy}{zx} = \dfrac{dz}{xy}$.

Taking the first two members, we have

$x\,dy - y\,dy = 0.$ $\therefore\ x^2 - y^2 = c_1$.

Similarly taking the first and the last members, we have

$x^2 - z^2 = c_2$.

$\therefore$ The general integral is $f(x^2 - y^2, x^2 - z^2) = 0$. **Ans.**

Example 9:

Solve

$(2x^2 + y^2 z^2 - 2yz - zx - xy)\,p + (x^2 + 2y^2 + z^2 - yz - 2zx - xy)\,q$
$= x^2 + y^2 + 2z^2 - yz - zx - 2xy.$ **(IAS, 92; Meerut, 96(BP))**

Solution:

We have $(2x^2 + y^2 z^2 - 2yz - zx - xy)\,p + (x^2 + 2y^2 + z^2 -$
$yz - 2zx - xy)\,q = x^2 + y^2 + 2z^2 - yz - zx - 2xy.$

The subsidiary equations are

$$\frac{dx}{2x^2+y^2+z^2-2yz-zx-xy} = \frac{dy}{x^2+2y^2+z^2-yz-2zx-xy}$$
$$= \frac{dz}{x^2+y^2+2z^2-yz-zx-2xy}$$

$$\therefore \quad \frac{dx - dy}{(x-y)(x+y+z)} = \frac{dy - dz}{(y-z)(x+y+z)} = \frac{dz - dx}{(z-x)(x+y+z)}$$

$$\therefore \quad \frac{dx - dy}{x - y} = \frac{dy - dz}{y - z}$$

Integrating, $\log(x - y) = \log(y - z) + \log c_1$

$$\Rightarrow \quad \frac{x-y}{y-z} = c_1 .$$

Similarly $\frac{x-y}{y-z} = c_2$.

$\therefore$ The solution is $f\left(\frac{x-y}{y-z}, \frac{z-x}{y-z}\right) = 0$. **Ans.**

Example 10:

Solve (mz – ny) p + (nx – lz) q = ly – mx.

(Delhi, Hons. 91; Meerut, 92, 96, 97, 98; Garhwal, 96)

Solution:

We have $(mz - ny)\,p + (nx - lz)\,q = ly - mx$.

The subsidiary equations are $\frac{dx}{mz - ny} = \frac{dy}{nx - lz} = \frac{dz}{ly - mx}$

Using x, y, z as multiples, we have

Each fraction $= \frac{x\,dx + y\,dy + z\,dx}{0}$.

$\therefore \quad x\,dx + y\,dy + z\,dz = 0$

Integrating, $x^2 + y^2 + z^2 = c_1$

Again using l, m, n as multipliers

Each fraction $= \frac{l\,dx + m\,dy + n\,dz}{0}$ $\therefore\ l\,dx + m\,dy + n\,dz = 0$.

Integrating, $lx + my + nz = c_2$

Hence the general integral is

$f(lx + my + nz,\ x^2 + y^2 + z^2) = 0$. **Ans.**

Example 11:

Solve $\frac{y^2 zp}{x} + zxq = y^2$.

Solution:

The given equation can be written as $y^2zp + zx^2q = xy^2$.

Thus, subsidiary equations are $\frac{dx}{y^2z} = \frac{dy}{zx^2} = \frac{dz}{xy^2}$.

Taking the first two members, we have

$$x^2dx = y^2dy \quad \therefore\ x^3 - y^3 = c_1.$$

Again taking the first and third members, we have

$$xdx = zdz \quad \therefore\ x^2 - z^2 = c_2.$$

$\therefore$ The general solution is $\quad f(x^3 - y^3, x^2 - y^2) = 0.$ **Ans.**

Example 12:

Solve p cos (x + y) + q sin (x + y) = z. **(Kanpur, 1996)**

Solution:

We have $p \cos(x + y) + q \sin(x + y) = z$.

The subsidiary equations are

$$\frac{dx}{\cos(x+y)} = \frac{dy}{\sin(x+y)} = \frac{dz}{z}.$$

$$\therefore \quad \frac{dx + dy}{\cos(x+y) + \sin(x+y)} = \frac{dz}{z}.$$

Putting, $x + y = u$ so that $dx + dy = du$

we have $\frac{du}{\cos u + \sin u} = \frac{du}{z}$

$$\Rightarrow \quad \frac{du}{\sqrt{2}\sin\left(u + \frac{\pi}{4}\right)} = \frac{dz}{z} \quad \Rightarrow \operatorname{cosec}\left(u + \frac{\pi}{4}\right) du = \sqrt{2}\,\frac{dz}{z}.$$

Integrating, $\log \tan\left(\frac{u}{2} + \frac{\pi}{8}\right) + \log c_1 = \sqrt{2} \log z$

$$\therefore \quad c_1 = z^{\sqrt{2}} \cot\left(\frac{x}{2} + \frac{y}{2} + \frac{\pi}{8}\right)$$

Also $\frac{dx + dy}{\cos(x+y) + \sin(x+y)} = \frac{dx - dy}{\cos(x+y) - \sin(x+y)}$

$$\Rightarrow \quad \frac{\cos(x+y)-\sin(x+y)}{\cos(x+y)+\sin(x+y)}(dx+dy) = dx - dy$$

Integrating, we have

$$\log[\cos(x+y)+\sin(x+y)] = (x-y) + \log c_2.$$

$\therefore \quad \cos(x+y) + \sin(x+y) = c_2 e^{x-y}$

$\therefore$ The general solution is given as

$$\cos(x+y) + \sin(x+y) = e^{x-y} f\left[z^{\sqrt{2}} \cot\left(\frac{x}{2}+\frac{y}{2}+\frac{\pi}{8}\right)\right]$$

$$= e^{x-y} f\left[z^{\sqrt{2}} \tan\left(\frac{3\pi}{8}-\frac{x+y}{2}\right)\right].$$ **Ans.**

Example 13:

Solve $z(xp - yq) = y^2 - x^2.$ **(G.N.U.A., 90)**

Solution:

The given equation is $zxp - zyq = y^2 - x^2$

The subsidiary equations are $\frac{dx}{zx} = \frac{dy}{-zy} = \frac{dz}{y^2 - x^2}$

Using x, y, z as multipliers, we have, each fraction $= \frac{x\,dx + y\,dy + z\,dz}{0}$

$\therefore \quad xdx + ydy + zdz = 0$

Integrating, $\quad x^2 + y^2 + z^2 = c_1$

From the first two members, we have

$y\,dx + x\,dy = 0 \quad$ Integrating $xy = c_2$

$\therefore$ The general solution is $\quad f(xy, x^2 + y^2 + z^2) = 0.$ **Ans.**

Example 14:

Solve $x\frac{\partial z}{\partial x} + y\frac{\partial z}{\partial y} + t\frac{\partial z}{\partial t} = az + \frac{xy}{t}.$

Solution:

The subsidiary equations are $\frac{dx}{x} = \frac{dy}{y} = \frac{dt}{t} = \frac{dz}{az + \frac{xy}{t}}$

Taking 1st and 2nd members, we have $\frac{dx}{x} = \frac{dy}{y}$

$\therefore \log x = \log y - \log c_1 \;\therefore\; \frac{y}{x} = c_1$

Similarly, taking 1st and 3rd members, we have $\frac{t}{x} = c_2$.

Again taking 1st and 4th members, we have

$$\frac{dx}{x} = \frac{dz}{ax + \frac{xy}{t}}$$

$$\Rightarrow \quad \frac{dz}{dx} - \frac{a}{x} z = \frac{y}{t} \qquad \Rightarrow \quad \frac{dz}{dx} - \frac{a}{x} z = \frac{c_1}{c_2}$$

which is a linear differential equation.

$$\text{I.F.} = e^{-\int \frac{a}{x} dx} = e^{-a \log x} = \frac{1}{x^a}$$

$$\therefore \; z.\frac{1}{x^a} = \frac{c_1}{c_2} \int x^{-a} dx + c_3 \quad \Rightarrow zx^{-a} = \frac{y}{t}.\frac{x^{1-a}}{1-a} + c_3$$

$$\Rightarrow \quad z.x^{-a} - \frac{y}{t}.\frac{x^{1-a}}{1-a} = c_3.$$

$\therefore$ The general solution is $f\left(\frac{y}{x}, \frac{t}{x}, zx^{-a} - \frac{y}{t}.\frac{x^{1-a}}{1-a}\right) = 0$. **Ans.**

Example 15:

Solve $\frac{(y-z)}{yz} + \frac{(z-x)}{zx} q = \frac{x-y}{xy}$.

Solution:

The given equation can be written as

$$(xy - zx)\, p\, (yz - xy)\, q = xz - yz.$$

The subsidiary equations are $\frac{dx}{xy - zx} = \frac{dy}{yz - xy} = \frac{dz}{xz - yz}$

Using 1, 1, 1 as multiples,

Each fraction $= \frac{dx + dy + dz}{0}$

$\therefore dx + dy + dz = 0$, Integrating $x + y + z = c_1$.

Again using $\frac{1}{x}, \frac{1}{y}, \frac{1}{z}$ as multiples,

Each fraction $\dfrac{\frac{1}{x}dx+\frac{1}{y}dy+\frac{1}{z}dz}{0}$

$\therefore \quad \dfrac{1}{x}dx+\dfrac{1}{y}dy+\dfrac{1}{z}dz=0$.

Integrating, $\log x + \log y + \log z = \log c_2$. $\therefore$ $xyz = c_2$.

$\therefore$ The general solution is

$$f(x + y + z, xyz) = 0.$$

Example 16:

Solve $x^2 (y - z) p + (z - x) y^2 q = z^2 (x - y)$.

(Rohilkhand, 1999)

Solution:

The subsidiary equations are

$$\frac{dx}{x^2(y-z)}=\frac{dy}{(z-x)y^2}=\frac{dz}{z^2(x-y)}$$

Using $\dfrac{1}{x^2}, \dfrac{1}{y^2}, \dfrac{1}{z^2}$ as multipliers,

Each fraction $= \dfrac{\frac{dx}{x^2}+\frac{dy}{y^2}+\frac{dz}{z^2}}{0}$ $\quad\therefore\quad \dfrac{dx}{x^2}+\dfrac{dy}{y^2}+\dfrac{dz}{z^2}=0.$

Integrating $\quad \dfrac{1}{x}+\dfrac{1}{y}+\dfrac{1}{z}=c_1$.

Again using $\dfrac{1}{x}, \dfrac{1}{y}, \dfrac{1}{z}$ as multipliers

Each fraction $= \dfrac{\frac{dx}{x}+\frac{dy}{y}+\frac{dz}{z}}{0}$ $\quad\therefore\quad \dfrac{dx}{x}+\dfrac{dy}{y}+\dfrac{dz}{z}=0$.

Integrating, $\log x + \log y + \log z = \log c_2$. $\therefore$ $xyz = c_2$.

Hence the solution is $f\left(xyz, \dfrac{1}{x}+\dfrac{1}{y}+\dfrac{1}{z}\right)=0$. **Ans.**

Example 17:

Solve $\quad z - xp - yq = a\sqrt{(x^2+y^2+z^2)}$. **(Meerut, 91, 96(P), 97(BP)**

Solution:

The given equation can be written as

$$xp + yq = z - a\sqrt{\left(x^2 + y^2 + z^2\right)}.$$

The subsidiary equations are

$$\frac{dx}{x} = \frac{dy}{y} = \frac{dz}{z - a\sqrt{\left(x^2 + y^2 + z^2\right)}}$$

$$= \frac{x\,dx + y\,dy + z\,dz}{x^2 + y^2 + z^2 - az\sqrt{\left(x^2 + y^2 + z^2\right)}}.$$

Putting, $x^2 + y^2 + z^2 = u^2$.

so that $x\,dx + y\,dy + z\,dz = u\,du$

$$\text{Each fraction} = \frac{dz}{z - au} = \frac{u\,du}{u^2 - azu} = \frac{du}{u - az} = \frac{du + dz}{(1 - a)(u + z)}.$$

$$\therefore \frac{dx}{x} = \frac{du + dz}{(1 - a)(u + z)}. \Rightarrow (1 - a)\frac{dx}{x} = \frac{du + dz}{u + z}.$$

Integrating, $(1 - a) \log x = \log (y + z) + \log c_1$

$$\Rightarrow \quad x^{1-a} = c_1 (u + z) \quad \Rightarrow x^{1-a} = c_1\left\{z + \sqrt{\left(x^2 + y^2 + z^2\right)}\right\}.$$

Again taking the first two members, we have $\dfrac{dx}{x} = \dfrac{dy}{y}$

$$\therefore \quad \log x = \log y + \log c_2 \quad \Rightarrow \quad \frac{x}{y} = c_2.$$

$\therefore$ The general solution is

$$x^{1-a} = \left\{z + \sqrt{\left(x^2 + y^2 + z^2\right)}\right\} f\left(\frac{x}{y}\right).$$

Ans.

Example 18:

Solve $x(y^2 + z)p - y(x^2 + z)q = z(x^2 - y^2).$

(Meerut, 90, 92, 93(P), 95(BP), Kanpur, 90, 97, 90; Agra, TDC 91)

Solution:

The subsidiary equations are

$$\frac{dx}{x\left(y^2+z\right)}=\frac{dy}{-y\left(x^2+z\right)}=\frac{dz}{z\left(x^2-y^2\right)}$$

Using $\frac{1}{x}, \frac{1}{y}, \frac{1}{z}$ as multipliers,

Each fraction = $\dfrac{\frac{dx}{x}+\frac{dy}{y}+\frac{dz}{z}}{0}$ $\quad\therefore\quad \frac{dx}{x}=\frac{dy}{y}+\frac{dz}{z}=0$.

Integrating, $\log x + \log y + \log z = \log c_1$

$\Rightarrow$ $\log xyz = \log c_1$ $\quad\therefore\ xyz = c_1$.

Again using x, y, – 1 as multipliers,

Each fraction = $\dfrac{xdx+ydx-dz}{0}$ $\quad\therefore\ 2x\,dx + 2y\,dy - 2dz = 0$

Integrating, $x^2 + y^2 - 2z = c_2$.

Hence the general solution is

$f(x^2 + y^2 - 2z, xyz) = 0$. **Ans.**

Example 19:

Find the surface whose tangent planes cut off an intercept of constant length k from the exist of z. **(IAS, 93)**

Solution:

Equation of the tangent plane at (x, y, z) is

$Z - z = p(X - x) + q(Y - y)$

$\because$ k is the intercept on the axis of z.

$\therefore$ When $X = 0 = Y$, $Z = k$.

$\therefore$ $k - z = p(-x) + q(-y)$ $\quad\Rightarrow xp + yq = z - k$

The subsidiary equations are $\frac{dx}{x}=\frac{dy}{y}=\frac{dz}{z-k}$.

Taking the first two members, we have $\frac{dx}{x}=\frac{dy}{y}$

$\therefore$ $\log x = \log y - \log c_1$ $\quad\Rightarrow \frac{y}{x}=c_1$.

Again taking the first and last members, we have

$$\frac{dx}{x}=\frac{dz}{z-k}.$$

$\therefore \log x = \log (z - k) - \log c_2 \Rightarrow \dfrac{z-k}{x} = c_2.$

$\therefore$ The general solution is $f\left(\dfrac{y}{x}, \dfrac{z-k}{x}\right) = 0$

which represent the required surface. **Ans.**

Example 20:

Solve $p + 3q = 5z + \tan (y - 3x)$. **(Meerut, 91 (P), 93, 94)**

Solution:

Subsidiary equations are

$$\frac{dx}{1} = \frac{dy}{3} = \frac{dz}{5z + \tan (y - 3x)}.$$

Taking the first two members, we have

$$dy - 3\,dx = 0$$

$$\therefore \quad y - 3x = c_1.$$

Again taking the first and the last members, we have,

$$\frac{dx}{1} = \frac{dz}{5z + \tan c_1}.$$

$$\therefore \quad 5x = \log (5z + \tan c_1) - \log c_2.$$

$$\therefore \quad c_2 = e^{-5x} [5z + \tan (y - 3x)].$$

The general solution is

$$f\,[y - 3x,\ e^{-5x} \{5z + \tan (y - 3x)\}] = 0.$$ **Ans.**

Example 21:

Solve $p_2 + p_3 = 1 + p_1$.

Solution:

The given equation can be written as $\quad -p_1 + p_2 + p_3 = 1.$

The subsidiary equations are

$$\frac{dx_1}{-1} = \frac{dx_2}{1} = \frac{dx_3}{1} = \frac{dz}{1}.$$

Taking the 1st and 4th members, we have

$$dx_1 + dz = 0 \therefore x_1 + z = c_1.$$

Similarly taking 1st and 2nd, 1st and 3rd members, we have

$x_1 + x_2 = c_2$; $\quad x_1 + x_3 = c_3$.

$\therefore$ The general solution is $f(x_1 + z, x_1 + x_2, x_1 + x_3) = 0$.

Example 22:

Solve $x_2x_3p_1 + x_3x_1p_2 + x_1x_2p_3 + x_1x_2x_3 = 0$.

Solution:

The given equation can be written as

$$x_2x_3p_1 + x_3x_1p_2 + x_1x_2p_3 = -x_1x_2x_3.$$

The subsidiary equations are $\dfrac{dx_1}{x_2x_3} = \dfrac{dx_2}{x_3x_1} = \dfrac{dx_3}{x_1x_2} = \dfrac{dz}{-x_1x_2x_3}$

Taking the 1st and 4th members, we have

$$x_1dx_1 + dz = 0 \qquad \therefore\ x_1^2 + 2z = c_1.$$

Similarly taking 1st and 2nd, 1st and 3rd members, we have

$$x_1^2 - x_2^2 = c_2 \quad \text{and} \quad x_1^2 - x_3^2 = c_3.$$

$\therefore$ The general integral is

$$f\left(x_1^2 + 2z,\ x_1^2 - x_2^2,\ x_1^2 - x_3^2\right) = 0.$$

Example 23:

Solve $(y + z + t)\dfrac{\partial t}{\partial x} + (z + x + t)\dfrac{\partial t}{\partial y}$

$$+ (x + y + t)\frac{\partial t}{\partial z} = x + y + z.$$ **(Agra, 92; IAS 95)**

Solution:

The subsidiary equations are

$$\frac{dx}{y+z+t} = \frac{dy}{z+x+t} = \frac{dz}{x+y+t} = \frac{dt}{x+y+z}$$

$$\therefore \frac{dx-dy}{-(x-y)} = \frac{dy-dz}{-(y-t)} = \frac{dz-dt}{-(z-t)} = \frac{dx+dy+dz+dt}{3(x+y+z+t)}$$

Taking 1st and 2nd, we have $\dfrac{dx-dy}{x-y} = \dfrac{dy-dz}{y-z}$

$\therefore \log(x-y) = \log(y-z) + \log c_1$., $\quad \dfrac{x-y}{y-z} = c_1$

Similarly, taking 2nd and 3rd we have $\dfrac{z-t}{y-z} = c_2$.

Again taking 3rd and 4th, we have

$$\frac{dz-dt}{z-t} + \frac{dx+dy+dz+dt}{3(x+y+z+t)} = 0.$$

$\therefore \log(z-t) + \dfrac{1}{3}\log(x+y+z+t) = \log c_3$

$\Rightarrow \quad (z-t)(x+y+z+t)^{1/3} = c_3$

The general Integral is $f\left[\dfrac{x-y}{y-z}, \dfrac{z-x}{y-z}, (z-t)(x+y+z+t)\, y_3\right] = 0.$

Ans.

3.9 Special Types of Equations

There is a general method for solving partial differential equation of first order but of any degree. Before giving this method, we shall deal with some special types of equations which can be solved easily by methods other than the general method.

3.10 Standard I. Equations Involving only p and q and no x, y, z

Let the equation be written as

$$f(p, q) = 0. \quad ...(1)$$

The complete integral is given be

$$z = ax + by + c. \quad ...(2)$$

Where a and b are connected by

$$f(a, b) = 0. \quad ...(3)$$

Since $p = \dfrac{\partial z}{\partial x} = a$ and $q = \dfrac{\partial z}{\partial y} = b$

which when substituted in (3), gives (1).

From (3) we may find b in terms of a

i.e., $b = \phi(a)$ say.

The complete integral of (1) is

$$z = ax\, \phi(a).y + c.$$

General Integral: The complete integral of (1) is

$$z = ax + \phi(a).\, y + c. \quad ...(4)$$

Taking c = ψ (a) where ψ is an arbitrary function, we have

$$z = ax + \phi(a).y + \psi(a) \qquad ...(5)$$

Differentiating (5) w.r.t. a, we have

$$0 = x + \phi'(a)\,y + \psi'(a).$$

Eliminating a between equation (5) and (6) we get the general integral.

Singular Integral. The complete integral of (1) is

$$z = ax + \phi(a).y + c.$$

In order to get the singular integral we have to find the envelope of (4) be eliminating arbitrary constants a and c.

Differentiating (4) partially, w.r.t. a and c, we have

$$0 = x + \phi'(a)y \quad \text{and} \quad 0 = 1.$$

2nd result is inadmissible which shows that such equations have no singular integral.

Note: Sometimes, change of variables reduces the questions to the form of standard 1.

Example 1:

Find the complete integral of

$(y - x)(qy - px) = (p - q)^2$.

(Agra, 98; Meerut, 90, 97(BP); Kanpur, 91, 97, 90; IAS 92)

Solution:

Let us put X = x + y and Y = xy

so that $$p = \frac{\partial z}{\partial x} = \frac{\partial z}{\partial X}.\frac{\partial X}{\partial x} + \frac{\partial z}{\partial Y}.\frac{\partial Y}{\partial x} = \frac{\partial z}{\partial X} + y\frac{\partial z}{\partial Y}$$

and $$q = \frac{\partial z}{\partial X}.\frac{\partial X}{\partial y} + \frac{\partial z}{\partial Y}.\frac{\partial Y}{\partial Y} = \frac{\partial z}{\partial X} + x\frac{\partial z}{\partial Y}.$$

Substituting in the given equation, we have

$$(y - x).(y - x)\frac{\partial z}{\partial X} = (y - x)^2\left(\frac{\partial z}{\partial Y}\right)^2 \Rightarrow \frac{\partial z}{\partial X} = \left(\frac{\partial z}{\partial Y}\right)^2$$

which of the form of standard I,

∴ The complete integral is given by

$$z = aX + bY + c. \quad \text{where } a = b^2$$

The complete integral is

$$z = b^2(x + y) + bxy = c.$$ **Ans.**

Example 2:

Solve $p^2 + q^2 = 1$. **(Meerut, 91)**

Solution:

The equation is of the form f (p, q) = 0,

The solution is given by

$$z = ax + by + c$$

where $a^2 + b^2 = 1 \Rightarrow b = \sqrt{(1-a^2)}$.

$\therefore$ the complete integral is

$$z = ax + \sqrt{(1-a^2)}\, y + c.$$

General Integral. Write $c = \psi(a)$.

The general integral is obtained by eliminating a from

$$z = ax + \sqrt{(1-a^2)}\, y + \psi(a) \text{ and } 0 = a + \frac{-a}{\sqrt{(1-a^2)}} y + \psi'(a) \qquad \textbf{Ans.}$$

Example 3:

Find the complete integral of

$$p^m \sec^{2m} x + z^l q^n \operatorname{cosec}^{2n} y = z^{lm\,(m\ n)}.$$

Solution:

We have $p^m \sec^{2m} x + z^l q^n \operatorname{cosec}^{2n} y = z^{lm/(m-n)}$.

The given equation can be written as

$$\left(\frac{z^{-l/(m-n)}}{\cos^2 x}\cdot\frac{\partial z}{\partial x}\right)^m + \left(\frac{z^{-l/(m-n)}}{\sin^2 y}\frac{\partial z}{\partial y}\right)^n = 1. \qquad ...(1)$$

Now putting

$$z^{-l/(m-n)}\, dz = dZ,\ \cos^2 x\, dx = dX \text{ and } \sin^2 y\, dy = dY,$$

so that $Z = \dfrac{(m-n)}{m-n-l} z^{(m-n-l)/(m-n)}$

$$X = \frac{1}{2}\left(x + \frac{1}{2}\sin 2x\right) \quad \text{and} \quad Y = \frac{1}{2}\left(y - \frac{1}{2}\sin 2y\right)$$

equation (1) reduces to

$$\left(\frac{\partial Z}{\partial X}\right)^m + \left(\frac{\partial Z}{\partial Y}\right)^n = 1,$$

which is of the form of standard I.

∴ The complete integral is given by

$Z = aX + bY + c$ where $a^m + b^n = 1 \Rightarrow b = (1 - a^m)^{1/n}$.

∴ The complete integral is

$$\frac{m-n}{(m-n-l)} z^{(m-n-l)/(m-n)} = a.\frac{1}{2}\left(x+\frac{1}{2}\sin 2x\right) + \left(1-a^m\right)^{1/n}.\frac{1}{2}\left(y-\frac{1}{2}\sin 2y\right)+c.$$ **Ans.**

Example 4:

$x^2p^2 + y^2q^2 = z^2$. **(Jiwaji, 92; Meerut, 90, 93, 97; Rohilkhand, 92)**

Solution:

We have $x^2p^2 + y^2q^2 = z^2$.

The given equation can be written as

$$\left(\frac{x}{z}.\frac{\partial z}{\partial x}\right)^2 + \left(\frac{y}{z}.\frac{\partial z}{\partial y}\right)^2 = 1. \quad ...(1)$$

Now, let us put $\frac{1}{z}dz = dZ$ i.e. $z = e^Z$

$\frac{1}{x}dx = dX$ i.e. $x = e^X$

and $\frac{1}{y}dy = dY$ i.e. $y = e^Y$.

Equation (1) becomes $\left(\frac{\partial Z}{\partial X}\right)^2 + \left(\frac{\partial Z}{\partial Y}\right)^2 = 1$

which is of the form of standard 1.

∴ The complete integral is given by

$Z\ aX + bY + c_1$ where $a^2 + b^2 = 1 \Rightarrow b = \sqrt{(1-a^2)}$.

∴ $Z = aX + \sqrt{(1-a^2)}\ Y + c_1$

⇒ $\log z = a \log x + \sqrt{(1-a^2)} \log y + c_1$

if we put $a = \cos\alpha$

$\log z = \cos\alpha. \log x + \sin\alpha \log y + \log c$

∴ $z = cx^{\cos\alpha}. y^{\sin\alpha}$

General Integral. It is obtained by eliminating α from

$$z = \phi(\alpha)\, x^{\cos\alpha}.\, y^{\sin\alpha}. \text{ Taking } c = \phi(\alpha)$$

and $$= \phi'(\alpha)\, x^{\cos\alpha}.\, y^{\sin\alpha} + \phi(\alpha)\, x^{\cos\alpha}.y^{\sin\alpha}\,(-\sin a).\log_e x$$
$$+ \phi'(\alpha)\, c^{\cos\alpha}.y^{\sin\alpha}.\cos\alpha.\log_e y$$

Singular Integral. It is obtained by eliminating α and c, from

$$z = cx^{\cos\alpha}.y^{\sin\alpha}$$

$$\frac{\partial z}{\partial \alpha} = -c\sin\alpha.x^{\cos\alpha}.y^{\sin\alpha}\log_e x + c\cos\alpha.x^{\cos\alpha}.y^{\sin\alpha}\log_e y = 0$$

and $\dfrac{\partial z}{\partial c} = x^{\cos\alpha}.y^{\sin\alpha} = 0.$

$\therefore$ Singular integral is $z = 0$. **Ans.**

Example 5:

Solve $(x^2 + y^2)(p^2 + q^2) = 1.$

(Meerut, 92(P), 95; Raj 91; Kanpur, 99; Rohilkhand, 94)

Solution:

We have $(x^2 + y^2)(p^2 + q^2) = 1.$

Let us put

$$x = r\cos\theta,\ y = r\sin\theta, \text{ i.e. } x^2 + y^2 = r^2,\ \theta = \tan^{-1}\frac{y}{x},$$

so that $$p = \frac{\partial z}{\partial x} = \frac{\partial z}{\partial r}.\frac{\partial r}{\partial x} + \frac{\partial z}{\partial \theta}.\frac{\partial \theta}{\partial x} = \frac{\partial z}{\partial r}\cos\theta - \frac{\sin\theta}{r}.\frac{\partial z}{\partial \theta}$$

and $$q = \frac{\partial z}{\partial y} = \frac{\partial z}{\partial r}.\frac{\partial r}{\partial y} + \frac{\partial z}{\partial \theta}.\frac{\partial \theta}{\partial y} = \frac{\partial z}{\partial r}\sin\theta + \frac{\cos\theta}{r}.\frac{\partial z}{\partial \theta}$$

$\therefore$ Given equation reduces to

$$r^2\left[\left(\frac{\partial z}{\partial r}\right)^2 + \frac{1}{r^2}\left(\frac{\partial z}{\partial \theta}\right)^2\right] = 1$$

$$r^2\left(\frac{\partial z}{\partial r}\right)^2 + \left(\frac{\partial z}{\partial \theta}\right)^2 = 1. \qquad ...(1)$$

Again putting $du = \dfrac{dr}{r}$, i.e. $u = \log r$ in (1), we have

$$\left(\frac{\partial z}{\partial u}\right)^2 + \left(\frac{\partial z}{\partial \theta}\right)^2 = 1$$

which is of the form of standard I.

$\therefore$ The complete integral is given by

$$z = au + b\theta + c \qquad \text{where } a^2 + b^2 = 1 \Rightarrow b = \sqrt{(1-a^2)}$$

$\therefore$ The complete integral is

$$z = a \log r + \sqrt{(1-a^2)}\,\theta + c$$

$$= \frac{a}{2}\log\left(x^2+y^2\right) + \sqrt{(1-a^2)}\tan^{-1}\frac{y}{x} + c.$$ **Ans.**

Example 6:

Solve $(x + y)(p + q)^2 + (x - y)(p - q)^2 = 1.$

(Kanpur, 98, 92; Meerut, 92, 97; IAS 91)

Solution:

We have $(x + y)(p + q)^2 + (x - y)(p - q)^2 = 1.$

Putting $x + y = X^2$, $x - y = Y^2$

so that $$p = \frac{\partial z}{\partial x} = \frac{\partial z}{\partial X}\cdot\frac{\partial X}{\partial x} + \frac{\partial z}{\partial Y}\cdot\frac{\partial Y}{\partial x} = \frac{1}{2X}\frac{\partial z}{\partial X} + \frac{1}{2Y}\cdot\frac{\partial z}{\partial Y}$$

and $$q = \frac{dz}{\partial y} = \frac{\partial z}{\partial X}\cdot\frac{\partial X}{\partial y} + \frac{\partial z}{\partial Y}\cdot\frac{\partial Y}{\partial y} = \frac{1}{2X}\frac{\partial z}{\partial X} - \frac{1}{2Y}\cdot\frac{\partial z}{\partial Y}$$

$$\therefore\ p + q = \frac{1}{x}\frac{\partial z}{\partial X} \text{ and } p - q = \frac{1}{Y}\frac{\partial z}{\partial X}.$$

Putting in the given equation, we have

$$\left(\frac{\partial z}{\partial X}\right)^2 + \left(\frac{\partial z}{\partial Y}\right)^2 = 1$$

which is of the form of standard I.

$\therefore$ The complete integral is given by

$$z = aX + bY + c$$

where $a^2 + b^2 = 1 \qquad \Rightarrow b = \sqrt{(1-a^2)}.$

$\therefore$ The complete integral is

$$z = a\sqrt{(x+y)} + \sqrt{(1-a^2)}\sqrt{(x-y)} + c.$$ **Ans.**

3.11 Standard II: Equations Involving only p, q and z i.e. Equations of the Form f (z, p, q) = 0. ...(1)

Let us assume z = f(x + ay) as a trial solution of given equation (1), where a is an arbitrary constant

$$\therefore \qquad z = f(x) \text{ where } X = x + ay$$

$$\therefore \qquad p = \frac{\partial z}{\partial x} = \frac{\partial z}{\partial X}\frac{\partial X}{\partial x} = \frac{\partial z}{\partial X} = \frac{\partial z}{\partial X}$$

and $$q = \frac{\partial z}{\partial y} = \frac{\partial z}{\partial X}\frac{\partial X}{\partial x} = a\frac{\partial z}{\partial X} = a\frac{\partial z}{\partial X}$$

$\therefore$ Equation (1) reduces to the form

$$f\left(z, \frac{dz}{dX}, a\frac{dz}{dX}\right) = 0$$

which is an ordinary differential equation of order one. Integrating it we may get the complete integral.

Method: To integrate the equation of the form of standard II i.e., of the form f (z, p, q) = 0, put $\frac{dz}{dX}$ and a $\frac{dz}{dX}$ for p and q respectively and then solve the equation so obtained. Then substitute x + ay for X.

General Integral: If F = 0 is the complete integral involving two constants a and b then replacing b by ϕ (a), the general is obtained by eliminating a from F = 0 and $\frac{dF}{da} = 0$.

Singular Integral: If F = 0 is the complete integral involving two arbitrary constants a and b then the singular integral is obtained by eliminating a and b from

$$F = 0, \ \frac{\partial F}{\partial a} = 0 \text{ and } \frac{\partial F}{\partial b} = 0.$$

Note: Sometimes, change of variable may reduce the equation to the form of standard II.

Example 1:

Solve $pz = 1 + q^2$. **(Meerut, 96 (P))**

Solution:

We have $pz = 1 + q^2$.

Putting z = f (x + ay) = f (X) where X = x + ay

so that $p = \dfrac{\partial z}{\partial x} = \dfrac{\partial z}{\partial X}$ and $q = \dfrac{\partial z}{\partial y} = a\dfrac{\partial z}{\partial X}$

The equation reduces to

$$z\frac{dz}{dX} = 1 + a^2\left(\frac{dz}{dX}\right)^2 \quad \Rightarrow a^2\left(\frac{dz}{dX}\right)^2 - z\frac{dz}{dX} + 1 = 0$$

$$\therefore \frac{dz}{dX} = \frac{z \pm \sqrt{(z^2 - 4a^2)}}{2a^2} \quad \Rightarrow \frac{dz}{z \pm \sqrt{(z^2 - 4a^2)}} = \frac{dX}{2a^2}$$

$$\Rightarrow \quad \frac{z \pm \sqrt{(z^2 - 4a^2)}}{4a^2}\, dz = \frac{dX}{2a^2} \Rightarrow \left\{z \pm \sqrt{(z^2 - 4a^2)}\right\} dz = 2dX.$$

Integrating, we have

$$\frac{z^2}{2} \pm \left[\frac{z}{2}\sqrt{(z^2 - 4a^2)} - \frac{4a^2}{2}\log\left\{z + \sqrt{(z^2 - 4a^2)}\right\}\right] = 2X + c$$

$$\Rightarrow \quad z^2 \pm \left[z\sqrt{(z^2 - 4a^2)} - 4a^2 \log\left\{z + \sqrt{(z^2 - 4a^2)}\right\}\right] = 4x + 4ay + 2c$$

which is the complete integral. **Ans.**

Example 2:

Find the complete integral of $p^2 = 2q$.

Solution:

We have $p^2 = 2q$.

Putting $z = f(x + ay) = f(X)$ in the given equation, we have

$$\left(\frac{dz}{dX}\right)^2 = za\frac{dz}{dX} \quad \Rightarrow \frac{dz}{dX} = za$$

$$\Rightarrow \quad \frac{dz}{z} = a\, dX.$$

Integrating we have $\log z = aX + \log b$

$\therefore z = be^{aX} \quad \Rightarrow \quad z = be^{a\,(x+ay)}$

$$\Rightarrow \quad z = be^{ax+a^2y}$$

which is the complete integral. **Ans.**

Example 3:

Solve $9(p^2z + q^2) = 4.$ **(Rohilkhand, 92; Meerut, 91(P), 98, 96(BP))**

Solution:

Putting $z = f(x + ay) = f(X)$ where $X = x + ay$

so that $p = \frac{\partial z}{\partial x} = \frac{\partial z}{\partial X}$ and $q = \frac{\partial z}{\partial q} = a\frac{\partial z}{\partial X}$

The equation reduces to

$$9\left[\left(\frac{dz}{dx}\right)^2 z + a^2\left(\frac{dz}{dX}\right)^2\right] = 4$$

$$\Rightarrow \quad 9\left(\frac{dz}{dX}\right)^2 \left(z + a^2\right) = 4$$

$$\Rightarrow \quad 3\sqrt{\left(z + a^2\right)}\frac{dz}{dX} = 2$$

$$\Rightarrow \quad 3\sqrt{\left(z + a^2\right)}\ dz = 2dX$$

Integrating, we get $(z + a^2)^{3/2} = X + b$

$\Rightarrow \quad (z + a^2)^{3/2} = x + ay + b$

$\Rightarrow \quad (a + a^2)^3 = (x + ay + b)^2$

which is the required solution. **Ans.**

Example 4:

Solve $q^2y^2 = z(z - px).$ **(Agra, 91, Meerut, 99, 92(P), 93, 97)**

Solution:

We have $q^2y^2 = z(z - px)$.

The given equation can be written as ...(1)

$$\left(y\frac{\partial z}{\partial y}\right)^2 = z\left(z - x\frac{\partial z}{\partial x}\right)$$

Putting $\frac{dx}{x} = dX$ so that $X = \log x$

and $\frac{dy}{y} = dY$ so that $X = \log y$

the given equation becomes

$$\left(\frac{\partial z}{\partial Y}\right)^2 = z\left(z - \frac{\partial z}{\partial Y}\right) \qquad ...(2)$$

which is of the form of standard II.

Now putting $z = f(X + aY) = f(u)$ where $u = X + aY$.

so that $\dfrac{\partial z}{\partial X} = \dfrac{dz}{du}\dfrac{\partial u}{\partial X} = \dfrac{dz}{du}$

and $\dfrac{\partial z}{\partial Y} = \dfrac{dz}{du}\dfrac{\partial u}{\partial Y} = a\dfrac{dz}{du}$.

The equation (2) reduces to $a^2\left(\dfrac{dz}{du}\right)^2 = z\left(z - \dfrac{dz}{du}\right)$.

$$\Rightarrow \quad a^2\left(\frac{dz}{du}\right)^2 = z\frac{dz}{du} - z^2 = 0$$

$$\frac{dz}{du} = \frac{-z \pm \sqrt{(z^2 + 4a^2z^2)}}{2a^2} = \frac{z}{2a^2}\left[-1 \pm \sqrt{(1 + 4a^2)}\right]$$

$$\Rightarrow \quad 2a^2\frac{dz}{z} = \left[-1 \pm \sqrt{(1 + 4a^2)}\right]du,$$

Integrating, $\quad 2a^2 \log z = \left[-1 \pm \sqrt{(1 + 4a^2)}\right](u + \log b)$

$$\Rightarrow \quad 2a^2 \log z = \left[-1 \pm \sqrt{(1 + 4a^2)}\right](X + aY + \log b)$$

$$\Rightarrow \quad 2a^2 \log z = \left[-1 \pm \sqrt{(1 + 4a^2)}\right](\log x + a \log y + \log b)$$

$$\Rightarrow \quad z^{2a^2/\left(-1 \pm \sqrt{(1+4a2)}\right)} = bxy^a.$$

which is the required integral. **Ans.**

Example 5:

Find the complete integral of

$$z^2(p^2z^2 + q^2) = 1. \qquad \textbf{(Meerut, 93(P), 96)}$$

Find also the singular integral if it exists.

Solution:

We have $z^2(p^2z^2 + q^2) = 1$.

Putting $z = f(x + ay) = f(X)$ where $X = x + ay$

so that $p = \frac{\partial z}{\partial x} = \frac{\partial z}{\partial X}$

and $q = \frac{\partial z}{\partial y} = a\frac{\partial z}{\partial X}$

The equation becomes

$$z^2\left[\left(\frac{dz}{dX}\right)^2 .z^2 + a^2\left(\frac{dz}{dX}\right)^2\right] = 1$$

$$\Rightarrow \quad z^2 (z^2 + a^2)\left(\frac{dz}{dX}\right)^2 = 1 \Rightarrow z\sqrt{(z^2 + a^2)}\, dz = dX$$

Integrating $\frac{1}{3}(z^2 + a^2)^{3/2} = X + b$

$$\Rightarrow \quad 9(x + ay + b)^2 = (z^2 + a^2)^3 \qquad ...(1)$$

Singular Integral. Differentiating (1) partially w.r.t a and b, we have

$$18 (x + ay + b) y = 6 (z^2 + a^2).a \qquad ...(2)$$

and $\quad 18 (x + ay + b) = 0 \qquad ...(3)$

From (2) and (3), we have $\quad a = 0 \qquad ...(4)$

which may be the singular integral.

From $z = 0$, we have $p = 0, q = 0$

which does not satisfy the given equation.

Hence, $z = 0$ is not singular integral i.e., given equation does not have singular integral.

Example 6:

Solve $p^2 = z^2(1 - pq)$. **(Meerut, 92, 94; Rohilkhand, 90, 93)**

Solution:

Putting $z = f(x + ay) = f(x)$ where $X = x + ay$

so that $p = \frac{\partial z}{\partial x} = \frac{dz}{dX}$ and $q = \frac{\partial z}{\partial y} = a\frac{dz}{dX}$

$$\left(\frac{dz}{dX}\right)^2 = z^2\left\{1 - \frac{dz}{dX}.a\frac{dz}{dX}\right\}$$

$$\Rightarrow \quad \left(1+az^2\right)\left(\frac{dz}{dX}\right)^2 = z^2$$

$$\Rightarrow \quad \frac{\sqrt{\left(1+az^2\right)}}{z}dz = dX$$

$$\Rightarrow \quad \frac{1+az^2}{z\sqrt{\left(1+az^2\right)}}dz = dX$$

$$\Rightarrow \quad \left(\frac{1}{z\sqrt{\left(1+az^2\right)}} + \frac{az}{\sqrt{\left(1+az^2\right)}}\right)dz = dX.$$

Integrating $\frac{1}{\sqrt{a}}\log\left[z\sqrt{a}+\sqrt{a\left(1+az^2\right)}\right]+\sqrt{\left(1+az^2\right)} = X + c$

which is the complete integral. **Ans.**

Example 7:

Find the complete integral of

$p^3 + q^3 = 27z.$ *where* $X = x + ay.$

Solution:

Putting $z = f(x + ay) = f(X)$

so that $p = \frac{\partial z}{\partial x} = \frac{\partial z}{\partial X}$ and $\frac{\partial z}{\partial y} = a\frac{dz}{dX}$

The equation reduces to $(1 + a^3)\left(\frac{dz}{dX}\right)^3 = 27z$

$$\Rightarrow \quad (1 + a^3)^{1/3}\frac{dz}{dX} = 3z^{1/3}$$

$$\Rightarrow \quad (1 + a^3)^{1/3}\frac{2}{3}z^{-1/3}\,dz = 2dX$$

Integrating, $z^{2/3}(1 + a^3) = 2X + c = 2(X + b)$

$\Rightarrow \quad (1 + a^3)z^2 = 8(x + ay + b)^3$ ...(1)

which is the required complete integral.

Singular Integral: Differentiating (1) partially w.r.t. a and b, we have

$3a^2z^2 = 24y(x + ay + b)^2$ ...(2)

and $= 24(x + ay + b)^2$

Eliminating a, b from (1), (2) and (3), we have

$$z = 0$$

Which is the singular solution. **Ans.**

Example 8:

Solve $p(1 + q^2) = q(z - a)$. **(Kanpur, 96; Meerut, 97(BP))**

Solution:

Putting $z = f(x + ay) = f(X)$ where $X = x + ay$

so that $p = \frac{\partial z}{\partial X} = \frac{\partial z}{\partial X}$ and $q = \frac{\partial z}{\partial y} = a\frac{\partial z}{\partial X}$.

The equation becomes

$$\frac{dz}{dX}\left[1 + a^2\left(\frac{dz}{dX}\right)^2\right] = a\,\frac{dz}{dX}(z - a)$$

$$\Rightarrow \quad a^2\left(\frac{dz}{dX}\right)^2 = (z - a) - 1$$

$$\Rightarrow \quad \frac{a\,dz}{\sqrt{[a(z-a)-1]}} = dX$$

Integrating, we have

$$2.\sqrt{[a(z-a)-1]} = X + b \Rightarrow 4a(z - a) - 4 = (x + ay + b)^2$$

$$\Rightarrow \quad 4a(z - a) = 4 + (x + ay + b)^2.$$ **Ans.**

Example 9:

Solve $z^2(p^2 + q^2 + 1) = c^2$. ...(1)

(Meerut, 97(P))

Solution:

Putting $z\,dz = dZ$ so that $X = z^2/2$

$$\frac{\partial Z}{\partial x} = \frac{\partial Z}{\partial z}.\frac{\partial z}{\partial x} = zp$$

$$\frac{\partial Z}{\partial y} = \frac{\partial Z}{\partial z}.\frac{\partial z}{\partial y} = zq.$$

The given equation reduces to

$$\left(\frac{\partial Z}{\partial x}\right)^2 + \left(\frac{\partial Z}{\partial y}\right)^2 + 2Z = c \qquad ...(2)$$

which is the form of Standard II.

$\therefore$ Let Z = f (x + ay) = f (X) where X = x + ay

so that $\frac{\partial Z}{\partial x} = \frac{\partial Z}{\partial X}.\frac{\partial X}{\partial x} = \frac{\partial Z}{\partial X}$

and $\frac{\partial Z}{\partial y} = \frac{\partial Z}{\partial X}.\frac{\partial X}{\partial y} = a\frac{\partial Z}{\partial X}$

The equation (2) becomes

$$\left(\frac{dZ}{dX}\right)^2 (1 + a^2) + 2Z = c.$$

$$\Rightarrow \quad \frac{\sqrt{(1+a^2)}}{\sqrt{(c-2Z)}} dZ = dX$$

Integrating $-\sqrt{(1+a^2)}.\sqrt{(c-2z)} = X + b$

$(1 + a^2)(c - z^2) = (x + ay + b)^2$. **Ans.**

Example 10:

Solve $pq = x^m y^n z^l$. **(IAS, 99, 94)**

Solution:

Putting $\frac{x^{m+1}}{m+1} = X, \frac{y^{n+1}}{n+1} = Y$

so that $p = \frac{\partial z}{\partial x} = \frac{\partial z}{\partial X}.\frac{\partial X}{\partial x} = x^m \frac{\partial z}{\partial X}$

and $q = \frac{\partial z}{\partial y} = \frac{\partial z}{\partial Y}.\frac{\partial Y}{\partial y} = y^n \frac{\partial z}{\partial Y}$

The given equation reduces to

$$\frac{\partial z}{\partial X}.\frac{\partial z}{\partial Y} = z^l \qquad ...(1)$$

which is the form of Standard II.

$\therefore$ Putting z = f (X + aY) = f (u) where X + aY = u

$$\frac{\partial z}{\partial X}=\frac{\partial z}{\partial u}.\frac{\partial u}{\partial X}=\frac{\partial z}{\partial u}$$

and $$\frac{\partial z}{\partial Y}=\frac{\partial z}{\partial u}.\frac{\partial u}{\partial Y}=a\frac{\partial z}{\partial u}.$$

equation (2) becomes

$$a\left(\frac{dz}{du}\right)^2=z^l. \qquad \Rightarrow z^{-l/2}\ dz=\frac{du}{\sqrt{a}}.$$

Integrating, we have $\dfrac{z^{-l/2+1}}{-l/2+1}=\dfrac{u}{\sqrt{a}}+b.$

$$\Rightarrow \quad \frac{z^{-l/2+1}}{-l/2+1}=\frac{1}{\sqrt{a}}\left(\frac{x^{m+1}}{n+1}+a\frac{y^{n+1}}{n+1}\right)+b.$$ **Ans.**

3.12 Standard III: i.e., Equation of the Form f (x, p) = F (y, q)

As a trial solution let us put each side equal to a arbitrary constant.

i.e. $f(x, p) = F(y, q) = a$

from which we obtain

$$p = f_1(x, a) \text{ and } q = f_2(y, a).$$

Now, from $dz = pdx + qdy$,

we have $dz = f_1(x, a)\,dx + f_2(y, a)\,dy$

$$z = \int f_1(x, a)\,dx + \int f_2(y, a)\,dy + b$$

which is the complete integral.

General Integral may be obtained as in other cases.

Singular Integral: As in standard I it may be shown that there will be no singular solution.

Method to Obtain Complete Integral: Put both the sides of the given equation of the above form equal to an arbitrary constant. Solving the we get values of p and q. Put the values of p and q in $dz = p\,dx + q\,dy$ and integrate to obtain the complete integral.

Example 1:

$z(p^2 - q^2) = x - y.$ **(IAS, 9l; PCS (UP), 92)**

Solution:

The given equation can be written as

$$\left(\sqrt{z}\frac{\partial z}{\partial x}\right)^2 - \left(\sqrt{z}\frac{\partial z}{\partial y}\right)^2 = x - y$$

Putting $\sqrt{z}\,dx = dZ$ so that $Z = \frac{2}{3}z^{3/2}$.

The equation becomes $\left(\frac{\partial Z}{\partial x}\right)^2 - \left(\frac{\partial Z}{\partial y}\right)^2 = x - y$.

$\Rightarrow$ $P^2 - Q^2 = x - y$ where $P = \frac{\partial Z}{\partial x}, Q = \frac{\partial Z}{\partial y}$

$\Rightarrow$ $P^2 - x\,Q^2 - y = a$

which is of the form of standard III.

$\therefore$ Let $P^2 - x = Q^2 - y = a$.

$\therefore$ $P = \sqrt{(x+a)}$ and $Q = \sqrt{(y+a)}$.

Putting in $dZ = Pdx + Qdy$, we have

$dZ = \sqrt{(x+a)}\,dx + \sqrt{(y+a)}\,dy$.

Integrating, $Z = \frac{2}{3}(x+a)^{3/2} + \frac{2}{3}(y+a)^{3/2} + b$

$\Rightarrow$ $z^{3/2} = (x + a)^{3/2} + (y + a)^{3/2} + c$. **Ans.**

Example 2:

Solve $\sqrt{p} + \sqrt{q} = 2x$. **(Rohilkhand, 91)**

Solution:

The given equation can be written as

$\sqrt{p} - 2x = -\sqrt{q}$

$\therefore$ Let $\sqrt{p} - 2x = a = -\sqrt{q}$

$\therefore$ $p = (a + 2x)^2$ and $q = a^2$

Putting these values in $dx = p\,dx + q\,dy$, we have

$$dz = (a + 2x)^2\,dx + a^2\,dy$$

Integrating $z = \frac{1}{6}(a + 2x)^3 + a^2y + b$,

which is the complete integral.

Example 3:

Solve $z^2 (p^2 + q^2) = x^2 + y^2$. **(Meerut, 91; Rohilkhand, 92)**

Solution:

Putting z dz = dZ

so that $\frac{z^2}{2} = Z$.

The given equation reduces to

$$P^2 + Q^2 = x^2 + y^2 \quad \text{where } P = \frac{dZ}{dx} \text{ and } Q = \frac{dZ}{dy}.$$

$$P^2 - x^2 = - Q^2 + y^2$$

which is of form of standard III.

$\therefore$ Let $P^2 - x^2 = - Q^2 + y^2 = a$

$\therefore P = \sqrt{(a + x^2)}$ and $Q = \sqrt{(y^2 - a)}$

Putting in dZ = Pdx + Qdy, we have

$$dZ = \sqrt{(a + x^2)}dx + \sqrt{(y^2 - a)}\, dy$$

Integrating, $z = \frac{x}{2}\sqrt{(a + x^2)} + \frac{a}{2}\log\left\{x + \sqrt{(a + x^2)}\right\}$

$$+ \frac{y}{2}\sqrt{(y^2 - a)} - \frac{a}{2}\log\left\{y + \sqrt{(y^2 - a^2)}\right\} + b$$

$$\Rightarrow \quad z^2 = x\sqrt{(a + x^2)} + a \log\left\{x + \sqrt{(a + x^2)}\right\}$$

$$+ y\sqrt{(y^2 - a)} - a \log\left\{y + (y^2 - a)\right\} + c.$$ **Ans.**

Example 4:

Solve $pq = xy$.

Solution:

Let $\frac{p}{x} = \frac{y}{q} = a$. $\therefore$ $p = ax$, $q = \frac{y}{a}$.

Putting in dz = pdx + q dy, we have

$$dz = ax\,dx + \frac{y}{a}.dy.$$

Integrating, we have $z = \frac{a}{2}x^2 + \frac{1}{2a}y^2 + b$

$\Rightarrow$ $$2az = a^2x^2 + y^2 + 2ab.$$ **Ans.**

Example 5:

Solve $pe^y = qe^x$.

Solution:

The given equation can be written as

$pe^{-x} = qe^{-y} = a$ (say)

$\therefore p = ae^x$ and $q = ae^y$.

Putting in $dz = pdx + qdy$, we have

$dz = ae^x + ae^y\,dy$.

Integrating, we get $z = ae^x + ae^y + b$. **Ans.**

3.13 Standard IV i.e., Equation of the Form

$$\mathbf{z = px + qy + f(p, q)} \quad ...(1)$$

(Analogous to Clairaut's Form)

We know that the solution of Clairaut's Equation

$y = px + f(p)$ where $p = \frac{dy}{dx}$ is $y = cx = f(c)$.

Similarly the complete integral of Clairaut's equation (1) is

$$z = ax + by + \ (a, b).$$

Method: To get the complete integral of the equation of this type replace p and q by a, b (two arbitrary constants) respectively.

General Integral is obtained as in other cases.

Singular Integral: The complete integral is

$$F = z - ax - by - f(a, b) = 0 \quad ...(1)$$

$$\frac{\partial F}{\partial a} = 0, \text{ gives } x + \frac{\partial f}{\partial a} = 0, \quad ...(2)$$

and $$\frac{\partial F}{\partial b} = 0, \text{ gives } y + \frac{\partial f}{\partial b} = 0. \quad ...(3)$$

Singular integral is obtained by eliminating a, b from (1), (2) and (3).

Example 1:

Solve $z = px + qy - \sqrt{(pq)}$.

Solution:

The complete integral is

$$z = ax + by - 2\sqrt{(ab)} . \quad \text{...(1)}$$

Singular Integral. Differentiating (1) partially w.r.t. a and b, we have

$$0 = x - \frac{2}{2\sqrt{(ab)}} b \text{ i.e. } x = \sqrt{\left(\frac{b}{a}\right)} \quad \text{...(1)}$$

and
$$0 = y - \frac{2}{2\sqrt{(ab)}} a \text{ i.e. } y = \sqrt{\left(\frac{a}{b}\right)}$$

Eliminating a, b, the singular solution is

$xy = 1$. **Ans.**

Example 2:

Solve $z = px + qy + c\sqrt{(1 + p^2 + q^2)}$.

(Meerut 99, 91, 91(P); IAS 99)

Solution:

This is of the form of standard IV.

∴ The complete integral is

$$z = ax + by + c\sqrt{(1 + a^2 + b^2)} \quad \text{...(1)}$$

Singular Integral. Differentiating equation (1) partially w.r.t. a and b, we have

$$0 = x + \frac{ac}{\sqrt{(1 + a^2 + b^2)}} \quad \text{...(2)}$$

and
$$0 = y + \frac{bc}{\sqrt{(1 + a^2 + b^2)}} \quad \text{...(3)}$$

$$\therefore x^2 + y^2 = \frac{(a^2 + b^2)c^2}{1 + a^2 + b^2}$$

$$\Rightarrow \quad c^2 - x^2 - y^2 = \frac{c^2}{1+a^2+b^2}$$

$$\Rightarrow \quad 1 + a^2 + b^2 = \frac{c^2}{c^2 - x^2 - y^2}$$

$\therefore$ From (2), (3) $\quad a = -\dfrac{x\sqrt{(1+a^2+b^2)}}{c} = \dfrac{-x}{\sqrt{(c^2-x^2-y^2)}}$

and $\quad b = -\dfrac{y\sqrt{(1+a^2+b^2)}}{c} = \dfrac{-y}{\sqrt{(c^2-x^2-y^2)}}$

Putting these values of a and b in (1), the singular integral is

$$z = -\frac{-x^2}{\sqrt{(c^2-x^2-y^2)}} - \frac{y^2}{\sqrt{(c^2-x^2-y^2)}} + \frac{c^2}{\sqrt{(c^2-x^2-y^2)}}$$

$$= \frac{(c^2-x^2-y^2)}{\sqrt{(c^2-x^2-y^2)}} = \sqrt{(c^2-x^2-y^2)}$$

$$\Rightarrow \quad x^2 + y^2 + z^2 = c^2.$$ **Ans.**

3.14 General Method of Solution

Now, we shall discuss general methods for the solution of the partial differential equations of order one but of any degree.

3.15 Two Independent Variables

Charpits Method. **(Rohilkhand, 90, 92, 97, 99, 91; Meerut, 93, 93(P), 96; Kanpur, 95, 98; I.A.S. 98, 95)**

When the given equation cannot be reduced to any of the standard of forms then this method is applied to solve such equations.

Let the given equation be

$$f(x, y, z, p, q) = 0 \quad \text{...(1)}$$

If we are able to find another relation

$$F(x, y, z, p, q) = 0 \quad \text{...(2)}$$

Then solving p, q from (1) and (2) substitute in

$$dz = p\,dx + q\,dy. \qquad ...(3)$$

Clearly the integral equation (3) will satisfy the given equation, for the values of p and q derived from it are the same as the values of p and q in (1).

Let equaiton (2) be the relation such that when the values of p and q derived from it and the given equation (1) are substituted in (3), it become integrable.

Thus, z, p, q may be expressed as functions of x and y. Since these values satisfy (1) and (2) identically, therefore their differential coefficients with respect to x and y will vanish.

Differentiating (1) and (2) with respect to x, and y, we have

$$\frac{\partial f}{\partial x}+\frac{\partial f}{\partial z}.p+\frac{\partial f}{\partial p}.\frac{\partial p}{\partial x}+\frac{\partial f}{\partial q}.\frac{\partial q}{\partial x}=0. \qquad ...(4)$$

$$\frac{\partial F}{\partial x}+\frac{\partial F}{\partial z}.p+\frac{\partial F}{\partial p}.\frac{\partial p}{\partial x}+\frac{\partial F}{\partial q}.\frac{\partial q}{\partial x}=0. \qquad ...(5)$$

$$\frac{\partial f}{\partial y}+\frac{\partial f}{\partial z}.q+\frac{\partial f}{\partial p}.\frac{\partial p}{\partial y}+\frac{\partial f}{\partial q}.\frac{\partial q}{\partial y}=0. \qquad ...(6)$$

$$\frac{\partial F}{\partial y}+\frac{\partial F}{\partial z}.q+\frac{\partial F}{\partial p}.\frac{\partial p}{\partial y}+\frac{\partial F}{\partial q}.\frac{\partial q}{\partial y}=0. \qquad ...(7)$$

Eliminating $\frac{\partial p}{\partial x}$ from (4) and (5), we have

$$\left(\frac{\partial f}{\partial x}+\frac{\partial f}{\partial z}p+\frac{\partial f}{\partial q}.\frac{\partial q}{\partial x}\right)\frac{\partial F}{\partial p}-\left(\frac{\partial F}{\partial x}+\frac{\partial F}{\partial z}p+\frac{\partial F}{\partial q}.\frac{\partial q}{\partial x}\right).\frac{\partial f}{\partial p}=0.$$

$$\Rightarrow \quad \left(\frac{\partial f}{\partial x}.\frac{\partial F}{\partial p}-\frac{\partial F}{\partial x}.\frac{\partial f}{\partial p}\right)+\left(\frac{\partial f}{\partial z}.\frac{\partial F}{\partial p}-\frac{\partial F}{\partial z}.\frac{\partial f}{\partial p}\right)p$$

$$+\left(\frac{\partial f}{\partial q}.\frac{\partial F}{\partial p}-\frac{\partial F}{\partial q}.\frac{\partial f}{\partial p}\right)\frac{\partial q}{\partial x}=0 \qquad ...(8)$$

Similarly eliminating $\frac{\partial q}{\partial y}$ from (6) and (7), we have

$$\left(\frac{\partial f}{\partial y}.\frac{\partial F}{\partial q}-\frac{\partial F}{\partial v}.\frac{\partial f}{\partial q}\right)+\left(\frac{\partial f}{\partial z}.\frac{\partial F}{\partial q}-\frac{\partial F}{\partial z}.\frac{\partial f}{\partial q}\right)q$$

$$\left(\frac{\partial f}{\partial p}.\frac{\partial F}{\partial q}-\frac{\partial F}{\partial p}.\frac{\partial f}{\partial q}\right)\frac{\partial p}{\partial y}=0 \qquad ...(9)$$

Since $\frac{\partial q}{\partial x}=\frac{\partial^2 z}{\partial x\,\partial y}=\frac{\partial^2 z}{\partial y\,\partial x}=\frac{\partial p}{\partial y}$.

Adding (9) and (9), we have

$$\left(\frac{\partial f}{\partial x}+p\frac{\partial f}{\partial z}\right)\frac{\partial F}{\partial p}+\left(\frac{\partial f}{\partial y}+q\frac{\partial f}{\partial z}\right)\frac{\partial F}{\partial q}+\left(-p\frac{\partial f}{\partial p}-q\frac{\partial f}{\partial q}\right)\frac{\partial F}{\partial z}$$

$$+\left(-\frac{\partial f}{\partial p}\right)\frac{\partial F}{\partial x}+\left(-\frac{\partial f}{\partial q}\right)\frac{\partial F}{\partial y}=0. \qquad ...(10)$$

which is linear equation of the first order with x, y, z, p, q as independent variables and F as dependent variable.

The auxiliary equations are

$$\frac{dp}{\frac{\partial f}{\partial x}+p\frac{\partial f}{\partial z}}=\frac{dq}{\frac{\partial f}{\partial y}+q\frac{\partial f}{\partial z}}=\frac{dz}{-p\frac{\partial f}{\partial p}-q\frac{\partial f}{\partial q}}$$

$$=\frac{dx}{\frac{\partial f}{\partial p}}=\frac{dy}{\frac{\partial f}{\partial q}}=\frac{\partial F}{0} \qquad ...(11)$$

Any integral of (11) will satisfy (10). Take the simplest relation involving at least one of p and q for F = 0. From f = 0 and F = 0, find the values of p and q and substitute in dz = p dx + q dy which on integration gives the solution.

Note: Charpits Method is to be applied only when other methods fail to solve the equation.

Example 1:

Solve px + qy = pq.

(Rohilkhand, 90, 94; Agra, 96; Meerut, 91, 96, 91(P), 92, 93(P)

Solution:

Here $f \equiv px + qy - pq = 0$.

$\therefore$ the Charpit's auxiliary equations are

$$\frac{dp}{p}=\frac{dq}{q}=\frac{dz}{-p(x-q)-q\,(y-p)}=\frac{dx}{-(x-q)}=\frac{dy}{-(y-p)}=\frac{dF}{0}$$

Taking the first two members, we have

$$\frac{dp}{p}=\frac{dq}{q}.$$

Integrating, log p = log q + log a

$$\Rightarrow \qquad p = aq. \qquad \ldots(2)$$

Putting p = 1q in (1), we have

$$aqx + qy = aq^2 \qquad \therefore \qquad q = \frac{y+ax}{a}$$

from (2) $p = aq = y + ax$.

Putting in $dz = p\,dx + q\,dy$, we have

$$dz = (y + ax)\,dx + \frac{y+ax}{a}dy$$

$$\Rightarrow \quad a\,dz = (y + ax)(dy + a\,dx).$$

Integrating, we get $az = \frac{1}{2}(y + ax)^2 + b$

which is the complete integral.

General Integral: Writing $b = \phi(a)$ we have

$$az = \frac{1}{2}(y + ax)^2 + \phi(a) \qquad \ldots(3)$$

Differentiating (3) partially w.r.t. a, we have

$$z = x(y + ax) = \phi'(a). \qquad \ldots(4)$$

General integral is obtained by eliminating a from equations (3) and (4).

Singular Integral: Differentiating the complete integral partially w.r.t. a and b, we have

$$z = x(y + ax) \qquad \text{and } 0 = 1.$$

Hence, there is no singular integral. **Ans.**

Example 2:

Solve $(p^2 + q^2)\,y = qz$. **(Rohilkhand, 96, 99; Meerut, 90, 94, 92, 95; Garhwal, 92, 94; PCS (UP) 90)**

Solution:

Here $f \equiv (p^2 + q^2)\,y - qz = 0$.

Now the Charpit's auxiliary equations are

$$\frac{dp}{\frac{\partial p}{\partial x} + p\frac{\partial f}{\partial z}} = \frac{dc_1}{\frac{\partial f}{\partial y} + q\frac{\partial f}{\partial z}} = \frac{dz}{-p\frac{\partial f}{\partial p} - q\frac{\partial f}{\partial q}} = \frac{dx}{\frac{\partial f}{\partial p}} = \frac{dy}{-\frac{\partial f}{\partial q}} = \frac{dF}{0}$$

$$\Rightarrow \quad \frac{dp}{-pq} = \frac{dq}{\left(p^2 + q^2\right) - q^2} = \frac{dz}{-2p^2y - 2q^2y + qz} = \frac{dx}{-2py} = \frac{dy}{-2qy + z} = \frac{dF}{0}$$

Taking the first two members, we have

$$\frac{dp}{-pq} = \frac{dq}{p^2} \quad \Rightarrow p\ dp + q\ dq = 0.$$

Integrating, $p^2 + q^2 = a^2$ (say)

Putting in equation (1), $q = \dfrac{a^2 y}{z}$

$\therefore$ From (2), $p = \sqrt{\left(a - q^2\right)} = \sqrt{\left\{a^2 - \dfrac{a^4 y^2}{z^2}\right\}} = \dfrac{a}{z}\sqrt{\left(z^2 - a^2 y^2\right)}.$

$\therefore$ From $dz = p\ dx + q\ dy$, we have

$$dz = \frac{a}{z}\sqrt{\left(z^2 - a^2 y^2\right)}\, dx + \frac{a^2 y}{z} dy,$$

$\Rightarrow \dfrac{z\ dz - a^2 y\ dy}{\sqrt{\left(z^2 - a^2 y^2\right)}} = a\ dx$. $\qquad \Rightarrow dx = a\ dx$ Putting $z^2 - a^2y^2 = t^2$

Integrating, we have $\quad t = \sqrt{\left(z^2 - a^2 y^2\right)} = ax + b$.

$\Rightarrow \qquad z^2 = a^2y^2 + (ax + b)^2.$ ...(3)

which is the complete solution.

General Integral: Writing $b = \phi$ (a) in the complete integral, we have

$$z^2 = a^2y^2 + [ax + \phi (a)]^2. \qquad ...(4)$$

Differentiating (4) partially w.r.t. 'a', we have

$$0 = 2a^2y + 2\ [ax + \phi (a)]\ [x + \phi' (a)] \qquad ...(5)$$

General integral is obtained by eliminating a from (4) and (5).

Singular Integral: Differentiating (3) partially w.r.t. and b, we have

$$0 = 2a^2y + 2\ (ax + b)\ x. \qquad ...(6)$$

and $\qquad 0 = 2\ (ax + b).$ ...(7)

Eliminating a and b from (3), (6) and (7), we get

$z = 0$ which clearly satisfy the given equation (1) and therefore is the singular integral.

Example 3:

Solve $p^2 + q^2 - 2px - 2qy + 1 = 0.$ ...(1)

(IAS, 90)

Solution:

$$f \equiv p + q^2 - 2px - 2px - 2qy + 1 = 0.$$

$$\frac{dp}{-2p} = \frac{dq}{-2q} = \frac{dz}{-p\,(2p-2x)-q\,(2q-2y)} = \frac{dx}{-(2p-2x)}$$

$$= \frac{dy}{-(2q-2y)} = \frac{dF}{0}$$

Taking the first two members, we have

$$\frac{dp}{p} = \frac{dq}{q}. \qquad \therefore\ p = aq$$

Putting in (1), $a^2q^2 + q^2 - 2\,aqx + 2qy + 1 = 0$

$$(a^2 + 1)\,q^2 - 2\,(ax + y)\,q + 1 = 0.$$

$$\therefore\ q = \frac{2\,(ax+y) + \sqrt{\left\{4\,(ax+y)^2 - 4\left(a^2+1\right)\right\}}}{2\left(a^2+1\right)}$$

$$\Rightarrow \qquad q = \frac{ax+y+\sqrt{\left\{(ax+y)^2 - \left(a^2+1\right)\right\}}}{a^2+1}$$

(Taking the +ve sign only)

and
$$p = aq = \frac{a\left[(ax+y) + \sqrt{\left\{(ax+y)^2 - \left(a^2+1\right)\right\}}\right]}{a^2+1}$$

Substituting in $dz = p\,dx + q\,dy$, we have

$$dz = \frac{(ax+y) + \sqrt{\left\{(ax+y)^2 - \left(a^2+1\right)\right\}}}{a^2+1}\left(a\,dx + dy\right)\cdot$$

Putting $ax + y = t$ so that $a\,dx + dy = dt$,

we have, $(a^2 + 1)\,dz = \left[t + \sqrt{\left\{t^2 - \left(a^2+1\right)\right\}}\right] dt$

Integrating, we get

$$(a^2 + 1)\,z = \frac{t^2}{2} + \frac{t}{2}\sqrt{\left\{t^2 - \left(a^2+a\right)\right\}} - \frac{a^2+1}{2}\log\left[t + \sqrt{\left\{t^2 - \left(a^2+1\right)\right\}}\right]$$

$+ b$ where $t = ax + y$

which is complete integral. **Ans.**

Example 4:

Apply Charpit's method to find complete integral of

$$z = px + qy + p^2 + q^2. \qquad ...(1)$$

Solution:

Here $f \equiv z - px - qy - p^2 - q^2 = 0$

$\therefore$ The Charpit's equations are

$$\frac{dp}{-p+p} = \frac{dq}{-q+q} = \frac{dz}{-p(-x-2p) - q(x-y-2p)} = \frac{dx}{-(-x-2p)}$$

$$= \frac{dy}{(-y-2q)} = \frac{dF}{0}$$

$\therefore$ $dp = 0$, and $dq = 0$.

Integrating, $p = a$ and $q = b$.

Putting in (1), $z = ax + by + a^2 + b^2$

Note: The given equation is of the form of Standard IV.

Example 5:

Apply Charpit's method to find complete integral of

$z^2 (p^2z^2 + q^2) = 1.$ **(Meerut 94, 96(P), 97)**

Solution:

Here $f \equiv p^2z^4 + q^2z^2 - 1 = 0$.

The Charpit's Equations are

$$\frac{dp}{0 + p.(4p^2z^3 + 2q^2z)} = \frac{dq}{q(4p^2z^3 + 2q^2z)} = \frac{dz}{-p(2pz^4) - q(2qz^2)}$$

$$= \frac{dx}{-2pz^4} = \frac{dy}{-2qz^2} = \frac{dF}{0}$$

Taking the first two members, we have

$$\frac{dp}{p} = \frac{dq}{q}$$

$\therefore$ $\log p = \log q + \log a, \quad \Rightarrow p = aq$

Putting in (1), $z^2 (a^2z^2q^2 + q^2) = 1$

$$\therefore \quad q^2 = \frac{1}{z^2(a^2z^2+1)} \quad \Rightarrow \quad q = \frac{1}{z\sqrt{(a^2z^2+1)}}$$

$$\therefore \quad p = \frac{1}{z\sqrt{(a^2z^2+1)}}$$

Substituting in $dz = p\,dx + q\,dy$, we have

$$dz = \frac{a}{z\sqrt{(a^2z^2+1)}}\,dx + \frac{1}{z\sqrt{(a^2z^2+1)}}\,dy$$

$$\Rightarrow \quad z\sqrt{(a^2z^2+1)}\,dz = a\,dx + dy. \quad \Rightarrow \frac{t^2}{a^2}\,dt = a\,dx + dy$$

Putting $a^2z^2 + 1 = t^2$.

Integrating, $\frac{1}{3a^2}t^3 = \frac{1}{3a^2}(a^2z^2+1)^{3/2} = ax + y + b$

$$\Rightarrow \quad (a^2z^2+1) = 9a^4(ax+y+b)^2.$$ **Ans.**

Note: The equation can be solved by method of Standard II.

Example 6:

Solve $p = (qy + z)^2$.

(Meerut, 93, 95(BP), 96(BP), 97(P); PCS (UP) 92)

Solution:

Here $f \equiv -p + (qy+z)^2 = 0$

Charpit's auxiliary equations are

$$\frac{dp}{2p(qy+z)} = \frac{dq}{4q(qy+z)} = \frac{dz}{(-p)(-1) - q.2(qy+z)\,y}$$

$$= \frac{dx}{-(-1)} = \frac{dy}{-2y(qy+z)} = \frac{dF}{0}$$

Taking the 1st and 5th members, we have

$$\frac{dp}{p} + \frac{dy}{y} = 0.$$

$$\therefore \quad \log p + \log y = \log a \Rightarrow p = \frac{a}{y}.$$

$$\therefore \quad \text{From (1), } \frac{a}{y} = (qy+z)^2$$

$$\sqrt{\left(\frac{a}{y}\right)} = qy + z. \qquad \therefore\ q = \frac{\sqrt{a}}{y^{3/2}} - \frac{z}{y}$$

Substituting these values in dz = p dx + q dy, we have

$$dz = \frac{a}{y}dx + \left(\frac{\sqrt{a}}{y^{3/2}} - \frac{z}{y}\right) dy$$

$$\Rightarrow \qquad y\,dz + z\,dy = a\,dx + \frac{\sqrt{a}}{\sqrt{y}}\,dy.$$

Integrating $yz = dx + 2\sqrt{(ay)} + b$

which is complete integral **Ans.**

Example 7:

Solve $2zx - px^2 - 2qxy + pq = 0$. ...(1)

(Rohilkhand, 93; Meerut, 95, 97 (R), 96; Kanpur, 92; IAS 91, 93)

Solution:

Here $f \equiv 2zx - px^2 - 2zxy + pq = 0$.

$\therefore$ The Charpit's auxiliary equations are

$$\frac{dp}{2z-2qy} = \frac{dq}{0} = \frac{dz}{px^2 - pq + 2xyq - pq} = \frac{dx}{x^2-q} = \frac{dy}{2xy-p} = \frac{dF}{0}$$

$\therefore dq = 0 \qquad \Rightarrow q = a.$

Putting in (1), we have $2zx - px^2 - 2axy + pa = 0$

$p(x^2 - a) = 2x(z - ay)$.

$$\therefore\ p = \frac{2x(z-ay)}{x^2-a}$$

Substituting these values in dz = p dx + q dy. We have

$$dz = \frac{2x(z-ay)}{x^2-a}\,dx + a\,dy \Rightarrow \frac{dz - a\,dy}{z-ay} = \frac{2x}{x^2-a}\,dx.$$

Integrating, $\log(z - ay) = \log(x^2 - a) + \log b$

$\therefore z - ay = b(x^2 - a) \Rightarrow z = ay + b(x^2 - a)$. **Ans.**

Example 59:

Find the complete integral of the equation

$2(z + xp + yq) = yp^2$. ...(1)

Solution:

Here $f \equiv 2(z + xp + yq) - yp^2 = 0$ ...(1)

∴ The Charpit's auxiliary equations are

$$\frac{dp}{2p+2p}=\frac{dq}{2q-p^2+2q}=\frac{dz}{-p\,(2x-2yp)-2qy}$$

$$=\frac{dx}{-2x+2yp}=\frac{dy}{-2y}$$

Taking the first and fifth members, we have

$$\frac{dp}{p}+2\,\frac{dy}{y}=0.$$

Integrating, we get $py^2 = a$ (say) ...(2)

Putting $p=\frac{a}{y^2}$ in (1), we have

$$z+\frac{ax}{y^2}+yq=\frac{a^2}{2y^3}.$$

$$\therefore\ q=-\frac{z}{y}-\frac{ax}{y^3}+\frac{a^2}{2y^4}.$$

∴ From $dz = p\,dx + q\,dy$, we have

$$dz=\frac{a}{y^2}dx=\frac{z}{y}dy-\frac{dx}{y^2}dy+\frac{a^2}{2y^4}dy$$

$$\Rightarrow\quad y\,dz+z\,dy=a.\left(\frac{y\,dx-x\,dy}{y^2}\right)+\frac{a^2}{2y^3}dy$$

Integrating, we have $yz-\frac{ax}{y}+\frac{a^2}{4y^2}=b$

$$\therefore\ z=\frac{ax}{y^2}-\frac{a^2}{4y^3}+\frac{b}{y}.$$

which is the complete integral. **Ans.**

Example 9:

Solve $p^2+q^2-2px-2qy+2xy=0.$ ...(1)

(Rohilkhand 99; Kanpur 92, 98; Garhwal, 95, 93; PCS (UP) 91)

Solution:

We have $p^2 + q^2 - 2px - 2qy + 2xy = 0.$

Here $f \equiv p^2 + q^2 - 2px - 2qy + 2xy = 0.$

Charpit's auxiliary equations are

$$\frac{dp}{-2p+2y} = \frac{dq}{-2q+2x} = \frac{dx}{2x-2p} = \frac{dy}{2y-2q}$$

i.e., $$\frac{dp}{-p+y} = \frac{dq}{-q+x} = \frac{dx}{x-p} = \frac{dy}{y-q}$$

$\Rightarrow$ $$\frac{dp}{x+y-p-q} = \frac{dx+dy}{x+y-p-q} \quad \Rightarrow \ dp + dq = dx + dy$$

$\therefore$ $(p - x) + (q - y) = a.$ ...(2)

Equation (1) can be written as

$$(p - x)^2 + (q - y)^2 = (x - y)^2 \qquad ...(3)$$

Putting the value of (q − y) from (2) in (3), we have

$$(p - x)^2 + [a - (p - x)]^2 = (x - y)^2$$

$\Rightarrow$ $2 (p - x)^2 - 2a (p - x) + \{a^2 - (x - y)^2\} = 0.$

$$\therefore p - x = \frac{2a + \sqrt{\left[4a^2 - 4.2\left\{a^2 - (x-y)^2\right\}\right]}}{4} \quad \text{(Taking + ve sign only)}$$

$$\frac{1}{2}\left[a + \sqrt{\left\{2(x-y)^2 - a^2\right\}}\right] \Rightarrow p = x + \frac{1}{2}\left[a + \sqrt{\left\{2(x-y)^2 - a^2\right\}}\right]$$

$$\therefore \text{From (2), } q - y = a - \frac{1}{2}\left[a + \sqrt{\left\{2(x-y)^2 - a^2\right\}}\right]$$

$\Rightarrow$ $$q = y + \frac{1}{2}\left[a - \sqrt{\left\{2(x-y)^2\right\} - a^2}\right]$$

Substituting these values in $dz = p\,dx + q\,dy$, we have

$$dz = x\,dx + y\,dy + \frac{a}{2}(dx+dy) + \frac{1}{2}\sqrt{\left[2(x-y)^2\right]}(dx - dy)$$

$$= x\,dx + y\,dy + \frac{a}{2}(dx+dy) + \frac{1}{\sqrt{2}}\sqrt{\left\{(x-y)^2 - \frac{a^2}{2}\right\}}(dx - dy).$$

Integrating, we have $z = \frac{x^2}{2} + \frac{y^2}{2} + \frac{a}{2}(x+y)$

$$+\frac{1}{\sqrt{2}}\left[\frac{x-y}{2}\sqrt{\left\{(x-y)^2-\frac{a^2}{2}\right\}}-\frac{a^2}{4}\log\left\{(x-y)+\sqrt{(x-y)^2-\frac{a^2}{2}}\right\}\right]$$

$\Rightarrow$ $2z = x^2 + y^2 + ax + ay$

$$+\frac{1}{\sqrt{2}}\left[(x-y)\sqrt{\left\{(x-y)^2-\frac{a^2}{2}\right\}}-\frac{a^2}{2}\log\left\{(x-y)+\sqrt{(x-y)^2-\frac{a^2}{2}}\right\}\right]$$

which is the complete integral. **Ans.**

Example 10:

Find a complete integral of the equation

$p^2x + q^2y = z.$ **(Garhwal, 92)**

Solution:

We have $p^2x + q^2y = z.$

Here $f \equiv p^2x + q^2y - z = 0$

$\therefore$ The Charpit's auxiliary equations are

$$\frac{dp}{-p+p^2}=\frac{dq}{-q+q^2}=\frac{dz}{-2\left(p^2x+q^2y\right)}=\frac{dx}{-2px}=\frac{dy}{-2qy}$$

from which, we have

$$\frac{p^2dx+2px\ dp}{p^2x}=\frac{q^2dy+2qy\ dq}{q^2y}$$

Integrating, we have $\log p^2x = \log q^2y + \log a$ $\therefore$ $p^2 = aq^2y$ ...(2)

where a is constant.

From equations (1) and (2), we have $aq^2y + q^2y = z$

$$\therefore \qquad q=\left\{\frac{z}{(1+a)\ y}\right\}^{1/2}$$

From equation (1), $p=\left\{\frac{az}{(1+a)\ x}\right\}^{1/2}$

Putting in $dz = p\ dx + q\ dy$, we have

$$dz=\left\{\frac{az}{(1+a)\ x}\right\}^{1/2}dx+\left\{\frac{z}{(1+a)\ y}\right\}^{1/2}dy$$

$$\Rightarrow \qquad \sqrt{(1+a)}\,\frac{dz}{\sqrt{z}} = \sqrt{a}\,\frac{dx}{\sqrt{x}} + \frac{dy}{\sqrt{y}}.$$

Integrating, we get $\sqrt{\{(1+a)\,z\}} = \sqrt{(ax)} + \sqrt{y} + b$

which is the complete integral.

Example 11:

Find the complete integral of

pxy + pq + qy = yz.

(Meerut, 93, 94(P); Raj., 98; Rohilkhand 94; Garhwal, 96)

Solution:

We have pxy + pq + qy = yz.

Here f ≡ pxy + qp + qy − yz = 0.

Charpit's auxiliary equations are

$$\frac{dp}{py + p(-y)} = \frac{dq}{(px+q) + q(y)} = \frac{dz}{-p(xy+q) - q(p+y)}$$

$$= \frac{dx}{-(xy+q)} = \frac{dy}{-(p+y)} = \frac{dF}{0}$$

$\therefore$ dp = 0, $\qquad \Rightarrow$ p = a.

Putting in (1), axy = aq + qy = yz

q (a + y) = y (z − ax)

$$\therefore\ q = \frac{y(z-ax)}{a+y}$$

Substituting these values of p and q in dz = p dx + q dy, we have

$$dz = a\,dx + \frac{y(z-ax)}{a+y}dy \ \Rightarrow\ \frac{dz - a\,dx}{z-ax} = \frac{y\,dy}{a+y}$$

$$\Rightarrow \qquad \frac{dz - a\,dx}{z-ax} = \left(1 - \frac{a}{a+y}\right)dy$$

Integrating, log (z − ax) = y − a log (a + y) + log b

$\Rightarrow \qquad z - ax = be^{y}(y+a)^{-a}.$ **Ans.**

Example 12:

Solve $(x^2 - y^2)\,pq - xy\,(p^2 - q^2) - 1 = 0.$ ...(1)

(Kanpur, 91; Raj. 90, 91; Rohilkhand, 92; Garhwal, 97)

Solution:

Here $f \equiv (x^2 - y^2)\, pq - xy\,(p^2 - q^2) - 1 = 0$

The Charpit's auxiliary equations are

$$\frac{dp}{2pqx - y\left(p^2 - q^2\right)} = \frac{dq}{-2y\,pq - x\left(p^2 - q^2\right)} = \frac{dx}{-\left(x^2 - y^2\right)q + 2pxy}$$

$$= \frac{dy}{-\left(x^2 - y^2\right)p - 2qxy} = \ldots.$$

Using x, y, p, q as multipliers, we have

$$\text{Each fraction} = \frac{x\,dp + y\,dq + p\,dx + q\,dy}{0}$$

$\therefore$ $(x\,dp + p\,dx) + (q\,dy + y\,dq) = 0.$

Integrating, $px + qy = a$, $\quad \therefore\ p = \dfrac{a - qy}{x}$.

$\therefore$ From (1), we have

$$(x^2 - y^2)\left(\frac{a - qy}{x}\right)q - xy\left[\frac{(a - qy)^2}{x^2} - q^2\right] - 1 = 0$$

$$\Rightarrow \quad \frac{a - qy}{x}\{(x^2 - y^2)\,q - (a - qy)\,y\} + xyq^2 - 1 = 0$$

$$\Rightarrow \quad \frac{a - qy}{x}(x^2q - ay) + x^2\,yq^2 - x = 0$$

$$\Rightarrow \quad (a - qy)\,(x^2q - ay) + x^2yq^2 - x = 0$$

$$\Rightarrow \quad ax^2q - a^2y - x^2yq^2 + ay^2q + x^2yq^2 - x = 0$$

$$\Rightarrow \quad qa\,(x^2 + y^2) = a^2y + x \quad \Rightarrow \quad q = \frac{a^2y + x}{a\left(x^2 + y^2\right)}$$

$$\therefore\ p = \left[a - \frac{\left(a^2y + x\right)y}{a\left(x^2 + y^2\right)}\right] = \frac{a^2x - y}{a\left(x^2 + y^2\right)}.$$

Substituting these values in $dz = p\,dx + q\,dy$, we have

$$dz = \frac{\left(a^2x - y\right)dx + \left(a^2y + x\right)dy}{a\left(x^2 + y^2\right)}$$

$$\Rightarrow \quad dz = a \cdot \frac{(x\,dx + y\,dy)}{x^2 + y^2} + \frac{x\,dy - y\,dx}{a\left(x^2 + y^2\right)}.$$

Integrating, we have

$$z = \frac{a}{2}\log\left(x^2 + y^2\right) + \frac{1}{a}\tan^{-1}\frac{y}{x} + b$$

which is the complete integral. **Ans.**

Example 13:

Solve $px + qy = z\,(1 + pq)^{1\,2}$ ...(1)

(Meerut, 91, 97 (BP); Kanpur, 95; IAS, 92)

Solution:

We have $px + qy = z\,(1 + pq)^{1/2}$

$f \equiv px + qy - z\,(1 + qp)^{1/2} = 0.$

The Charpit's auxiliary equations are

$$\frac{dp}{p - p\,(1 + pq)^{1/2}} = \frac{dq}{q - q\,(1 + pq)} = \ldots$$

Taking the first two members, we have

$$\frac{dp}{p} = \frac{dq}{q}$$

Integrating, we have $\log p = \log q + \log a.$ $\therefore\ p = aq.$

Putting in (1), we have $aqx + qy = z\,(1 + aq^2)^{1/2}$

$$\Rightarrow \quad q^2\,(ax + y)^2 = z^2\,(1 + aq^2)$$

$$\Rightarrow \quad q^2 = \frac{z^2}{(ax + y)^2 - az^2}$$

$$\Rightarrow \quad q = \frac{z}{\sqrt{\left\{(ax + y)^2 - az^2\right\}}} \qquad \therefore \quad p = \frac{az}{\sqrt{\left\{(ax + y)^2 - az^2\right\}}}$$

Substituting, these values in $dz = p\,dx + q\,dy$, we have

$$dz = \frac{(az\,dx + z\,dy)}{\sqrt{\left\{(ax + y)^2 - az^2\right\}}} \quad \Rightarrow \quad \frac{dz}{z} = \frac{a\,dx + dy}{\sqrt{\left\{(ax + y)^2 - az^2\right\}}}$$

we get $$\frac{dz}{z} = \frac{\sqrt{a}\,du}{\sqrt{\left(au^2 - az^2\right)}}$$

$$\Rightarrow \quad \frac{du}{dz} = \frac{1}{z}\sqrt{\left(u^2 - z^2\right)}.$$

Again putting $u = vz$

so that $$v + z\frac{dv}{dz} = \frac{1}{z}\sqrt{\left(v^2 z^2 - z^2\right)}$$

$$\Rightarrow \quad v + z\frac{dv}{dz} = \sqrt{\left(v^2 - 1\right)}.$$

$$\Rightarrow \quad z\frac{dv}{dz} = \sqrt{\left(v^2 - 1\right)} - v$$

$$\Rightarrow \quad \frac{dz}{z} = \frac{dv}{\sqrt{\left(v^2 - 1\right)} - v}$$

$$\Rightarrow \quad \frac{dz}{z} = -\left\{\sqrt{\left(v^2 - 1\right)} + v\right\} dv.$$

Integrating, $\log z = = \left[\frac{v}{2}\sqrt{\left(v^2 - 1\right)} - \frac{1}{2}\log\left\{v + \sqrt{\left(v^2 - 1\right)}\right\}\right] - \frac{v^2}{2} + b$

$$\Rightarrow \quad -\log z + \frac{v^2}{2} + \frac{v}{2}\sqrt{\left(v^2 - 1\right)} - \frac{1}{2}\log\left\{v + \sqrt{\left(v^2 - 1\right)}\right\} = b.$$

which is the complete integral, where $v = \frac{u}{z} = \frac{ax + y}{z\sqrt{a}}$. **Ans.**

3.14 Three or more Independent Variables

Jacobi's Method **(Meerut, 96)**

Let us consider the partial differential equation with three independent variables x_1, x_2, x_3, as

$$F(x_{11}\ x_{29}\ x_{39}\ p_1, p_2, p_3) = 0 \qquad \text{...(1)}$$

z is the dependent variable not occurring in the equation except in its partial derivatives p_1, p_2, p_3 w.r.t. x_1, x_2, x_3 respectively.

The idea of Jacobi's method is to find two additional equations

$$F_1\ (x_1, x_2, x_3, p_1, p_2, p_3) = a_1, \qquad \text{...(2)}$$

and $$F_2\ (x_1, x_2, x_3, p_1, p_2, p_3) = a_2, \qquad \text{...(3)}$$

where a_1, a_2 are arbitrary constants such that (1), (2), (3) may be solved for p_1, p_2, p_3 as functions of x_1, x_2, x_3, which when substituted in

$$dz = p_1 dx_1 + p_2 dx_2 + p_3 dx_3. \qquad ...(4)$$

it becomes integrable, for which conditions are

$$\frac{\partial p_2}{\partial x_1} = \frac{\partial^2 z}{\partial x_1 \partial x_2} + \frac{\partial p_1}{\partial x_2}, \frac{\partial p_3}{\partial x_2} = \frac{\partial p_2}{\partial x_3}, \frac{\partial p_1}{\partial x_3} = \frac{\partial p_3}{\partial x_1}$$

Now differentiating equations (1) and (2), partially w.r.t. x_1, we have

$$\frac{\partial f_2}{\partial x_1} + \frac{\partial f_2}{\partial p_1} . \frac{\partial p_1}{\partial x_1} + \frac{\partial f}{\partial p_2} . \frac{\partial p_2}{\partial x_1} + \frac{\partial f}{\partial p_3} . \frac{\partial p_3}{\partial x_1} = 0 \qquad ...(5)$$

$$\frac{\partial F_1}{\partial x_1} + \frac{\partial F_1}{\partial x_1} + \frac{\partial p_1}{\partial x_1} + \frac{\partial F_1}{\partial p_2} . \frac{\partial p_2}{\partial x_1} + \frac{\partial F_1}{\partial p_3} . \frac{\partial p_3}{\partial x_1} = 0. \qquad ...(6)$$

Eliminating $\frac{\partial p_1}{\partial x_1}$ from equations (5) and (6), we have

$$\left(\frac{\partial f}{\partial x_1}\frac{\partial F_1}{\partial p_1} - \frac{\partial f}{\partial p_1}\frac{\partial F_1}{\partial x_1}\right) + \left(\frac{\partial f}{\partial p_2}\frac{\partial F_1}{\partial p_1} - \frac{\partial f}{\partial p_1}\frac{\partial F_1}{\partial p_2}\right)\frac{\partial p_2}{\partial x_1}$$

$$\left(\frac{\partial f}{\partial p_3}\frac{\partial F_1}{\partial p_1} - \frac{\partial f}{\partial p_1}\frac{\partial F_1}{\partial p_3}\right)\frac{\partial p_3}{\partial x_1} = 0. \qquad ...(7)$$

Similarly, we have

$$\left(\frac{\partial f}{\partial x_2}\frac{\partial F_1}{\partial p_2} - \frac{\partial f}{\partial p_2}\frac{\partial F_1}{\partial x_2}\right) + \left(\frac{\partial f}{\partial p_1}\frac{\partial F_1}{\partial p_2} - \frac{\partial f}{\partial p_2}\frac{\partial F_1}{\partial p_1}\right)\frac{\partial p_2}{\partial x_2}$$

$$+\left(\frac{\partial f}{\partial p_3}\frac{\partial F_1}{\partial p_2} - \frac{\partial f}{\partial p_2}\frac{\partial F_1}{\partial p_3}\right)\frac{\partial p_3}{\partial x_2} = 0. \qquad ...(8)$$

and $$\left(\frac{\partial f}{\partial x_3}\frac{\partial F_1}{\partial p_3} - \frac{\partial f}{\partial p_3}\frac{\partial F_1}{\partial p_3}\right) + \left(\frac{\partial f}{\partial p_1}\frac{\partial F_1}{\partial p_3} - \frac{\partial f}{\partial p_3}\frac{\partial F_1}{\partial p_1}\right)\frac{\partial p_1}{\partial x_3}$$

$$+\left(\frac{\partial f}{\partial p_2}\frac{\partial F_1}{\partial p_3} - \frac{\partial f}{\partial p_2}\frac{\partial F_1}{\partial p_2}\right)\frac{\partial p_2}{\partial x_3} = 0. \qquad ...(9)$$

Adding equations (7), (8) and using $\frac{\partial p_2}{\partial x_1} = \frac{\partial p_1}{\partial x_2}, \frac{\partial p_3}{\partial x_2} = \frac{\partial p_2}{\partial x_3}$

and $\frac{\partial p_1}{\partial x_3} = \frac{\partial p_3}{\partial x_1}$, we have

$$\left(\frac{\partial f}{\partial x_1}\frac{\partial f}{\partial p_1} - \frac{\partial f}{\partial p_1}\frac{\partial F_1}{\partial x_1}\right) + \left(\frac{\partial f}{\partial x_2}\frac{\partial F_1}{\partial p_3} - \frac{\partial f}{\partial p_2}\frac{\partial F_1}{\partial x_2}\right)$$

$$+\left(\frac{\partial f}{\partial x_3}\frac{\partial F_1}{\partial p_3}-\frac{\partial f}{\partial p_3}\frac{\partial F_1}{\partial x_3}\right)=0. \qquad ...(10)$$

i.e., $$(f, F_1) = \sum_{r=1}^{3}\left(\frac{\partial f}{\partial x_3}\frac{\partial F_1}{\partial p_3}-\frac{\partial f}{\partial p_3}\frac{\partial F_1}{\partial x_3}\right)=0.$$

Similarly, we have

$$(f, F_2) = \sum_{r=1}^{3}\left(\frac{\partial f}{\partial x_r}\frac{\partial F_2}{\partial p_r}-\frac{\partial f}{\partial p_r}\frac{\partial F_2}{\partial x_r}\right)=0$$

and $$(F_1, F_2) = \sum_{r=1}^{3}\left(\frac{\partial F_1}{\partial x_r}\frac{\partial F_2}{\partial p_r}-\frac{\partial F_1}{\partial p_r}\frac{\partial F_2}{\partial x_r}\right)=0.$$

Since equation (10) is a linear equation of the first order with x_1, x_2, x_3 p_1, p_2, p_3 as independent variables and F_1 as dependent variable.

$\therefore$ The auxiliary equations are

$$\frac{dx_1}{-\dfrac{\partial f_1}{\partial p_1}}=\frac{dp_1}{\dfrac{\partial f_1}{\partial x_1}}=\frac{dx_2}{-\dfrac{\partial f}{\partial p_2}}=\frac{dp_2}{\dfrac{\partial f}{\partial x_2}}=\frac{dx_3}{-\dfrac{\partial f}{\partial p_3}}=\frac{dp_3}{\dfrac{\partial f}{\partial p_3}} \qquad ...(11)$$

Any integral of (11) will satisfy (10):

Working Method: Find two independent integrals $F_1 = a_1$ and $F_2 = a_2$ of the subsidiary equation (11).

If these satisfy the condition

$$(F_1, F_2) = \sum_{r=1}^{3}\left(\frac{\partial F_1}{\partial x_r}\frac{\partial F_2}{\partial x_r}-\frac{\partial F_1}{\partial p_r}\frac{\partial F_2}{\partial x_r}\right)=0. \qquad ...(12)$$

and if p_1, p_2, p_3, can be found as functions of x_1, x_2, x_3 from

$F = 0$, $F_1 = a_1$ and $F_2 = a_2$

then substitute these values in

$$dz = p_1\, dx_1 + p_2\, dx_2 + p_3\, dx_3$$

and integrate to get the complete integral.

Example 1:

Solve $p_1p_2p_3 = z^3x_1x_2x_3$ ...(1)

(IAS, 95)

Solution:

This equation is not of the form considered, as it involves the dependent variable z also.

The equation can be written as

$$\left(\frac{1}{z}\frac{\partial z}{\partial x_1}\right)\left(\frac{1}{z}\frac{\partial z}{\partial x_2}\right)\left(\frac{1}{z}\frac{\partial z}{\partial x_3}\right) = x_1x_2x_3$$

Putting, $\frac{dz}{z} = dZ$ so that $\log z = Z$. We have

$$\frac{\partial Z}{\partial x_1}.\frac{\partial Z}{\partial x_2}.\frac{\partial Z}{\partial x_3} = x_1x_2x_3$$

$\Rightarrow$ $P_1P_2P_3 = x_1x_2x_3$. ...(2)

$\therefore$ Now $f \circ P_1P_2P_3 - x_1x_2x_3 = 0$.

$$\frac{dx_1}{-P_2\,P_3} = \frac{dP_1}{-x_2\,x_3} = \frac{dx_2}{-P_1\,P_3} = \frac{dP_2}{-x_1\,x_3} = \frac{dx_3}{-P_1\,P_2} = \frac{d\,P_3}{-x_1\,x_2}$$

Taking the first two members, we have

$$\frac{P_1\,dx_1}{-P_1\,P_2\,P_3} = \frac{x_1\,dP_1}{-x_1\,x_2\,x_3}$$

$\Rightarrow$ $P_1dx_1 = x_1dP_1$ from (2) $\Rightarrow \frac{dP_1}{P_1} = \frac{dx_1}{x_1}$.

$\Rightarrow$ $\log P_1 = \log x_1 + \log a_1$

$F_1 \equiv P_1 - a_1x_1 = 0$. ...(3)

Similarly, $F_2 \equiv P_2 - a_2x_2 = 0$. ...(4)

Now, $(F_1, F_2) = \sum_{r=1}^{3}\left(\frac{\partial F_1}{\partial x_r}\frac{\partial F_2}{\partial P_r} - \frac{\partial F_1}{\partial P_r}\frac{\partial F_2}{\partial x_r}\right) = 0$.

Therefore (3), (4) may be taken as two additional equations.

Substituting $P_1 = a_1x_1$, $P_2 = a_2x_2$, $P_2 = \frac{x_3}{a_1a_2}$ in

$dZ = P_1dx_1 + P_2dx_2 + P_3dx_3$, we have

$$dZ = a_1x_1dx_1 + a_2x_2dx_2 + \frac{x^3}{a_1a_2}dx_3.$$

Integrating, we get $2Z = a_1\,a_1x_1^2 + a_2x_2^2 + \frac{x_3^2}{a_1a_2} + a_3$

$$\Rightarrow \quad \frac{2}{a_3}\log z = \frac{a_1}{a_3}x_1^2 + \frac{a_2}{a_3}x_2^2 + \frac{1}{a_1a_2a_3}x_3^2 + 1$$

which is the complete integral.

Note: If $\frac{a_1}{a_3} = A_1, \frac{a_2}{a_3} = A_2, \frac{1}{a_1 a_2 a_3} = A_3$ then $A_1A_2A_3 = \frac{1}{a_3^3}$.

$\therefore \frac{1}{a^3} = (A_1A_2A_3)^{1/3}$.

$\therefore$ The complete integral is

$$2\,(A_1A_2A_3)^{1/3} \log z = A_1\, A_1x_1^2 + A_2x_2^2 - A_3x_3^2 + 1$$

where A_1, A_2, A_3 are arbitrary constants. **Ans.**

Example 2:

Apply Jacobi's method to find complete integral of

$$p_1^3 + p_2^2 + p_3 - 1 = 0. \qquad ...(1)$$

Solution:

Here $\quad f \equiv p_1^3 + p_2^3 + p_3 - 1 = 0.\ p_1^3 + p_2^3 + p_3 - 1 = 0$

The subsidiary equations are

$$\frac{dx_1}{-3p_1^2} = \frac{dp_1}{0} = \frac{dx_1}{-2p_2} = \frac{dp_2}{0} = \frac{dx_3}{-1} = \frac{dp_3}{0}$$

$\therefore dp_1 = 0 \qquad$ and $\quad dp_2 = 0$.

$\Rightarrow \quad F_1 \circ p_1 - a_1 \quad$ and $\Gamma_2 \circ p_2 = a_2$.

Now $(F_1, F_2) = 0$. Therefore these may be taken as additional equations.

$\therefore$ From (1) $p_3 = 1 - a_1^3 - a_2^2$.

Substituting these values in $dz = p_1dx_1 + p_2dx_2 + p_3dx_3$, we have

$$dz = a_1dx_1 + a_2dx_2 + \left(1 - a_1^3 - a_2^2\right)dx_3.$$

Integrating $\qquad z = a_1x_1 + a_2x_2 + \left(1 - a_1^3 - a_2^2\right)x_3 + a_3,$

which is the complete integral. **Ans.s**

Example 3:

Solve $(x_2 + x_3)(p_2 + p_3)^2 + z\,p_1 = 0. \qquad ...(1)$

Solution:

This equation is not of the form considered, as it involves the dependent variable z.

Putting $z = x_4$.

so that $$p_1 = \frac{\partial z}{\partial x_1} = \frac{\partial z}{\partial x_4}\cdot\frac{\partial x_4}{\partial x_1} = \frac{\partial x_4}{\partial x_1} = -\frac{\frac{\partial u}{\partial x_1}}{\frac{\partial u}{\partial x_4}} = -\frac{P_1}{P_4} \text{ (say)}$$

where $u = 0$ is an integral of (1).

Similarly, $p_2 = -\frac{P_2}{P_4}$ and $p_3 = -\frac{P_3}{P_3}$.

The given equation (1) reduces to

$$(x_2 + x_3)\left(-\frac{P_2}{P_4} - \frac{P_3}{P_4}\right)^2 + x_4\left(-\frac{P_1}{P_4}\right) = 0$$

$$\Rightarrow \quad (x_2 + x_3)(P_2 + P_3)^2 - x_4 P_1 P_4 = 0. \qquad \text{...(2)}$$

i.e. $\quad f \equiv (x_2 + x_3)(P_2 + P_3)^2 - x_4 P_1 P_4 = 0$

is an equation in four independent variables x_1, x_2, x_3, x_4 and not involving the dependent variable i.e., except by its partial derivative P_1, P_2, P_3, P_4.

The subsidiary equations are

$$\frac{dx_1}{x_4P_4} = \frac{dP_1}{0} = \frac{dx_2}{2(x_1+x_3)(P_2+P_3)} = \frac{dP_2}{(P_2+P_3)^2}$$

$$= \frac{dx_3}{-2(x_2+x_3)(P_2+P_3)} = \frac{dP_3}{(P_2+P_3)^2} = \frac{dx_4}{x_4P_1} = \frac{dP_4}{-P_1P_4}$$

$\therefore dP_1 = 0$

$$\Rightarrow \quad F_1 \equiv P_1 = a_1. \qquad \text{...(3)}$$

Similarly taking 4th and 6th members, we have

$$d P_2 - d P_3 = 0,$$

$$\therefore \quad F_2 \equiv P_2 - P_3 = a_2, \qquad \text{...(4)}$$

Taking last two members, we have

$$\frac{dx_4}{x_4} + \frac{dP_4}{P_4} = 0.$$

$$\therefore \quad F_3 \equiv x_4 P_4 = a_3. \qquad \text{...(5)}$$

Now, $(F_1, F_2) = \sum_{r=1}^{4}\left(\frac{\partial F_1}{\partial x_r}\frac{\partial F_2}{\partial P_r} - \frac{\partial F_1}{\partial P_r}\frac{\partial F_2}{\partial x_r}\right) = 0$.

Similarly, $(F_2, F_3) = 0$ and $(F_1, F_3) = 0$.

Therefore, equation (3), (4) and (5) may be taken as three additional equations.

Solving equations (3) and (5), we have

$P_1 = a_1$, $P_4 = \frac{a_3}{x_4}$, putting in (2), we have

$$(P_2 + P_3)^2 = a_1 a_3/(x_2 + x_3).$$

$$\therefore P_2 + P_3 = \pm \sqrt{\frac{a_1 a_3}{(x_2 + x_3)}}. \qquad ...(6)$$

Solving (4) and (6), we have $P_2 = \frac{a_2}{2} \pm \frac{1}{2}\sqrt{\frac{a_1 a_3}{(x_2 + x_3)}}$

and $\qquad P_3 = -\frac{a_2}{2} \pm \frac{1}{2}\sqrt{\frac{a_1 a_3}{x_2 + x_3}}$

Substituting these values in

$$du = P_1\, dx_1 + P_2\, dx_2 + P_3 dx_3 + P_4 dx_4.$$

We have $du = a_1 dx_1 + \left[\frac{a_2}{2} \pm \frac{1}{2}\sqrt{\frac{a_1 a_3}{x_2 + x_3}}\right] dx_2$

$$+ \left[-\frac{a_2}{2} \pm \frac{1}{2}\sqrt{\frac{a_1 a_3}{(x_2 + x_3)}}\right] dx_3 + \frac{a_3}{x_4}\, dx_4.$$

$$\Rightarrow \quad du = a_1 dx_1 + \frac{a_2}{2}(dx_2 - dx_3) \pm \sqrt{(a_1\, a_3)}\cdot\frac{dx_2 + dx_3}{\sqrt{(x_2 + x_3)}} + \frac{a_3}{x_4} dx_4.$$

Integrating, we have

$$u = a_1 x_1 + \frac{a_2}{2}\,(x_2 - x_3) \pm \sqrt{\{a_1 a_3\,(x_2 + x_3)\}} + a_3 \log x_4 + a_4$$

$\therefore$ u = 0 gives

$$a_1 x_1 + \frac{a_2}{2}(x_2 - x_3) \pm \sqrt{\{a_1 a_3\,(x_2 + x_3)\}} + a_3 \log z + a_4 = 0$$

$$\Rightarrow \quad \log z + \frac{a_1}{a_3}x_1 + \frac{a_2}{2a_3}(x_2 - x_3) \pm \sqrt{\left\{\frac{a_1}{a_3}(x_2 + x_3)\right\}} + \frac{a_4}{a_3} = 0.$$

If $\frac{a_1}{a_3} = A_1, \frac{a_2}{2a_3} = A_2, \frac{a_4}{a_3} = A_3.$

The complete integral of equation (1) is

$$\log z + A_1x_1 + A_2(x_2 - x_3) \pm \sqrt{\{A_1(x_2+x_3)\}} + A_3 = 0. \qquad \textbf{Ans.}$$

Example 4:

Find the complete integral of

$$p_3x_3(p_1 + p_2) + x_1 + x_2 = 0. \qquad ...(1)$$

(Delhi, Hons. 94)

Solution:

Here $f \equiv p_3x_3(p_1 + p_2) + x_1 + x_2 = 0$.

The subsidiary equations are

$$\frac{dx_1}{-p_3x_3} = \frac{dp_1}{1} = \frac{dx_2}{-p_3x_3} = \frac{dp_2}{1} = \frac{dx_3}{-x_3(p_1+p_2)} = \frac{dp_3}{p_3(p_1+p_2)}$$

Taking 2nd and 4th members, we have

$$dp_1 = dp_2$$

$$\therefore \quad F_1 \equiv p_1 - p_2 = a_1. \qquad ...(2)$$

Again taking 5th and 6th members, we have

$$\frac{dx_3}{x_3} + \frac{dp_3}{p_3} = 0$$

$$\therefore \log x_3 + \log p_3 = \log a_2$$

$$F_2 \equiv p_3x_3 = a_2 \qquad ...(3)$$

$$\text{Now, } (F_1, F_2) = \sum_{r=1}^{3}\left(\frac{\partial F_1}{\partial x_r}\frac{\partial F_2}{\partial p_r} - \frac{\partial F_1}{\partial p_r}\frac{\partial F_2}{\partial x_r}\right) = 0.$$

$\therefore$ Equation (2) and (3) may be taken as additional equations.

From equation (1) and (3), we have

$$a_2(p_1 + p_2) + x_1 + x_2 = 0$$

$$p_1 + p_2 = -\frac{x_1+x_2}{a_2} \qquad ...(4)$$

From equation (2) and (4) we have

$$p_1 = \frac{a_1}{2} - \frac{x_1+x_2}{2a_2} \quad \text{and} \quad p_2 = -\frac{a_1}{2} - \frac{x_1+x_2}{2a_2}$$

From equation (3) $\qquad p_3 = \dfrac{a_2}{x_3}$

Substituting these values in

$dz = p_1\, dx_1 + p_2\, dx_2 + p_3\, dx_3$, we have

$$dz = \frac{a_1}{2}(dx_1 - dx_2) - \frac{(x_1 + x_2)}{2a_2}(dx_1 + dx_2) + \frac{a_3}{x_3}dx_3$$

Integrating, we have

$$z = \frac{a_1}{2}(x_1 - x_2) - \frac{1}{4a_2}(x_1 + x_2)^2 + a_2 \log x_3 + a_3$$

$\Rightarrow$ $4a_2z = 2a_1a_2(x_1 - x_2) - (x_1 + x_2)^2 + 4a_2^2 \log x_3 + 4a_2a_3.$ **Ans.**

Example 5:

Find the complete integral of

$$p_1p_2p_3 + a_4^3x_1, x_2x_3a_4^3 = 0.$$

Solution:

Here $f \equiv p_1p_2p_3 + a_4^3\, x_1x_2x_3\, x_4^3 = 0.$ ...(1)

The subsidiary equations are

$$\frac{dx_1}{-p_2p_3} = \frac{dp_1}{p_4^3x_2x_3x_4^3} = \frac{dx_2}{-p_1p_3} = \frac{dp_2}{p_4^3x_1x_3x_4^3} = \frac{dx_3}{-p_1p_2}$$

$$= \frac{dp_3}{p_4^3x_1x_2x_4^3} = \frac{dx_4}{-3p_4^2x_1x_2x_3x_4^3} = \frac{dp_4}{3p_4^2x_1x_2x_3x_4^2}$$

Taking the first two members, we have

$$\frac{p_1dx_1}{-p_1p_2p_3} = \frac{x_1dp_1}{p_4^3x_1x_2x_3x_4^3}$$

$\Rightarrow$ $p_1dx_1 = x_1dp_1$ from (1)

$\Rightarrow$ $\dfrac{dp_1}{p_1} = \dfrac{dx_1}{x_1} \Rightarrow \log p_1 = \log x_1 + \log a_1$

$\therefore$ $F_1 \equiv p_1 - a_1x_1 = 0$ ...(2)

Similarly $F_2 \equiv p_2 - a_2x_2 = 0$ ...(3)

and $F_3 \equiv p_3 - a_3x_3 = 0$ ...(4)

Now $(F_1, F_2) = \displaystyle\sum_{r=1}^{4}\left(\frac{\partial F_1}{\partial x_r}\frac{\partial F_2}{\partial p_r} - \frac{\partial F_1}{\partial p_r}\frac{\partial F_2}{\partial x_r}\right) = 0.$

Similarly, $(F_2, F_3) = 0$ and $(F_1, F_3) = 0$.

Therefore, equation (2), (3) and (4) may be taken as three additional equations.

Solving equation (1), (2), (3) and (4), we have

$$p_1 = a_1x_2,\ p_2 = a_2x_2,\ p_3 = a_3x_3$$

$$p_4 = -\frac{(a_1a_2a_3)^{1/3}}{x_4}$$

Substituting these values in

$dz = p_1dx_1 + p_2dx_2 + p_3dx_3 + p_4dx_4$, we have

$$dz = a_1x_1dx_1 + a_2x_2\,dx_2 + a_3x_3dx_3 - \frac{(a_1a_2a_3)^{1/3}}{x_4}dx_4.$$

Integrating, we have

$$2z = a_1\,x_1^2 + a_2\,x_2^2 + a_3\,x_3^2 - (a_1a_2a_3)^{1/3}\log x_4 + a_4$$

which is the complete integral.

Example 6:

Find the complete integral of

$$2p_1x_1x_3 + 3p_2x_3^2 + p_2^2p_3 = 0.$$

Solution:

Here $f \equiv 2p_1x_1x_3 + 3p_2\,x_3^2 + p_2^2p_3 = 0.$ ...(1)

The subsidiary equations are

$$\frac{dx_1}{-2x_1x_3} = \frac{dp_1}{2px_3} = \frac{dx_2}{-3x_3^2 + 2p_2p_3} = \frac{dp_2}{0} = \frac{dx_3}{-p_2^2} = \frac{dp_3}{2p_1x_1 + 6p_2x_3}$$

Taking the first two members, we have

$$\frac{dx_1}{x_1} + \frac{dp_1}{p_1} = 0$$

$\therefore$ $F_1 \equiv p_1x_1 = a_1.$ ...(2)

Again taking 1st and 4th members, we have

$$dp_2 = 0$$

$\therefore$ $F_1 \equiv p_2 = a_2.$

Now, $$(F_1, F_2) = \sum_{r=1}^{3}\left(\frac{\partial F_1}{\partial x_r}\frac{\partial F_2}{\partial p_r} - \frac{\partial F_1}{\partial p_r}\frac{\partial F_2}{\partial x_r}\right) = 0.$$

Therefore (2) and (3) may be taken as additional equations.

$$\therefore \qquad p_1 = \frac{a_1}{x_1},\ p_2 = a_2$$

and from (1), $2a_1x_3\ 3a_2x_3^2 + x_2^2p_3 = 0$

$$\therefore \qquad p_3 = \frac{-1}{a_2^2}\left(2a_1x_3 + 3a_2x_3^2\right).$$

Substituting in $dz = p_1dx_1 + p_2dx_2 + p_2\ dx_2 + p_3\ dx_3$, we have

$$dz = \frac{a_1}{x_1}dx_1 + a_2dx_2 - \frac{1}{a_2^2}\left(2a_1x_3 + 3a_2x_3^2\right)dx_3.$$

Integrating, we have

$$z = a_1 \log x_1 + a_2 x_2 - \frac{1}{a_2^2}\left(a_1x_3^2 + a_2x_3^3\right) + a_3.$$

which is the complete integral. **Ans.**

EXERCISES

1. $lx + my + nz = \phi(x^2 + y^2 + z^2)$. **(Kanpur, 91)**

   [Ans. $(l + np)y + z(lq - mp) = (m + nq)x$]

2. $f(x + y + z, x^2 + y^2 - z^2) = 0$. (PCS (UP) 93)

   [Ans. $(y + z)p - (z + x)q = x - y$]

3. $z = f(x) + xg(y)$. **(Kanpur, 90)**

   [Ans. $\dfrac{\partial^2 z}{\partial y^2} = x\dfrac{\partial^2 z}{\partial x \partial y}$]

4. $z = f(x + ay) + F(x - ay)$.

   [Ans. $\dfrac{\partial^2 z}{\partial y^2} = a^2\dfrac{\partial^2 z}{\partial x^2}$]

5. $z = e^{my}\phi(x - y)$. **[Ans.** $p + q = mz$]

6. $z = F(x^2 + y^2)$. **(Meerut, 95 (BP))**

   [Ans. $yp - xq = 0$]

7. $z = f\left(\dfrac{y}{x}\right)$. **[Ans.** $yp - xq = 0$]

8. $z = a(x + y) + b$. **[Ans.** $p = q$]

9. $(x-h)^2+(y-k)^2+z^2=c^2$. [**Ans.** $z^2(p^2+q^2+1)=c^2$]

10. $z=axe^y+\frac{1}{2}a^2e^{2y}+b$. [**Ans.** $q=px=p^2$]

11. $az+b=a^2x+y$. [**Ans.** $pq=1$]

12. $z=ax+by+ab$. [**Ans.** $z=px+qy+pq$]

13. $z=ax+a^2y^2+b$. **(Kanpur 91)**

[**Ans.** $q=2yp^2$]

14. $(y^2+z^2)p-xyq=-3x$. (IAS 90)

[**Ans.** $\phi(y/z, x^2+y^2+z^2)=0$]

15. $y^2p-xyq=x(z-2y)$. **(Meerut, 94, Delhi, Hons. 95)**

[**Ans.** $f(x^2+y^2, zy-y^2)=0$]

16. $p_1+x_1p_2+x_1x_2p_3=x_1x_2x_3\sqrt{z}$.

[**Ans.** $f\left(4\sqrt{z}-x_3^2, 2x_3-x_2^2, 2x_2-x_1^2\right)=0$]

17. $p+q=1$. [**Ans.** $f(x-z, y-z)=0$]

18. $(x+2z)p+(4zx-y)q=2x^2+y$.
[**Hint.** Form subsidiary equations and use, y, x – 2z and 2x. – 1, – 1 as multipliers].

[**Ans.** $f(xy-z^2, x^2-y-z)=0$]

19. $p_1+p_2+p_3=4z$. [**Ans.** $f\left(\frac{z}{e^{3x_1}}, \frac{z}{e^{4x_2}}, \frac{z}{e^{4x_3}}\right)=0$]

20. $(x_1p_1+2x_2p_2+3x_3p_3+4x_4p_4)=0$.

[**Ans.** $f\left(z, x_1^2x_2^{-1}, x_1^3x_3^{-1}, x_1^4x_4^{-1}\right)=0$]

21. $\frac{y^2z}{x}p-xzq=y^2$. [**Ans.** $f(x^2-z^2, x^3+y^3)=0$]

22. $x^2p+y^2q=z^2$. [**Ans.** $\frac{1}{x}-\frac{1}{y}=f\left(\frac{1}{x}-\frac{1}{z}\right)$]

23. $p\tan x+q\tan y=\tan z$. **(Kanpur, 91(P); PCS (UP) 90)**

[**Ans.** $\frac{\sin z}{\sin y}=f\left(\frac{\sin x}{\sin y}\right)$]

24. $x_2x_3z\,p_1+x_3x_1z\,p_2+x_1x_2z\,p_3=x_1x_2x_3$. **(Kanpur, 92)**

[**Ans.** $f\left(x_2^2-x_1^2, x_3^2-x_1^2, z^2-x_1^2\right)=0$]

25. $xzp = yzp = xy.$ [**Ans.** $xy - z^2 = f\left(\frac{x}{y}\right)$]

26. $p - q = z/(x + y).$ [**Ans.** $(x + y) \log z = x + f(x + y)$]

27. $z = px + qy + pq.$ **(Meerut, 90)**
[**Ans.** $z = ax + by + ab$]

28. $z = px + qy - p^2q.$ [**Ans.** $z = ax + by - a^2b$]

29. $z = px + qy - 2p - 3q.$ [**Ans.** $z = ax + by - 2a - 3b$]

30. $x(1 + y)p = y(1 + x)q.$ **(Agra, (TDC) 91)**
[**Ans.** $z = ax + by - 2a - 3b$]

31. $p^2 = q + x.$ [**Ans.** $3z = \pm 2(x + a)^{3/2} + b$]

32. $p(z + p) + q = 0.$ **(Kanpur, 92)**
[**Ans.** $(z + a)e^{x+ay} = b$]

33. $p - 3x^2 = q^2 - y.$ **(Meerut, 96)**
[**Ans.** $z = x^3 + ax \pm \frac{2}{3}(y + a)^{3/2} + b$]

34. $p(1 + q) = qz.$ [**Ans.** $az - 1 = be^{x+ay}$]

35. $pq = p + q.$ [**Ans.** $z = ax + \frac{a}{a-1}y + c$]

36. $p^2 - q^2 = \lambda.$ [**Ans.** $z = ax + \sqrt{(a^2 - \lambda)}y + c$]

37. $p = e^q.$ [**Ans.** $z = ax + y \log a + c$]

38. $p^2 + q^2 = m^2.$ **(Meerut, 93 (P); Kanpur, 90)**
[**Ans.** $z = ax + \sqrt{(m^2 - a^2)}y + c$]

39. Apply Jacobi's Method to find the complete integral of

$$p_1x_1 + p_2x_2 = p_3^2.$$

[**Ans.** $z = z_1 \log x_1 + a_2 \log x_2 + x_3 \sqrt{(a_1 + a_2)} + a_3$]

40. $pq + 4(px + qy + z) = 0.$ **(Meerut, 95)**
[**Ans.** $(t - u)(t + 2u)^2 =$ where $t^2 = (ax - y)^2 - az$ and $u = ax + y$]

41. $zpq = p + q.$ **(G.N.U.A., 91)**
[**Ans.** $z^2 = 2(a + 1)(x + y/a) + b$]

42. $q = px + p^2$. **(Meerut 94)**

[**Ans.** $z = -\frac{x^2}{4} \pm \frac{1}{2}\left[\frac{x}{2}\sqrt{(x^2+4a)} + 2a \log\left\{x + \sqrt{(x^2+4a)}\right\}\right] + ay + b$]

43. $z = pq$. **(Meerut, 94)**

[**Ans.** $2\sqrt{z} = \sqrt{ax} + \frac{1}{\sqrt{a}}y + b$]

44. $p^2 - y^2q = y^2 - x^2$. **(Meerut, 94)**

[**Ans.** $z = \frac{x}{2}\sqrt{(a^2 - x^2)} + \frac{a^2}{2}\sin^{-1}\frac{x}{a} - \frac{a^2}{y} - y + b$]

45. $2z + p^2 + qy + 2y^2 = 0$.

(Rohilkhand, 93; Madras, 93(P); B.H.U. 97, 98)

[**Ans.** $y^2\{(x-a)^2 + y^2 + 2z\} = b$]

46. $q = 3y^2$. [**Ans.** $z = ax + y^3 + b$]

47. $2(pq + py + qx) + x^2 + y^2 = 0$.

(Raj., 93; Kanpur, 93; Rohilkhand, 95)

[**Ans.** $2z = ax - x^2 + ay - y^2 + \frac{1}{2}(x-y)\sqrt{\{2(x-y)^2 + a^2\}}$

$+ \frac{a^2}{2^{3/2}}\log\left[\left\{\sqrt{2}(x-y)\right\} + \sqrt{\{2(x-y)^2 + a^2\}}\right] + b$]

48. $yzp^2 - q = 0$. **(Rohilkhand, 90)**

[**Ans.** $z^2(a - y^2) = (x + b)^2$]

49. $f(p, q) = 0$. **(G.N.U.A., 91)**

[**Ans.** $z = ax + f(a)y + b)$]

50. Find the general integral of the equation

$$(x - y)p + (y - x - z)q = z$$

and the particular solution through the circle $z = 1$, $x^2 + y^2 = 1$.

[**Ans.** $(x - y + z)^2 + z^4(x + y + z)^2 - 2z^2(x - y + z) - 2z^4(x + y + z) = 0$

51. Find the equation of the integral surface of the differential equation

$2y(z - 3)p(2x - z)q = y(2x - 3)$

which passes through the circle $z = 0,\ x^2 + y^2 = 2x.$

[**Ans.** $x^2 + y^2 - z^2 - 2x + 4z = 0$]

52. Find the general integral of the partial differential equation

$$(2xy - 1)\,p + (z - 2x^2)\,q = 2\,(x - yz)$$

and also the particular integral which passes through the line

$x = 1,\ y = 0.$ [**Ans.** $x^2 + y^2 - xz - y - z - 1 = 0$]

53. Find the integral surface of the linear partial differential equation

$$x\,(y^2 + z)\,p - y\,(x^2 + z)\,q = (x^2 - y^2)\,z$$

which contains the straight line $x + y = 0,\ z = 1.$ **(Kanpur, 97)**

[**Ans.** $x^2 + y^2 + 2xyz - 2z + 2 = 0$]

54. $(p_1 + x_1)^2 + (p_2 + x_2)^2 + (p_3 + x_3)^2 = 3\,(x_1 + x_2 + x_3).$

$$= -(a_1 + a_2)\,x_1 + (2a_1 - a_2)\,y_2 + (-a_1 + 2a_2)\,x_3 - \frac{1}{2}\left(x_1^2 + x_2^2 + x_3^2\right)$$

$$\pm \frac{2}{3}\left(x_1 + x_2 2a_1^2 + 2a_1^2 + 2a_1 a_3 - 2a_2^2\right)^{3/2} + a_3\,]$$

55. $p_1^2 + p_2p_3 - z\,(p_2 + p_3) = 0.$

[**Ans.** $(1 + a_1 a_2)\log z = (a_1 + a_2)\,(x_1 + a_1 x_2 + a_2 x_3 + a_3).$

56. $x_3^2\ p_1^2\ p_2^2\ p_3^2 + p_1^2\ p_2^2 - p_3^2 = 0.$

[**Ans.** $z = a_1x_1 + a_2x_2 \pm \sin^{-1}(a_1\, a_2\, a_3) + a_3$]

57. Apply Charpit's Method to solve $(p + q)\,(px + qy) = 1.$

Apply Jacobi's Method to find the complete integrals of the following:

[**Ans.** $\sqrt{(1+a)}z = 2\sqrt{(x+ay)} + b$]

58. $z = px + qy + \sqrt{(\alpha p^2 + \beta q^2 + \lambda)}.$

[**Ans.** $z = ax + by + \sqrt{(\alpha a^2 + \beta b^2 + \gamma)}$

59. $q\,(p - \cos x) = \cos y.$ [**Ans.** $az = a^2x + a\sin x + \sin y + ab$]

60. $4z = pq.$ [**Ans.** $az = (x + ay + b)^2$]

61. $p^3 + q^3 - 3\,pqz = 0.$ **[Ans.** $3a\,(x + ay + b) = (1 + a^3)\log z$]

62. $(x_3 - x_2)\,p_1 + x_2p_2 - x_3p_3 + x_2^2 - (x_2x_1 + x_2x_3) = 0.$

[Ans. $f\,(z - x_1x_2,\ x_1 + x_2 + x_3,\ x_2x_3) = 0$]

63. $px^2 + qy^2 = nxy.$ **[Ans.** $f\left[\dfrac{y-x}{xy}, \dfrac{nxy}{-x+y}\log\dfrac{y}{x} - x\right] = 0$]

64. $xy^2p - y^3q + axz = 0.$ **[Ans.** $\log z = -\dfrac{ax}{3y^2} + \phi\,(x, y)$]

65. $(x^3 + 3xy^2)\,p + (y^3 + 3x^2y)\,q = 2\,(x^2 + y^2)\,z.$

[Ans. $\dfrac{1}{(x-y)^2} - \dfrac{1}{(x+y)^2} = f\left(\dfrac{z^2}{xy}\right).$

4

Partial Differential Equations with Constant Coefficients

4.1 Introduction

In this chapter, these differential equation will be considered which are linear w.r.t the dependent variables and all its differential coefficient and in which the various terms are multiplied by constants only the general form of the equation is

$$f\left(\frac{\partial}{\partial x}, \frac{\partial}{\partial y}\right)z = V(x, y) \text{ where f is}$$

a rational integral algebrical function with constant coefficient and "V" may be any function of the independent variable x and y.

4.2 Homogeneous Linear Equations with Constant Coefficients

A partial differential equation, which is linear with respect to the dependent variable and its partial differential coefficients, and in which the coefficients are not functions of the dependent variable but merely constants is called a **Linear Partial Differential Equation with Constant Coefficients,** such as

$$\left(\frac{\partial^n z}{\partial x^n} + A_1 \frac{\partial^n z}{\partial x^{n-1}\partial t} + A_2 \frac{\partial^n z}{\partial x^{n-2}\partial y^2} + \ldots + A_n \frac{\partial^n z}{\partial y^n}\right)$$

$$+\left(B_0 \frac{\partial^{n-1} z}{\partial x^{n-1}} + \ldots\right) + \ldots + \left(M\frac{\partial z}{\partial x} + N\frac{\partial z}{\partial y}\right) + Bz = f(x, y)$$

where $A_1,...,A_n$, $B_0,...$, M, N, B are all constants.

A Linear Homogeneous Partial Differential Equation is of the form where all the derivatives are of the same order, such as

$$\frac{\partial^n z}{\partial x^n} + A_1 \frac{\partial^n z}{\partial x^{n-1} \partial y} +...+ A_n \frac{\partial^n z}{\partial y^n} = f(x, y).$$

Denoting $\frac{\partial}{\partial x}$ by D and $\frac{\partial}{\partial y}$ by D', it can be written as

$$(D_n + A_1 D^{n-1} D' + A_2 D^{n-2} D'^2 +...+ A_n D'^n) z = f(x, y)$$

$$\Rightarrow \quad \phi (D D') z = f(x, y) \qquad ...(1)$$

4.3 Solution of Linear Partial Differential Equation

As in the case of ordinary linear differential equations the complete solution of (1) will consist of two parts, (a) the complementary function (C.F) and (b) the **Particular Integral (P.I).**

The complementary function is the solution of

$$\phi (D, D') z = 0$$

4.4 To Find the Complementary Function (Meerut, 90)

Let the given equation be

$$\phi (D, D') z = f(x, y) \qquad ...(1)$$

Let z = F(y + mx) be its complementary function.

i.e., the solution of ϕ (D, D') z = 0 ...(2)

Now, $Dz = \frac{\partial z}{\partial x} = mF'(y + mx)$

$$D^2 z = \frac{\partial^2 z}{\partial x^2} = m^2 F''(y + mn)$$

... ...

... ...

$$D^n z = \frac{\partial^2 z}{\partial x^n} = m^n F^n (y + mx)$$

and $D' z = \frac{\partial z}{\partial y} = F'(y + mx)$

... ...

... ...

$$D'z = \frac{\partial^n z}{\partial y^n} = F^n(y + mx).$$

Thus, $D^r D'^{n-r} z = \dfrac{\partial^n z}{\partial y^r\, \partial y^{n-r}} = m^r F^n(y - mx).$

$\therefore$ From (2), we have

$$\phi(D, D') = (m^n + A_1 m^{n-1} + A_2 m^{n-2} + \ldots + A_n)\, F^n\,(y + mx) = 0,$$

which is satisfied if

$$m^n + A_1 m^{n-1} + A_2 m^{n-2} + \ldots + A_n = 0$$

The equation is known as the auxiliary equation.

If $m_1, m_2, \ldots, m_n$ are the distinct roots of the auxiliary equation, then C.F. of (1) is

$$z = \psi_1(y + m_1 x) + \psi_2(y + m_2 x) + \ldots + \psi_n(y + m_n x).$$

Note: The auxiliary equation is obtained by putting D = n an D' = 1 in ϕ (D, D') = 0.

4.5 When the Auxiliary Equation has Equal (Repeated) Roots

Let us consider the equation having two roots equal as

$$(D - mD')^2 z = 0 \qquad \ldots(1)$$

Putting $(D - mD')\, z = u$, ...(2)

we have $(D - nD')\, u = 0$,

$\therefore$ $u = f(y + mu)$

$\Rightarrow$ $(D - mD')\, z = f(y + mx)$

$\Rightarrow$ $p - mq = f(y + mx)$

The subsidiary equations are

$$\frac{dx}{1} = \frac{dy}{-m} = \frac{dz}{f(y + mx)}.$$

Taking the first two members, we have

$$dy + m\, dx = 0.$$

$\therefore$ $y - mz = a.$

Again taking the first and third members, we have

$dz = f(y + mx)\, dx \quad \Rightarrow \quad dz = f(a)\, dx$

$\therefore\ z = x\, f(a) + b \quad \Rightarrow \quad z = x\, f(y + mx) + \phi(y + mx).$

Thus, in general if the auxiliary equation has r roots equal, then

$$z = f_1(y + mx) + x\, f_2(y + mx) + \ldots + x^{r-1}\, f_r(y + mx)$$

Example 1:

Solve $\frac{\partial^4 z}{\partial x^4} - \frac{\partial^4 z}{\partial y^4} = 0.$

Solution:

We have $\frac{\partial^4 z}{\partial x^4} - \frac{\partial^4 z}{\partial y^4} = 0.$

The given equation can be written as

$(D^4 - D'^4)\, z = 0.$

A.E. is $m^4 - 1 = 0$

$\Rightarrow \quad (m - 1)(m + 1)(m^2 + 1) = 0.$

$\therefore m = 1, -1, \pm i.$

Hence the solution is

$z = f_1(y + x) + f_2(y - x) + f_3(y + ix) + f_4(y - ix).$ **Ans.**

Example 2:

Solve $2r + 5s + 2t = 0$ **(Meerut, 95)**

Solution:

We have $2r + 5s + 2t = 0$

The given equation can be written as

$(2D^2 + 5DD' + 2D'^2)\, z = 0$

A.E. is $2m^2 + 5m + 2 = 0$

$\Rightarrow \quad (2m + 1)(m + 2) = 0 \quad \therefore \quad m = \frac{1}{2}, -2$

$\therefore z = f\left(y - \frac{x}{2}\right) + \psi(y - 2x).$

or $z = \phi(2y - x) + \psi(y - 2x).$ **Ans.**

Example 3:

Solve $(D^3 - 4D^2D' + 4DD'^2)z = 0.$ **(Meerut, 91 (P))**

Solution:

We have $(D^3 - 4D^2D' + 4DD'^2)z = 0.$

Here A.E. is $m^3 - 4m^2 + 4m = 0$

$\Rightarrow \quad (m - 2)^2 = 0 \quad \therefore \quad m = 0, 2, 2.$

Hence the solution is

$z = \phi(y) + f_1(y + 2x) + x\, f_2(y + 2x).$ **Ans.**

Example 4:

Solve $r = a^2t$. **(Meerut, 91 (P))**

Solution:

We have $r = a^2t$

The given equation can be written as

$$(D^2 - a^2 D'^2) z = 0.$$

AE. is $m^2 - a^2 = 0.$

$\therefore$ $(m - a)(m + a) = 0 \Rightarrow m = -a, a,$

Hence the solution is

$$y = f_1(y + ax) + f_2(y - ax)$$ **Ans.**

Example 5:

Solve $\dfrac{\partial^3 z}{\partial x^3} + \dfrac{\partial^3 z}{\partial x\, \partial y^2} + 6\dfrac{\partial^3 z}{\partial y^3} = 0$

Solution:

The given equation can be written be

$$(D^3 - 7DD'^2 + 6D'^3) z = 0.$$

A.E is $m^3 - 7m + 6 = 0$

or $(m + 1)(m - 2)(m + 3) = 0.$ $\therefore$ $m = 1, 2. -3.$

Hence the solution is

$$z = f_1(y + x) + f_2(y + 2x) + f_3(y - 3x).$$ **Ans.**

Example 6:

Solve $(D^4 - 2D^3D' + 2DD'^3 - D'^4) z = 0.$

Solution:

We have $(D^4 - 2D^3D' + 2DD'^3 - D'^4) z = 0.$

A.E. is $(m^4 - 2m^3 + 2m - 1) = 0$

$\Rightarrow$ $(m + i)(m - 1)^3 = 0$ $\therefore$ $m = -1, 1, 1, 1.$

Hence the solution is

$$z = f_1(y - x) + f_2(y + x) + xf_3(y + x) + x^2f_4(y + x).$$ **Ans.**

Example 7:

Solve $\dfrac{\partial^3 z}{\partial x^3} - 3\dfrac{\partial^3 z}{\partial y^2 \partial y} + 2\dfrac{\partial^3 z}{\partial x\, \partial y^2} = 0.$ **(Meerut, 99)**

Solution:

We have $\frac{\partial^3 z}{\partial x^3} - 3\frac{\partial^3 z}{\partial y^2 \partial y} + 2\frac{\partial^3 z}{\partial x\, \partial y^2} = 0.$

The given equation can be written as

$$(D^3 - 3D^2 D' + 2DD'^2)\, z = 0.$$

A.E is $m^3 - 3m^2 + 2m = 0$

$\Rightarrow$ $m(m - 1)(m - 2) = 0 \quad \therefore \quad m = 0, 1, 2.$

Hence the solution: is

$z = f_1(y) + f_2(y + x) + f_3(y + 2x).$ **Ans.**

Example 8:

Solve $(D^3 - 6D^2D' + 11\, DD'^2 - 6D'^3)\, z = 0.$

Solution:

We have $(D^3 - 6D^2D' + 11\, DD'^2 - 6D'^3)\, z = 0.$

Here A. e. is $m^3 - 6m^2 + 11m - 6 = 0.$

$\therefore$ $(m - 1)(m - 2)(m - 3) = 0$

$\Rightarrow$ $m = 1, 2, 3.$

Hence the solution is

$z = f_1(y + x) + f_2(y + 2x) + f_3(y + 3x).$ **Ans.**

Example 9:

Solve $25r - 40s + 16t = 0$ **(Kanpur, 92)**

Solution:

We have $25r - 40s + 16t = 0$

The given equation can be written as

$$(25D^2 - 40DD' + 16D'^2)\, z = 0.$$

A.E. is $25m^2 - 40m + 16 = 0$

$\Rightarrow$ $(5m - 4)^2 = 0 \quad \therefore \quad m = \frac{4}{5}, \frac{4}{5}.$

Hence the solution is

$$z = f_1\left(y + \frac{4}{5}x\right) + xf_2\left(y + \frac{4}{5}x\right)$$

$\Rightarrow$ $z = \phi_1(5y + 4x) + x\phi_2(5y + 4x).$

4.6 The Particular Integral

Let, the differential equation by

$$f(D, D')\, z = \phi(x, y).$$

We know that, the general value of z is the sum of complementary function and the particular integral.

Particular integral may be written as

$$\frac{1}{f(D, D')} \cdot \phi(x, y)$$

We may treat that symbolic function D and D' as we did that of D alone, that is, factorising it, resolving interpartial fractions, or expanding in an infinite series.

Example 1:

Solve $(D^2 + 3DD' + 2D'^2)\, z = x + y.$

Solution:

A.E. is $m^2 + 3m + 2 = 0$

$\Rightarrow \quad (m + 1)(m + 2) = 0 \quad \therefore \quad m = -1, -2$

C.F. $= f_1(y - x) + f_2(y - 2x)$

Now, P.I. $= \dfrac{1}{D^2 + 3DD' + 2D'^2}(x + y)$

$$= \frac{1}{D^2}\left(1 + \frac{3D'}{D} + 2\frac{D'^2}{D^2}\right)^{-1} \cdot (x + y)$$

$$= \frac{1}{D^2}\left(1 + \frac{3D'}{D} \ldots\right) \cdot (x + y) = \frac{1}{D^2}(x + y) - \frac{3}{D^3} D'.(x + y)$$

$$= \frac{x^3}{6} + \frac{x^2}{2} \cdot y - \frac{3}{D^2} \cdot 1 = \frac{x^3}{6} + \frac{1}{2}x^2 y - 3\frac{x^3}{6}$$

$$= \frac{1}{3}x^3 + \frac{1}{2}x^2 y.$$

$\therefore$ The general solution is

$$z = f_1(y - x) + f_2(y - 2x) - \frac{1}{3}x^3 + \frac{1}{2}x^2 y.$$ **Ans.**

Example 2:

Solve $(2D^2 - 5DD' + 2D'^2)\, z = 24\,(y - x).$ **(IAS, 97; Meerut, 97)**

Solution:

We have $(2D^2 - 5DD' + 2D'^2)\, z = 24\,(y - x)$

Here A.E. is $2m^2 - 5m + 2 = 0$.

$\Rightarrow \quad (2m - 1)(m - 2) = 0. \quad \therefore \quad m = \frac{1}{2}, 2$

C.F. $= f_1(2y + x) + f_2(y + 2x)$.

Now, P.I. $= \frac{1}{2D^2 - 5DD' + 2D'^2} \cdot 24\,(y = x)$

$$= \frac{1}{2D^2}\left(1 - \frac{5D'}{2D} + \frac{D'^2}{D^2}\cdots\right)^{-1} . 24(y - x)$$

$$= \frac{1}{2D^2}\left(1 - \frac{5D'}{2D}\cdots\right) . 24(y = x)$$

$$= \frac{1}{2D^2} . 24\,(y - x) + \frac{5}{4D^3} . 24$$

$$= 12\left(\frac{x^2}{2}y - \frac{x^3}{6}\right) + \frac{5}{4} . 24\frac{x^3}{6} = 6x^2y + 3x^3.$$

$z = f_1(2y + x) + f_2(y + 2x) + 6x^2y + 3x^3$. **Ans.**

Example 3:

Solve $\frac{\partial^3 z}{\partial x^3} - \frac{\partial^3 z}{\partial y^3} = x^3y^3$. **(Kanpur, 94; Agra, 98)**

Solution:

We have $\frac{\partial^3 z}{\partial x^3} - \frac{\partial^3 z}{\partial y^3} = x^3y^3$.

The given equation can be written as

$$(D^2 - D'^3)z = x^3y^3.$$

A.E. is $m^3 - 1 = 0$

$$\Rightarrow \quad (m - 1)(m - \omega)(m - \omega^2) = 0$$

where ω is one of the cube roots of unity

$\therefore \quad m = 1, \omega, \omega^2$.

C.F. $= f_1(y + x) + f_2(y + \omega x) + f_3(y + \omega^2 x)$.

$$\text{Now P. I.} = \frac{1}{D^3 - D'^3} x^3y^3 = \frac{1}{D^3}\left(1 - \frac{D'^3}{D^3}\right)^{-1} \cdot x^3y^3$$

$$= \frac{1}{D^3}\left(1 + \frac{D'^3}{D^3} + \ldots\right). x^3y^3 = \frac{1}{D^3} x^3y^3 + \frac{1}{D^6}.(6x^3)$$

$$= \frac{x^6y^3}{4.5.6} + \frac{6x^9}{4.5.6.7.8.9} = \frac{x^6y^3}{120} + \frac{x^3}{10080}$$

Hence the solution is

$$z = f_1(y + x) + f_2(y + \omega x) + f_3(y = \omega^2 x) + \frac{1}{120} x^6y^3 + \frac{x^9}{10080}.$$

Example 4:

Solve $r + (a + b) s + adt = xy$.

(Kanpur, 91, 95; Agra, 91; Meerut, 99, 93 (P)

Solution:

We have $r + (a + b) s + adt = xy$.

The given equation can be written as

$$\{D^2 + (a + b) DD' + ab D'^2\} z = xy.$$

A.E. is $m^2 + (a + b) m + ab = 0$

$\Rightarrow$ $(m = a)) (m + b) = 0$ $\therefore$ $m = -a, -b$

C.F. $= f_1(y - ax) + f_2(y - bx)$.

$$\text{Now P.I.} = \frac{1}{D^2 + (a + b) DD' + ab D'^2}. xy$$

$$= \frac{1}{D^2}\left\{1 + (a + b)\frac{D'}{D} + ab\frac{D'^2}{D^2}\right\}^{-1}. xy$$

$$= \frac{1}{D^2}\left\{1 - (a + b)\frac{D'}{D}\ldots\right\}. xy$$

$$\frac{1}{D^2} xy - (a + b). = \frac{1}{D^2} x = \frac{1}{6}x^3y - \frac{1}{24}(a + b)x^4.$$

Hence the solution is

$$z = f_1(y - ax) + f_2(y = bx) + \frac{1}{6}x^3y - \frac{1}{24}(a + b)x^4.$$ **Ans.**

Example 5:

Find a real function V of x and y, reducing to zero when y = 0 and satisfying

$$\frac{\partial^2 V}{\partial x^2} + \frac{\partial^2 V}{\partial y^2} = -4\pi(x^2 + y^2).$$ **Ans.**

Solution:

The real function V will be given by the P.I. of the given equation.

Note P.I. $= \frac{1}{D^2 + D'^2} \{-4\pi(x^2 + y^2)\}.$

$$= \frac{1}{D^2}\left[1 + \frac{D'^2}{D^2}\right]^{-1} . \{-4\pi(x^2 + y^2)\}.$$

$$= \frac{1}{D^2}\left[1 - \frac{D'^2}{D^2} \cdots\right] . \{-4\pi(x^2 + y^2)\}.$$

$$= \frac{1}{D^2}\left\{-4\pi\,(x^2 + y^2)\right\} = \frac{1}{D^2} \cdot \{-4\pi.\ 2\}$$

$$= -4\pi\left[\frac{x^4}{12} + \frac{x^2y^2}{2}\right] - \left[-8\pi.\ \frac{x^4}{24}\right]$$

$$= -\frac{\pi x^4}{3} - 2\pi x^2 y^2 + \frac{\pi x^4}{3}$$

$= -2\pi x^2y^2. \quad \therefore \quad V = -2\pi x^2y^2.$ **Ans.**

Example 6:

Solve $(D^2 - 2DD' + D'^2)\, z = 12xy.$ **(Meerut, 91)**

Solution:

We have $(D^2 - 2DD' + D'^2)\, z = 12xy.$

Here A.E. is $m^2 - 2m + 1 = 0,$

$\Rightarrow \quad (m - 1)^2 = 0 \quad \therefore \quad m = 1, 1.$

$\therefore \quad$ C.F. $= f_1(y + x) + xf_2(y + x).$

Now P.I. $= \frac{1}{D^2 - 2DD' + D'^2}\, 12\,xy$

$$= \frac{1}{(D - D')^2} . 12xy = \frac{1}{D^2}\left(1 - \frac{D'}{D}\right)^{-2} . 12xy$$

$$= \frac{1}{D^2}\left(1 + \frac{2D'}{D} + \ldots\right). 12xy$$

$$= \left(\frac{1}{D^2} + \frac{2D'}{D^3} + \ldots\right) 12xy = 2x^3 y + \frac{2}{D^3} 12x = 2x^3y + x^4.$$

$\therefore$ The general solutions is

$z = f_1(x + y) + xf_1(y + x) + 2x^3y + x^4.$ **Ans.**

Example 7:

Solve $(D - 6DD' + 9D'^2)\, z = 12x^2 + 36xy$

Solution:

We have $(D - 6DD' + 9D'^2)\, z = 12x^2 + 36xy.$

Here A.E. is $m^2 = 6m + 9 = 0$

$\Rightarrow \quad (m - 3)^2 = 0 \quad \therefore \quad m = 3, 3.$

C.F. $= f_1(y + 3x) + xf_2(y + 3x),$

$$\text{Now, P.I.} = \frac{1}{D^2 - 6DD' + 9D'^2} \cdot (12x^2 + 36xy)$$

$$= \frac{1}{D^2}\left(1 - \frac{3D'}{D}\right)^{-2} \cdot (12x^2 + 36xy)$$

$$= \frac{1}{D^2}\left(1 - \frac{6D'}{D} + 27\frac{D'2}{D^2} + \ldots\right) (12x^2 + 36xy)$$

$$= \frac{1}{D^2} (12x^2 + 36xy) + \frac{6}{D^3} D' (12x^2 + 36xy)$$

$$= x^4 + 6x^3y = \frac{6}{D^3} . 36x = x^4 + 6x^3y + 9x^4 = 10x^4 + 6x^3y.$$

$\therefore$ The general solution is

$z = f_1(y + 3x)\, xf_2(y + 3x) + 10x^4 + 6x^3y.$ **Ans.**

4.7 Short Method

When $\phi(x, y)$ is a function of $ax + by$, shorter method may be used.

Now, $D\psi(ax + by) = a\psi'(ax + by)$

$D^2\psi(ax + by) = a^2\psi''(ax + by)$

$D'\psi(ax + by) = b\psi'(ax + by)$

$D'^2\psi(ax + by) = b^2\psi''(ax + by)$

where $\psi', \psi'', \ldots,$ are the derivatives of ψ w.r.t. $ax + by$

Hence, $f(D, D')\ \psi\ (ax + by) = f(a, b)\ \psi^n\ (ax + by)$

when n is the degree of $f(D, D')$

and ψ^n is the nth derivative of ψ w.r.t. $ax + by$

Consequently $\dfrac{1}{f(D, D')}\ \psi^n(ax + by) = \dfrac{1}{f(a, b)}\ \psi(ax + by)$

Provided $f(a, b) \neq 0$.

Working Method: To find the P.I. $\dfrac{1}{f(D, D')}\ y^n(ax + by)$

where $f(D, D')$ is a rational integral homogeneous function of degree n, integrate $\psi^n(ax + y)$ n times w.r.t. $ax + by$ considered as one variable and then divide the result by $f(a, b)$.

4.8 Exceptional Case. When f(a, b) = 0

Let us consider the equation

$$(bD - aD')\ z = x^r\ \psi(ax + by)$$

$$\Rightarrow \quad bp - aq = x^r\ \psi(ax + by).$$

$\therefore$ subsidiary equations are

$$\frac{dx}{b} = \frac{dy}{-a} = \frac{dz}{x^r \psi(ax + by)}.$$

First two members give $ax + by = c$ (const.)

From the first and last members, we have

$$dz = \frac{x^r}{b}\ y(ax + by)\ dx = \frac{x^r}{b}\ y(c)\ dx$$

$$\therefore z = \frac{x^{r+1}}{b(r + 1)}\ \psi(c) \qquad \Rightarrow z = \frac{x^{r+1}}{b(r + 1)}\ \psi(ax + by)$$

$$\therefore z = \frac{1}{(bD - aD')}\ x^r\ \psi(ax + by) = \frac{x^{r+1}}{b(r + 1)}\ \psi(ax + by) \qquad \ldots(1)$$

Hence, $z = \dfrac{1}{(bD - aD')^n}\ \psi(ax + by)$

$$z = \frac{1}{(bD - aD')^{n-1}} \frac{1}{(bD - aD')} \cdot \psi(ax + by)$$

$$= \frac{1}{(bD - aD')^{n-1}} \cdot \frac{x}{b}\ \psi(ax + by) \text{ form (1).}$$

$$= \frac{1}{(bD - aD')^{n-2}} \frac{1}{(bD - aD')} \frac{x}{b}\ \psi(ax + by)$$

$$= \frac{1}{(bD - aD')^{n-2}} \frac{1}{b} \cdot \frac{x^2}{2b} \ \psi(ax + by)$$

... ...

... ...

$$= \frac{x^n}{b^n \ n!} \ \psi(ax + by)$$

$$\therefore \ \frac{1}{(bD - aD')^n} \ y(ax + by) = \frac{x^n}{b^n \ n!} \ y(ax + by).$$

Example 1:

Find P.I. of

$$(D^3 - 10D^2D' + D^3) \ z = \cos(2x + 3y).$$

Solution:

We have $(D^3 - 10D^2D' + D^3) \ z = \cos(2x + 3y)$.

$$\text{Here P.I.} = \frac{1}{D^3 - 10D^2D' + D'^3} \ \cos(2x + 3y)$$

$$= \frac{1}{2^3 \ 10D^2 . 3 + 3^3} \cdot \{-\sin(2x + 3y)\}$$

$$= \frac{1}{85} \ \sin(2x + 3y).$$ **Ans.**

Example 2:

Solve $(D^2 - 5DD' + 4D^2) \ z = \sin(4x + y)$. **(Garhwal, 92)**

Solution:

We have $(D^2 - 5DD' + 4D^2) \ z = \sin(4x + y)$.

Here A.E. is $m^2 - 5m + 4 = 0$

$\Rightarrow \quad (m - 1)(m - 4) = 0 \qquad \therefore \quad m = 1, 4.$

C.F. $= f_1(y + x) + f_2(y + 4x)$.

$$\text{P.I.} = \frac{1}{D^2 - 5DD' + 4D'^2} \ \sin(4x + y) \qquad \because \ f(a, b) = 0$$

$$= \frac{1}{(D - 4D')} \cdot \frac{1}{(D - D')} \ \sin(4x + y)$$

$$= \frac{1}{(D - 4D')} \cdot \frac{1}{4 - 1} \ \{-\cos(4x + y)\}$$

$$= \frac{1}{(D - 4D')} \cdot \left\{-\frac{1}{3}\cos(4x + y)\right\} \qquad \because \ f(a, b) = 0$$

$$= \frac{x}{3} \cos(4x + y),$$

The solution is $z = f_1(y + x) + f_2(y + 4x) - 1/3\ x \cos(4x + y)$. **Ans.**

Example 3:

Solve $4r - 4s + t = 16 \log(x + 2y)$.

(Meerut, 94 (P), 98; Kanpur, 86; Rohilkhand, 93, 95, 98)

Solution:

We have $4r - 4s + t = 16 \log(x + 2y)$.

The equation can be written as

$$(4D^2 - 4DD' + D'^2)\, z = 16 \log(x + 2y)$$

A.E. is $\quad 4m^2 - 4m + 1 = 0$

$$\Rightarrow \qquad (2m - 1)^2 = 0$$

$$\text{C.F.} = f_1(2y + x) + x f_2(2y + x)$$

$$\text{P.I.} = \frac{1}{4D^2 - 4DD' + D'^2}\ 16 \log(x + 2y)$$

$$= \frac{1}{(2D - D')^2} \cdot 16 \log(x + 2y)$$

$$\frac{x^2}{2^2\ 2!} \cdot 16 \log(x + 2y) = 2x^2 \log(x + 2y). \qquad \because\ f(a, b) = 0$$

$\therefore$ The solution is

$$z = f_1(2y + x) + x f_2(2y + x) + 2x^2 \log(x + 2y).$$ **Ans.**

Example 4:

Solve $\dfrac{\partial^2 z}{\partial x^2} + \dfrac{\partial^2 z}{\partial y^2} = 12(x + y)$.

Solution:

We have $\dfrac{\partial^2 z}{\partial x^2} + \dfrac{\partial^2 z}{\partial y^2} = 12(x + y)$.

The given equation can be written as

$$(D^2 + D'^2)\, z = 12(x + y).$$

A.E is $\quad m^2 + 1 = 0, \qquad \therefore \quad m = \pm i.$

$$\text{C.F.} = f_1(y + ix) + f_2(y - ix).$$

$$\text{P.I.} = \frac{1}{D^2 + D'^2}\ 12(x + y)$$

$$= \frac{1}{1^2 + 1^2} . 12 \frac{(x + y)^3}{6} = (x + y)^3.$$

$\therefore$ The equation is

$z = f_1(y + ix) + f_2\ y - ix) + (x + y)^3.$ **Ans.**

Example 5:

Solve $2r - s - 3t = 5e^x/e^y$. **(Rohilkhand, 94)**

Solution:

We have $2r - s - 3t = 5e^x/e^y$.

The given equation can be written as

$(2D^2 - DD' - 3D'^2)\ z = 5e^{x-y}$.

A.E. is $\quad 2m^2 - m - 3 = 0$

$\Rightarrow \quad (2m - 3)(m + 1) = 0$

$\therefore$ C.F. $= f_1(2y + 3x) + f_2(y - x)$

$$\text{P.I.} = \frac{1}{(2D^2 - DD' - 3D'2)} \cdot 5e^{x-y}$$

$$= \frac{1}{(D + D')} \cdot \frac{1}{(D - 3D')} .5e^{x-y}$$

$$= \frac{1}{D + D'} \frac{1}{2 + 3} 5e^{xy} = \frac{1}{D + D'} e^{x-y}$$

$$= \frac{x}{1!} e^{x-y} = xe^{x-y} \quad \because \quad f(a, b) = 0$$

$\therefore$ The solution is $z = f_1(2y + 3x) + f_2(y - x) + xe^{x-y}$. **Ans.**

Example 6:

Solve $(D^3 - 4D^2D' = 4DD'^2)\ z = 4 \sin (2x + y)$.

(Kanpur, 93, 92; Agra, 95)

Solution:

We have $(D^3 - 4D^2D' = 4DD'^2)\ z = 4 \sin (2x + y)$.

Here A.E. is $m^3 - 4m^2 + 4m = 0$

$\Rightarrow \quad m(m - 2)^2 = 0 \quad \therefore \quad m = 0, 2, 2.$

C.F. $= f_1(y) + f_2(y + 2x) + xf_3(y + 2x)$

$$\text{P.I.} = \frac{1}{D^3 - 4D^2D' + 4DD'^2} 4 \sin (2x + y)$$

$$= \frac{1}{D(D-2D')^2} \cdot 4 \sin(2x + y)$$

$$= \frac{1}{(D-2D')^2} \cdot \frac{1}{D} \cdot 4 \sin(2x + y)$$

$$= \frac{1}{(D-2D')^2} \{-2 \cos(2x + y)$$

$$= \frac{x^2}{1!} \{-2 \cos(2x + y)\} = - x^2 \cos(2x + y)$$

∴ The solution is

$$z = f_1(y) + f_2(y + 2x) + xf_3(y + 2x) - x^2 \cos(2x + y).$$ **Ans.**

Example 7:

Solve $(D^2 - 2DD' + D'^2)\, z = e^{x+2y} + x^3.$ **(Meerut, 92, 96)**

Solution:

Here A. E. is $m^2 - 2m + 1 = 0$

$\Rightarrow \quad (m - 1)^2 = 0 \quad \therefore \quad m = 1, 1$

C.F. $= f_1(y + x) + f_2(y + x)$

$$\text{P.I.} = \frac{1}{D^2 - 2DD' + D'^2} e^{x+2y} + \frac{1}{D^2 - 2DD' + D'^2} x^3$$

$$= \frac{1}{1^2 - 2.1.2 + 2^2} . e^{x+2y} + \frac{1}{D^2}\left(1 - \frac{D'}{D}\right)^{-2} x^3$$

$$= e^{x+2y} + \frac{1}{D^2}\left(1 + \frac{2D'}{D} ...\right) x^3 = e^{x+2y} + \frac{1}{20} x^5.$$

∴ The solution is $y = f_1(y + x) + xf_2(y + x) + e^{x+2y} + \frac{1}{20} x^5.$ **Ans.**

Example 8:

Solve $(D^2 - 6DD' - 9D'2)\, z = 6x + 2y.$

Solution:

We have $(D^2 - 6DD' - 9D'2)\, z = 6x + 2y.$

Here A.E. is m^2 is $m^2 - 6m + 9 = 0.$

$\Rightarrow \quad (m - 3)^2 = 0 \quad \Rightarrow \quad m = 3, 3.$

C.F. $= f_1(y + 3x) + xf_2(y + 3x).$

$$\text{P.I.} = \frac{1}{(D - 3D')^2}.(6x + 2y) = 2.\frac{1}{(D - 3D')^2}(3x + y)$$

$$2.\frac{x^2}{2!}.\frac{(3x + y)}{1^2} = x^2(3x + y) \qquad \because f(a, b) = 0$$

The solution is $z = f_1(y + 3x) + xf_2(y + 3x) + x^2(3x + y)$. **Ans.**

Example 9:

Solve $\frac{\partial^2 z}{\partial^2 x^2} + \frac{\partial^2 z}{\partial^2 y^2} = \cos mx \cos ny$. **(PCS (UP), 94)**

Solution:

The given equation can be written as

$$(D^2 + D'^2)\, z = \frac{1}{2}[\cos(mx + ny) + \cos(mx - ny)]$$

A.E. is $m^2 + 1 = 0 \quad \therefore \quad m = \pm i$

C.F. $= f_1(y + ix) + f_2(y - ix)$

$$\text{P.I.} = \frac{1}{2}\frac{1}{D^2 + D'^2}\cos(mx + ny) + \frac{1}{2}\frac{1}{D^2 + D'^2}\cos(mx - ny)$$

$$= \frac{1}{2}\frac{1}{m^2 + n^2}\{-\cos(mx + ny)\} + \frac{1}{2}\frac{1}{m^2 + n^2}\{-\cos(mx - ny)\}$$

$$= \frac{1}{2}\frac{1}{m^2 + n^2}[\cos(mx + ny) + \cos(mx - ny)]$$

Hence the required solution is

$$z = f_1(y + ix) + f_2(y - ix) - \frac{1}{2(m^2 + n^2)}[\cos(mx + ny) + \cos(mx - ny) \qquad \textbf{Ans.}$$

4.9 General Method

Now, we proceed to find the general method for getting a P.I.

Let us consider the equation

$$(D - mD')\, z = \phi(x, y)$$

$$\Rightarrow \quad p - mq = \phi(x, y).$$

The subsidiary equations are

$$\frac{dx}{1} = \frac{dy}{-m} = \frac{dz}{\phi(x, y)}$$

First two members give $y + mx = a$ (constant)

From the first and the last members, we have

$dz = \phi(x, y)\, dx = \phi(x, a - mx)\, dx$

$\therefore z = \int \phi(x, a - mx)\, dx$

$\therefore \text{ P.I.} = \dfrac{1}{(D - mD')}\, \phi(x, y) = \int \phi(x, a - mx)\, dx.$

After integrating a is replaced by y + mx.

Working Method : Take the P.I. corresponding to

$$\frac{1}{(D - mD')}\, \phi(x, y) \quad \text{as} \quad \int \phi(x, a - mx)\, dx,$$

and replace a by y = mx after integration.

Example 1:

Solve $(D^2 - DD' - 2D'^2)\, z = (y - 1)e^x$.

Solution:

We have $(D^2 - DD' - 2D'^2)\, z = (y - 1)e^x$.

Here A.E. is $m^2 - m - 2 = 0$

$\Rightarrow \quad (m - 2)(m + 1) = 0. \qquad \therefore \quad m = 2, -1.$

$\text{C.F} = f_1(y + 2x) + f_2(y - x).$

$$\text{P.I.} = \frac{1}{D^2 - DD' - 2D'^2}.\,(y - 1)e^x$$

$$= \frac{1}{(D - 2D')}\cdot\frac{1}{(D + D')}.\,(y - 1)e^x$$

$$= \frac{1}{(D - 2D')}\cdot\int (x - a - 1)\, e^x\, dx \text{ where } y - x = a.$$

$$= \frac{1}{(D - 2D')}.\left[(a - 1)\int e^x\, dx + \int xe^x\, dx\right]$$

$$= \frac{1}{(D - 2D')}\cdot\, [(a + 1)\, e^x + (x + 1)\, e^x]$$

$$= \frac{1}{(D - 2D')}\cdot\, (a + x - 2)\, e^x = \frac{1}{(D - 2D')}\cdot\, (y - 2)e^x$$

$$= \int (b - 2x - 2)\, e^x.dx \quad \text{where} \quad y + 2x = b$$

$$= (b - 2)\int e^x\, dx - 2\int xe^x\, dx$$

$= (b - 2)\, e^x - 2(x - 1)\, e^x = (b - 2x)\, e^x = ye^x.$

$\therefore$ The solution is

$z = f_1(y + 2x) + f_2(y - x) + ye^x.$ **Ans.**

Example 2:

Solve $(D^2 - 2DD' - 15D'^2)\, z = 12xy.$

Solution:

A.F is $m^2 - 2m - 15 = 0$

$\Rightarrow \quad (m - 5)(m + 3) = 0. \quad \therefore \quad m = 5, -3$

C.F. $= f_1(y + 5x) + f_2(y - 3x)$

$$\text{P.I.} = \frac{1}{D^2 - 2DD' - 15D'^2}.\,12xy = \frac{1}{(D + 3D')}.\,\frac{1}{(D - 5D')}.12xy$$

$$= \frac{12}{(D + 3D')} \int x(a - 5x)\, dx \quad \text{where } y + 5x = a$$

$$= \frac{12}{(D + 3D')} \left(\frac{1}{2}ax^2 - \frac{5}{3}x^3\right)$$

$$= \frac{12}{(D + 3D')} \left\{\frac{1}{2}(y + 5x)\, x^2 - \frac{5}{3}x^3\right\}$$

$$= \frac{2}{(D + 3D')}\cdot (3x^2y + 5x^3)$$

$$= 2 \int \left[3x^2\, (3x + b) + 5x^3\right] dx \qquad (\text{where } y - 3x = b)$$

$$= 2 \int (14x^3 + 3x^2 b)\, dx = 2\left(\frac{14}{4}\, x^4 + x^3 b\right)$$

$= 7x^4 + 2x^3(y - 3x) = x^4 + 2x^3y.$

$\therefore$ The solution is

$z = f_1(y + 5x) + f_2(y - 3x) + x^4 + 2x^3y.$

Aliter. $\text{P.I} = \dfrac{1}{D^2 - 2DD' - 15D'^2} 12xy$

$$= \frac{1}{D'}\left(1 - \frac{2D'}{D} - \frac{15D'^2}{D^2}\right)^{-1}.\,12xy$$

$$= \frac{1}{D^2}\left(1 - \frac{2D'}{D} + \ldots\right).\,12xy$$

$$= \frac{1}{D^2}(12xy) + \frac{2}{D^3} \cdot 12x = 2x^3y = x^4.$$ **Ans.**

Example 3:

Solve (r + s – 6t) = y cos x.

Solution:

We have (r + s – 6t) = y cos x.

The given equation can be written as

$$(D^2 + DD' - 6D'^2)\, z = y \cos x$$

A.E. is $m^2 + m - 6 = 0$

$\Rightarrow$ $(m + 3)(m - 2) = 0.$ $\therefore$ $m = -3, 2.$

C.E. $= f_1(y - 3x) + f_2(y + 2x)$

$$\text{P.I.} = \frac{1}{D^2 + DD' - 6D'^2} \cdot y \cos x = \frac{1}{(D - 2D')} \cdot \frac{1}{(D + 3D')} \cdot y \cos x$$

$$= \frac{1}{(D - 2D')} \cdot \int (3x + a) \cos x\, dx \text{ where } y - 3x = a$$

$$= \frac{1}{(D - 2D')} \cdot [a \sin x + 3x \sin x + 3 \cos x]$$

$$= \frac{1}{(D - 2D')} \cdot [(y - 3x) \sin x + 3x \sin x + 3\cos x]$$

$$= \frac{1}{(D - 2D')} \cdot [y \sin x + 3 \cos x].$$

$$= \int \left[(b - 2x) \sin x + 3\cos x\right] dx \quad \text{where } y + 2x = b$$

$$= -b \cos x - 2(-x \cos x + \sin x) + 3 \sin x$$

$$= -(y = 2x) \cos x + 2x \cos x + \sin x = -y \cos x + \sin x.$$

$\therefore$ The required solution is

$$z = f_1(y - 3x) + f_2(y + 2x) - y \cos x + \sin x.$$ **Ans.**

Example 4:

Solve $(D^2 + 2DD' + D'^2)\, z = 2 \cos y - x \sin y$. **(Meerut, 91 (P))**

Solution:

We have $(D^2 + 2DD' + D'^2)\, z = 2 \cos y - x \sin y.$

A.E. is $m^2 + 2m + 1 = 0.$

$\Rightarrow \quad (m+1)^2 = 0 \quad \therefore \quad m = -1, -1,$

$$\text{C.F.} = f_1(y-x) + xf_2(y-x)$$

$$\text{P.I.} = \frac{1}{(D+D')} \cdot \frac{1}{(D+D')} \cdot (2\cos y - x \sin y)$$

$$= \frac{1}{(D+D')} \int [2\cos(x+a) - x\sin(x+a)]\, dx \quad \text{where } y - x = a$$

$$= \frac{1}{(D+D')} \cdot [2\sin(x+a) - \{-x\cos(x+a) + \sin(x+a)\}]$$

$$= \frac{1}{(D+D')} \cdot [\sin(x+a) + x\cos(x+a)]$$

$$= \frac{1}{(D+D')} \cdot (\sin y + x \cos y)$$

$$= \frac{1}{(D+D')} \cdot (\sin y + x \cos y)$$

$$= \int [\sin(x+b) + x\cos(x+b)]\, dx \quad \text{where } y - x = b.$$

$$= -\cos(x+b) + x\sin(x+b) + \cos(x+b)$$

$$= x\sin(x+b) = x \sin y.$$

$\therefore$ The solution is $z = f_1(y-x) + xf_2(y-x) + x \sin y$. **Ans.**

4.10 Non Homogeneous Linear Equations With Constant Coefficients

Let us consider simplest case.

$$(D - mD' - \alpha)\, z = 0 \quad \Rightarrow \quad p - mq = \alpha z$$

The subsidiary equations are $\frac{dx}{1} = \frac{dy}{-m} = \frac{dz}{\alpha z}$.

The first two members give $y + mx = a$ (const.)

Again from the first and the third relations, we have

$$\frac{dz}{z} = \alpha\, dx.$$

$\therefore \log z = \alpha x + \log b \quad \Rightarrow \quad z = be^{\alpha x}$

$\therefore z = e^{\alpha x} f(y + mx)$.

Similarly it can easily be shown that the integral of

$$(D - m_1D' - \alpha_1)(D - m_2D' - \alpha_2)\ldots(D - m_nD' - \alpha_n)\, z = 0$$

is $z = e^{\alpha_1 x} f_1(y + m_1x) + e^{\alpha_2 x} f_2(y + m_2x) + \ldots + e^{\alpha_n x} f_n(y = m_nx)$.

Also in case of repeated factors

e.g., $(D - mD' - \alpha) z = 0$

the integral is

$$z = e^{\alpha x} f_1(y + mx) + xe^{\alpha x} f_2(y + mx) + ... + x^{r-1}e^{\alpha x} f_r(y + mx).$$

Note: In case f(D, D') cannot be resolved into factors linear in D and D', the equation cannot be integrated by the above methods. In such cases a trial method is used to find solutions.

4.11 Particular Integrals

The methods of obtaining particular integrals of non homogeneous partial differential equations are very similar to those used in solving linear equations with constant coefficients.

Here we are considering few cases only.

Case I: *When the right hand side of differential equation is of form $e^{ax\ by}$.*

The $\dfrac{1}{f(D, D')} \cdot e^{ax+by} = \dfrac{2}{f(a, b)} e^{ax+by}$ provided $f(a, b) \neq 0$. i.e.

Put D = a and D' = b.

Case II: *When the right hand side of differential equation is of the form sin (ax + by or cos (ax + by).*

Then $\dfrac{1}{f(D, D')}$ sin (ax + dy) is obtained by putting

$D^2 = - a^2$, $DD' = - ab$, $D'^2 = - b^2$

provided the denominator is not zero.

Case III: *When the right hand side of a differential equation is of the form $x^m y^n$, where m and n are positive integers.*

$$\text{Then } \frac{1}{f(D, D')} x^m y^n = [f(D, D')]^{-1} x^m y^n$$

Case IV: *When the right hand side of a differential equation is of the form e^{ax+by} V.*

$$\text{Then } \frac{1}{f(D, D')} (e^{ax+by} v) = e^{ax+by} \frac{1}{f(D + a, D' + b)} . V.$$

Example 1:

Solve $(D - 3D' - 2)^2 z = 2e^{2x} \tan (y + 3x)$.

(Rohilkhand, 93; Meerut, 90, 94)

Solution:

We have $(D - 3D' - 2)^2 z = 2e^{2x} \tan (y + 3x)$.

Here C.F. = $e^{2x} f_1(y + 3x) + x.e^{2x} f_2(y + 3x)$,

$$\text{P.I.} = \frac{1}{(D - 3D' - 2)^2} 2e^{2x} \tan(y + 3x)$$

$$= 2e^{2x} . \frac{1}{(D + 2 - 3D' - 2)^2} \tan(y + 3x)$$

$$= 2e^{2x} . \frac{1}{(D - 3D')^2} \tan(y + 3x)$$

$$= 2e^{2x} . \frac{1}{D - 3D'} x \tan(y + 3x)$$

$$= 2e^{2x} . \frac{x^2}{2} \tan(y + 3x) = x^2 e^{2x} \tan(y + 3x)$$

∴ To solution is

$$z = e^{2x} f_1(y + 3x) + xe^{2x} f_2(y + 3x) + x^2e^{2x} \tan(y + 3x).$$ **Ans.**

Example 2:

Solve $(D^2 - DD' + D' - 1) z = \cos(x + 2y) + e^y$

(Kanpur, 99; Agra, 93; T.D.C. 91; Raj., IAS 92)

Solution:

The given equation is

$$(D - 1)(D - D' + 1) z = \cos(x + 2y) + e^y$$

∴ C.F. = $e^x f_1(y) + e^{-x} f_2(y + x)$.

P.I. corresponding to $\cos(x + 2y)$

$$= \frac{1}{D^2 - DD' + D' - 1} \cos(x + 2y)$$

$$= \frac{1}{1^2 - (-12) + D' - 1} \cos(x + 2y)$$

$$= \frac{1}{D'} \cos(x + 2y) = \frac{D'}{D'^2} \cos(x + 2y)$$

$$= \frac{D'}{D'^2} \cos(x + 2y) = -\frac{1}{4} \{-2\sin(x + 2y)\} = \frac{1}{2} \sin(x + 2y).$$

P.I. corresponding to e^y

$$= \frac{1}{(D - 1)(D - D' + 1)} . e^y = \frac{1}{(0 - 1)(D - D' + 1)} e^y$$

$$= - e^y . \frac{1}{D - (D + 1) + 1} . 1 = - e^y . \frac{1}{D - D'} .1$$

$$= - \frac{1}{D}\left(1 - \frac{D'}{D}\right)^{-1} . 1 = - e^y . \frac{1}{D} . 1 = - xe^y .$$

∴ The solution is

$$z = e^x f_1(y) + e^{-x} f_2(y + x) + \frac{1}{2} \sin (x + 2y) - xe^y. \qquad \textbf{Ans.}$$

Example 3:

Solve $(D^3 - 3DD' + D' + 1) z = e^{2x+3y}$. **(Meerut, 95)**

Solution:

We have $(D^3 - 3DD' + D' + 1) z = e^{2x+3y}$.

Here $D^2 - 3DD' + D' + 1$ cannot be resolved into factors linear in D and D'.

∴ C.F. $= \sum Ae^{hx+ky}$ where $h^3 - 3hk + k + 1 = 0$.

$$\text{P.I.} = \frac{1}{D^3 - 3DD' + D' + 1} . e^{2x+3y}$$

$$= \frac{e^{2x+3y}}{2^3 - 3.2.3 + 3 + 1} = - \frac{1}{6} e^{2x+3y} .$$

∴ The required solution is $z = - = \frac{1}{6} e^{2x+3y} + \sum Ae^{hx+ky}$

where $h^3 - 3hk + k + 1 = 0$. **Ans.**

Example 4:

Solve $(D^2 - D'^2 + D - D') z = 0$.

Solution:

We have $(D^2 - D'^2 + D - D') z = 0$.

The given equation as

$(D - D') (D + D' + 1) z = 0$.

∴ $z = f_1(y + x) + e^{-x} f_2(y - x)$. **Ans.**

Example 5:

Solve $DD'(D - 2D' - 3) z = 0$.

Solution:

Here $z = e^{0.x} f_1(y) + e^{0.x} f_2(0 - x) + e^{0.x} f_3(y + 3x)$

$= f_1(y) + \phi(x) + e^{3x} f_3(y + 3x).$ **Ans.**

Example 6:

Solve $r + 2s + t + 2p + 2q + z = 0.$ **(Meerut, 93 (P))**

Solution:

We have $r + 2s + t + 2p + 2q + z = 0.$

The given equation can be written as

$$(D^2 + 2DD' + D'^2 + 2D + 2D' + 1)\, z = 0$$

$$\Rightarrow \qquad (D + D' + 1)^2 = 0$$

$\therefore z = e^{-x} f_1(y - x) + xe^{-x} f_2(y - x).$ **Ans.**

Example 7:

Solve $(D - 2D' - 1)(D - 2D'^2 - 1)\, z = 0$

Solution:

C.F. Corresponding to the first factor is

$$z = e' f_1(y + 2x)$$

and C.F. corresponding to first factor is

$$\sum Ae^{hx+ky}$$

where $h - 2k^2 - 1 = 0$ or $h = 2k^2 + 1.$

$$z = e^x f_1(y + 2x) + \sum Ae^{ky+(2k^2+1)x.}$$ **Ans.**

Example 8:

Solve $(D - D' - 1)(D - D' - 2)\, z = e^{2x-y} + x.$ **(Rohilkhand, 95)**

Solution:

We have $(D - D' - 1)(D - D' - 2)\, z = e^{2x-y} + x.$

Here C. F. $e^x f_1(y + x) + e^{2x} f_2(y + x)$

and P.I. corresponding to e^{2x-y}

$$= \frac{1}{(D - D' - 1)(D - D' - 2)} . e^{2x-y}$$

$$= \frac{1}{(2 + 1 - 1)(2 + 1 - 2)} . e^{2x-y} = \frac{1}{2} . e^{2x-y}.$$

P.I. Corresponding to $x = \frac{1}{(D - D' - 1)(D - D' - 2)} . x$

$$= \frac{1}{2}(1 - D + D')^{-1}\left(1 - \frac{1}{2}D + \frac{1}{2}D'\right)^{-1} .x$$

$$= \frac{1}{2}(1 + D - D'...)\left(1 + \frac{D}{2} - \frac{D'}{2}...\right).x$$

$$= \frac{1}{2}\left(1 + D + \frac{D}{2} + ...\right). x$$

$$= \frac{1}{2} x + \frac{1}{2} . \frac{2}{3} Dx = \frac{1}{2}x + \frac{3}{4}.$$

$\therefore$ the solution is

$$z = e^x f_1(y + x) + e^{2x} f_2(y + x) + \frac{x}{2} + \frac{3}{4} + \frac{1}{2}e^{2x-y}.$$ **Ans.**

Example 9:

Solve $s + p - q = z + xy$. **(IAS 91; Delhi, Hons. 91, 95; Agra, 93)**

Solution:

We have $s + p - q = z + xy$

The given equation can be written as

$$(DD' + D - D' - 1) z = xy \quad \Rightarrow \quad (D - 1)(D' + 1) z = xy$$

$\therefore$ C.F. $= e^x f_1(y) + e^{-y} f_2(x)$

$$\text{P.I.} = \frac{1}{(D - 1)(D' + 1)}.xy$$

$$= - (1 - D)^{-1} (1 + D')^{-1} xy = - (1 + D + ...) (1 - D' + ...) xy$$

$$= - (1 + D - D' - DD'..) xy = - (xy + y - x + 1)$$

$\therefore$ The required solution is $z = e^x f_1(y) + e^{-y} f_2(x) - xy - y + x + 1.$

Ans.

Example 10:

Solve $(2D^4 - 3D^2D' + D'^2) z = 0.$ **(Meerut, 97 (R))**

Solution:

We have $(2D^4 - 3D^2D' + D'^2) z = 0.$

The given equation can be written as

$$(2D^2 - D')(D^2 - D') z = 0$$

Let $z = \sum Ae^{hx+ky}$

be the C.F. corresponding to $(D^2 - D')z$ ...(1)

$\therefore (D^2 - D')z = \sum Ah^2 e^{hx+ky} - \sum Ak\, e^{hx+ky} = \sum A(h^2 - k)e^{hx+ky}$

$\therefore$ The given equation (1) will be satisfied by this substitution if $h^2 - k = 0$.

$\therefore$ The C.F. corresponding to $(D^2 - D)z$ is

$$\sum Ae^{hx+h^2y}$$

Similarly C.F. corresponding to $(2D^2 - D')z$ is

$$\sum Be^{h'x+k'y} \qquad \text{where } 2h'^2 - k' = 0$$

i.e., $\sum Be^{h'x+2k'2y}$

$\therefore$ The most general solution of the given equation is

$$z = \sum Ae^{hx+h^2y} + \sum Be^{h'x+2k'2y}.$$ **Ans.**

4.12 Equations Reducible to Homogeneous Linear Form

An equation in which the coefficient of derivative of any order is a multiple of the variables of the same degree, may be transformed into the partial differential equations with constant coefficients.

For this we substitute $x = e^x$, $y = e^y$

so that, $X = \log x$ and $Y = \log y$,

$$\frac{\partial z}{\partial x} + \frac{\partial z}{\partial X} \cdot \frac{\partial X}{\partial x} = \frac{1}{x}\frac{\partial z}{\partial X}$$

$$\Rightarrow \quad x\frac{\partial z}{\partial x} + \frac{\partial z}{\partial X}. \qquad \therefore x\frac{\partial}{\partial x} \equiv \frac{\partial}{\partial X} \equiv D \text{ (say)}$$

Now, $$x\frac{\partial}{\partial x}\left(x^{n+1}\frac{\partial^{n-1}z}{\partial x^{n-1}}\right) = x^n \frac{\partial^n z}{\partial x^n} + (n-1)\, x^{n-1}\frac{\partial^{n-1}z}{\partial x^{n-1}}$$

$$x^2\frac{\partial^n z}{\partial x^n}\left(x\frac{\partial}{\partial x} - n + 1\right) x^{n-1}\frac{\partial^{n-1}z}{\partial x^{n-1}}.$$

Putting $n = 2, 3, \ldots,$ we have

$$x^2\frac{\partial^2 z}{\partial x^2} = (D-1)\, x\frac{\partial z}{\partial x} = D(D-1)z$$

$$x^3 \frac{\partial^3 z}{\partial x^3} = (D-2)\, x^2 \frac{\partial^2 z}{\partial x^2} = (D-2)(D-1)\, Dz, \text{ etc.}$$

Similarly, $y \frac{\partial z}{\partial y} = \frac{\partial z}{\partial Y} = D'z$

$$y^2 \frac{\partial^2 z}{\partial y^2} = D'(D'-1)\, z, \text{ etc.}$$

and $xy \frac{\partial^2 z}{\partial x \partial y} = DD'\, z$ etc.

Substituting in the given equation, it reduces to the form

$$F(D, D')\, z = V$$

which is an equation having constant coefficients and can easily by solved by the methods discussed above.

Example 1:

Solve $x^2r - 3xys + 2y^2t + px + 2qy = x + 2y$.

Solution:

Given equation can be written as

$$x^2 \frac{\partial^2 z}{\partial x^2} - xy \frac{\partial^2 z}{\partial x \partial y} + 2y^2 \frac{\partial^2 z}{\partial y^2} + x \frac{\partial z}{\partial x} + 2y \frac{\partial z}{\partial y} = x + 2y$$

Substituting $x = e^x$, $y = e^x$ and denoting $\frac{\partial}{\partial X}$ and $\frac{\partial}{\partial Y}$ by D and D' the given equation reduces to

$$[D(D-1) - 3DD' + 2D'(D'-1) + D + 2D']\, z = e^x + 2e^y$$

$\Rightarrow \quad (D - D')(D - 2D')\, z = e^x + 2e^y.$

$\therefore \quad$ C.F. $f_1(Y + X) + f_2(Y + 2X)$

$= f_1(\log y + \log x) + f_2(\log y + 2\log x)$

$= f_1(\log xy) + f_2(\log x^2y)$

$= \phi_1(xy) + \phi_2(x^2y).$

$$\text{P.I.} = \frac{1}{(D-D')(D-2D')}.\, e^x + \frac{1}{(D-D')(D-2D')}.2e^y$$

$$= \frac{c^X}{(1-0)(1-0)} + \frac{2e^Y}{(0-1)(0-2)} = e^X + e^Y = x + y$$

The complete solution is

$z = \phi_1(xy) + \phi_2(x^2y) + x + y.$ **Ans.**

Example 2:

Solve $x^2 \frac{\partial^2 z}{\partial x^2} + 2xy \frac{\partial^2 z}{\partial x \partial y} + y^2 \frac{\partial^2 z}{\partial y^2} = 0.$

(IAS 97; Delhi, Hons. 90, 94; Kanpur, 92)

Solution:

Substituting $x = e^X$, $y = e^Y$

so the $X = \log x$, $Y = \log y$

and denoting $\frac{\partial}{\partial X}$ and $\frac{\partial}{\partial Y}$ by D and D' respectively the given equation reduces to

$$[D(D - 1) + 2DD' = D'(D' - 1)]\, z = 0$$

$$\Rightarrow \quad (D + D')(D + D' - 1)\, z = 0$$

Hence the solution is

$$z = f_1(Y - X) + e^x F_2(Y - X)$$

$$= f_1(\log y - \log x) + x f_2(\log y - \log x)$$

$$= f_1\left(\log \frac{y}{x}\right) + x f_2\left(\log \frac{y}{x}\right) = \phi_1\left(\frac{y}{x}\right) + x\phi_2\left(\frac{y}{x}\right).$$ **Ans.**

Example 3:

Solve $x^2 \frac{\partial^2 z}{\partial x^2} - y^2 \frac{\partial^2 z}{\partial y^2} - y \frac{\partial z}{\partial x} + x \frac{\partial z}{\partial x} = 0.$

Solution:

Substituting $x = e^X$, $y = e^Y$

so that $X = \log x$, $Y = \log y$,

and denoting $\frac{\partial}{\partial X}$ and $\frac{\partial}{\partial Y}$ by D and D' respectively the given equation reduces to

$$[D(D - 1) - D'(D' - 1) - D' + D]\, z = 0$$

$$\Rightarrow \quad (D^2 - D'^2)\, z = 0$$

$$\therefore z = f_1(Y + X) + f_2(Y + X)$$

$$= f_1(\log y + \log x) + f_2(\log y - \log x)$$

$$= f_1(\log xy) + f_2\left\{\log \frac{y}{x}\right\}$$

$$= \phi_1(xy) + \phi_2(y/x).$$ **Ans.**

Example 4:

Solve $x^2 \dfrac{\partial^2 z}{\partial x^2} - 4xy \dfrac{\partial^2 z}{\partial x \partial y} + 4y^2 \dfrac{\partial^2 z}{\partial y^2} + 6y \dfrac{\partial z}{\partial y} = x^3 y^4.$

(Rohilkhand, 82; Meerut, 92; Delhi, Hons. 95)

Solution:

Substituting $x = e^X$, $y = e^Y$

and denoting $\dfrac{\partial}{\partial X}$ and $\dfrac{\partial}{\partial Y}$ by D and D' the equation reduces to

$$[D(D - 1) - 4DD'(D' - 1) + 6D']\, z = e^{3X+4Y}$$

$$\Rightarrow \quad (D - 2D')(D - 2D' - 1)\, z = e^{2X+4Y}$$

$$\begin{aligned} \text{C.F.} &= f_1(Y + 2X) + e^X f_2(Y + 2X) \\ &= f_1(\log y + 2\log x) + f_2(\log y + 2\log x) \\ &= f_1(\log yx^2) + xf_2(\log yx^2) \\ &= \phi_1(yx^2) + x\phi_2(yx^2). \end{aligned}$$

$$\text{P.I.} = \frac{1}{(D - 2D')(D - 2D' - 1)} \cdot e^{3X+4Y}$$

$$= \frac{e^{3X+4Y}}{(3 - 8)(3 - 8 - 1)} = \frac{1}{30} x^3 y^4$$

$\therefore$ The solution is $z = f_1(yx^2) + xf_2(yx^2) + \dfrac{1}{30} x^3 y^4.$ **Ans.**

Example 5:

Solve $x^2 \dfrac{\partial^2 z}{\partial x^2} + 2xy \dfrac{\partial^2 z}{\partial x \partial y} + y^2 \dfrac{\partial^2 z}{\partial y^2}$

$$- nx \frac{\partial z}{\partial x} - ny \frac{\partial z}{\partial y} + nz = x^2 + y^2.$$

Solution:

Substituting $x = e^X$, $y = e^Y$,

if D and D' denote $\dfrac{\partial}{\partial X}$ and $\dfrac{\partial}{\partial Y}$, the given equation reduces to

$$[D(D - 1) + 2DD' + D'(D' - 1) - nD - nD' + n]\, z = e^{2X} + e^{1Y}$$

$\Rightarrow \quad (D + D - 1)(D + D' - n) z = e^{2X} + e^{eY}$

$\therefore \quad$ C.F. $= e^{X} f_1(Y - X) + e^{nX} f_2(Y - X)$

$$= xf_1(\log y - \log x) + x^n f_2(\log y - \log x)$$

$$= xf_1\left(\log \frac{y}{x}\right) + x^n f_2\left(\log \frac{y}{x}\right)$$

$$= x\phi_1\left(\frac{y}{x}\right) + x^n\phi_2\left(\frac{y}{x}\right).$$

$$\text{P.I.} = \frac{1}{(D + D' - 1)(D + D' - n)}.e^{2X} + \frac{1}{(D + D' - 1)(D + D' - n)}.e^{2Y}$$

$$= \frac{e^{2X}}{(2 + 0 - 1)(2 + 0 - n)} + \frac{e^{2Y}}{(0 + 2 - 1)(0 + 2 - n)}$$

$$= \frac{e^{2X} + e^{2Y}}{2 - n} = \frac{x^2 + y^2}{2 - n}$$

$\therefore$ The solution is $z = x\phi_1\left(\frac{y}{x}\right) + x^2\phi_2\left(\frac{y}{x}\right) + \frac{x^2 + y^2}{2 - n}$. **Ans.**

Example 6:

Find a surface passing through the two lines $z = x = 0$. $z - 1 = x - y = 0$, satisfying $r - 4s + 4t = 0$.

Solution:

The given equation can be written as

$$(D^2 - 4DD' - 4D'^2) z = 0.$$

A.E is $m^2 - 4m + 4 = 0$.

$\Rightarrow \quad (m - 2)^2 = 0. \quad \therefore \quad m = 2, 2$ ...(1)

$$z = f(y + 2x) + xf(y + 2x).$$

Since the surface passes through the lines

$z = x = 0$ and $z = 1 = x - y = 0$.

$\therefore 0 = f(y + x)$ ...(2)

and $1 = f(y + 2x) + x\phi(y + 2x)$ ...(3)

From equation (2) and (3), we have

$$\phi(y + 2x) = \frac{1}{x} = \frac{3}{3x} = \frac{3}{2x + x} = \frac{3}{2x + y} \quad \because \quad y - x = 0.$$

Hence the required solution is $z = x. \dfrac{3}{2x + y}$

$\Rightarrow \quad z(2x + y) = 3x.$ **Ans.**

Example 7:

Solve $r - s - 2t = (2x^2 + xy - y^2) \sin xy - \cos xy.$

Solution:

We have $r - s - 2t = (2x^2 + xy - y^2) \sin xy - \cos xy.$

The given equation can be written as

$$(D' - DD' - 2D'^2) z = (2x^2 + xy - y^2) \sin xy - \cos xy.$$

A.E. is $\quad m^2 - m - 2 = 0.$

$\Rightarrow \quad (m - 2)(m - 1) = 0 \qquad \therefore \quad m = 2, -1.$

C.F. $= f_1(y + 2x) + f_2(y - x)$

$$\text{P.I.} = \frac{1}{(D^2 - DD' - 2D'^2)} \cdot \{(2x^2 + xy - y^2) \sin xy - \cos xy\}$$

$$= \frac{1}{(D - D')} \cdot \frac{1}{(D - 2D)} \cdot \{(2x^2 + xy - y^2) \sin xy - \cos xy\}$$

$$= \frac{1}{(D + D')} \int \left[\left\{ 2x^2 + x(a - 2x) - (a - 2x)^2 \right\} \sin x(a - 2x) - \cos x(a - 2x) \right] dx \text{ where } y + 2x = a$$

$$= \frac{1}{(D + D')} [(a - 4x)(x - a) \sin (ax - 2x^2) - \cos (ax - 2x^2)] dx$$

$$= \frac{1}{(D + D')} \left[\int (-a + x). \left\{ (a - 4x) \sin (ax - 2x^2) \right\} dx - \int \cos (ax - 2x^2)\, dx \right]$$

$$= \frac{1}{(D + D')} \left[(-a + x) \cos (ax - 2x^2) + \int \cos (ax - 2x^2)\, dx - \int \cos (ax - 2x^2)\, dx \right]$$

$$= \frac{1}{(D + D')} (-x + a) \cos (ax - 2x^2)$$

$$= \frac{1}{(D + D')} (x + y) \cos xy$$

$$= \int (2x + b) \cos x (x + b)\, dx \qquad \text{where } y - x = b$$

$$= \int (2x + b) \cos x (x^2 + xb)\, dx = \sin (x^2 + xb) = \sin xy.$$

∴ The solution is

$z = f_1(y = 2x) + f_2(y + x) + \sin xy.$ **Ans.**

Example 8:

Solve $\frac{\partial^3 u}{\partial x^3} + \frac{\partial^3 u}{\partial y^3} + \frac{\partial^3 u}{\partial z^3} - 3\frac{\partial^3 u}{\partial x\, \partial y\, \partial z} = x^3 + y^3 + z^3 - 3xyz$ ...(1)

Solution:

The given equation can be written as

$$(D_1^3 + D_2^3 + D_3^3 - 3D_1D_2D_3)\, u = x^3 + y^3 + z^3 - 3xyz$$

$$\Rightarrow (D_1 + D_2 + D_3)(D_1^2 + D_2^2 + D_3^2 - D_1D_2 - D_2D_3 - D_3D_1)\, u = x^3 + y^3 + z^3 - 3xyz$$

$$\Rightarrow (D_1 + D_2 + D_3)(D_1 + \omega D_2 + \omega^2 D_3)(D_1 + \omega^2 D_2 + \omega D_3)\, u = x^3 + y^3 + z^3 - 3xyz$$

where w is imaginary cube root of unity.

Now let us consider

$(D_1 + \omega D_2 + \omega^2 D_3)\, u = 0$ ...(2)

Subsidiary equations are $\frac{dx}{1} = \frac{dy}{\omega} = \frac{dz}{\omega^2}$.

∴ $y - xw = a,\ z = \omega^2 x = b.$

∴ Solution of (2) is given by $f_1(y - \omega^2 x, z - \omega x)$.

Similarly other factors give

$f_2(y - x, z - x)$ and $f_3(y - \omega^2 x, z - \omega x)$.

∴ $\text{C.F.} = f_1(y - \omega x, z - \omega^2 x) + f_2(y - x, z - x) + f_3(y - \omega^2 x, z - \omega x)$

Now, P.I corresponding to x^3 is

$$\frac{1}{D_1^3 + D_2^3 + D_3^3 - 3D_1D_2D_3} x^3$$

$$= \frac{1}{D_1^3}\left(1 + \frac{D_2^3}{D_1^3} + \ldots\right)^{-1} . x^3 = \frac{1}{D_1^3} . x^3 = \frac{x^6}{4.5.6} = \frac{x^6}{120}.$$

Similarly, particular integrals corresponding to y^3 and z^3 are $\frac{y^6}{120}$ and $\frac{z^6}{120}$ and P.I. corresponding to $-3xyz$

$$= \frac{1}{D_1^3 + D_2^3 + D_3^3 - 3D_1D_2D_3}(-3xyz)$$

$$= \frac{1}{3D_1D_2D_3}.\left(1 - \frac{D_1^2}{3D_2D_3}...\right)^{-1}.(-2xyz)$$

Hence the solution is

$$u = f_1(y - \omega x, z - \omega^2 x) + f_2(y - x, z - x) + f_3(y - \omega^2 x, z - \omega x)$$
$$+ \frac{x^2 + y^6 + z^6}{120} + \frac{x^2y^2z^2}{8}. \qquad \textbf{Ans.}$$

Example 9:

Solve $(D^3 - 4D^2D' + 4DD'^2)\, z = \cos(y + 2x)$.

Solution:

The given equation is

$$D(D^2 - 4DD' + 4D'^2)\, z = \cos(y + 2x)$$

$\Rightarrow \quad D(D - 2D')^2\, z = \cos(y + 2x).$

$\therefore \quad C.F. = f_1(y) + f_2(y + 2x) + xf_3(y + 2x)$

$$P.I. = \frac{1}{(D - 2D')^2}.\frac{1}{D}\cos(y + 2x)$$

$$= \frac{1}{(D - 2D')^2}.\frac{\sin(y + 2x)}{2}$$

$$= \frac{x^2}{2!}.\frac{\sin(y + 2x)}{2} \qquad \because f(a, b) = 0$$

$$= \frac{1}{4}x^2 \sin(y + 2x).$$

$\therefore$ The required solution is

$$z = f_1(y) + f_2(y + 2x) + xf_3(y + 2x) + \frac{1}{4}x^2 \sin(y + 2x). \qquad \textbf{Ans.}$$

Example 10:

Solve $\frac{\partial^2 z}{\partial x^2} - 4\frac{\partial^2 z}{\partial y^2} = \frac{4x}{y^2} - \frac{y}{x^2}$. **(Rohilkhand, 91; Meerut, 92)**

Solution:

The given equation can be written as

$$(D^2 - 4D'^2)\, z = \frac{4x}{y^2} - \frac{y}{x^2} \quad \Rightarrow \quad (D + 2D')(D - 2D')\, z = \frac{4x}{y^2} - \frac{y}{x^2}.$$

$\therefore$ C.F $= f_1(y - 2x) + f_2(y + 2x)$.

$$\text{P.I.} = \frac{1}{(D + 2D')} \cdot \frac{1}{(D - 2D')} \cdot \left(\frac{4x}{y^2} - \frac{y}{x^2}\right)$$

$$= \frac{1}{(D + 2D')} \int \left\{\frac{4x}{(a - 2x)^2} - \frac{a - 2x}{x^2}\right\} dx \quad \text{where } y + 2x = a.$$

$$= \frac{1}{(D + 2D')} \int \left[-\frac{2}{(a - 2x)} + \frac{2a}{(a - 2x)^2} - \frac{a}{x^2} + \frac{2}{x}\right] dx$$

$$= \frac{1}{(D + 2D')} \left[\log (a - 2x) + \frac{a}{(a - 2x)} + \frac{a}{x} + 2 \log x\right]$$

$$= \frac{1}{(D + 2D')} \left[\log y + 2 \log x \, \frac{y + 2x}{y} + \frac{y + 2x}{x}\right]$$

$$= \frac{1}{(D + 2D')} \left[\log y + 2 \log x \, \frac{2x}{y} + \frac{y}{x} + 3\right]$$

$$= \int \left[\log(b + 2x) + 2 \log x + \frac{2x}{b + 2x} + \frac{b + 2x}{x} + 3\right] dx$$

where $y - 2x + b$

$$= \int \left[\log(b + 2x) + 2 \log x + \frac{2x}{b + 2x} + \frac{b}{x} + 5\right] dx$$

$$= x. \log(b + 2x) - \int \frac{2x}{(b + 2x)}\, bx + 2x \log x - 2 \int \frac{1}{x}. xdx$$

$$+ \int \frac{2x}{(b + 2x)}\, dx + b \log x + 5x$$

$= x \log (b + 2x) + (2x + b) \log x + 3x = x \log y + y \log x + 3x.$

$\therefore$ The solution is

$z = f_1(y - 2x) = f_2(y + 2x) + x \log y + y \log x + 3x.$ **Ans.**

Example 11:

Solve $\frac{1}{x^2}\frac{\partial^2 z}{\partial x^2} - \frac{1}{x^3}\frac{\partial z}{\partial x} = \frac{1}{y^2}\frac{\partial^2 z}{\partial y^2} - \frac{1}{x^3}\frac{\partial z}{\partial x}.$

Solution:

We have $\frac{1}{x^2}\frac{\partial^2 z}{\partial x^2} - \frac{1}{x^3}\frac{\partial z}{\partial x} = \frac{1}{y^2}\frac{\partial^2 z}{\partial y^2} - \frac{1}{x^3}\frac{\partial z}{\partial x}.$

Putting $\frac{x^2}{2} = X, \frac{y^2}{2} = Y$

so that x dx = dX, y dy = dY

$$\therefore \frac{\partial z}{\partial X} = \frac{\partial z}{\partial x}.\frac{\partial x}{\partial X} = \frac{1}{x}\frac{\partial z}{\partial x}$$

$$\frac{\partial^2 z}{\partial X^2} = \frac{\partial}{\partial x}.\left(\frac{1}{x}\frac{\partial z}{\partial x}\right).\frac{\partial x}{\partial X} = -\frac{1}{x^3}\frac{\partial z}{\partial x} + \frac{1}{x^2}\frac{\partial^2 z}{\partial x^2}.$$

Similarly, $\frac{\partial^2 z}{\partial Y^2} = -\frac{1}{x^3}\frac{\partial z}{\partial y} + \frac{1}{y^2}\frac{\partial^2 z}{\partial y^2}.$

The given equation reduces to

$$\frac{\partial^2 z}{\partial X^2} - \frac{\partial^2 z}{\partial Y^2} = 0$$

$$\Rightarrow \quad (D^2 - D'^2)\, z = 0$$

$$\Rightarrow \quad (D - D')(D + D')\, z = 0$$

$$\therefore z = f_1(Y + X) + f_2(Y - X)$$

$$= f_1\left(\frac{y^2}{2} + \frac{x^2}{2}\right) + f_2\left\{\frac{y^2}{2} - \frac{x^2}{2}\right\}$$

$$= \phi_1(y^2 + x^2) + \phi_2(y^2 - x^2).$$

Example 12:

Solve $yt - q = xy.$

Solution:

We have $yt - q = xy.$

The given equation can bw written as

$$y\frac{\partial^2 z}{\partial y^2} - \frac{\partial z}{\partial y} = xy \Rightarrow y^2\frac{\partial^2 z}{\partial y^2} - y\frac{\partial z}{\partial z} = xy^2. \quad ...(1)$$

Substituting $x = e^X$, $y = e^Y$

and denoting $x = \dfrac{\partial z}{\partial X}$ and $\dfrac{\partial}{\partial Y}$ by D and D', the given equation (1) reduces to

$$[D'(D' - 1) - D']\, z = e^{X+2Y}$$

$$\Rightarrow \quad D'(D' - 2)\, z = e^{X+2Y}$$

$$\therefore \quad C.F. = f_1(X) + e^{2Y} f_2(X)$$

$$= f_1(\log x) + y^2 f_2(\log x) = \phi_1(x) =+ y^2 \phi_2(x)$$

and $$P.I. = \frac{1}{D'(D'-2)} . e^{X+2Y} = \frac{1}{2} . \frac{1}{D'-2} e^{X+2Y}$$

$$- \frac{1}{2} e^{X+2Y} = \frac{1}{D' + 2 - 2} . 1 = \frac{1}{2} e^{X+2Y} . \frac{1}{D'} .1$$

$$= \frac{1}{2} Y e^{X+2Y} = \frac{1}{2} xy^2 \log y.$$

$\therefore$ The complete solution is

$$z = \phi_1(x) + y^2\phi_2(x) + \frac{1}{2} xy^2 \log y.$$ **Ans.**

Example 13:

Solve

$$\frac{\partial^2 z}{\partial x^2} - \frac{\partial^2 z}{\partial x \partial y} - 2\frac{\partial^2 z}{\partial y^2} + 2\frac{\partial z}{\partial x} + 2\frac{\partial z}{\partial x} \quad e^{2x+3y} + \sin(2x + y) + xy.$$

Solution:

The given equation can be written as

$$(D^2 - DD' - 2D'^2 + 2D + 2D')\, z = e^{2x+3y} + \sin(2x + y) + xy$$

$$\Rightarrow \quad (D + D')(D - 2D' + 2)\, z = e^{2x+3y} + \sin(2x + y) + xy.$$

$\therefore$ C.F. $= f_1(y - x) + e^{-2x} f_2(y + 3x)$

P.I. corresponding to e^{2x+3y}

$$= \frac{1}{D^2 - DD' - 2D'^2 + 2D + 2D} . e^{2x+3y}$$

$$= \frac{e^{2x+3y}}{2^2 - 2.3 - 2.3^3 + 2.2 + 2.3} = \frac{1}{10} e^{2x+3y}$$

P.I. corresponding to sin (2x + y)

$$= \frac{1}{D^2 - DD' - 2D'^2 + 2D + 2D'} . \sin(2x + y)$$

$$= \frac{1}{2^2 - (-2.1) - 2.(-1)^2 + 2D + 2D'} \sin(2x + y)$$

$$= \frac{1}{2(D + D')} \sin(2x + y) = \frac{(D - D')}{2(D^2 - D'^2)} \sin(2x + y)$$

$$= \frac{(D - D')}{2(-2^2 + 1^2)} \sin(2x + y)$$

$$= \frac{1}{6}[2\cos(2x + y) - \cos(2x + y)] = -\frac{1}{6}\cos(2x + y)$$

and P.I. corresponding to xy $= \dfrac{1}{(D + D')(D - 2D' + 2)} . xy$

$$= \frac{1}{2(D + D')} . \left(1 + \frac{D}{2} - D'\right)^{-1} . xy$$

$$= \frac{1}{2(D + D')} . \left(1 + \frac{D - 2D')}{2}\right)^{-1} . xy$$

$$= \frac{1}{2(D + D')} . \left\{1 + \frac{(D - 2D'}{2} + \left(\frac{D - 2D')}{2}\right)^2 \ldots\right\} . xy$$

$$= \frac{1}{2(D + D')} . \left(1 - \frac{D}{2} + D' - DD' \ldots\right) . xy$$

$$= \frac{1}{2(D + D')} . \left(xy - \frac{y}{2} + x - 1\right)$$

$$= \frac{1}{2D}\left(1 + \frac{D'}{2}\right)^{-1} . \left(xy - \frac{y}{2} + x - 1\right)$$

$$= \frac{1}{2D}\left(1 - \frac{D'}{2} \ldots\right) . \left(xy - \frac{y}{2} + x - 1\right)$$

$$= \frac{1}{2D}\left(xy - \frac{1}{2}y + x - 1 - \frac{1}{D}x + \frac{1}{D} . \frac{1}{2}\right)$$

$$= \frac{1}{2D}\left(xy - \frac{1}{2}y + \frac{3}{2}x - 1 - \frac{x^2}{2}\right)$$

$$= \frac{1}{2}\left(\frac{x^2}{2}y - \frac{1}{2}xy + \frac{3}{4}x^2 - x - \frac{1}{6}x^3\right)$$

$$= \frac{x}{24}(6xy - 6y + 9x - 12 - 2x^2).$$

$\therefore$ The complete solution is

$$z = f_1(y - x) + e^{-x} f_2(y + 2x) - \frac{1}{10}e^{2x+3y} - \frac{1}{6}(2x + y)$$

$$= \frac{x}{24}(6xy - 6y + 9x - 12 - 2x^2).$$ **Ans.**

Example 14:

A surface is drawn satisfying $r + t = 0$ and touching $x^2 + z^2 = 1$ along its section by $y = 0$. Obtain its equation in the form

Solution:

The given equation is

$$r + t = 0$$

$$\Rightarrow \quad (D^2 + D'^2)\, z = 0.$$

$$\Rightarrow \quad (D - iD')(D + iD)\, z = 0.$$

$$\therefore \quad z = f(y + ix) + \phi(y - ix), \qquad ...(1)$$

Now, $\quad p = \dfrac{\partial z}{\partial x} = if'(y + ix) - i\phi'(y - ix)$

and $\quad q = \dfrac{\partial z}{\partial y} = f'(y + ix) + \phi'(y + ix),$

Also from $\quad x^2 + x^2 = 1. \qquad ...(2)$

$$z = \sqrt{(1 - x^2)}$$

$$\therefore \; p = \frac{\partial z}{\partial x} = \frac{-x}{\sqrt{(1 - x^2)}} \quad \text{and} \quad q = \frac{\partial z}{\partial y} = 0.$$

Now, under the given conditions for the surface the values of p and q for any point on $y = 0$, should be equal.

$$\therefore if'(y + ix) - i\phi'(y - ix) = \frac{-x}{\sqrt{(1 - x^2)}} \qquad ...(3)$$

$$f'(y + ix) + \phi'(y - ix) = 0. \qquad ...(4)$$

and $\quad y = 0. \qquad ...(5)$

From (3) and (4), we have

$$2if'(y + ix) = -\frac{x}{\sqrt{(1 - x^2)}}$$

$$\Rightarrow \quad f'(y + ix) = \frac{ix}{\sqrt{\{1 + (ix)^2\}}}$$

$$\Rightarrow \quad f'(y + ix) = \frac{(y + ix)}{2\sqrt{\{1 + (y + ix)^2\}}} \qquad \because \; y = 0 \quad \text{from (5)}$$

$$\therefore \quad f(y + ix) = \frac{1}{2}\sqrt{\{1 + (y + ix)^2\}} + K_1$$

from equation (4) $\phi'(y - ix) = - f'(y + ix) = \dfrac{-ix}{2\sqrt{(1 - x^2)}}$

$$= \frac{(y - ix)}{2\sqrt{\{1 + (y - ix)^2\}}} \qquad \because \; y = 0 \quad \text{from (5)}$$

$$\therefore \quad \phi(y - ix) = \frac{1}{2}\sqrt{\{1 + (y - ix)^2\}} + K_2$$

$\therefore$ from (1), we have

$$z = \frac{1}{2}\left\{\sqrt{[1 + (y - ix)^2]} + \sqrt{[1 + (y - ix)^2]}\right\} + K \qquad ...(6)$$

where $K_1 + K_2 = K$.

Equating the two values of z from (2) and (6) when y = 0, we have

$$\sqrt{(1 - x^2)} = \sqrt{(1 - x^2)} + K \qquad \therefore \; K = 0$$

$\therefore$ from (6), we have

$$2z = \sqrt{\{1 + (y + ix)^2\}} + \sqrt{\{1 + (y + ix)^2\}}.$$

$$\Rightarrow \quad 2z - \sqrt{\{1 + (y + ix)^2\}} = \sqrt{\{1 + (y + ix)^2\}}.$$

Squaring both sides, we have

$$4z^2 + 1 + (y + ix)^2 - 4z\sqrt{\{1 + (y + ix)^2\}} = 1 + (y - ix)^2$$

$$z^2 + ixy = z\sqrt{\{1 + (y + ix)^2\}}.$$

Again squaring both sides, we have

$$z^4 + 2ixyz^2 + i^2x^2y^2 = z^2\{1 + (y + ix)^2\}$$

$$\Rightarrow z^4 - x^2y^2 = z^2 + z^2y^2 - z^2x^2$$

$$\Rightarrow z^2(z^2 + x^2 - 1) = y^2(z^2 + x^2).$$

Ans.

Example 15:

Solve $\dfrac{\partial^2 z}{\partial x^2} - \dfrac{\partial^2 z}{\partial y^2} + \dfrac{\partial z}{\partial x} + 3\dfrac{\partial z}{\partial y} - 2z = e^{x-y} - x^2y.$

Solution:

We have $\dfrac{\partial^2 z}{\partial x^2} - \dfrac{\partial^2 z}{\partial y^2} + \dfrac{\partial z}{\partial x} + 3\dfrac{\partial z}{\partial y} - 2z = e^{x-y} - x^2y.$

The given equation can be written as

$$[D^2 - D'^2 + D + 3D' - 2]\, z = e^{x-y} - x^2y$$

$$\Rightarrow \quad [(D - D')(D + D') + 2(D + D') - (D - D' + 2)]\, z = e^{x-y} - x^2y$$

$$\Rightarrow \quad (D - D' + 2)(D + D' - 1)\, z = e^{x-y} - x^2y$$

$\therefore$ C.F. $= e^{-2x} f_1(y + x) + e^x f_2(y - x)$

Now, P.I. Corresponding to e^{x-y}

$$= \frac{1}{(D - D' + 2)(D + D' - 1)} . e^{x-y}$$

$$\frac{1}{(1 + 1 + 2)(1 - 1 - 1)} . e^{x-y} = -\frac{1}{4} e^{x-y}$$

and P.I. Corresponding to $(-x^2y) = = \dfrac{1}{(D - D' + 2)(D + D' - 1)} . x^2y$

$$= \frac{1}{2}\left(1 + \frac{D - D'}{2}\right)^{-1} \{1 - (D + D')\}^{-1}. x^2y$$

$$= \frac{1}{2}\left\{\left(\frac{D - D'}{2}\right) + \left(\frac{D - D'}{2}\right)^2 - \left(\frac{D - D'}{2}\right)^3 ...\right\}$$

$$\times\{1 + D + D') + (D + D')^2 + (D + D')^3 + ...\}. x^2y$$

$$= \frac{1}{2}\left\{1 - \frac{D}{3} + \frac{D'}{2} + \frac{D^2}{4} - \frac{DD'}{2} + \frac{D'2}{4} + \frac{3D^2D'}{8} ...\right\}$$

$$\times\{1 + D + D' + D^2 + 2DD' + D'^2 + 3D^2D' ...\}. x^2y$$

$$= \frac{1}{2}\left(1 - \frac{D}{2} + \frac{D'}{2} + \frac{D^2}{4} - \frac{DD'}{8} + \frac{3D^2D'}{8} + D - \frac{D^2}{2} + \frac{DD'}{2} - \frac{D^2D'}{2}\right.$$

$$D' - \frac{DD'}{2} + \frac{D^2D'}{4} + D^2 + \frac{D^2D'}{2}$$

$$\left. + 2DD' - D^2D' + 3D^2D' - ...\right) x^2y$$

$$= \frac{1}{2}\left(1 + \frac{1}{2}D + \frac{3}{2}D' + \frac{3}{4}D^2 + \frac{3}{2}DD' + \frac{21}{8}D^2D' \ldots\right)x^2y$$

$$= \frac{1}{2}\left(x^2y + xy + \frac{3}{2}x^2 + \frac{3}{2}y + 3x + \frac{21}{4}\right).$$

$\therefore$ The solution is

$$y = e^{-2x} f_1(y + x) + e^x f_2(y - x)$$
$$+ \frac{1}{2}\left(x^2y + xy + \frac{3}{2}x^2 + \frac{3}{2}y + 3x + \frac{21}{4}\right).$$ **Ans.**

Example 16:

Solve $\frac{\partial^2 z}{\partial x^2} + \frac{\partial^2 z}{\partial x \partial y} - 6\frac{\partial^2 z}{\partial y^2} = x^2 \sin(x + y)$

(Garhwal, 94; Rohilkhand, 90; Agra, 89; Meerut, 93, 94, 96(BP)

Solution:

We have $\frac{\partial^2 z}{\partial x^2} + \frac{\partial^2 z}{\partial x \partial y} - 6\frac{\partial^2 z}{\partial y^2} = x^2 \sin(x + y)$

The given equation can be written as

$$(D^2 + DD' - 6D'^2)\, z = x^2 \sin(x + y)$$

$$\Rightarrow \quad (D - 2D')(D + 3D')\, z = x^2 \sin(x + y)$$

$\therefore$ C.F $= f_1(y + 2x) + f_2(y = 3x)$.

Now, $\text{P.I.} = \frac{1}{D^2 + DD' - 6D'^2} x^2 \sin(x + y)$

$$= \text{I.P. of } \frac{1}{D^2 + DD' - 6D'^2} e^{i(x+y)} x^2$$

$$= \text{I.P. of } e^{i(x+y)} \frac{1}{(D + i)^2 + (D + i)(D + i) - 6(D' + i)^2} . x^2$$

$$= \text{I.P. of } e^{i(x+y)} \frac{1}{D^2 + 3iD + DD' - 11D'i - 6D'^2 + 4} x^2$$

$$= \text{I.P. of } e^{i(x+y)} \frac{1}{4}\left[1 + \left(\frac{D^2}{4} + \frac{3iD}{4} + \frac{DD'}{4} - \frac{11D'i}{4} - \frac{9D'^2}{4}\right)\right]^{-1} . x^2$$

$$= \text{I.P. of } e^{i(x+y)} \cdot \frac{1}{4}\left[1 - \frac{D^2}{4} - \frac{3iD}{4} - \frac{DD'}{4} + \frac{11D}{4}i + \frac{9D'^2}{4} + \frac{9i^2D^2}{16} \ldots\right] \cdot x^2$$

$$= \text{I.P. of } e^{i(x+y)} \cdot \frac{1}{4}\left[x^2 - \frac{1}{2} - \frac{3ix}{2} - \frac{9}{8}\right]$$

$$= \text{I.P. of } \frac{1}{4} \; [\cos(x+y) + i\sin(x+y)] \left(x^2 - \frac{13}{8} - \frac{3ix}{2}\right)$$

$$= \frac{1}{4}\left(x^2 = \frac{13}{8}\right)\sin(x+y) - \frac{3}{8}x\cos(x+y)$$

∴ The required solution is

$$z = f_1(y + 2x) + f_2(y - 3x)$$

$$\frac{1}{4}\left(x^2 = \frac{13}{8}\right)\sin(x+y) - \frac{3}{8}x\cos(x+y).$$ **Ans.**

EXERCISES

1. $(D^2 - 2aDD' + a^2D^2)\, z = f(y + ax)$.

   Ans. $z = f_1(y + ax) + xf_2(y + ax) + \frac{1}{2}\, x^2 f(y + ax$

2. $(D^2 - D')\, z = xe^a\, x + a^2y$.

   Ans. $z = \sum Ae^{hx+h^2y} - e^{ax+a^2y}\left(\frac{x^2}{4a} - \frac{x}{4a^2}\right)$.

3. $(4D^2 + 12D.D' + 9D'^2) = 0$.

   Ans. $z = f_1(2y - 3x) + f_2(2y - 3x)$.

4. $\frac{\partial^2 z}{\partial x^2} - 3\frac{\partial^2 z}{\partial x\,\partial y^2} - 2\frac{\partial^2 z}{\partial y^2} = 0$. **Ans.** $z = f_1(2y - x) + f_2(y - 2x)$.

5. $\frac{\partial^4 z}{\partial x^4} + \frac{\partial^4 z}{\partial y^4} = 2\frac{\partial^4 z}{\partial x^2\,\partial y^2}$ **(Meerut, 93)**

   Ans. $z = f_1(y + x) + xf_2(y + x) + f_3(y - x) + xf_4(y - x)$.

6. $R - 4s + 4t = 0$. **Ans.** $z = f_1(y + 2x) + xf_2(y + 2x)$.

7. $(D^2 - a^2 D'^2)\, z = x^2$. **Ans.** $z = f_1(y + ax) + f_2(y = ax) + \frac{1}{12}x^4$.

8. $\dfrac{\partial^2 z}{\partial x^2} + 3\dfrac{\partial^2 z}{\partial x \partial y} + 2\dfrac{\partial^2 z}{\partial y^2} = 2x + 3y.$ **(Agra, 92)**

Ans. $z = f_1(y - x) + f_2(y - 2x) - \dfrac{7}{6}x^3 + \dfrac{3}{2}x^2 y$

9. $\dfrac{\partial^2 z}{\partial x^2} + 3\dfrac{\partial^2 z}{\partial x \partial y} + 2\dfrac{\partial^2 z}{\partial y^2} = 6(x + y).$ **(Agra, 95)**

Ans. $z = f_1(y - x) + f_2(y - 2x) - 2x^3 + 3x^2 y.$

10. $(D^2 + 3DD' + 2D'^2)\, z = 12xy.$ **(Rohilkhand, 96)**

Ans. $z = f_1(y - x) + f_2(y - 2x) + 2x^3 y - \dfrac{3}{2}x^4$

11. $\dfrac{\partial z}{\partial x} + \dfrac{\partial z}{\partial y} = \sin x.$ **Ans.** $z = f(y - x) - \cos x.$

12. $r - 2s + t = \sin(2x + 3y).$

Ans. $z = f_1(x + y) + x f_2(x + y) - \sin(2x + 3y).$

13. $(D^3 - 4D^2D' + 4DD'^2\, z = \cos(2x + y).$ **(Agra, 93)**

Ans. $z = f_1(y) + f_2(y + 2x) + x f_3(y + 2x) + \dfrac{1}{4}\, x^2 \sin(2x + y).$

14. $\dfrac{\partial^3 z}{\partial x^3} - 4\dfrac{\partial^3 z}{\partial x^2 \partial y} + 4\dfrac{\partial^3 z}{\partial x \partial y^2} = \sin(2x + y)$ **(PCS (UP) 92; Agra, 95)**

Ans. $z_1 = f_1(y) + f_2(y + 2x) + x f_3(y + 2x) - \dfrac{x^2}{4} \cos(2x + y).$

15. $(2D^2 - 5DD' + 2D'^2)\, z = 5 \sin(2x + y).$

Ans. $z = f_1(2y + x) + f_2(y + 2x) - (5x/3)\cos(2x + y)$

16. $\dfrac{\partial^2 z}{\partial x^2} - 4\dfrac{\partial^3 z}{\partial x \partial y} - 6\dfrac{\partial^2 z}{\partial y^2} = y \sin x.$

Ans. $z = f_1(y - 3x) + f_2(y + 2x) - (y \sin x + \cos x).$

17. $t + s + q = 0.$ **(Meerut, 91(P), 93(P); Kanpur, 90)**

Ans. $z = f_1(x) + e^{-x} f_2(y - x)$

18. $(D + D' - 1)(D + 2D' - 2)\, z = 0.$

Ans. $z = e^x f_1(y - x) + e^{2x} f_2(y = 2x).$

19. $(D^2 + DD' + D' - 1)\, z = \sin(x + 2y).$

(Rohilkhand, 94; Agra, 90, 91)

Ans. $z = e^x f_1(y) + e^x f_2(y - x) - \dfrac{1}{20}[2\cos(x + 2y) + 4 \sin(x + 2y)$

20. $(D^2 - DD' - 2D)\,z = \sin(3x + 4y) - e^{2x+y}$. **(Meerut, 95(BP)**

Ans. $z = f_1(y) + e^{2x} f_2(y + x) + \frac{1}{15}\sin(3x + 4y)$
$+ \frac{2}{15}\cos(3x + 3y) + \frac{1}{2}e^{2x+y}$

21. $x^2\frac{\partial^2 z}{\partial x^2} - y^2\frac{\partial^2 z}{\partial y^2} = xy$. **Ans.** $z = x\phi_1(y/x) + \phi_2(xy) + xy\log x$

22. $x^2\frac{\partial^2 z}{\partial x^2} - y^2\frac{\partial^2 z}{\partial y^2} = x^2y$. **Ans.** $z = x\phi_1(y/x) + \phi_2(xy) + \frac{1}{2}x^2y$

23. $x^2r - y^2t + xp - rq = \log x$. **Ans.** $z = f_1(xy) + f_2(y/x) + \frac{1}{6}$
$(\log x)^3$

24. $x^2D^2 + 2xy\,DD' + y^2D'^2 = x^my^n$.

Ans. $z = \phi_1\left(\frac{y}{x}\right) + x\phi_2\left(\frac{y}{x}\right) + \frac{x^my^{n1}}{(m + n)(m + n - 1)}$

25. Solve $(D^2 - D'^2 + D - D')\,z = e^{2x+3y}$.

Ans. $z = f_1(y + x) + e^{-x}f_2(y - x) - \frac{1}{6}e^{2x+3y}$

26. $(D^2 - 5\,DD'^2 - 6D'^3)\,z = \frac{1}{(y - 2x)}$. **(IAS, 91)**

Ans. $z = \phi_1(y - 2x) + \phi_2(y - 3x) + x\log(y - 2x)$

27. $(D^2 - 3aDD' + 2a^2D'^2)\,z = 0$. **Ans.** $z = f_1(y + ax) + f_2(y + 2zx)$

28. $(D^2 - 2D^2D' + DD'^2)\,z = 0$. **Ans.** $z = f_1(y) + x\,f_2(y + x) + f_3(y + x)$

29. $(D^2 - DD' - 6D'^2)z = xy$.

Ans. $z = f_1(y + 3x) + f_2(y - 2x) + \frac{1}{6}x^3y + \frac{1}{24}x^4$.

30. $\frac{\partial^2 z}{\partial x^2} - \frac{\partial^2 z}{\partial y^2} = x - y$ **(Meerut, 96)**

Ans. $z = f_1(y + x) + f_2(y - x) + \frac{1}{6}x^3 + \frac{1}{2}x^2y$.

31. $r + t + 2s = 0$. **Ans.** $z = f_1(y - x) + xf_2(y - x)$.

5

Partial Differential Equations of Order Two

5.1 Introduction

In this chapter, the partials differential equations of order two with varriable co-efficients will be considered. Only two independent variables and one dependent variable will be taken Second Order partial derivatives will be dentoted by r, s and t so that

$$r = \frac{\partial^2 z}{\partial x^2}, \quad t = \frac{\partial^2 z}{\partial y^2}, \quad s = \frac{\partial^2 z}{\partial x \partial y} .$$

An equation is said to be of order two, if it is involves atleast one of differential co-efficient r, s, t but non of higher order the quantities p and q may also enter into the equation.

Thus, the general form of equation of order two is

$$f(x, y, z, p, q, r, s, t) = 0.$$

Example 1:

Solve $t - xq = x^2$.

Solution:

The given equation can be written as

$$\frac{\partial q}{\partial y} - xq = x^2$$

which is linear in q. I.F. $= e^{\int -x dy} = e^{-xy}$

$\therefore \quad qe^{-xy} = \int x^2 e^{-xy} dy + f(x) = -xe^{-xy} + f(x)$

$\Rightarrow \quad q = \dfrac{\partial z}{\partial y} = -x + f(x).e^{xy}$

$\therefore \quad z = -xy + f(x) . \int e^{xy} dy + F(x)$

$\Rightarrow \quad z = -xy + \dfrac{1}{x} f(x) e^{xy} + F(x).$

$\Rightarrow \quad z = -xy + \phi(x) e^{xy} + F(x).$ **Ans.**

Example 2:

Solve xys = 1.

Solution:

The given equation can be written as

$$\frac{\partial^2 z}{\partial x \partial y} = \frac{1}{xy}.$$

Integrating w.r.t. 'y' we have

$$\frac{\partial z}{\partial x} = \frac{1}{x} \log y + f(x).$$

Again integrating w.r.t. x, we have

$$z = \log x \log y = \int f(x) dx + F(y).$$

$\Rightarrow \quad z = \log x \log y + f(x) + F(y).$ **Ans.**

Example 3:

Solve yt – q = xy. **(Meerut, 97)**

Solution:

The given equation can be written as

$$\frac{\partial q}{\partial y} \frac{\partial q}{\partial y} - \frac{1}{y} q = x$$

which is linear, I.F. $= e^{\int -1/y \, dy} = e^{-\log y} = 1/y$

$\therefore \quad q \dfrac{1}{y} = \int x . \dfrac{1}{y} dy + f(x) = x \log y + f(x)$

$\Rightarrow \quad q = \dfrac{\partial z}{\partial y} = xy \log y + yf(x).$

Integrating, we have w.r.t. y, we have

$$z = x \int y \log y \, dy + f(x) \int y \, dy + F(x)$$

$$\Rightarrow \quad z = x \left[\frac{y^2}{2} . \log y - \int \frac{1}{y} . \frac{y^2}{2} \, dy \right] + \frac{y^2}{2} f(x) + F(y)$$

$$\Rightarrow \quad z = \frac{1}{2} xy^2 \log y - \frac{1}{4} xy^2 + \frac{y^2}{2} f(x) + F(x).$$ **Ans.**

Example 4:

$r = 2y^2.$

Solution:

The given equation can be written as

$$\frac{\partial p}{\partial x} = 2y^2 .$$

Integrating w.r.t. x, we have

$$p = 2xy^2 + f(y)$$

$$y \frac{\partial z}{\partial y} + z \frac{\partial z}{\partial x} = 2xy^2 + f(y).$$

$$\therefore z = x^2y^2 + xf(y) + F(y).$$ **Ans.**

Example 5:

Solve $t = \sin xy$.

Solution:

The given equation can be written as

$$\frac{\partial^2 z}{\partial y^2} = \sin xy.$$

Integrating w.r.t. y, we have $\frac{\partial z}{\partial y} = -\frac{1}{x} \cos xy + f(x).$

Again integrating w.r.t. y, we have

$$z = -\frac{1}{x^2} \sin xy + yf(x) + F(x).$$ **Ans.**

Example 6:

Solve $s = \frac{x}{y} + a.$

Solution:

The given equation can be written as

$$\frac{\partial^2 z}{\partial x\,\partial y} = \frac{x}{y} + a.$$

Integrating w.r.t. x, we have

$$\frac{\partial z}{\partial y} = \frac{1}{y}\frac{x^2}{2} + ax + f(y).$$

Again integrating w.r.t. y, we have

$$Z = \frac{x^2}{2}\log y + axy + \phi(y) + F(x).$$ **Ans.**

Example 7:

Solve yx + p = cos (x + y) – sin (x + y). **(Meerut, 95 (BP)**

Solution:

The given equation can be written as

$$y\frac{\partial q}{\partial x} + \frac{\partial z}{\partial x} = \cos(x + y) - y\sin(x + y).$$

Integrating w.r.t. x, we have

$$yq + z = \sin(x + y) + y\cos(x + y) + f(y).$$

$$\Rightarrow y\frac{\partial z}{\partial y} + z = \sin(x + y) + y\cos(x + y) + f(y).$$

Integrating w.r.t. y, we have

$$yz = \sin(x + y) + \phi(y) + F(x).$$ **Ans.**

Example 8:

Solve p + r + s = 1. **(Meerut, 93, 97 (P))**

Solution:

We have $p + r + s = 1$.

The given equation can be written as

$$\frac{\partial z}{\partial x} + \frac{\partial p}{\partial x} + \frac{\partial q}{\partial x} = 1.$$

Integrating w.r.t. x, we have

$$z + p + q = f(y) + x$$

$$\Rightarrow \quad p + q = x + f(y) - z.$$

$\therefore$ Subsidiary equations are

$$\frac{dx}{1} = \frac{dy}{1} = \frac{dz}{x + f(y) - z}$$

Taking the first two members, we have

$dx = dy, \qquad \therefore x - y = a.$

Again taking the last two members, we have

$$\frac{dz}{dy} + z = x + f(y) \Rightarrow \frac{dz}{dy} + z = a + y + f(y)$$

which is linear.

$\therefore$ I.F. $= e^{\int dy} = e^y.$

$$\therefore z.e^y = \int\left[a e^y + \{y + f(y)\}e^y\right]dy + b$$

$$= ae^y + \int\{y + f(y)\} e^y\, dy + b$$

$\Rightarrow \qquad ze^y = ae^y + \phi(y) + b$

$\therefore z = a + e^{-y} f(y) + be^{-y}$

$\Rightarrow \qquad z = x - y + e^{-y} \phi(y) + e^{-y} F(x - y).$ **Ans.**

Example 9:

Solve rx = (n – 1) p.

Solution:

We have $rx = (n - 1)p$.

The given equation can be written as

$$\frac{\partial p}{\partial x} = \frac{n-1}{x}p \qquad \Rightarrow \qquad \frac{\partial p}{\partial x} - \frac{n-1}{x}p = 0.$$

Integrating w.r.t. x, we have

$$p.e^{-\int \frac{n-1}{x}dx} = f(y) \qquad \Rightarrow \qquad P.E.^{-(n-1)\log x} = f(y)$$

$$\Rightarrow \qquad p.\frac{1}{x^{n-1}} = f(y) \qquad \Rightarrow \qquad p = \frac{\partial z}{\partial x}\ x^{n-1} f(y).$$

Again integrating w.r.t. x, we have

$$z = \frac{1}{n}x^n\ f(y) + F(y).$$

Example 10:

Solve $s - t = \dfrac{x}{y^2}$. **(IAS 88)**

Solution:

The equation can be written as

$$\frac{\partial p}{\partial y} - \frac{\partial q}{\partial y} = \frac{x}{y^2}.$$

Integrating w.r.t. y, we have

$$p - q = -\frac{x}{y} + f(x).$$

Applying Lagrange's Method

$$\frac{dx}{1} = \frac{dy}{-1} = \frac{dz}{-\frac{x}{y} + f(x)}$$

Taking first two members, we have

$$dx + dy = 0 \quad \therefore \; x + y = a,$$

Again taking the firs and the last members, we have

$$dz = -\frac{x}{y}\, dx + f(x)\, dx$$

$$\Rightarrow \quad dz = -\frac{x}{a - x}\, dx + f(x)\, dx$$

$$\Rightarrow \quad dz = \left(1 - \frac{a}{a - x}\right) dx + f(x)\, dx$$

Integrating, we have

$$z = x + a \log(a - x) + \phi(x) + b$$

$$\Rightarrow \quad z = a \log(a - x) + \psi(x) + b$$

$$\Rightarrow \quad z = (x + y) \log y + \psi(x) + F(x + y).$$

Ans.

Example 11:

Show that a surface of revolution satisfying the differential equation $r = 12x^2 + 4y^2$ *and touching the plane* $z = 0$ *is* $z = (x^2 + y^2)^2$.

(Agra, 88; Meerut, 93)

Solution:

The given differential eqn. is

$$r = \frac{\partial^2 z}{\partial x^2} = 12x^2 + 4y^2$$

Integrating partially w.r.t. 'x', $\frac{\partial z}{\partial x} = p = 4x^3 + 4xy^2 + f(y)$ ...(1)

Again integrating partially w.r.t. 'x',

$z = x^4 + 2x^2y^2 + xf(y) + F(y)$ ...(2)

$\therefore$ The surface touches the plane $z = 0$

$\therefore$ for $z = 0$, $\frac{\partial z}{\partial x} = p = 0 \quad \therefore \; 4x^3 + 4xy^2 + f(y) = 0$

But $4x^3 + 4xy^2$ is not a function of y only,

$\therefore$ we have $f(y) = 0$.

$\therefore 4x^3 + 4xy^2 = 0 \quad \Rightarrow \quad y^2 = -x^2$ ...(3)

Putting $z = 0$ and $x^2 = -y^2$ in (2), we get

$0 = y^4 - 2y^4 + 0 + F(y)$. $\therefore$ $F(y) = y^4$.

Hence from (2), the required surface is

$z = x^4 + 2x^2y^2 + y^4 = (x^2 + y^2)$. **Ans.**

Example 12:

Find the surface passing through the parabolas

$$z = 0,\ y^2 = 4ax \text{ and } z = 1,\ y^2 = -4ax$$

and satisfying the equation

$$xr + 2p = 0.$$

(Agra, 91; IAS 92; Meerut 89, 92(P), 93(P))

Solution:

The given equation can be written as

$$x\frac{\partial p}{\partial x} + 2p = 0$$

$$\Rightarrow x^2\frac{\partial p}{\partial x} + 2xp = 0. \quad \Rightarrow \quad \frac{\partial}{\partial x}(x^2 p) = 0$$

Integrating w.r.t. x, we have

$$x^2p = f(y) \quad \Rightarrow \quad p = \frac{\partial z}{\partial x} = \frac{1}{x^2}f(y)$$

$\therefore z = -\frac{1}{x}f(y) + F(y)$. ...(1)

Now, putting $z = 0$, $x = \frac{y^2}{4a}$, we have

$$0 = -\frac{4a}{y^2} f(y) + F(y). \qquad ...(2)$$

Again putting $z = 1$ and $x = -\frac{y^2}{4a}$ in (1), we have

$$1 = \frac{4a}{y^2} f(y) + F(y).$$

Adding (2) and (3), we have

$$2F(y) = 1 \quad \Rightarrow \quad f(y) = \frac{1}{2}$$

Putting $F(y) = \frac{1}{2}$ in (2), we have $f(y) = y^2/8a$.

Putting these values in (1), the required surface is

$$z = -\frac{y^2}{8ax} + \frac{1}{2}$$

$$\Rightarrow \quad 8axz = 4ax - y^2.$$ **Ans.**

Example 13:

Show that a surface satisfying $r = 6x + 2$ and touching $z = x^3 + y^3$ along its section by the plane $x + y + 1 = 0$ is

$$z = x^3 + y^3 + (x + y + 1)^2.$$

Solution:

Here $r = 6x + 2$

$$\Rightarrow \quad \frac{\partial p}{\partial x} = 6x + 2.$$

Integrating w.r.t. x, we have

$$p = 3x^2 + 2x + f(y)$$

$$\Rightarrow \quad \frac{\partial z}{\partial x} = 3x^2 + 2x + f(y)$$

$$\therefore \quad z = x^3 + x^2 + xf(y) + F(y) \qquad ...(1)$$

$$p = \frac{\partial z}{\partial x} = 3x^2 + 2x + f(y)$$

and $$q = \frac{\partial z}{\partial y} = xf'(y) + F'(y).$$

Also from $z = x^3 + y^3$, ...(2)

$$p = 3x^2, \; q = 3y^2.$$

If equation (1) touches (2) along its section by the plane $x + y + 1 = 0$, then the values of p and q for any point on this plane should be equal.

i.e., $3x^2 + 2x + f(y) = 3x^2$ ...(3)

$xf'(y) + F'(y) = 3y^2,$ ...(4)

and $x + y + 1 = 0$ ...(5)

from equations (3) and (5), we have

$$f(y) = -2x = 2(y + 1)$$

$\therefore$ $f'(y) = 2.$

Putting in (4), we have

$$F'(y) = 3y^2 - 2x$$

$\Rightarrow$ $F'(y) = 3y^2 + 2(y + 1).$

Integrating $F(y) = y^3 + y^2 + 2y + k.$

Substituting in (1), we have

$$z = x^3 + x^2 + x.\ 2(y + 1) + y^3 + y^2 + 2y + k.$$

Now equating the values of z from (2) and (6) when

$y = -(x + 1)$, we have

$$x^3 - (x + 1)^3 = x^3 + x^2 + 2x.(-x) - (x + 1)^3 + (x + 1)^2 - 2(x + 1) + k.$$

$\therefore$ $k = 1.$

$\therefore$ The required surface is

$$z = x^3 + x^2 + 2x(y + 1) + y^3 + y^2 + 2y + 1$$

$\Rightarrow$ $z = x^3 + y^3 + (x + y + 1)^2.$ **Ans.**

Example 14:

Find the surface satisfying $t = 6x^3y$, containing the two lines,

$$y = 0 = z,\ y = 1 = z.$$

(Meerut, 92; Agra 99; Kanpur, 96)

Solution:

The given equation can be written as

$$\frac{\partial q}{\partial y} = 6x^3y.$$

Integrating w.r.t. y, we have $q = 3x^3y^2 + f(x)$

$$\Rightarrow \quad \frac{\partial z}{\partial y} = 3x^3y^2 + f(x) \qquad ...(1)$$

$$\therefore \quad z = x^3y^3 + yf(x) + F(x). \qquad ...(2)$$

If equation (1) contains the line y = 0 = z and y = 1 = z

then $\qquad 0 = F(x)$

and $\qquad 1 = x^3 + f(x) + F(x)$

$\Rightarrow \qquad f(x) = 1 - x^3. \qquad$ ...(4)

$\therefore$ The required surface is

$$z = x^3y^3 + y(1 - x^3).$$ **Ans.**

Example 15:

Find a surface satisfying $r + s = 0$ and touching the elliptic paraboloid $z = 4x^2 + y^2$ along its section by the plane $y = 2x + 1$.

(Meerut, 97 (R); IAS 94)

Solution:

Here $r + s = 0$.

$$\Rightarrow \qquad \frac{\partial p}{\partial x} + \frac{\partial q}{\partial x} = 0.$$

Integrating w.r.t. x, we have $p + q = f(y)$.

Now, the subsidiary equations are

$$\frac{dx}{1} = \frac{dy}{1} = \frac{dz}{f(y)}$$

Taking the first two members, we have

$$dy = dx \qquad \therefore\ y - x = a$$

Again taking the last two members, we have

$$dz = \phi(y)\, dy$$

$\therefore \qquad z = \phi(y) + b$

$\therefore \qquad$ The solution is

$$z = \phi(y) + F(y - x). \qquad ...(2)$$

Now, $\qquad p = \dfrac{\partial z}{\partial x} = -F'(y - x)$

and $\qquad q = \dfrac{\partial z}{\partial y} = f'(y) + F'(y - x)$

Also from $z = 4x^2 + y^2 \qquad$ ...(3)

$$p = \frac{\partial z}{\partial x} = 8x \text{ and } q = \frac{\partial z}{\partial y} = 2y.$$

If equation (2) touches (3) along its section by the plane $y = 2x + 1$, then the values of p and q for any point $y = 2x + 1$ should be equal.

i.e., $\quad - F'(y - x) = 8x$ ...(4)

and $\quad \phi'(y) + F'(y - x) = 2y$ ...(5)

$y = 2x + 1.$ ...(6)

From equations (4) and (6), we have

$$- F'(y - x) = 8(y - x - 1)$$

$$\therefore \quad - F(y - x) = 8, [\frac{1}{2}(y - x)^2 - (y - x)] + a_1$$

$$\Rightarrow \quad F(y - x) = - 4(y - x)^2 + 8(y - x) + c$$

Again from (4) and (5), we have

$$\phi'(y) = 8x + 2y$$

$$\Rightarrow \quad \phi'(y) = 4(y - 1) + 2y$$

$$\Rightarrow \quad \phi'(y) = 6y - 4$$

Integrating $\phi(y) = 3y^2 - 4y + b_1$.

Substituting in (2), the required surface is

$$z = 3y^2 - 44 + b_1 - 4(y - x)^2 + 8(y - x) + c_2$$

$$\Rightarrow \quad z + yx^2 + y^2 - 8xy + 8x - 4y - c = 0 \qquad ...(1)$$

Now equate the values of z, from c_3 and (7)

When the required surface is

$\therefore z + 4x^2 + y^2 - 8x + 8x - 44 + 2 = 0.$ **Ans.**

Example 16:

Solve $xr + p = 9x^2y^3$.

Solution:

The given equation can be written as

$$\frac{\partial p}{\partial x} + \frac{1}{x}p = 9xy^3$$

which is linear in p.

Here $\quad$ I.F. $= e^{\int 1/x \, dx} = e^{\log x} = x.$

$$\therefore \quad px = \int 9x^2 y^3 \, dx + f(y)$$

$$\Rightarrow \quad px = 3x^3y^3 + f(y)$$

$$\Rightarrow \quad p = \frac{\partial z}{\partial x} = 3x^2y^3 + \frac{1}{x}f(y)$$

Integrating, we have w.r.t. x, $z = x^3y^3 + f(y) \log x + F(y)$. **Ans.**

Example 17:

Solve log s = x + y.

Solution:

The given equation can be written as

$$\frac{\partial^2 z}{\partial x\,\partial y} = e^{x+y}$$

Integrating w.r.t. x, we have

$$\frac{\partial z}{\partial y} = e^{x+y} + f(y).$$

Again integrating w.r.t. y, we have

$$z = e^{x+y} + \int + (y)dy + F(x).$$

$\Rightarrow \quad z = e^{x+y} + \phi\,(y) + F\,(x).$ **Ans.**

Example 18:

Solve s = 2x + 2y.

Solution:

The given equation can be written as

$$\frac{\partial^2 z}{\partial x \partial y} = 2x + 2y.$$

Integrating w.r.t. x, we have $\frac{\partial z}{\partial y} = x^2 + 2xy + f\,(y).$

Now integrating w.r.t. y, we have

$$z = x^2\,y + xy^2 + \int f\,(y)\,dy + F\,(x).$$

$\Rightarrow \quad z = x^2y + xy^2 + f\,(y) + F\,(x).$ **Ans.**

5.2 Canonical Forms

We now consider the equations of the type

$$Rr + Ss + Tt + f\,(x, y, z, p, q) = 0 \qquad ...(1)$$

in which R, S, T are continuous functions of x and y possessing continuous partial derivatives of as high order as necessary.

By suitable change of independent variables, we shall show that the equation (1) can be transformed into one of the three *canonical forms* which can be integrated easily.

Let the independent variables x, y be changed to u, v the transformation equations

$$u = u\ (x, y),\ v = v\ (x, y). \qquad \text{...(2)}$$

Now, we have

$$p = \frac{\partial z}{\partial x} = \frac{\partial z}{\partial u}.\frac{\partial u}{\partial x} + \frac{\partial z}{\partial v}.\frac{\partial v}{\partial x},\ q = \frac{\partial z}{\partial y} = \frac{\partial z}{\partial u}.\frac{\partial u}{\partial y} + \frac{\partial z}{\partial v}.\frac{\partial v}{\partial y}$$

$$\therefore\ \frac{\partial}{\partial x} = \frac{\partial u}{\partial x}.\frac{\partial}{\partial x} + \frac{\partial v}{\partial x}.\frac{\partial}{\partial v},\ \text{and}\ \frac{\partial}{\partial y} \equiv \frac{\partial u}{\partial y}.\frac{\partial}{\partial u} + \frac{\partial v}{\partial y}.\frac{\partial}{\partial v}$$

$$r = \frac{\partial^2 z}{\partial u^2} = \frac{\partial}{\partial x}\left(\frac{\partial z}{\partial x}\right) = \left(\frac{\partial u}{\partial x}.\frac{\partial v}{\partial x} + \frac{\partial v}{\partial x}.\frac{\partial}{\partial v}\right)\left(\frac{\partial u}{\partial x}.\frac{\partial z}{\partial u} + \frac{\partial v}{\partial x}.\frac{\partial z}{\partial v}\right)$$

$$= \frac{\partial^2 z}{\partial u^2}\left(\frac{\partial u}{\partial x}\right)^2 + 2\frac{\partial^2 z}{\partial u\,\partial v}.\frac{\partial u}{\partial x}.\frac{\partial v}{\partial x} + \frac{\partial^2 u}{\partial v^2}\left(\frac{\partial v}{\partial x}\right)^2 + \frac{\partial z}{\partial u}.\frac{\partial^2 u}{\partial x^2}$$

$$+ \frac{\partial z}{\partial v}.\frac{\partial^2 u}{\partial x^2}$$

$$s = \frac{\partial^2 z}{\partial x\,\partial y} + \frac{\partial}{\partial x}\left(\frac{\partial z}{\partial y}\right) = \left(\frac{\partial u}{\partial x}\frac{\partial}{\partial v} + \frac{\partial v}{\partial x}\frac{\partial}{\partial v}\right)\left(\frac{\partial z}{\partial u}\frac{\partial u}{\partial y} + \frac{\partial z}{\partial v}\frac{\partial v}{\partial y}\right)$$

$$= \frac{\partial^2 z}{\partial u^2}\frac{\partial u}{\partial x}\frac{\partial u}{\partial y} + \frac{\partial^2 z}{\partial u\,\partial v}\left(\frac{\partial u}{\partial x}\frac{\partial v}{\partial y} + \frac{\partial u}{\partial y}\frac{\partial v}{\partial x}\right)$$

$$+ \frac{\partial^2 u}{\partial v^2}\frac{\partial v}{\partial x}\frac{\partial v}{\partial y} + \frac{\partial z}{\partial u}\frac{\partial^2 u}{\partial y\,\partial x} + \frac{\partial z}{\partial v}\frac{\partial^2 v}{\partial y\,\partial x}$$

and $$t = \frac{\partial^2 z}{\partial y^2} = \frac{\partial}{\partial y}\left(\frac{\partial z}{\partial y}\right)\left(\frac{\partial u}{\partial y}\frac{\partial}{\partial u} + \frac{\partial v}{\partial y}\frac{\partial}{\partial v}\right)\left(\frac{\partial u}{\partial y}\frac{\partial z}{\partial u} + \frac{\partial z}{\partial u} + \frac{\partial z}{\partial v}\frac{\partial v}{\partial y}\right)$$

$$= \frac{\partial^2 z}{\partial u^2}\left(\frac{\partial u^2}{\partial y}\right)^2 + 2\ \frac{\partial^2 z}{\partial u\,\partial v}\ \frac{\partial u}{y}\ \frac{\partial u}{\partial u\,\partial y} + \frac{\partial^2 u}{\partial v^2}\left(\frac{\partial v}{\partial y}\right)^2$$

$$+ \frac{\partial z}{\partial u}\frac{\partial^2 u}{\partial y^2} + \frac{\partial z}{\partial v}\frac{\partial^2 v}{\partial y^2}.$$

Putting these values of p, q, r, s, t in (1) and simplifying, we have

$$A\frac{\partial^2 z}{\partial u^2} + 2B\frac{\partial^2 z}{\partial u\,\partial v} + C\frac{\partial^2 z}{\partial v^2} + F\left(u, v, z, \frac{\partial z}{\partial u}, \frac{\partial z}{\partial v}\right) = 0 \qquad \text{...(3)}$$

where $$A = R\left(\frac{\partial u}{\partial x}\right)^2 + S\frac{\partial u}{\partial x}\frac{\partial v}{\partial y} + T\left(\frac{\partial u}{\partial y}\right)^2 \qquad \text{...(4)}$$

$$B = R\frac{\partial u}{\partial x}.\frac{\partial v}{\partial x}+\frac{1}{2}S\left(\frac{\partial u}{\partial x}\frac{\partial v}{\partial y}+\frac{\partial u}{\partial y}\frac{\partial u}{\partial x}\right)+T\frac{\partial u}{\partial y}\frac{\partial v}{\partial y} \quad ...(5)$$

$$C = R\left(\frac{\partial u}{\partial x}\right)^2+S\frac{\partial v}{\partial x}\frac{\partial v}{\partial y}+T\left(\frac{\partial v}{\partial y}\right)^2 \quad ...(6)$$

and $F\left(u, v, w, \frac{\partial z}{\partial u}, \frac{\partial z}{\partial v}\right)$ is the transformed form of f (x, y, z, p, q).

Now, we determine u and v so that the equation (3) takes the simplest possible form. The equation (3) reduces to the simple integrable form when the discriminant $S^2 - 4RT$ of the quadratic equation

$$R\lambda^2 + S\lambda + T = 0 \quad ...(7)$$

is either positive, negative or zero everywhere.

The three cases are discussed separately as follows.

Case I. $S^2 - 4RI > 0$: In this case the two roots λ_1, λ_2 of equation (7) would be real and distinct.

We choose u and v such that

$$\frac{\partial u}{\partial x} = \lambda_1\frac{\partial u}{\partial y} \quad ...(8)$$

and $$\frac{\partial v}{\partial x} = \lambda_2 . \frac{\partial v}{\partial y} \quad ...(9)$$

$\therefore$ $$A = \left(R\lambda_1^2 + S\lambda_1 + T\right)\left(\frac{\partial u}{\partial y}\right)^2 = 0,$$

$\because$ $R\lambda_1^2 + S\lambda_1 + T = 0$ as λ is the root of equation (7).

Similarly C = 0.

For differential equation (9) the auxiliary equations are

$$\frac{dx}{1} = \frac{dy}{-\lambda_1} = \frac{du}{0}$$

Given du = 0 $\therefore$ u = c_1 (constant)

and $$\frac{dy}{dx}+\lambda_1 = 0. \quad ...(10)$$

Let f_1 (x, y) = c_2 (constant) be the solution of equation (10).

$\therefore$ Solution of equation (8) can be taken as

$$u = f_1\ (x, y) \qquad ...(11)$$

Similarly, if $f_2\ (x, y)$ = constant, is a solution of

$$\frac{dy}{dx} + \lambda_1 = 0,$$

the solution of (9) can be taken as

$$v = f_1\ (x, y).$$

Now, it can be shown that

$$AC - B^2 = \frac{1}{4}\left(4\ RT - S^2\right)\left(\frac{\partial u}{\partial x}\frac{\partial v}{\partial y} - \frac{\partial u}{\partial y}\frac{\partial v}{\partial x}\right)^2$$

$$\Rightarrow B^2 = \frac{1}{4}\left(S^2 - 4\ RT\right)\left(\frac{\partial u}{\partial x}\frac{\partial v}{\partial y} - \frac{\partial u}{\partial y}\frac{\partial v}{\partial x}\right)^2, \qquad ...(13)$$

$$\because \quad A = C = 0.$$

$S^2 - 4RT > 0 \quad \therefore B^2 > 0,$

Therefore we can divide both sides of the equation by it. Hence the equation (1) is reduced to the form

$$\frac{\partial^2 z}{\partial u\ \partial v} = \phi\left(u, v, z, \frac{\partial z}{\partial u}, \frac{\partial u}{\partial v}\right) \qquad ...(14)$$

which is the *canonical form* of the equation (1).

Case II $S^2 - 4RT = 0$: In this case the two roots of equation (7) would be equal (real).

Here we choose u, as in case I

i.e., $$\frac{\partial u}{\partial x} = \lambda_1 \frac{\partial u}{\partial y}.$$

Giving, $u = f\ (x, y)$.

Also we take v to be any function of x and y, which is independent of u.

$\therefore$ As in Case I, $A = 0$.

Also from (13), $B^2 = 0 \quad \because \quad S^2 - 4RT = 0$

i.e., $B = 0$.

Putting $A = 0$, $B = 0$, in equation (3), and dividing by $C\ (\neq 0)$, the equation (1) takes the form

$$\frac{\partial^2 z}{\partial v^2} = \phi\left(u, v, z, \frac{\partial z}{\partial u}, \frac{\partial z}{\partial v}\right) \qquad ...(15)$$

which is the other *canonial form* of the equation (1).

Case III. $S^2 - 4RT < 0$: In case the two roots of equation (7) would be complex.

Proceeding as in case I, here the equation (1) will reduce to the same canonical form equation (14)] as in case I but here the variables u and v are not real but are infect the complex conjugates.

To get a real canonical form.

Let $\quad u = \alpha + i\beta, \quad v = \alpha - i\beta$

so that $\quad \alpha = \frac{1}{2}(u + v)$ and $\beta = \frac{1}{2i}(v - u)$.

Now, we transform the independent variables u and v to α and β with the help of these relations.

$$\frac{\partial z}{\partial u} = \frac{\partial z}{\partial \alpha}\cdot\frac{\partial \alpha}{\partial u} + \frac{\partial z}{\partial \beta}\cdot\frac{\partial \beta}{\partial u} = \frac{1}{2}\left(\frac{\partial z}{\partial \alpha} - i\frac{\partial z}{\partial \beta}\right)$$

Similarly, $$\frac{\partial z}{\partial v} = \frac{1}{2}\left(\frac{\partial z}{\partial \alpha} - i\frac{\partial z}{\partial \beta}\right)$$

$$\therefore \frac{\partial^2 z}{\partial u\,\partial v} = \frac{\partial}{\partial u}\left(\frac{\partial z}{\partial v}\right) = \frac{1}{4}\left(\frac{\partial}{\partial \alpha} - i\frac{\partial}{\partial \beta}\right)\left(\frac{\partial z}{\partial \alpha} + i\frac{\partial z}{\partial \beta}\right)$$

$$= \frac{1}{4}\left(\frac{\partial^2 z}{\partial \alpha^2} + \frac{\partial^2 z}{\partial \beta^2}\right)$$

Substituting in equation (14), the transformed *canonical form* of the given equation is

$$\frac{\partial^2 z}{\partial \alpha^2} + \frac{\partial^2 z}{\partial \beta^2} = \psi\left(\alpha, \beta, z, \frac{\partial z}{\partial \alpha^2}, \frac{\partial z}{\partial \beta^2}\right).$$

Example 1:

Reduce the equation

$$\frac{\partial^2 z}{\partial x^2} = x^2\frac{\partial^2 z}{\partial y^2}, \text{ to canonical form.}$$

Solution:

The given equation can be written as

$$r - x^2t = 0 \qquad ...(1)$$

Comparing (1) with

$$Rr + Ss + Tt + f(x, y, z, p, q) = 0, \text{ we have}$$

$\therefore$ The quadratic equation, $R\lambda^2 + S\lambda + T = 0$, becomes

$$\lambda^2 - x^2 = 0, \text{ giving } \lambda = \pm x$$

i.e., $\lambda_1 = x$ and $\lambda_2 = -x$. (Real and distinct roots)

$\therefore$ The equations $\frac{dy}{dx} + \lambda_1 = 0$ and $\frac{dy}{dx} + \lambda_2 = 0$, becomes

$$\frac{dy}{dx} + x = 0 \text{ and } \frac{dy}{dx} - x = 0.$$

Integrating them, we get,

$$y + \frac{1}{2}x^2 = \text{constant and } y - \frac{1}{2}x^2 = \text{constant.}$$

$\therefore$ To change the independent variable x, y, to u, v, we take

$$u = y^2 + \frac{1}{2}x^2 \text{ and } y - \frac{1}{2}x^2.$$

which give

$$p = \frac{\partial z}{\partial x} = \frac{\partial z}{\partial u}.\frac{\partial u}{\partial x} + \frac{\partial z}{\partial v}.\frac{\partial v}{\partial x} = x.\frac{\partial z}{\partial u} - x\frac{\partial z}{\partial v}$$

$$q = \frac{\partial z}{\partial y} = \frac{\partial z}{\partial u}.\frac{\partial u}{\partial y} + \frac{\partial z}{\partial v}.\frac{\partial v}{\partial xy} = \frac{\partial z}{\partial u} + \frac{\partial z}{\partial v}$$

$$r = \frac{\partial^2 z}{\partial x^2} = \frac{\partial}{\partial x}\left(\frac{\partial z}{\partial x}\right) = \frac{\partial}{\partial x}\left\{x.\left(\frac{\partial z}{\partial u} - \frac{\partial z}{\partial v}\right)\right\}$$

$$= \left(\frac{\partial z}{\partial u} - \frac{\partial z}{\partial v}\right) + 1.\left(\frac{\partial z}{\partial u} - \frac{\partial z}{\partial v}\right)$$

$$= x\left[\frac{\partial}{\partial u}\left(\frac{\partial z}{\partial u} - \frac{\partial z}{\partial v}\right)\frac{\partial u}{\partial x} + \frac{\partial}{\partial v}\left(\frac{\partial z}{\partial u} - \frac{\partial z}{\partial v}\right).\frac{\partial v}{\partial x}\right] + \frac{\partial z}{\partial u} - \frac{\partial z}{\partial v}$$

$$=x^2\left(\frac{\partial^2 z}{\partial y^2}-2\frac{\partial^2 z}{\partial u\,\partial v}+\frac{\partial^2 z}{\partial v^2}\right)+\frac{\partial z}{\partial u}-\frac{\partial z}{\partial v}$$

and $t=\dfrac{\partial^2 z}{\partial y^2}=\dfrac{\partial}{\partial y}\left(\dfrac{\partial z}{\partial y}\right)=\left(\dfrac{\partial}{\partial u}+\dfrac{\partial}{\partial v}\right)\left(\dfrac{\partial z}{\partial u}+\dfrac{\partial z}{\partial v}\right)$

$$=\frac{\partial^2 z}{\partial u^2}+\frac{\partial^2 z}{\partial u\,\partial v}+\frac{\partial^2 z}{\partial v^2}.$$

∴ Substituting in (1), we

$$x^2\left(\frac{\partial^2 z}{\partial u^2}-2\frac{\partial^2 z}{\partial u\,\partial v}+\frac{\partial^2 z}{\partial v^2}\right)+\frac{\partial z}{\partial u}-\frac{\partial z}{\partial v}$$

$$-x^2\left(\frac{\partial^2 z}{\partial u^2}+2\frac{\partial^2 z}{\partial u\,\partial v}+\frac{\partial^2 z}{\partial v^2}\right)=0$$

$$\Rightarrow \quad \frac{\partial^2 v}{\partial u\,\partial v}=\frac{1}{4x^2}\left(\frac{\partial z}{\partial u}-\frac{\partial z}{\partial v}\right)$$

$$\Rightarrow \quad \frac{\partial^2 z}{\partial u\,\partial v}=\frac{1}{4(u-v)}\left(\frac{\partial z}{\partial u}-\frac{\partial z}{\partial v}\right).$$

which is the required-canonical form of the equation.

Example 2:

Reduce the equation

$(y - 1)\, r - (y^2 - 1)\, s + y\,(y - 1)\, t = p - q = 2ye^{2x}\,(1 - y)^3$ ...(1)

to canonical form and hence solve it. **(Rohilkhand, 92)**

Solution:

Comparing the given equation (1) with

$Rr + Ss + Tt + f(x, y, z, p, q) = 0$, we have

$R = y - 1,\ S = -(y^2 - 1),\ T = y\,(t - 1)$

∴ The quadratic equation. $R\lambda^2 + S\lambda + T = 0$, becomes

$(y - 1)\,\lambda^2 - (y^2 - 1)\,\lambda + y\,(y - \lambda) = 0$

⇒ $\lambda^2 - (y + 1)\, 1 + y = 0$

⇒ $(\lambda - 1)\,(\lambda - y) = 0$, giving $\lambda = 1, y$

i.e., $\lambda_1 = 1$ and $\lambda_2 = y$ (Real and distinct roots)

$\therefore$ The equations $\dfrac{dy}{dx}+\lambda_1=0$ and $\dfrac{dy}{dx}+\lambda_2=0$ becomes

$\dfrac{dy}{dx}+1=0$ and $\dfrac{dy}{dx}+y=0$

Integrating these equations, we get

$$x + y = \text{Constant and } ye^x = \text{Constant.}$$

$\therefore$ The change the independent variables x, y to u, v, we take

$$u = x + y \text{ and } v = ye^x,$$

$$\therefore\ p=\frac{\partial z}{\partial x}=\frac{\partial z}{\partial u}.\frac{\partial u}{\partial x}+\frac{\partial z}{\partial v}\frac{\partial v}{\partial x}=\frac{\partial z}{\partial u}+ye^x\frac{\partial z}{\partial x}=\frac{\partial z}{\partial u}+v\frac{\partial z}{\partial v}$$

$$q=\frac{\partial z}{\partial y}=\frac{\partial z}{\partial u}.\frac{\partial u}{\partial xy}+\frac{\partial z}{\partial v}\frac{\partial v}{\partial y}=\frac{\partial z}{\partial u}+e^x\frac{\partial z}{\partial v}$$

$$r=\frac{\partial}{\partial x}=\left(\frac{\partial z}{\partial x}\right)=\left(\frac{\partial}{\partial u}+v\frac{\partial}{\partial v}\right)\left(\frac{\partial z}{\partial u}+v\frac{\partial z}{\partial v}\right)$$

$$=\frac{\partial^2 z}{\partial u^2}+2v\frac{\partial^2 z}{\partial u\,\partial v}=v^2\frac{\partial^2 z}{\partial v^2}+v\frac{\partial z}{\partial v}$$

$$s=\frac{\partial}{\partial x}\left(\frac{\partial z}{\partial y}\right)=\frac{\partial}{\partial x}\left(\frac{\partial z}{\partial u}+e^x\frac{\partial z}{\partial v}\right)$$

$$=\frac{\partial}{\partial x}\left(\frac{\partial z}{\partial u}\right)=e^x\frac{\partial}{\partial x}\left(\frac{\partial z}{\partial v}\right)+e^x\frac{\partial z}{\partial v}$$

$$=\left(\frac{\partial}{\partial u}+v\frac{\partial}{\partial v}\right)\left(\frac{\partial z}{\partial u}\right)+e^x\left(\frac{\partial}{\partial u}+v\frac{\partial}{\partial v}\right)\left(\frac{\partial z}{\partial v}\right)+e^x\frac{\partial z}{\partial v}$$

$$=\frac{\partial^2 z}{\partial u^2}+\left(e^x+v\right)\frac{\partial^2 z}{\partial u\,\partial v}+ve^x\frac{\partial^2 z}{\partial v^2}+e^x\frac{\partial z}{\partial v}$$

and $$t=\frac{\partial}{\partial y}\left(\frac{\partial z}{\partial y}\right)=\frac{\partial}{\partial y}\left(\frac{\partial z}{\partial u}e^x\frac{\partial z}{\partial v}\right)=\frac{\partial}{\partial y}\left(\frac{\partial z}{\partial u}\right)+e^x\frac{\partial}{\partial y}\left(\frac{\partial z}{\partial v}\right)$$

$$=\frac{\partial}{\partial u}\left(\frac{\partial z}{\partial u}\right)\frac{\partial u}{\partial y}+\frac{\partial}{\partial v}\left(\frac{\partial z}{\partial u}\right)\frac{\partial v}{\partial y}+e^x\left[\frac{\partial}{\partial u}\left(\frac{\partial z}{\partial v}\right)\frac{\partial u}{\partial y}+\frac{\partial}{\partial v}\left(\frac{\partial z}{\partial v}\right)\frac{\partial v}{\partial y}\right]$$

$$= \frac{\partial^2 z}{\partial u^2} + 2e^x \frac{\partial^2 z}{\partial u \partial v} + e^{2x} \frac{\partial^2 z}{\partial v^2}$$

Substituting in (1), we have

$$(y-1)\left(\frac{\partial^2 z}{\partial u^2} + 2v\frac{\partial^2 z}{\partial v^2} + v^2 \frac{\partial^2 z}{\partial v^2} + v\frac{\partial z}{\partial v}\right)$$

$$-(y^2-1)\left[\frac{\partial^2 z}{\partial u^2} + \left(e^x + v\right)\frac{\partial^2 z}{\partial u \partial v} + ve^x \frac{\partial^2 z}{\partial v^2} + e^x \frac{\partial z}{\partial v}\right]$$

$$+(y-1)\left[\frac{\partial^2 z}{\partial u^2} + 2e^x \frac{\partial^2 z}{\partial u \partial v} + e^{2x}\frac{\partial^2 z}{\partial v^2}\right] + \left(v - e^x\right)\frac{\partial z}{\partial v}$$

$$= 2ye^{2x}(1-y)^3$$

$$\Rightarrow \quad (1-y)^3 e^x \frac{\partial^2 z}{\partial u \partial v} = 2ye^{2x}(1-y)^3$$

$$\Rightarrow \quad \frac{\partial^2 z}{\partial u \partial v} = 2ye^x \quad \Rightarrow \quad \frac{\partial^2 z}{\partial u \partial v} = 2v \qquad ...(2)$$

which is the canonical form.

Integrating (2), w.r.t. v, we have

$$\frac{\partial z}{\partial u} = v^2 + \phi_1(u). \qquad ...(3)$$

where ϕ_1 (u) is an arbitrary function of u.

Now integrating (3) w.r.t. u, we have

$$z = uv^2 + \psi_1(u) + \psi_2(v)$$

Where y_1 (u) is integral of ϕ_1 (u) w.r.t. u and ψ_2 (v) is an arbitrary function of v

$$\Rightarrow \quad z = (x+y)\, y^2 e^{2x} + \psi_1(x+y) + \psi_2(ve^x). \qquad \textbf{Ans.}$$

Example 3:

Reduce the equation

$$\frac{\partial^2 z}{\partial x^2} + x^2 \frac{\partial^2 z}{\partial y^2} = 0,$$

to canonical form. **(Delhi, Hons. 95; Kanpur, 92)**

Solution:

The given equation can be written as

$$r + x^2t = 0 \qquad ...(1)$$

Comparing it with

$Rr + Ss + Tt + f(x, y, z, p, q) = 0$, we have

$R = 1, S = 0, T = x^2.$

$\therefore$ The quadratic equation, $R\lambda^2 + S\lambda + T = 0$, becomes

$\lambda^2 + x^2 = 0$, giving $\lambda = \pm ix$

i.e., $\lambda_1 + ix\ \lambda_2 = -ix$ (Complex roots)

$\therefore$ The equations, $\dfrac{dy}{dx}+\lambda_1 = 0$ and $\dfrac{dy}{dx}+\lambda_2 = 0$. becomes

$$\frac{dy}{dx}+ix = 0 \text{ and } \frac{dy}{dx}-ix = 0$$

Integrating them, we have

$y + \dfrac{i}{2}x^2 =$ constant and $y - \dfrac{i}{2}x^2 =$ constant.

$\therefore$ To change the independant variable x, y to u, v we take

$u = y+\dfrac{i}{2}\ x^2 = \alpha + i\beta$ (say)

and $v = y - \dfrac{i}{2}x^2 = \alpha - i\beta$

so that $a = y$ and $\beta = \dfrac{1}{2}x^2$

Now, we transform the independent variables x and y to α and β. With the help of these relations

$$p=\frac{\partial z}{\partial x}=\frac{\partial z}{\partial \alpha}.\frac{\partial \alpha}{\partial x}+\frac{\partial z}{\partial \beta}.\frac{\partial \beta}{\partial x}=x\ \frac{\partial z}{\partial \beta}$$

$$q=\frac{\partial z}{\partial y}=\frac{\partial z}{\partial \alpha}.\frac{\partial \alpha}{\partial y}+\frac{\partial z}{\partial \beta}.\frac{\partial \beta}{\partial y}=\frac{\partial z}{\partial \alpha}$$

$$r=\frac{\partial}{\partial x}=\left(\frac{\partial z}{\partial x}\right)=\frac{\partial}{\partial x}\left(x.\frac{\partial z}{\partial \beta}\right)=1.\frac{\partial z}{\partial \beta}+x.\frac{\partial}{\partial x}\left(\frac{\partial z}{\partial \beta}\right)$$

$$=\frac{\partial z}{\partial \beta}+x\left[\frac{\partial}{\partial \alpha}\left(\frac{\partial z}{\partial \beta}\right)\frac{\partial \alpha}{\partial x}+\frac{\partial}{\partial \beta}\left(\frac{\partial z}{\partial \beta}\right).\frac{\partial \beta}{\partial x}\right]=\frac{\partial z}{\partial \beta}+x^2\ \frac{\partial^2 z}{\partial \beta^2}.$$

$$t=\frac{\partial}{\partial y}\left(\frac{\partial z}{\partial y}\right)=\frac{\partial}{\partial y}\left(\frac{\partial z}{\partial \alpha}\right)=\frac{\partial^2 z}{\partial \alpha^2}$$

Substituting in (1), we have

$$\left(\frac{\partial}{\partial\beta}+x^2\frac{\partial^2 z}{\partial\beta^2}\right)+x^2\frac{\partial^2 z}{\partial\alpha^2}=0$$

$$\Rightarrow \quad \frac{\partial^2 z}{\partial\alpha^2}+\frac{\partial^2 z}{\partial\beta^2}=-\frac{1}{x^2}\frac{\partial z}{\partial\beta}$$

$$\Rightarrow \quad \frac{\partial^2 z}{\partial\alpha^2}+\frac{\partial^2 z}{\partial\beta^2}=-\frac{1}{2\beta}\frac{\partial z}{\partial\alpha}$$

which is the canonical form. **Ans.**

Example 4:

Reduce the equation

$$y^2\frac{\partial^2 z}{\partial x^2}-2xy\frac{\partial^2 z}{\partial x\,\partial y}+x^2\frac{\partial^2 z}{\partial y^2}+\frac{y^2}{x}\frac{\partial z}{\partial x}+\frac{x^2}{y}\frac{\partial z}{\partial y}$$

to canonical form, and hence solve it.

Solution:

The given equation can be written as

$$y^2r-2xys+x^2t-\frac{y^2}{x}p-\frac{x^2}{y}q=0. \qquad ...(1)$$

Comparing with, $Rr + Ss + Tt + f(x, y, z, p, q) = 0$, we have

$$R = y^2,\ S = -2xy,\ T = x^2.$$

∴ The quadratic equation $R\lambda^2 + S\lambda + T = 0$, becomes

$$y^2\lambda^2 - 2xy\lambda + x^2 = 0$$

$$\Rightarrow \quad (y\lambda - x)^2 = 0$$

Giving $\lambda=\frac{x}{y},\frac{x}{y}$ i.e., $\lambda_1=\lambda_2=\frac{x}{y}$ (equal roots).

∴ The equation $\frac{dy}{dx}+\lambda=0$, becomes $\frac{dy}{dx}+\frac{x}{y}=0$.

$\Rightarrow$ $y\,dy + x\,dx = 0$. Integrating $x^2 + y^2 =$ constant.

∴ To change the independent variables x, y to u, v, we take

$$u = x^2 + y^2.$$

Also we have to take v as some function of x and y independent of u,

∴ let $v = x^2 - y^2$.

$$\therefore\ p=\frac{\partial z}{\partial x}=\frac{\partial z}{\partial u}.2x+\frac{\partial z}{\partial v}.2x=2x\left(\frac{\partial z}{\partial u}+\frac{\partial z}{\partial v}\right)$$

$$q=\frac{\partial z}{\partial y}=2y\left(\frac{\partial z}{\partial u}-\frac{\partial z}{\partial v}\right)$$

$$r=\frac{\partial}{\partial x}\left(\frac{\partial}{\partial x}\right)=\frac{\partial}{\partial x}\left\{2x\left(\frac{\partial z}{\partial u}+\frac{\partial z}{\partial v}\right)\right\}$$

$$=2\left(\frac{\partial z}{\partial u}+\frac{\partial z}{\partial v}\right)+2x\frac{\partial}{\partial x}\left(\frac{\partial z}{\partial u}+\frac{\partial z}{\partial v}\right)$$

$$=2\left(\frac{\partial z}{\partial u}+\frac{\partial z}{\partial v}\right)+2x.\left[\frac{\partial}{\partial u}\left(\frac{\partial z}{\partial u}+\frac{\partial z}{\partial v}\right).\frac{\partial u}{\partial x}+\frac{\partial}{\partial v}\left(\frac{\partial z}{\partial u}+\frac{\partial z}{\partial v}\right).\frac{\partial u}{\partial x}\right]$$

$$=2\left(\frac{\partial z}{\partial u}+\frac{\partial z}{\partial v}\right)+4x^2\left[\frac{\partial^2 z}{\partial u^2}+2\frac{\partial^2 z}{\partial u\,\partial v}+\frac{\partial^2 z}{\partial v^2}\right]$$

$$s=\frac{\partial}{\partial x}\left(\frac{\partial z}{\partial y}\right)+\frac{\partial}{\partial x}\left[2y\left(\frac{\partial z}{\partial u}-\frac{\partial z}{\partial v}\right)\right]$$

$$=2y\,\frac{\partial}{\partial x}\left(\frac{\partial z}{\partial u}-\frac{\partial z}{\partial v}\right)$$

$$=2y\left[\frac{\partial}{\partial u}\left(\frac{\partial z}{\partial u}-\frac{\partial z}{\partial v}\right)\cdot\frac{\partial u}{\partial x}+\frac{\partial}{\partial v}\left(\frac{\partial z}{\partial u}-\frac{\partial z}{\partial v}\right)\cdot\frac{\partial v}{\partial x}\right]$$

$$=4xy\left[\frac{\partial^2 z}{\partial u^2}-\frac{\partial^2 z}{\partial v^2}\right]$$

and $t=\frac{\partial}{\partial y}\left(\frac{\partial z}{\partial y}\right)=\frac{\partial}{\partial y}\left[2y\left(\frac{\partial z}{\partial u}-\frac{\partial z}{\partial v}\right)\right]$

$$=2\left(\frac{\partial z}{\partial u}-\frac{\partial z}{\partial v}\right)=2y\,\frac{\partial z}{\partial y}\left(\frac{\partial z}{\partial u}-\frac{\partial z}{\partial v}\right)$$

$$=2y\left(\frac{\partial z}{\partial u}-\frac{\partial z}{\partial v}\right)+2y\left[\frac{\partial}{\partial u}\left(\frac{\partial z}{\partial u}-\frac{\partial z}{\partial v}\right)\cdot\frac{\partial u}{\partial y}+\frac{\partial}{\partial v}\left(\frac{\partial z}{\partial u}-\frac{\partial z}{\partial v^2}\right)\frac{\partial v}{\partial y}\right]$$

$$=2\left(\frac{\partial z}{\partial u}-\frac{\partial z}{\partial v}\right)+4y^2\left[\frac{\partial^2 z}{\partial u^2}-2\,\frac{\partial^2 z}{\partial u\,\partial v}+\frac{\partial^2 z}{\partial v^2}\right]$$

Substituting in (1) and simplifying, we have

$$\frac{\partial^2 z}{\partial v^2} = 0, \text{ which is the canonical form.}$$

Integrating w.r.t. v, $\frac{\partial z}{\partial v} = (u)$

Again integrating w.r.t. v, we have

$$z = v\,\phi_1(u) + \phi_2(u)$$

where $\phi_1(u)$, $\phi_2(u)$ are arbitrary functions of u.

$\therefore$ the solution is

$$z = (x^2 - y^2)\,\phi_1(x^2 + y^2) + \phi_2(x^2 + y^2).$$ **Ans.**

Example 5:

Reduce the equation

$$(n-1)^2 \frac{\partial^2 z}{\partial x^2} - y^{2n} \frac{\partial^2 z}{\partial y^2} = n\, y^{2n-1} \frac{\partial z}{\partial y}$$

to canonical form, and find its general solution.

Meerut, 94; Delhi Hons., 94)

Solution:

The given equation can be written as

$$(n-1)^2 r - y^{2n} t - n\, y^{n-1} q = 0 \qquad \text{...(1)}$$

Comparing with $Rr + Ss + Tt + f(x, y, z, p, q) = 0$, we have

$$R = (n-1)^2,\ S = 0,\ T = -y^{2n}.$$

$\therefore$ The quadratic equation $R\lambda^2 + S\lambda + T = 0$, becomes

$$(n-1)^2 l^2 - y^{2n} = 0$$

$$\Rightarrow \qquad \lambda^2 = \frac{1}{(n-1)^2} y^{2n}$$

Giving $\lambda = \pm \frac{1}{(n-1)} y^n$ i.e., $\lambda_1 = \frac{1}{(n-1)} y^n$, $\lambda_2 = \frac{-1}{n-1} y^n$

(Two real distinct roots)

$\therefore$ The equations $\frac{dy}{dx} + \lambda_1 = 0$ and $\frac{dy}{dx} + \lambda_2 = 0$ becomes

$$\frac{dy}{dx} + \frac{1}{(n-1)} y^n = 0 \quad \text{and} \quad \frac{dy}{dx} + \frac{1}{(n-1)} y^n = 0$$

$$\Rightarrow \qquad (n-1)y^{-n}\,dy + dx = 0 \text{ and } (n-1)\,y^{-n}\,dy - dx = 0$$

Integrating, we have

$x - y^{-n+1}$ = constant and $x + y^{-n+1}$ constant

∴ To change the independent variables x, y to u, v we take

$u = x - y^{-n+1}$ and $v = x + y^{-n+1}$

$$\therefore\ p = \frac{\partial z}{\partial x} = \frac{\partial z}{\partial u} + \frac{\partial z}{\partial v}$$

$$q = \frac{\partial z}{\partial y} = (n-1)\, y^{-n}\left(\frac{\partial z}{\partial u} - \frac{\partial z}{\partial v}\right)$$

$$r = \frac{\partial}{\partial x}\left(\frac{\partial z}{\partial x}\right) = \left(\frac{\partial}{\partial u} + \frac{\partial}{\partial v}\right)\left(\frac{\partial z}{\partial u} + \frac{\partial z}{\partial v}\right) = \frac{\partial^2 z}{\partial u^2} + 2\frac{\partial^2 z}{\partial u\, \partial v} + \frac{\partial^2 z}{\partial v^2}$$

and $$t = \frac{\partial}{\partial y}\left(\frac{\partial z}{\partial y}\right) = \frac{\partial}{\partial y}\left[(n-1)\, y^{-n}\left(\frac{\partial z}{\partial u} - \frac{\partial z}{\partial v}\right)\right]$$

$$= -n(n-1)\, y^{-n-1}\left(\frac{\partial z}{\partial u} - \frac{\partial z}{\partial v}\right) + (n-1)\, y^{-n}\frac{\partial}{\partial y}\left(\frac{\partial z}{\partial u} - \frac{\partial z}{\partial v}\right)$$

$$= -n(n-1)\, y^{-n-1}\left(\frac{\partial z}{\partial u} - \frac{\partial z}{\partial v}\right)$$

$$+ (n-1)\, y^{-n}\left[\frac{\partial}{\partial u}\left(\frac{\partial z}{\partial u} - \frac{\partial z}{\partial v}\right)\frac{\partial u}{\partial y} + \frac{\partial}{\partial v}\left(\frac{\partial z}{\partial u} - \frac{\partial z}{\partial v}\right)\frac{\partial v}{\partial y}\right]$$

$$= -n(n-1)\, y^{-n-1}\left(\frac{\partial z}{\partial u} - \frac{\partial z}{\partial v}\right) + (n-1)^2\, y^{-2n}\left[\frac{\partial^2 z}{\partial u^2} - 2\frac{\partial^2 z}{\partial u\, \partial v} + \frac{\partial^2 z}{\partial v^2}\right]$$

Substituting these values in (1) and simplifying, we have

$$\frac{\partial^2 z}{\partial u\, \partial v} = 0,$$ which is the canonical form.

Integrating, w.r.t. v, we have

$\frac{\partial z}{\partial u} = \phi_1(u)$, where $\phi_1(u)$ is an arbitrary function of u.

Integrating again, w.r.t. u, we have

$z = \psi_1(u) + \psi_2(v)$

Where $\psi_1(u)$ is the integral of $\phi_1(u)$ and $\psi_2(v)$ is arbitrary function of v.

Hence the solution is

$z = \psi_1 (x - y^{-n+1}) + \psi_2 (x + y^{-n+1}).$ **Ans.**

Example 6:

Reduce the equation

$$\frac{\partial^2 z}{\partial x^2} + 2\frac{\partial^2 z}{\partial u\,\partial v} + \frac{\partial^2 z}{\partial y^2} = 0$$

to canonical form and hence solve it.

(G.N.D.U. Amritsar 87; Raj. 83; B.H.U. 87; Delhi Hons. 94)

Solution:

The given equation can be written as

$$r + 2s + t = 0 \qquad ...(1)$$

Comparing this equation with

$Rr + Ss + Tt + f(x, y, z, p, q) = 0$, we have

$R = 1$, $S = 2$ and $T = 1$.

$\therefore$ The quadratic equation $R\lambda^2 + S\lambda + T = 0$, becomes

$\lambda^2 + 2\lambda + 1 = 0 \Rightarrow (\lambda + 1)^2 = 0.$

Giving $\lambda = -1$, i.e. $\lambda_1 = \lambda_2 = -1$. (Equal roots)

$\therefore$ The equation $\frac{dy}{dx} + \lambda = 0$, becomes $\frac{dy}{dx} - 1 = 0$

Integrating, we have $x - y$ = constant.

$\therefore$ To change the independent variables x, y to u, v we take $u = x - y$.

Also we have to take v as some function of x and y independent of u

$\therefore$ Let $v = x + y$.

$$\therefore\ p = \frac{\partial z}{\partial x} = \frac{\partial z}{\partial u}\cdot\frac{\partial u}{\partial x} + \frac{\partial z}{\partial v}\cdot\frac{\partial v}{\partial x} = \frac{\partial z}{\partial u} + \frac{\partial z}{\partial v}$$

$$\therefore\ q = \frac{\partial z}{\partial y} = \frac{\partial z}{\partial u}\cdot\frac{\partial u}{\partial y} + \frac{\partial z}{\partial v}\cdot\frac{\partial v}{\partial y} = -\frac{\partial z}{\partial u} + \frac{\partial z}{\partial v}$$

$$r = \frac{\partial}{\partial x}\left(\frac{\partial z}{\partial x}\right) = \left(\frac{\partial}{\partial u} + \frac{\partial}{\partial v}\right)\left(\frac{\partial z}{\partial u} + \frac{\partial z}{\partial v}\right)$$

$$= \frac{\partial^2 z}{\partial u^2} + 2\frac{\partial^2 z}{\partial u\,\partial v} + \frac{\partial^2 v}{\partial v^2}$$

$$s = \frac{\partial}{\partial x}\left(\frac{\partial z}{\partial y}\right) = \left(\frac{\partial}{\partial u} + \frac{\partial}{\partial v}\right)\left(-\frac{\partial z}{\partial u} + \frac{\partial z}{\partial v}\right) = -\frac{\partial^2 z}{\partial u^2} + \frac{\partial^2 z}{\partial v^2}$$

and $t = \frac{\partial}{\partial y}\left(\frac{\partial z}{\partial y}\right) = \left(-\frac{\partial}{\partial u} + \frac{\partial}{\partial v}\right)\left(-\frac{\partial z}{\partial u} + \frac{\partial z}{\partial v}\right)$

$$= \frac{\partial^2 z}{\partial u^2} - 2\frac{\partial^2 z}{\partial u\,\partial v} + \frac{\partial^2 z}{\partial v^2}.$$

Substituting in equation (1), we have $\frac{\partial^2 z}{\partial v^2} = 0$.

Which is the canonical form.

Integrating w.r.t. v, we have

$$\frac{\partial z}{\partial v} = \phi_1(u)$$

where $\phi_1(u)$ is some arbitrary function of u.

Integrating again w.r.t. v, we have

$$z = v\,\phi_1(u) + \phi_2(u).$$

where $\phi_2(u)$ is another arbitrary function of u.

Hence, the complete solution is

$$z = (x + y)\,\phi_1(x - y) + \phi_2(x - y).$$ **Ans.**

EXERCISES

1. $xr + 2p = 0.$ **[Ans.** $z = \frac{1}{x} f(y) + F(y)$]
2. $ar = xy.$ **(Meerut, 94)**
   [Ans. $az = \frac{1}{6}x^3y + x\,f(y) + F(y)$]
3. $xs + q = 4x + 2y + 2.$
   [Ans. $zx = 2x^2y + y^2x + 2xy + f(y) + F(x)$]
4. $r = 6x.$ **(Meerut, 95)**
   [Ans. $z = x^3 + x\,f(y) + F(y)$]
5. $s = 0.$ **[Ans.** $z = f(x) + F(y)$]
6. $xr = p.$ **[Ans.** $z = \frac{x^2}{2} f(y) + F(y)$]
7. $2yq + y^2t = 1.$ **(Meerut, 97(BP))**
   [Ans. $yz = y\log y - f(x) + y\,F(x)$]

8. $t + s + q = 0.$ **(Meerut, 93 (P), 94(P))**

[**Ans.** $ze^x = \phi(x) + F(x - y)$]

9. $xys - qy = x^2.$ **(Delhi, Hons. 92, 93)**

[**Ans.** $z = x^2 \log y + x f_1(y) + f_2(y)$]

10. $st = x/y^2.$ **(IAS, 97)**

[**Ans.** $z = (x + y) \log y + \phi_1(x) + \phi_2(x + y)$]

11. Show that a surface through the circle

$$z = 0,\ x^2 + y = 1$$

and satisfying the differential equation $s = 8xy$ is

$$z = (x^2 + y^2) - 1.$$ **(Meerut, 91, 94)**

12. $x(xy-1)r - (x^2y^2 - 1)s + y(xy - 1)t + (x - 1)p + (y - 1)q = 0.$

[**Ans.** $\dfrac{\partial^2 z}{\partial u\, \partial v} = 0,\ z = \phi_1(ye^x) + \phi_2(xe^y)$]

13. $x^2r - 2xys + y^2t - xp + 3yq = 8y/x.$

[**Ans.** $v\dfrac{\partial^2 z}{\partial v^2} + 2\dfrac{\partial z}{\partial v} = 2,\ z = \dfrac{y}{x} + x^2\ \phi_1(xy) + \phi_2(xy)$]

14. $xy\,r - (x^2 - y^2)s - xyt + py - qx = 2(x^2 - y^2).$

[**Ans.** $\dfrac{\partial^2 z}{\partial u\, \partial v} = \dfrac{v^2 - 1}{(v^2 + 1)^2},\ z = -xy + \phi_1(x^2 + y^2) + \phi_2\left(\dfrac{y}{x}\right)$].

6

Monge's Methods of Integration

6.1 Monge's Method of Integrating Rr + Ss + Tt = V

Where r, s, t have their usual meaning and R, S, T, V are functions of x, y, z, p and q.

(Meerut, 95, 95,(BP), 96,(P) 96(BP), 97(BP), 97(BP); Delhi, Hons. 92, 94; Garhwal, 92, 94; Rohilkhand, 91)

The given equation is

$$Rr + Ss + Tt = V. \qquad ...(1)$$

We know that $dp = \frac{\partial p}{\partial x} dx + \frac{\partial p}{\partial y} dy = r\,dx + s\,dy$

and $dq = \frac{\partial q}{\partial x} dx + \frac{\partial q}{\partial y} dy + s\,dx + t\,dy$

$\therefore\ r = \frac{dp - s\,dy}{dx}$ and $t = \frac{dq - s\,dx}{dy}$

Substituting these values in equation (1), we get

$$R\left(\frac{dp - s\,dy}{dx}\right) + Ss + T\left(\frac{dq - s\,dx}{dy}\right) = V$$

$\Rightarrow\quad R\,dy(dp - s\,dy) + Ss\,dx\,dy + T\,(dq - s\,dx) = V\,dx\,dy$

$\Rightarrow\quad (R\,dp\,dy + T\,dq\,dx - V\,dx\,dy) - s(R\,dy^2 - s\,dx\,dy + T\,dx^2) = 0 \qquad ...(2)$

If some relation between x, y, z, p, q makes each of the bracketed expressions vanish the relation will satisfy equation (2),

$$R\,dy^2 - S\,dx\,dy + T\,dx^2 = 0. \quad ...(3)$$

$$R\,dp\,dy + T\,dq\,dx - V\,dx\,dy = 0 \quad ...(4)$$

The equations (3) and (4) are called **Monge's Subsidiary Equations.**

Let (3) be resolved into two linear equations in dx and dy such that

$$dy - m_1\,dx = 0. \quad ...(5)$$

and $$dy - m_2\,dx = 0 \quad ...(6)$$

Now, from equation (4) and (5), combined if necessary with $dz = p\,dx + q\,dy$, obtain two integrals

$$u_1 = a \text{ and } v_1 = b, \text{ say.}$$

Then the relation $u_1 = f_1(v_1)$. ...(7)

is the solution and is called an **Intermediate Integral.**

Similarly from (4) and (6), obtain another intermediate integral

$$u_2 = f_2(v_2). \quad ...(8)$$

From equations (7) and (8) find the values of p and q in terms of x and y. Substituting these value in

$dz = p\,dx + q\,dy$ and integrating it, the complete integral of the equation (1) is obtained.

Example 1:

Solve $(b + cq)^2 r - 2(b + cq)(a + cp)s + (a + cp)^2 t = 0.$

Solution:

We have $(b + cq)^2 r - 2(b + cq)(a + cp)s + (a + cp)^2 t = 0.$

Putting $r = \dfrac{dp - s\,dy}{dx}$ and $t = \dfrac{dq - s\,dx}{dy}$, the given equation reduces to

$$(b + cq)^2\left(\frac{dp - s\,dy}{dx}\right) - 2(b + cq)(a + cp)s + (a + cp)^2\left(\frac{dq - s\,dx}{dy}\right) = 0$$

$$\Rightarrow \quad \{(b + cq)^2\,ep\,dy + (a + cp)^2\,dq\,dx\} - s\,\{(b + cq)^2\,dy^2 + 2(b + cq)(a + cp)\,dx\,dy + (a + cp)^2\,dx^2\} = 0.$$

The Monge's Subsidiary Equations are

$$(b + cq)^2\,dp\,dy + (a + cp)^2\,dq\,dx = 0 \quad ...(1)$$

and $(b + cq^2\,dy^2 + 2\,(b + cq)(a + cp)\,dx\,dy + (a + cp)^2\,dx^2 = 0$ ...(2)

Equation (2) gives $[(b + cq)\,dy + (a + cp)\,dx]^2 = 0$ ...(3)

$$\Rightarrow \quad b\,dy + a\,dx + c(p\,dx + q\,dy) = 0$$

$$\Rightarrow \qquad a\,dx + b\,dy + c\,dz = 0.$$

$$\therefore \qquad ax + by + cz = A.$$

From (1) and (3), we have

$$(b + cq)\,dp - (a + cp)\,dq = 0$$

$$\Rightarrow \qquad \frac{dq}{a + cp} - \frac{dq}{b + cq} = 0.$$

Integrating, $\dfrac{a + dp}{b + cq} = B.$

$$\Rightarrow \qquad (a + cp) = (b + cq)\,B.$$

$\therefore$ The intermediate integral is

$$a + cp = (b + cq)\,f(ax + by + cz)$$

$$\Rightarrow \qquad cp - cf(ax + by + cz)\,q = -a + bf(ax + by + az).$$

$\therefore$ Lagrange's Subsidiary Equations are

$$\frac{dx}{c} = \frac{dy}{-cf(ax + by + cz)} - \frac{dz}{-a + bf(ax + by + cz)}$$

using a, b, c as multipliers.

Each fraction $= \dfrac{a\,dx + b\,dy + c\,dz}{0}$

$\therefore a\,dx + b\,dy + c\,dz = 0$

$$\Rightarrow \qquad ax + by + cz = A.$$

Again taking the first two members, we have

$$dx = \frac{dy}{-f(A)}$$

$$\Rightarrow \qquad dy + f(A)\,dx = 0$$

$$\therefore \qquad y + f(A)\,x = A' \text{ (cons)}$$

$$\Rightarrow \qquad y + xf(ax + by + cz) = F(ax + by + cz).$$

which is the required solution. **Ans.**

Example 2:

Solve $y^2r - 2ys + t = p + 6y.$ **(Meerut, 91, 94, 96, 97(P); Kanpur, 93; Delhi Hons 94)**

Solution:

We have $y^2r - 2ys + t = p + 6y.$

Putting $r = \dfrac{dp - s\,dy}{dx}$ and $t = \dfrac{dq - s\,dx}{dy}$,

in the given equation, we have

$$y^2 . \frac{dp - s\,dy}{dx} - 2ys + \frac{dq - s\,dx}{dy} = p + 6y.$$

$\Rightarrow$ $[y^2\,dp\,dy - (p + 6y)\,dx\,dy + dq\,dx] - s[y^2dy^2 + 2ydx\,dy + dx^2] = 0$

Hence the Monge's subsidiary equations are

$$y^2\,dp\,dy - (p + 6y)\,dx\,dy + dq\,dx = 0. \qquad ...(1)$$

and $\quad y^2\,dy^2 + 2y\,dx\,dy + dx^2 = 0 \qquad ...(2)$

From (2), we have $(y\,dy + dx)^2 = 0$

$\therefore y\,dy + dx = 0 \qquad ...(3)$

Integrating, $y^2 + 2x = A \qquad ...(4)$

$\therefore$ From (1) and (3), we have

$y\,dp + (p + 6y)\,dy - dq = 0$

$\Rightarrow$ $(y\,dp + p\,dy) + 6y\,dy - dq = 0$

$\therefore py + 2y^3 - q = B. \qquad ...(5)$

$\therefore$ From (4) and (5) the intermediate integral is

$py + 3y^2 - q = f(y^2 + 2x)$

$\Rightarrow$ $py - q = = 3y^2 + f(y^2 + 2x)$

Lagrange's Subsidiary Equations are

$$\frac{dx}{y} = \frac{dy}{-1} = \frac{dz}{-3y^2 + f(y^2 + 2x)}$$

From the first two members, we have

$y\,dy + dx = 0 \qquad \therefore \quad y^2 + 2x = A.$

Again from the second and third members, we have

$dz = [3y^2 - f(y^2 + 2x)]\,dy$

$\Rightarrow$ $dz = 3y^2\,dy - f(A)\,dy$

$\therefore z = y^3 - yf(A) + B$

$\therefore z = y^3 - yf(y^2 + 2x) + F(y^2 + 2x).$

which is the required solution. **Ans.**

Example 3:

Solve $r = a^2t$. **(Agra, 94; Meerut, 88, 90 (P), 96 (P), 97; IAS 88; Garhwal 92; Delhi Hons. 95; Rohilkhand, 89, 94)**

Solution:

We have $r = a^2t$.

We know that

$$dq = \frac{\partial q}{\partial x} dx + \frac{\partial p}{\partial y} dy = r\,dx + s\,dy,$$

and $$dq = \frac{\partial q}{\partial x} dx + \frac{\partial p}{\partial y} dy = s\,dx + t\,dy,$$

$$\therefore\ r = \frac{dp - s\,dy}{dx} \text{ and } t = \frac{dq - s\,dx}{dy}$$

Substituting these values in the given equation we have

$$\frac{dp - s\,dy}{dx} = a^2 \frac{dq - s\,dx}{dy}$$

$$\Rightarrow \quad (dp\,dy - a^2\,dq\,dx) - s\,(dy^2 - a^2\,dx^2) = 0$$

$\therefore$ The Monge's subsidiary equations are.

$$dp\,dy - a^2\,dq\,dx = 0 \qquad ...(1)$$

and $$dy^2 - a^2\,dx^2 = 0 \qquad ...(2)$$

From equation (2), we have

$$dy - a\,dx = 0 \qquad ...(3)$$

and $$dy + a\,dx = 0. \qquad ...(4)$$

From equations (1) and (3), we have

$$dp\,(a\,dx) = a^2\,dq\,dx = 0$$

$$\Rightarrow \quad dp - a\,dx = 0. \qquad ...(5)$$

From equation (3) and (5), we obtain

$$y - ax + a_1 \text{ and } p - aq = b_1$$

$$\therefore \quad p - aq = f_1(y = ax) \qquad ...(6)$$

is an intermediate integral.

Similarly, from equations (4) and (1) we obtain the second intermediate integral

$$p + aq = f_2(y + ax). \qquad ...(7)$$

Solving equation (6) and (7) we have

$$p = \frac{1}{2} f_1(y - ax) + \frac{1}{2} f_2(y + ax)$$

and $$q = \frac{1}{2a} f_2(y + ax) - \frac{1}{2a} f_1(y - ax).$$

Substituting these values in dz = p dx + q dy, we have

$$dz = \left[\frac{1}{2} f_1(y - ax) + \frac{1}{2} f_2(y + ax)\right] dx$$

$$+ \left[\frac{1}{2a} f_2(y + ax) - \frac{1}{2a} f_1(y - ax)\right] dy$$

$$= \frac{1}{2a} f_2(y = ax)\ (dy + a\ dx) - \frac{1}{2a} f_1(y - ax)\ (dy - a\ dx).$$

Integrating, we have $z = \dfrac{1}{2a} f_2(y + ax) - \dfrac{1}{2a} f_1(y - ax)$

$\Rightarrow$ $$z = F_2(y + ax) = F_1(y = ax).$$ **Ans.**

Example 4:

Solve $(1 + q)^2 r - 2(1 + p + q + pq) s + (1 + p)^2 t = 0$. **(Kanpur, 94)**

Solution:

We have $(1 + q)^2 r - 2(1 + p + q + pq) s + (1 + p)^2 t = 0$

Putting $r = \dfrac{dp - s\ dy}{dx}$ and $t = \dfrac{dq - s\ dx}{dy}$ in the given equation we have

$$(1 + q)^2 . \frac{dp - s\ dy}{dx} - 2(1 + p + q + pq) s + (1 + p)^2 . \frac{dq - s\ dx}{dy} = 0$$

$\Rightarrow$ $$\{(1 + q)^2\ dp\ dy + (1 + p)^2\ dq\ dx\} - s\ \{(1 + p)^2\ dy^2$$
$$+ 2(1 + p)(1 + q)\ dx\ dy + (1 + p)^2\ dx^2\} = 0.$$

Hence the Monge's subsidiary equations are

$$(1 + q)^2\ dp\ dy + (1 + p)^2\ dq\ dx = 0 \qquad ...(1)$$

and $$(1 + q)^2\ dy^2 + 2(1 + p)(1 + q)\ dx\ dy + (1 + p)^2\ dx^2 = 0 \qquad ...(2)$$

From equation (2), we have

$$[(1 + q)\ dy + (1 + p)\ dx]^2 = 0$$

$\Rightarrow$ $$(1 + q)\ dy + (1 + p)\ dx = 0 \qquad ...(3)$$

$\Rightarrow$ $$dx + dy + p\ dx + q\ dy = 0$$

$\Rightarrow$ $$dx + dy + dx = 0 \qquad \because\ dz = p\ dx + q\ dy$$

$\therefore$ $$x + y + z = A. \qquad ...(4)$$

Again from (1) and (3), we have

$$(1 + q)\ dp - (1 + p)\ dq = 0$$

$\Rightarrow$ $$\frac{dp}{1 + p} - \frac{dq}{1 + q} = 0.$$

Integrating $\log(1+p) - \log(1+q) = \log B$.

$\Rightarrow \quad (1+p) = B(1+q).$ ...(5)

From (4) and (5) the intermediate integral is

$(1+p) = (1+q) f(x+y+z)$

$\Rightarrow \quad p - q\, fx + y + z) = f(x+y+z) - 1$

$\therefore$ The Lagrange's subsidiary equations are

$$\frac{dx}{1} = \frac{dy}{-f(x+y+z)} = \frac{dz}{f(x+y+z)-1} = \frac{dx+dy+dz}{0}$$

$\therefore \quad dx + dy + dz = 0 \quad \Rightarrow \quad x + y + z = A.$

Taking the first two members, we have

$dy = -f(A)\,dx$

$\therefore y = -xf(A) + B.$

$\therefore y + xf(x+y+z) = F(x+y+z)$

Which is the complete integral. **Ans.**

Example 5:

Solve $t - r\sec^4 y = 2q\tan y$.

(Meerut, 91(P) 92(P), 93, 95(BP); Agra, TDC 91; Delhi, Hons, 95)

Solution:

We have $t - r\sec^4 y = 2q\tan y$.

Putting $r = \dfrac{dp - s\,dy}{dx}$ and $t = \dfrac{dq - s\,dy}{dy}$, the given equation gives

$$\frac{dp - s\,dy}{dy} - \frac{dq - s\,dy}{dx}\sec^4 y = 2q\tan y$$

$\Rightarrow \quad (dq\,dx - dp\,dy\sec^4 y = 2q\tan y\,dx\,dy) - s(dx^2 - dy^2\sec^4 y) = 0.$

The Monge's subsidiary equations are

$dq\,dx - dp\,dy\sec^4 y - 2q\tan y\,dx\,dy = 0$ ...(1)

and $\quad dx^2 - dy^2\sec^4 y = 0$ ...(2)

from equation (2), we have $\quad dx - dy\sec^2 y = 0$ ...(3)

and $\quad dx + dy\sec^2 y = 0$ ...(4)

from equations (1) and (3), we have

$dq - dp\sec^2 y - 2q\tan y\,dy = 0$

$\Rightarrow \quad dp - (\cos^2 y\,dq - 2q\sin y\cos y\,dy) = 0.$

Integrating, $p - q \cos^2 y = A$

Also integrating equation (3), we have $x - \tan y = B$.

$$\therefore \quad p - q \cos^2 y = f_1(x - \tan y) \qquad ...(5)$$

is an intermediate integral.

Similarly from (1) and (4), the other intermediate integral is

$$p + q \cos^2 y = f_2(x + \tan y). \qquad ...(6)$$

Solving equations (5) and (6), we have

$$p = \frac{1}{2} [f_1(x - \tan y) + f_2(x + \tan y)]$$

and $$q = \frac{1}{2} \sec^2 y \, [f_1(x + \tan y) - f_1(x - \tan y)].$$

Substituting these values in

$dz = p\,dx + q\,dy$, we have

$$dz = \frac{1}{2} [f_1(x - \tan y) + f_2(x + \tan y)]\, dx$$
$$+ \frac{1}{2} \sec^2 y \, [f_2(x + \tan y) - f_1(x = \tan y)]\, dy$$
$$= \frac{1}{2} f_1(x - \tan y)(dx - \sec^2 y\, dy)$$
$$+ \frac{1}{2} f_2(x + \tan y)(dx + \sec^2 y\, dy)$$

$$\therefore \quad z = f_1(x + \tan y) + f_2(x + \tan y).$$ **Ans.**

Example 6:

Solve $xy(t - r) + (x^2 - y^2)(s + 2) = py - qx.$

Solution:

We have $xy(t - r) + (x^2 - y^2)(s + 2) = py - qx$.

The given equation can be written as

$$xyr - (x^2 - y^2)s - xyt = 2(x^2 - y^2) - py + qx.$$

Putting $r = \dfrac{dp - s\,dy}{dx}$ and $t = \dfrac{dq - s\,dx}{dy}$, we have

$$xy\frac{dp - s\,dy}{dx} - (x^2 - y^2)s - xy\frac{dq - s\,dx}{dy}$$
$$= 2(x^2 - y^2) - py + qx$$

$$\Rightarrow \quad xy\,dp\,dy - xy\,dq\,dx - \{2(x^2 - y^2) + py - qx\}\,dx\,dy$$

$$-s\{xy\ dy^2 + (x^2 - y^2)\ dx\ dy - xy\ dx^2\} = 0$$

Hence, Monge's Subsidiary Equations are

$$xy\ dp\ dy - xy\ dq\ dx - \{2\{x^2 - y^2) + py - qx\}\ dx\ dy = 0. \quad ...(1)$$

and $\quad xy\ dy^2 + (x^2 - y^2)\ dx\ dy - xy\ dx^2 = 0. \quad ...(2)$

equation (2) give $\quad x\ dx + y\ dy = 0 \quad ...(3)$

and $\quad x\ dy - y\ dx = 0. \quad ...(4)$

From equations (3), we have $x^2 + y^2 = A$.

From equations (1) and (3), we have

$$x\ dp + y\ dp - 2x\ dy - 2y\ dx + p\ dx + q\ dy = 0$$

$$\Rightarrow \quad (x\ dp + p\ dx) + (y\ dq + q\ dy) - 2(x\ dy + y\ dx) = 0$$

$\therefore$ $px + qy - 2xy = B$.

$\therefore$ One intermediate integral is

$$px + qy - 2xy = f_1(x^2 + y^2). \quad ...(5)$$

Similarly from (1) and (4) the second intermediate integral is

$$-py + qx + x^2 - y^2 = f_2\left(\frac{y}{x}\right).$$

Solving (5) and (6), we have

$$p = y + \frac{1}{(x^2 + y^2)}\left\{xf_1\ (x^2 + y^2) - yf_2\left(\frac{y}{x}\right)\right\}$$

and $$q = x + \frac{1}{(x^2 + y^2)}\left\{yf_1\ (x^2 + y^2) - xf_2\left(\frac{y}{x}\right)\right\}.$$

Substituting these values in

$dz = p\ dx + q\ dy$, we have

$$dz = \left[y + \frac{1}{(x^2 + y^2)}\left\{yf_1\ (x^2 + y^2) - yf_2\left(\frac{y}{x}\right)\right\}\right] dx$$

$$+ \left[x + \frac{1}{(x^2 + y^2)}\left\{yf_1\ (x^2 + y^2) + xf_2\left(\frac{y}{x}\right)\right\}\right] dy$$

$$= (y\ dx + x\ dy) + f_1(x^2 + y^2).\ \frac{x\ dx + y\ dy}{x^2 + y^2}$$

$$+ f_2\left(\frac{y}{x}\right).\ \frac{x\ dx + y\ dy}{x^2 + y^2}$$

$$= (y\,dx + x\,dy) + \frac{1}{2}\,f_1(x^2 + y^2).\,d\,\{\log(x^2 + y^2)\}$$

$$+ f_2\left(\frac{y}{x}\right).d\left\{\tan^{-1}\frac{y}{x}\right\}$$

$$\therefore z = xy + \frac{1}{2}\,f_1(x^2 + y^2)\,d\,\{\log(x^2 + y^2)\} + f_2\left(\frac{y}{x}\right)d\left\{\tan^{-1}\frac{y}{x}\right\}$$

$$\Rightarrow \quad z = xy + f_1(x^2 + y^2) + \phi_2\left(\frac{y}{x}\right)$$

which is the required solution. **Ans.**

Example 7:

Solve $pt - qs = q^3$.

(Meerut, 93(P), 94(P), 96(BP), 97(BP); Rohilkhand, 94; Agra, 89; Kanpur, 87)

Solution:

We have $pt - qs = q^3$.

Putting $t = \dfrac{dq - s\,dx}{dy}$, the given equation becomes

$$\frac{p(dq - s\,dx)}{dy} - qs = q^3$$

$$\Rightarrow \quad (p\,dq - q^3\,dy) - s(p\,dx + q\,dy) = 0.$$

The Monge's subsidiary equations are

$$p\,dq - q^3\,dy = 0. \quad ...(1)$$

and

$$p\,dx + q\,dy = 0. \quad ...(2)$$

From equations (2), we have $dz = 0$. $\therefore$ $z = A$. ...(3)

Combining equations (1) and (2), we have

$$dq + q^2\,dx = 0$$

$$\Rightarrow \quad \frac{dq}{q^2} + dx = 0 \qquad \therefore \quad -\frac{1}{q} + x = B$$

$$\Rightarrow \quad -\frac{1}{q} + x = f(z)$$

$$\Rightarrow \quad \frac{\partial y}{\partial x} - x = -f(z)$$

Integrating, $y = xz - \int f(z)\, dz + c.$

$\therefore \quad y = xz + f_1(z) + c.$

and $\quad y + xz = f_1(z) + f_2(x)$

Since c is a function of x which is regarded constant at the time of integration.

Example 8:

Obtain the integral of $q^2r - 2pqs + p^2t = 0$ in the form

$$y + x\, f(z) = F(z)$$

and show that this represents a surface generated by straight lines that are parallel fixed plane.

(Meerut, 89, 94, 95; Rohilkhand, 90; Kanpur, 88)

Solution:

Putting $r = \dfrac{dp - s\, dy}{dx}$ and $t = \dfrac{dq - s\, dx}{dy}$

in the given equation, we have

$$q^2 \frac{dp - s\, dy}{dx} - 2p\, qs + p^2 \frac{dq - s\, dx}{dy} = 0$$

$\Rightarrow \quad (q^2\, dp\, dy + p^2\, dq\, dx) - s\,(q^2\, dy^2 + 2pq\, dx\, dy + p^2\, dx^2) = 0.$

$\therefore$ The Monge's subsidiary equations are

$$q^2\, dy^2 + 2p^2\, 2pq\, dx = 0 \qquad ...(1)$$

and $$q^2\, dp\, dy + p^2\, dq\, dx = 0 \qquad ...(2)$$

(1) gives, $(q\, dy + p\, dx)^2 = 0$

$\Rightarrow \quad p\, dx + q\, dy = 0 \qquad ...(3)$

$\therefore \quad dz = p\, dx + q\, dy = 0$

$\Rightarrow \quad z = A. \qquad ...(4)$

from (2) and (3), we have

$q\, dp - p\, dq = 0$

$\Rightarrow \quad \dfrac{dp}{p} = \dfrac{dq}{q}$

$\Rightarrow \quad \log p = \log q + \log B$

$\therefore \quad p = pB. \qquad ...(5)$

Hence, the intermediate integral is

$$p - q\, f(z) = 0$$

$\therefore$ Lagrange's subsidiary equations are

$$\frac{dx}{1} = \frac{dy}{-f(z)} = \frac{dz}{0}.$$

The last member gives, $dz = 0$ $\therefore z = c$

Again from the first two members, we have

$$dy + f(z)\, dx = 0$$

$$\Rightarrow \quad dy + f(c)\, dx = 0$$

Integrating, we have $y + xf(c)\, dx = 0$

$$\Rightarrow \quad y + xf(z) = F(z) \qquad ...(6)$$

which is the required integral.

The integral of the given differential equation is the surface (6) which is the locus of the straight lines given by the intersection of planes.

$$y + x\, f(c) = F(c) \text{ and } z = c.$$

These lines are all parallel to the planes $z = 0$ as they lie on the plane $z = 0$ for varying values of c. **Ans.**

Example 9:

Solve $q(1 + q)\, r - (p + q + 2pq)\, s + p(1 + p)\, t = 0.$

Solution:

We have $q(1 + q)\, r - (p + q + 2pq)\, s + p(1 + p)\, t = 0.$

Putting $r = \dfrac{dp - s\,dy}{dx}$ and $t = \dfrac{dq - s\,dx}{dy}$ in the given equation, we have

$$q(1 + q).\frac{dp - s\,dy}{dx} - (p + q + 2qp)\, s + p\,(1 + p).\frac{dq - s\,dx}{dy} = 0$$

$$\Rightarrow \quad \{q(1 + q)\, dp\, dy + p(1 + p)\, dq\, dx\} - s\,\{q(1 + q)\, dy^2 + (p + q + 2qp)\, dx\, dy + p(1 + p)\, dx^2\} = 0.$$

Hence Monge's Subsidiary Equations are

$$q(1 + q)\, dp\, dy + p(1 + p)\, dq\, dx = 0 \qquad ...(1)$$

and $$q(1 + q)\, dy^2 + (p + q + 2qp)\, dx\, dy + p(1 + p)\, dx^2 = 0 \qquad ...(2)$$

equation (2) gives, $$p\, dx + q\, dy = 0 \qquad ...(3)$$

and $$(1 + p)\, dx + (1 + q)\, dy = 0 \qquad ...(4)$$

from equation (3), we have $dz = p\, dx + q\, dy = 0$ $\quad\therefore\quad z = A.$

Also from (1) and (3), we have

$$p(1 + 1)\, dp - p(1 + p)\, dq = 0$$

$$\Rightarrow \quad \frac{dp}{1+p} - \frac{dq}{1+q} = 0.$$

$$\therefore \quad \log(1 + p) - \log(1 + q) = \log B$$

$$\Rightarrow \quad (1 + p) = (1 + q)\, f_1(z) \qquad ...(5)$$

which is one intermediate integral.

Now from (4) we have $dx + dy + (p\, dx + a\, dy) = 0$

$$\Rightarrow \quad dx + dy + dz = 0. \quad \therefore \quad x + y + z = A'$$

from (1) and (4) we have

$$q\, dp - p\, dq = 0$$

$$\Rightarrow \quad \frac{dp}{p} = \frac{dq}{q}$$

$$\Rightarrow \quad \log p = \log q + \log B'$$

$$\therefore \quad p = qf_2(x + y + z). \qquad ...(6)$$

Solving (5) and (6), we have

$$p = \frac{(f_1 - 1)\, f_2}{f_2 - f_1} \text{ and } q = \frac{f_1 - 1}{f_2 - f_1}.$$

Substituting these values in $dz = p\, dx + q\, dy$, we have

$$dz = \frac{(f_1 - 1)\, f_2}{f_2 - f_1} dx + \frac{f_1 - 1}{f_2 - f_1} dy$$

$$\Rightarrow \quad (f_2 - f_1)\, dz = (f_1 - 1)\, f_2\, dx + (f_1 - 1)\, dy$$

$$\Rightarrow \quad (f_2 - f_1)\, dz = (f_1 - 1)\, f_2\, dx + (f_1 - 1)(dx + dy + dz) - (f_1 - 1)(dx + dz)$$

$$\Rightarrow \quad (f_2 - 1)\, dz = (f_1 - 1)(f_2 - 1)\, dx + (f_1 - 1)(dx + dy + dz)$$

$$\Rightarrow \quad \frac{dz}{f_1 - 1} = dx + \frac{dx + dy + dz}{f_2 - 1}$$

$$\Rightarrow \quad \frac{dz}{f_1\, z - 1} = dx + \frac{dx + dy + dz}{f_2(x + y + z) - 1}$$

Integrating, we have

$$\phi_1(z) = x + \phi_2(x + y + z)$$

which the required solution. **Ans.**

Example 10:

Solve (q = 1) s = (p + 1) t. **(Meerut, 91)**

Solution:

We have (q = 1) s = (p + 1) t.

Putting $t = \dfrac{dq - s\,dx}{dy}$, in the given equation, we have

$$(q + 1)\,s = (p = 1)\,\frac{(dq - s\,dx)}{dy}$$

$\Rightarrow$ $[(p + 1)\,dq] - s\,[(q + 10\,dy + (p + 1)\,dx] = 0$

Hence, Monge's Subsidiary Equations are

$$(p + 1)\,dq = 0 \qquad ...(1)$$

and $(q + 1)\,dy + (p + 1)\,dx = 0$ ...(2)

From equation (2) we have

$$dx + dy + (p\,dx + q\,dy) = 0$$

$\Rightarrow$ $dx + dy + dz = 0, \quad \because \quad dz = p\,dx + q\,dy$

Integrating, $x + y + z = A$ ...(3)

from equation (1) we have $dq = 0$

Integrating, $q = B$ ...(4)

$\therefore$ One intermediate integral is

$$q = f(x + y + z)$$

$\Rightarrow$ $\dfrac{\partial z}{\partial y} = f(x + y + z)$ ...(5)

Integrating (5) partially w.r.t y, taking x as a constant, we have

$$z = F(x + y + z) + \phi(x).$$

where F and ϕ are arbitrary functions.

Example 11:

Solve $r - t\cos^2 x + p\tan x = 0.$

(Meerut, 93(P), 93(P); B.H.U. 88; Rohilkhand, 86, 91; Agra, 91)

Solution:

We have Solve $r - t\cos^2 x + p\tan x = 0.$

Putting $r = \dfrac{dp - s\,dy}{dx}$ and $t = \dfrac{dq - s\,dx}{dy}$, in the given equation, we have

$$\frac{dp - s\,dy}{dx} - \frac{dq - s\,dx}{dy} \cos^2 x + \tan x = 0$$

$\Rightarrow$ $(dp\,dy - \cos^2 x\,dq\,dx + p \tan x\,dx\,dy) - s(dy^2 - \cos^2 x\,dx^2) = 0.$

Hence the Monge's subsidiary equations are

$$dp\,dy - \cos^2 x\,dq\,dx + p \tan x\,dx\,dy = 0 \qquad ...(1)$$

and $$dy^2 - \cos^2 x\,dx^2 = 0 \qquad ...(2)$$

(2) gives $dy - \cos x\,dx = 0$ ...(3)

and $dy + \cos x\,dx = 0$ ...(4)

From (3), we have $y - \sin x = A$.

From (1) and (4), we have

$$dp - \cos x\,dq + p \tan x\,dx = 0$$

$\Rightarrow$ $\sec x\,dp + p \tan x \sec x\,dx - dq = 0.$

Integrating $p \sec x - q = B$.

$\therefore$ $p \sec x - q = f_1(y - \sin x)$ ...(5)

which is one intermediate integral is

Similarly the other intermediate integral is

Solving (4) and (6), we have

$$p = \frac{1}{2} \cos x\,[f_1(y - \sin x) + f_2(y + \sin x)].$$

and $$q = \frac{1}{2} [-f_1(y - \sin x) + f_2(y + \sin x)].$$

Substituting these values in $z = p\,dx + q\,dy$, we have

$$dz = \frac{1}{2} \cos x\,[f_1(y - \sin x) + f_2(y + \sin x)]\,dx + \frac{1}{2} \cos x\,[f_1(y - \sin x) + f_2(y + \sin x)]\,dy$$

$$dz = \frac{1}{2} f_1(y - \sin x).\,(dy - \cos x\,dx) + \frac{1}{2} f_1(y + \sin x).\,(dy + \cos x\,dx)$$

$\Rightarrow$ $$dz = \frac{1}{2} f_1(y - \sin x).\,(y - \sin x) + \frac{1}{2} f_1(y + \sin x).\,(y + \sin x)$$

$\therefore \quad z = f_1(y - \sin x) + f_2(y + \sin x)$

which is the complete solution. **Ans.**

Example 12:

Solve $y^2r + 2xys + x^2t + px + qy = 0.$ **(Rohilkhand, 97)**

Solution:

We have $y^2r + 2xys + x^2t + px + qy = 0.$

Putting $r = \dfrac{dp - s\,dy}{dx}$ and $t = \dfrac{dq - s\,dx}{dy}$ in the given equation, we have

$$y^2 \frac{dp - s\,dy}{dx} 2xys + x^2 \frac{dq - s\,dx}{dy} + px + qy = 0$$

$$\Rightarrow \quad \{y^2\,dp\,dy + x^2\,dq\,dx + (px + qy)\,dx\,dy\}$$
$$-s\{y^2\,dy^2 - 2xy\,dx\,dy + x^2\,dx^2\} = 0$$

Hence Monge's subsidiary equations are

$$y^2\,dp\,dy + x^2\,dq\,dx + (pz + qy)\,dx\,dy = 0 \quad ...(1)$$

and $\quad y^2\,dy^2 - 2xy\,dx\,dy + x^2\,dx^2 = 0, \quad ...(2)$

equation (2) gives $\quad (y\,dy - x\,dx)^2 = 0.$

$\Rightarrow \quad y\,dy - x\,dx = 0.$

$\therefore \quad -y^2 + x^2 = A.$

From (1) and (3) we have

$y\,dp + x\,dq + p\,dy + q\,dx = 0$

$\Rightarrow \quad (y\,dp + p\,dy) + (x\,dq + q\,dx) = 0.$

$\therefore \quad yp + xq = B.$

$\therefore$ One intermediate integral is

$yp + xq = f(x^2 - y^2).$

Hence Lagrange's subsidiary equations are

$$\frac{dx}{y} = \frac{dy}{x} = \frac{dz}{f(x^2 - y^2)}.$$

From the first two members, we have

$x\,dx - y\,dy = 0. \qquad \therefore x^2 - y^2 = A.$

Again from the last two members, we have

$$\frac{dx}{y} = \frac{dy}{f(A)} \quad \Rightarrow \quad dz = f(A)\frac{dy}{\sqrt{(y^2 + A)}}$$

$\therefore \quad z = f(A).\log[\log + \sqrt{(A + y^2)}] + A'$

$\Rightarrow \quad z = f(x^2 - y^2)\log(y + x) + F(x^2 - y^2)$

which is the complete integral. **Ans.**

Example 13:

Solve $y^2r - 2ys + t = p + 6y.$ **(Meerut, 91, 94, 96, 97(P); Agra, 92 (P); Kanpur, 83; Delhi, Hons. 94)**

Solution:

We have $y^2r - 2ys + t = p + 6y.$

Putting $r = \dfrac{dp - s\,dy}{dx}$ and $t = \dfrac{dq - s\,dx}{dy}$,

in the given equation, we have

$$y^2 . \frac{dp - s\,dy}{dx} - 2ys + \frac{dq - s\,dx}{dy} = p + 6y.$$

$\Rightarrow \quad [y^2\,dp\,dy - (p + 6y)\,dx\,dy + dq\,dx] - s[y^2dy^2 + 2ydx\,dy + dx^2] = 0$

Hence the Monge's subsidiary equations are

$$y^2\,dp\,dy - (p + 6y)\,dx\,dy + dq\,dx = 0. \qquad ...(1)$$

and $\quad y^2\,dy^2 + 2y\,dx\,dy + dx^2 = 0 \qquad ...(2)$

From (2), we have $(y\,dy + dx)^2 = 0$

$\therefore \quad y\,dy + dx = 0 \qquad ...(3)$

Integrating, $y^2 + 2x = A \qquad ...(4)$

$\therefore$ From (1) and (3), we have

$$y\,dp + (p + 6y)\,dy - dq = 0$$

$\Rightarrow \quad (y\,dp + p\,dy) + 6y\,dy - dq = 0$

$\therefore \quad py + 2y^3 - q = B. \qquad ...(5)$

$\therefore$ From (4) and (5) the intermediate integral is

$$py + 3y^2 - q = f(y^2 + 2x)$$

$\Rightarrow \quad py - q = = 3y^2 + f(y^2 + 2x)$

Lagrange's subsidiary equations are

$$\frac{dx}{y} = \frac{dy}{-1} = \frac{dz}{-3y^2 + f(y^2 + 2x)}$$

From the first two members, we have

$y\,dy + dx = 0 \qquad \therefore \quad y^2 + 2x = A.$

Again from the second and third members, we have

$$dz = [3y^2 - f(y^2 + 2x)]\ dy$$

$$\Rightarrow \quad dz = 3y^2\ dy - f(A)\ dy$$

$$\therefore z = y^3 - yf(A) + B$$

$$\therefore z = y^3 - yf(y^2 + 2x) + F(y^2 + 2x).$$

which is the required solution. **Ans.**

Example 14:

Solve $(x - y)(xr - xs - ys + yt) = (x + y)(p - q)$.

Solution:

We have $(x - y)(xr - xs - ys + yt) = (x + y)(p - q)$.

Putting r $r = \dfrac{dp - s\,dy}{dx}$ and $t = \dfrac{dq - s\,dx}{dy}$, is the given equation, we have

$$x(x - y)\frac{dp - s\,dy}{dx} - (x - y)(x + y)s + y(x - y)\frac{dq - s\,dx}{dy}$$

$$= (x + y)(p - q \quad)$$

$$\Rightarrow \quad \{x(x - y)\,dp\,dy + y(x - y)dq\,dx - (x + y)(p - q)\,dx\,dy\}$$

$$-s\,\{x - y)\,dy^2 + (x - y)(x + y)\,dx\,dy + y(x - y)\,dx^2\} = 0.$$

Hence the Monge's subsidiary equation are

$$x(x - y)\,dp\,dy + y(x - y)\,dq\,dx - (x + y)(p - q)\,dx\,dy = 0 \quad ...(1)$$

and $\quad x\,dy^2 + (x + y)\,dx\,dy + y\,dx^2 = 0 \quad ...(2)$

Equation equation (2) gives $\quad x\,dy + y\,dx = 0 \quad ...(3)$

and $\quad dx + dy = 0 \quad ...(4)$

From equation (3), $xy = A$.

From equations (1) and (3), we have

$$-y(x - y)\,dp + y(x - y)\,dq - (p - q)[-ydx + ydy] = 0$$

$$\Rightarrow \quad (x - y)(dp - dq) - (p - q)(dx - dy) = 0$$

$$\Rightarrow \quad \frac{dp - dq}{p - q} = \frac{dx - dy}{x - y}$$

$$\therefore \log(p - q) = \log(x - y) + \log B$$

$$\Rightarrow \quad p - q = (x - y)B.$$

$\therefore$ One intermediate integral is

$$p - q = (x - y)\, f(xy). \qquad ...(5)$$

The Lagrange's subsidiary equations are

$$\frac{dx}{1} = \frac{dy}{-1} = \frac{dz}{(x - y)\, f(xy)}.$$

From the first two members, we have

$$dx + dy = 0 \qquad \therefore \qquad x + y = A_1.$$

Again we have

$$\text{each fraction} = \frac{(y\,dx + x\,dy)\, f(xy) + dz}{0}$$

$$\therefore \qquad (y\,dx + x\,dy)\, f(xy) + dz = 0$$

$$\Rightarrow \qquad dz + f(xy)\, d\,(xy) = 0$$

Integrating, we get $z = \phi_1(xy) + B_1$

$$\Rightarrow \qquad z = \phi_1(xy) + \phi_2(x + y)$$

which is the complete solution. **Ans.**

6.2 Monge's Method of Integrating

$Rr + Ss + Tt = U(rt - s^2) = V$ **(Rohilkhand, 93; Meerut, 97)**

where R, S, T, U, V are functions of x, y, z, p, q.

We have $dp = \frac{\partial q}{\partial x}.\, dx + \frac{\partial q}{\partial y}\, dy = r\,dx + s\,dy$

and $dq = \frac{\partial q}{\partial x}.\, dx + \frac{\partial q}{\partial y}\, dy = s\,dx + t\,dy$

$$\therefore \qquad r = \frac{dp - s\,dy}{dx} \text{ and } t = \frac{dq - s\,dx}{dy}$$

Putting these values in the equation, we have

$$R.\frac{dp - s\,dy}{dx} + Ss + T.\frac{dq - s\,dx}{dy}$$

$$+ U\left\{\frac{(dp - s\,dy)\,(dq - s\,dx)}{dx\,dy}\right\} = V$$

$$\Rightarrow \qquad (R\,dp\,dy + T\,dq\,dx + U\,dp\,dq - V\,dx\,dy)$$

$$-s(R\,dy^2 - S\,dx\,dy + T\,dx^2 + U\,dp\,dx + U\,dq\,dy) = 0$$

Hence Monge's Subsidiary Equation are

$$L \equiv R\,dp\,dy + T\,dq\,dx + U\,dp\,dq - V\,dx\,dy = 0. \qquad ...(1)$$

and $L \equiv R\,dy^2 - S\,dx\,dy + T\,dx^2 + U\,dp\,dx + U\,dq\,dy = 0$ ...(2)

Now, equation (2) may not be factorised, on account of the presence of the term

$$U\,dp\,dx + U\,dq\,dy.$$

$\therefore$ Let us try to factorise $M + \lambda L = 0$

i.e., $(R\,dy^2 - S\,dx\,dy + T\,dx^2 + U\,dp\,dx + U\,dq\,dy)$
$+ \lambda(R\,dp\,dy + T\,dq\,dx + U\,dp\,dq - V\,dx\,dy) = 0.$ ...(3)

where λ is some multiplier to be determined.

Let the factors of (3) be

$$(R\,dy + mT\,dx + KU\,dp)\left(dy + \frac{1}{m}\,dx + \frac{\lambda}{K}\,dq\right) = 0. \quad ...(4)$$

$\therefore$ Comparing equations (3) and (4), we have

$$\frac{R}{M} + mT = -(S + \lambda V) \quad ...(5)$$

$$K = m \text{ and } \frac{R\lambda}{K} = U.$$

From the last two, we have $m = \dfrac{R\lambda}{U}$.

$\therefore$ From equation (5), we have

$$\frac{U}{\lambda} + \frac{P\lambda T}{U} = -(S + \lambda V)$$

$\Rightarrow$ $\lambda^2(UV + RT + lUS + U^2 = 0.$ ...(6)

Let λ_1 and λ_2 be the values of λ from (6).

When $\lambda = \lambda_1$, $m = \dfrac{R\lambda_1}{U}$.

$\therefore$ From (4), we have

$$\left(R\,dy = \frac{R\lambda_1}{U}\,T\,dx + R\lambda_1\,dp\right)\left(dy + \frac{U}{R\lambda_1}\,dx + \frac{U}{R}\,dq\right) = 0$$

$\Rightarrow$ $(U\,dy + \lambda_1\,T\,dx + \lambda_1\,U\,dp)(U\,dx + \lambda_1\,R\,dy + \lambda_1\,U\,dq) = 0$ ...(7)

Similarly, when $l = l_2$, $m = \dfrac{R\lambda_2}{U}$, from (4), we have

$(U\,dy + \lambda_2\,T\,dx + \lambda_2\,U\,dp)(U\,dx + \lambda_2\,R\,dy + \lambda_2\,U\,dq) = 0$ ...(8)

Now, one factor of (7) is combined with one factor of (8) to give an intermediate integral and similarly other pair gives another intermediate integral.

Now, let us suppose that two integrals $u_1 = a_1$, $v_1 = b_1$ are obtained from the equations.

$$\left.\begin{aligned} &\mathbf{U\ dy + \lambda_1\ T\ dx + \lambda_1\ U\ dp = 0} \\ \text{and} \quad &\mathbf{U\ dx + \lambda_2\ R\ dy + \lambda_2\ U\ dq = 0} \end{aligned}\right\} \quad ...(9)$$

and similarly let $u_2 = a_2$, $v_2 = b_2$ be two integrals obtained from the equations

$$\left.\begin{aligned} &\mathbf{U\ dy + \lambda_2\ T\ dx + \lambda_2\ U\ dp = 0} \\ \text{and} \quad &\mathbf{U\ dx + \lambda_1\ R\ dy + \lambda_1\ U\ dq = 0} \end{aligned}\right\} \quad ...(10)$$

Thus, two intermediate integrals are

$$u_1 = f_1(v_1)$$

and $$u_2 = f_2(v_2).$$

Solving them obtain the values of p and q and substitute in

$$dz = p\ dx + q\ dy.$$

The complete integral is obtained by solving the above relation.

Notes:

1. When the two values of λ are equal we shall get only one intermediate integral say u_1 f_1 (v) which together with one of the integrals $u_1 = a_1$ and $v_1 = b_1$ will give values of p and q suitable to solve
   $$dz = p\ dx + q\ dy.$$
2. If suitable values of p and q may not be obtained from the two intermediate integrals $u_1 = f_1(v_1)$ and $u_2 = f_2(v_2)$, for integration in $dz = p\ dx + q\ dy$, we may take one of the intermediate integrals say $u_1 = f_1\ (v_1)$ and one of the integrals from $u_2 = a_2$ and $v_2 = b_2$.
3. An integral of a more general form can be obtained by supposing linear. Supposing
   $$u_1 = mv_1 = n,$$
   the integrating by Lagrange's method we get the solution of the given equation.

Ans.

Example 1:

$5r + 6s + 3t + 2(rt - s^2) + 3 = 0.$ **(Kanpur, 90)**

Solution:

We have $5r + 6s + 3t + 2(rt - s^2) + 3 = 0.$

Comparing it with the equation

$Rr + Ss + Tt + U(rt - s^2) = V$, we have

$R = 3, S = 6m\ T = 3, U = 2, V = -3.$

From λ equation $\lambda^2 (UV + RT) + \lambda US + U^2 = 0$, we have

$$9\lambda^2 + 12\lambda + 4 = 0$$

$$(3\lambda + 2)^2 = 0 \qquad \therefore \qquad \lambda_1 = \lambda_2 = -\frac{2}{3}.$$

Putting these values in

$$U\ dy + \lambda T\ dx + \lambda U\ dp = 0$$

and $\quad U\ dx + \lambda R\ dx + \lambda U\ dq = 0.$

and simplifying, we have

$$3dy - 3dx - 2dp = 0 \text{ and } -5dy + 3dx - 2dq = 0.$$

$$\therefore \qquad 3y - 3x - 2p = A$$

and $\quad -5y + 3x - 2q = B$

$$\therefore \quad p = \frac{1}{2}(3y - 3x - A)$$

and $\quad q = \frac{1}{2}(-5y - 3x - B) = A$

Substituting in $dz = p\ dx + q\ dy$, we have

$$dz = \frac{1}{2}(3y - 3x - A)\ dx + \frac{1}{2}(-5y + 3x - B)\ dy$$

$$\Rightarrow \qquad 2dz = 3(y\ dx + x\ dy) - 3x\ dx - 5y\ dy - A\ dx - B\ dy$$

$$\therefore \quad 2z = 3xy - \frac{3}{2}x^2 - \frac{5}{2}y^2 - Ax - By + c$$

$$\Rightarrow \qquad 4z = 6xy - 3x^2 - 5y^2 - 2Ax - 2By + c.$$

which is the required general solution.

Aliter. To get the more general integral, let us take

$$3y - 3x - 2p = (-5y + 3x - 2q)\ m + n \text{ where } m, n \text{ are constants}$$

$$\Rightarrow \qquad 2p - 2qm = 3y - 3x + 5ym - 3mx - n.$$

Lagrange's Auxiliary Equations are

$$\frac{dx}{2} = \frac{dy}{-2m} = \frac{dz}{3y - 3x + 4ym - 3mx - n}$$

From the first two members, we have $y + mx = a$.

Again we have $\dfrac{dx}{2} = \dfrac{\frac{3x}{2}dx + \frac{5}{2}dy + dz}{3y - 3mx - n}$

$$\Rightarrow \qquad 3x\ dx + 5y\ dy + 2dz = (3y - 3mx - n)\ dx$$

$$\Rightarrow \quad 2dz = -3x\,dx - 5y\,dy + [3(a - mx) - 3mx - n]\,dx$$
$$\Rightarrow \quad 2dz = -3x\,dx - 5y\,dy + (3a - 6mx - n)\,dx$$
$$\therefore \quad 4z = -3x^2 - 5y^2 + 6ax - 6mx^2 - 2nx + b$$
$$\Rightarrow \quad 4z = -3x^2 - 5y^2 + 6x(y + mx) - 6mx^2 - 2nx + b$$
$$\Rightarrow \quad 4z = 6xy - 3x^2 - 5y^2 - 2nx + \phi(y + mx).$$ **Ans.**

Example 2:

Solve $r + 3s + t\ 3s + t\ (rt - s^2) = 1$.

(Rohilkhand, 92, 85; Meerut, 93(P), 87; Garhwal, 97)

Solution:

We have $r + 3s + t\ 3s + t\ (rt - s^2) = 1$.

Comparing it with the equation.

$$Rr + Ss + Tt + U(rt - s^2) = V.$$

we have $R = 1$, $S = 3$, $T = 1$, $U = 1$, $V = 1$.

From λ equation, $\lambda^2 (UV + RT) + Us + U^2 = 0$ we have

$$2\lambda^2 + 3\lambda + 1 = 0$$

$$\Rightarrow \quad (2\lambda + 1)(\lambda + 1) = 0 \qquad \therefore \quad \lambda_1 = -1, \lambda_2 = -\frac{1}{2}.$$

Hence first intermediate integral is give by

$$U\,dy + \lambda_1 T\,dx + \lambda_1 U\,dp = 0$$

and
$$U\,dx + \lambda_2 R\,dy + \lambda_2 U\,dq = 0$$

$$\Rightarrow \quad dy - dx - dp = 0$$

and
$$dx - \frac{1}{2}dy - \frac{1}{2}dq = 0$$

$$\Rightarrow \quad dy - dx - dp = 0$$

and
$$-2dx + dy + dq = 0$$

$$\therefore \quad y = x - p = A \text{ and } -2x + y\ q = B.$$

so the second intermediate integral is

$$2y - x - p = f_2(-y + x - q) = f_2(\beta), \qquad ...(2)$$

$$\therefore \quad \alpha + \beta = -x \qquad \text{where } \alpha = y - 2x + q.$$

$$\Rightarrow \quad d\alpha + d\beta = -dx$$

and from (2) and (1),

$$y = f_2(\beta) - f_1(\alpha)$$

$\Rightarrow \quad dy = f_2'(\alpha)\, d\beta - f_1'(\alpha)\, d\alpha$

and $\quad p = y - x - f_1(\alpha),\ q = x - y - \beta.$

Substituting in $dz = p\, dx + q\, dy$, we have

$$dz = [y - x - f_1(\alpha)]\, dx + (x - y - \beta)\, dy$$

$$\Rightarrow \quad dz = (y - x)\, dx - f_1(\alpha)\, dx + (x - y)\, dy - \beta\, dy$$

$$= -(x - y)(dx - dy) - f_1(\alpha)(-d\alpha - d\beta)$$

$$-\beta[f_2'(\beta)\, d\beta - f_1'(\alpha)\, d\alpha)$$

$$= -(x - y)(dx - dy) + f_1(\alpha)\, d\alpha - bf_2'(\beta)\, d\beta$$

$$+\{f_1(\alpha)\, d\beta + bf_1'(\alpha)\, d\alpha$$

Integrating, we have

$$z = -\frac{1}{2}(x - y)^2 + \int f_1(\alpha)\, d\alpha - \int \beta f'_2(\beta)\, d\beta + \beta f_1(\alpha)$$

$$= \frac{1}{2}(x - y)^2 + \phi_1(\alpha) - bf_2(\beta) + \int f_2(\beta)\, d\beta + \beta f_1(\alpha)$$

$$= \frac{1}{2}(x - y)^2 + \phi_1(\alpha) - \beta f_2(\beta) + \phi_2(\beta) + bf_1(\alpha)$$

which is the required solution. **Ans.**

Example 3:

Solve $qr + (p + x)s + yt + y(rt - s^2) + q = 0.$

(Rohilkhand, 92)

Solution:

We have $qr + (p + x)s + yt + y(rt - s^2) + q = 0.$

Comparing it with the equation

$$Rr + Ss + Tt + U(rt - s^2) = V.$$

we have $R = q.\ S = p + x,\ T = y,\ U = y,\ V = -q.$

$\therefore$ From λ equation, $\lambda^2(UV + RT) + \lambda US + U^2 = 0$

$$\Rightarrow \quad \lambda = \frac{y}{(p + x)}$$

Substituting this in equation

$$Udy + \lambda Tdx + \lambda\, Udp = 0$$

we have $y\,dy - \dfrac{y^2}{p + x}\, dx - \dfrac{y^2}{p + x}\, dp = 0$

$$\Rightarrow \qquad \frac{dy}{y} = \frac{dx + dp}{x + p} \qquad ...(1)$$

$$\therefore \qquad \log y + \log (p + x) + \text{const.}$$

$$\therefore \qquad \frac{(p + x)}{y} = a \text{ (const.)} \qquad ...(2)$$

Also from $Rdy^2 - Sdx\, dy + Tdx^2 + Udp\, dx + Udq\, dy = 0$

we have $q\, dy^2 - (p + x)\, dx\, dy + ydx^2 = y\, dp\, dx + y\, dq\, dy = 0$

$$\Rightarrow \qquad dy\,(q\, dy + y\, dq) = dx\,[(p + x)\, dy - y\,(dx + dp)]$$

$$\Rightarrow \qquad dy\,(q\, dy + y\, dq) = 0 \text{ using (1)}$$

$$\Rightarrow \qquad q\, dy + y\, dq = 0 \quad \therefore \quad qy = b.$$

$\therefore$ The intermediate integral is

$$qy = f\left(\frac{p + x}{y}\right).$$

Charpit's auxiliary equations are

$$\frac{dx}{\frac{1}{y}. f'\left(\frac{p + x}{y}\right)} = \frac{dy}{y} = \frac{dp}{\frac{1}{y}. f'\left(\frac{p + x}{y}\right)}$$

Taking the first and the last members, we have

$$dx + dp + 0 \quad \therefore \; p + x = c.$$

From $dz = p\, dx + q\, dy$, we have

$$dz = (c - x)\, dx + (1/y)\, f(c/y)\, dy$$

$$\therefore \; z = cx - \frac{1}{2}\, x^2 + \phi(c/y) + \psi(c)$$

which is the required solution. **Ans.**

Example 4:

Solve $2s + (rt - s^2) = 1$.

(Rohilkhand, 94; Meerut 92; Garhwal, 95)

Solution:

We have $2s + (rt - s^2) = 1$.

Comparing the given equation with the equation

$$Rr + Ss + Tt + U(rt - s^2) = V$$

we have $R = 0,\; S = 2,\; T = 0,\; U = 1,\; V = 1$.

$\therefore$ From λ equation $\lambda^2(UV + RT) + \lambda US + U^2 = 0$.

we have $\lambda^2 + 2\lambda + 1 = 0$

$\therefore \qquad \lambda_1 = -1$ and $\lambda_2 = -1$.

Putting $\lambda_1 = -1 = \lambda_2$ in one of pair of equations.

$$Udy + \lambda_1 Tdx + \lambda_1\ Udp = 0$$

and $\qquad Udx + \lambda_2\ Rdy + \lambda_2\ udq = 0.$

we have $dy - dp = 0$ and $dx - dq = 0$

which give $\quad y - p = a$, and $x - q = b$.

$\therefore$ The intermediate integral is

$$y - p = f(x - q)$$

Also $p = y - a$ and $q = x - b$.

substituting these values of p and q in $dz = p\ dx + q\ dy$, we have

$$dz = (y = a)\ dx + (x - b)\ dy$$

$$= (y\ dx + x\ dy) - a\ dx - b\ dy$$

$\therefore \quad z = xy - ax - by + c$

which is the complete integral.

Aliter. Taking $y - p = m\ (x - q) + n$ where m and n are constants.

$\Rightarrow \qquad p - mq = y\ mx - n.$

$\therefore$ Lagranges's auxiliary equations are

$$\frac{dx}{1} = \frac{dy}{-m} = \frac{dz}{y - mx - n}$$

Taking first two members, we have

$dy + mdx = 0 \qquad\qquad \therefore\ y + mx = a.$

Again taking the first and last members, we have

$$dz = (y - mx - n)\ dx$$

$$= (a - 2mx - n)\ dx$$

$\therefore\ z = \ ax - mx^2 - nx + b$

$\Rightarrow \quad z = (y + mx)\ x - mx^2 - nx + b$

$\Rightarrow \quad z = xy\ nx + \phi(y + mx).$ **Ans.**

Example 5:

Solve $(rt - s^2) - s\ (\sin x + \sin y) = \sin x \sin y.$

Solution:

We have $(rt - s^2) - s(\sin x + \sin y) = \sin x \sin y$.

Comparing with the equation

$$Rr + Ss + Tt + U(rt - s^2) = V, \text{ we have}$$

$R = 0$, $S = -(\sin x + \sin y)$, $T = 0$ $U = 1$, $V = \sin x \sin y$.

From λ equation

$$\lambda^2 (UV + RT) + \lambda US + U^2 = 0, \text{ we have}$$

$$\sin x \sin y.\lambda^2 - (\sin x + \sin y)\lambda + 1 = 0$$

$$\Rightarrow \quad (\lambda \sin x - 1)(\lambda \sin y - 1) = 0$$

$$\therefore \quad \lambda_1 \operatorname{cosec} y \text{ and } \lambda_2 = \operatorname{cosec} x.$$

Substituting these values in

$$U\,dy + \lambda_1 T\,dx + \lambda_1 U\,dp = 0$$

and $\quad U\,dx + \lambda_1 R\,dy + \lambda_1 U\,dq = 0$

We have

$$dy + \operatorname{cosec} y\,dp = 0$$

and $\quad dx + \operatorname{cosec} x\,dq = 0$

$$\Rightarrow \quad \sin y\,dy + dp = 0$$

and $\quad \sin x\,dx + dq = 0$

$$\therefore \quad -\cos y + p = A \text{ and } -\cos x + q = B$$

$\therefore$ Let $p - \cos y = m(q - \cos x) + n$, where m and n are constants.

$$\Rightarrow \quad p - mq = \cos y - m\cos x + n.$$

$\therefore$ Lagrange's auxiliary equations are

$$\frac{dx}{1} = \frac{dy}{-m} = \frac{dz}{\cos y - m\cos x + n}$$

Taking the first two members, we have

$$dy = m\,dx = 0 \qquad \therefore y + mx = a$$

and from the first and the last members, we have

$$dz = (\cos y - m\cos x + n)\,dx$$

$$= \{\cos(a - mx) - m\cos x + n\}\,dx$$

$$\therefore z = -\frac{1}{m}\sin(a - mx) - m\sin x + nx + b$$

$$\Rightarrow \quad mz + \sin y + m^2 \sin x - mn\,x + mf(y + mx).$$ **Ans.**

Example 6:

Solve $2r + te^x - (rt - s^2) = 2e^x$. **(Raj., 91)**

Solution:

We have $2r + te^x - (rt - s^2) = 2e^x$.

Here we proceed directly.

Putting $r = \dfrac{dp - s\,dy}{dx}$ and $t = \dfrac{dq - s\,dx}{dy}$, in the given equation we have

$$2\frac{dp - s\,dy}{dx} + \frac{dq - s\,dx}{dy}e^x - \left(\frac{dp - s\,dy}{dx}.\frac{dq - s\,dx}{dy} - s^2\right) = 2e^x.$$

$\Rightarrow$ $2dp\,dy + e^x\,dq\,dx - dp\,dq - 2e^x\,dx\,dy$

$$-s(2dy^2 + e^x\,dx^2 - dp\,dx - dq\,dy) = 0$$

Hence, the Monge's Equations are

$$2dp\,dy + e^x\,dq\,dx - dp\,dq - 2e^x\,dx\,dy = 0 \quad ...(1)$$

and $$2dy^2 + e^x\,dx^2 - dp\,dx\,dq\,dy = 0 \quad ...(2)$$

(1) gives, $(2dy - dq)\,(dp = e^x\,dx) = 0.$

$\therefore$ $2dy - dq = 0$

$\Rightarrow$ $dp - e^x\,dx = 0$

Integrating $2y - q = A$ and $p - e^x = B$

$\Rightarrow$ $q = 2y - A$ and $p = e^x = B$.

Substituting in $dz = p\,dx + q\,dy$, we have

$dx = (e^x + B)\,dx + (2y - A)\,dy$

Integrating, we get $z = e^x + Bx + y^2\,Ay + C$.

Aliter. Taking $p - e^x = (-2y + q)\,m + n$,

where m, n are const.

$\Rightarrow$ $p - qm = e^x - 2ym + n$.

$\therefore$ Lagrange's Auxiliary Equations are

$$\frac{dx}{1} = \frac{dy}{-m}\;\frac{dz}{e^x - 2ym + n} = \frac{-e^x\,dx - 2y\,dy + dz}{n}$$

From the first two members, we have $y + mx = a$.

Again from the first and the last members, we have

$-e^x\,dx - 2y\,dy + dx = n\,dx$

$\Rightarrow \quad dz = e^x\, dx + 2y\, dy + n\, dx$

$\therefore \quad z = e^x + y^2 + nx + b$

$\Rightarrow \quad z = e^x + y^2 + nx + \phi(y + mx)$

Example 7:

Solve $(q^2 - 1)\, z\, r - 2pq\, zs + (p^2 - 1)\, zt + z^2\, (rt - s^2) = p^2 + q^2 - 1.$

Solution:

We have $(q^2 - 1)zr - 2pq\, zs + (p^2 - 1)zt + z^2(rt - s^2) = p^2 + q^2 - 1.$

Here $\quad R = (q^2 - 1)\, z,\ S = = 2pq\, z,\ T = (p^2 - 1)\, z,$

$U = z^2 \quad \text{and} \quad V = p^2 + q^2 - 1.$

From λ equation $\lambda^2\, (UV + RT) + USl + U^2 = 0$, we have

$$[z^2(p^2 - q^2 - 1) + (q^2 - 1)\, (p^2 - 1)\, z^2]\, \lambda^2 - 2pq\, z^3\lambda + z^4 = 0$$

$\Rightarrow \quad p^2q^2\lambda^2 - 2pq\, z\lambda + z^2 = 0$

$\Rightarrow \quad (pq\, \lambda - z)^2 = 0 \qquad \therefore \quad \lambda = \dfrac{z}{pq}$

$\therefore$ Intermediate integral is given by

$$U\, dy + \lambda\, T\, dx + \lambda\, U\, dp = 0$$

and $\quad U\, dx = \lambda\, R\, dy + \lambda\, U\, dq = 0$

$$z^2\, dy + \frac{z^2\, (p^2 - 1)}{pq}\, dx + \frac{z^3}{pq\, dp}\, dp = 0$$

and $\quad z^2\, dy + \dfrac{z^2\, (p^2 - 1)\, dx + z\, dp}{pq}\, dy + \dfrac{z^3}{pq}\, dq = 0$

$\Rightarrow \quad dy + \dfrac{(p^2 - 1)\, dx + z\, dp}{pq} = 0$

and $\quad dx + \dfrac{(q^2 - 1)\, dy + z\, dq}{pq} = 0$

$\Rightarrow \quad pq\, dy + p^2\, dx - dx + z\, dp = 0 \qquad ...(1)$

and $\quad pq\, dx + q^2\, dy - dy + z\, dq = 0. \qquad ...(2)$

From (1), we have

$$p(p\, dx + q\, dy) - dx + z\, dp = 0$$

$\Rightarrow \quad p\, dz + z\, dp - dx = 0 \qquad \therefore \quad dz = p\, dx + q\, dy.$

$\therefore \quad pz - x = A. \qquad ...(3)$

Similarly from equation (2), we have

$dz - y = B.$ ...(4)

Substituting the values of p, q from (3) and (4) in

$dz = p\,dx + q\,dy$, we have

$$dz = \frac{A + x}{z}dx + \frac{B + y}{z}dy$$

$\Rightarrow$ $z\,dz = x\,dx + y\,dy + A\,dx + B\,dy.$

$\therefore$ $z^2 = x^2 + y^2 + 2Ax + 2By + c.$

Aliter. A more general integral may be obtained by taking

$pz - x = (qz - y)\,m + n$ where m and n are constants.

$\Rightarrow$ $zp - zmq = x - ym + n.$

$\therefore$ Lagrange's auxiliary equation are

$$\frac{dx}{z} = \frac{dy}{-zm} = \frac{dz}{x - ym + n}$$

Taking the first two members, we have

$dy + m\,dx = 0 \quad \therefore \quad y + mx = 0$

Now, using $-\frac{x}{z}, -\frac{y}{z}$ and 1 as multipliers.

$$\text{each fraction} = \frac{dx}{z} = \frac{-\frac{x}{z}dx = \frac{y}{z}dy + dz}{n}$$

$\therefore$ $n\,dx = -x\,dx - y\,dy + z\,dz$

$\Rightarrow$ $z\,dz = x\,dx + y\,dy + n\,dx.$

$\therefore$ $z^2 = x^2 + y^2 + 2nx + b$

$\Rightarrow$ $z^2 = x^2 + y^2 + 2nx + f(y - mx).$ **Ans.**

Example 8:

Solve $qx\,r + (x + y)\,s + pyt + xy\,(rt - s^2) = 1 - pq.$

Solution:

We have $qx\,r + (x + y)\,s + pyt + xy\,(rt - s^2) = 1 - pq.$

Here $R = qx$, $S = x + y$, $T = py$, $U = xy$ and $V = 1 - pq.$

From λ equation $\lambda^2\,(UV + RT) + US\lambda + U^2 = 0$ we have

$[xy\,(1 - pq) + qx\,py]\,\lambda^2 + xy\,(x + y)\,\lambda + x^2y^2 = 0$

$\Rightarrow$ $\lambda^2 + (x + y)\,\lambda + xy = 0$

$\Rightarrow$ $(\lambda + x)(\lambda + y) = 0 \quad \therefore \quad \lambda_1 = -x, \lambda_2 = -y.$

Hence one intermediate integral is given by

$$U\,dy + \lambda_1 T\,dx - \lambda_1 U\,dp = 0$$

and $$U\,dx + \lambda_2 R\,dx - \lambda_2 U\,dq = 0$$

$\Rightarrow$ $$xy\,dy - xpy\,dx - x.xy\,dp = 0$$

and $$xy\,dx - y\,dx\,dy - yxy\,dq = 0$$

$\Rightarrow$ $$dy - p\,dx - x\,dp = 0$$

and $$dx - q\,dy - y\,dq = 0$$

$\therefore$ $$-y + px = A$$

and $$-x + yq = B.$$

$\therefore$ Let $px - y = m(qy - n) + n$ where m and n are constants

$\Rightarrow$ $$px - m\,qy = y - mx + n.$$

$\therefore$ Lagrange's auxiliary equations are

$$\frac{dx}{x} = \frac{dy}{-my} = \frac{dz}{y - mx + n}$$

Taking the first two members, we have

$$m\frac{dx}{x} + \frac{dy}{y} = 0.$$

$\therefore$ $m \log x + \log y = \log a \quad \Rightarrow \quad x^m y = a.$

Again using m, $\frac{1}{m}$ and 1 as multipliers, we have

$$\text{Each fraction} = \frac{m\,dx + \frac{1}{m}dy + dz}{n}$$

$\therefore$ $$\frac{dx}{x} = \frac{m\,dx + \frac{1}{m}dy + dz}{n}$$

$\Rightarrow$ $$dz + m\,dx + \frac{1}{m}dy - n\frac{dx}{x} = 0.$$

Integrating, we have $z = mx + \frac{y}{m} - n \log x = b$

$\Rightarrow$ $$a = mx + \frac{y}{m} - n \log x = f(x^m y).$$ **Ans.**

Example 9:

Solve $z(1 + q^2)\,r - 2pq\,zs + z(1 + p^2).\,t + z^2(rt - s^2) + (1 + p^2 + q^2) = 0.$

Solution:

We have $z(1 + q^2)\, r - 2pq\, zs + z(1 + p^2).\, t$
$+ z^2(rt - s^2) + (1 + p^2 + q^2) = 0.$

Comparing it with the equation

$$Rr + Ss + Tt + U(rt - s^2) = V, \text{ we have}$$

$$R = z(1 + q^2),\ S = -2pqz,\ T = z(1 + p^2),\ U = z^2$$

and $V = -(1 + p^2 + q^2)$.

$\therefore$ From λ equation $\lambda^2 (UV + RT) + \lambda US + U^2 = 0$, we have

$$\{-z^2(1 + p^2 + q^2) + z^2(1 + p^2)(1 + q^2)\}\ \lambda^2 - 2pqz^3\ \lambda + z^4 = 0$$

$$\Rightarrow \quad p^2q^2\lambda^2 - 2pqz\lambda + z^2 = 0$$

$$\Rightarrow \quad (pql - z)^2 = 0. \qquad \therefore \qquad \lambda_1 = \lambda_2 = \frac{z}{pq}.$$

$\therefore$ The intermediate integral is give by

$$U\, dy + lT\, dx + lU\, dp = 0$$

and

$$U\, dx + lR\, dy + lU\, dq = 0$$

i.e., by $\quad z^2\, dy + \frac{z}{pq}.\, z\,(1 + p^2)\, dx + \frac{z}{pq}\, .\, z^2 — dp = 0$

$$\Rightarrow \quad pq\, dy + (1 + p^2)\, dx + z\, dp = 0. \qquad ...(1)$$

and

$$pq\, dx + (1 + q^2)\, dx + z\, dq = 0. \qquad ....(2)$$

From (1), we have

$$p(p\, dx + q\, dy) + z\, dp + dx = 0.$$

$$\Rightarrow \quad p\, dz + z\, dp + dx = 0 \quad \text{since } dz = p\, dx + q\, dy$$

$$\therefore \quad pz + x = A. \qquad ...(3)$$

From (2), we have

$$q(p\, dx + q\, dy) + z\, dq + dy = 0$$

$$p\, dz + z\, dq + dy = 0.$$

$$\therefore \quad qz + y = B \qquad ...(4)$$

Substituting the values of p and q from (3) and (4) in

$dz = p\, dx + q\, dy$, we have

$$dz = \frac{A - x}{z}\, dx + \frac{B - y}{z}\, dy$$

$$\Rightarrow \quad z\, dz = (A - x)\, dx + (B - y)\, dy.$$

Integrating, we have

$$z^2 = -(A - x)^2 - (B - y)^2 + c.$$

$$\Rightarrow \quad z^2 + (A - x^2) + (B - y^2) = c,$$

which is the general solution.

Aliter. Taking $pz + x = (qz + y)\,m + n.$

where m and n are constants,

$$\Rightarrow \quad zp - z\,mq = -x + ym + n.$$

$\therefore$ Lagrange's auxiliary equations are

$$\frac{dx}{z} = \frac{dy}{-zm} = \frac{dz}{-x + ym + n}$$

first two members give $y + mx = a$.

Using $\frac{x}{z}, \frac{y}{z}, 1,$ as multipliers, we have

$$\text{Each fraction} = \frac{dx}{z} = \frac{\frac{x}{z}\,dx + \frac{y}{z}\,dy + dz}{n}$$

$$\Rightarrow \quad n\,dx = x\,dx + y\,dy + z\,dz$$

$$\Rightarrow \quad z\,dz + x\,dx + y\,dy - n\,dx = 0.$$

$$\therefore \quad z^2 + x^2 + y^2 - 2nx = b.$$

$$\Rightarrow \quad z^2 + x^2 + y^2 - 2nx = \phi(y + mx).$$ **Ans.**

Example 10:

Solve $ar + bs + ct + e\,(rt - s^2 = h$, where a, b, c, e and h are constant.

Solution:

We have $ar + bs + ct + e\,(rt - s^2 = h$.

Here $R = a$, $S = b$, $T = c$, $U = e$ and $V = h$.

$\therefore$ From λ equation

$\lambda^2(RT + UV) + \lambda US + U^2 = 0$, we have

$$(ac + eh)\,\lambda^2 + \lambda be + e^2 = 0. \qquad ...(1)$$

Putting $\lambda = -e/m$, ...(2)

(1) becomes $\frac{e^2}{m^2}(ac + eh) - \frac{e^2 h}{m} + e^2 = 0.$

$$\Rightarrow \quad m^2 - bm + (ac + eh) = 0. \qquad ...(3)$$

Let m_1, m_2 be the roots of (3).

$\therefore$ The first intermediate integral is given by

$$U\ dy + \lambda_1\ T\ dx + \lambda_1\ U\ dp = 0$$

and $$U\ dx + \lambda_2\ R\ dy + \lambda_2\ U\ dq = 0$$

$$\Rightarrow \quad e\,dy - \frac{e}{m_1} a\ dy - \frac{e}{m_1}\ e\ dp = 0$$

and $$e\,dx - \frac{e}{m_2} a\ dy - \frac{e}{m_2}\ e\ dq = 0$$

$$\Rightarrow \quad c\ dx + e\ dp - m_1\ dy = 0$$

and $$a\ dy + e\ dq - m_2\ dx = 0$$

$\therefore$ $cx + eq - m_2x = B'$

$\therefore$ One intermediate integral is

$$cx + ep - m_1y = f_2(ay + eq - m_2x) \qquad ...(4)$$

Similarly the second intermediate integral is

$$cx + ep + m_2y \ = f_2(ay = eq - m_1x) \qquad ...(5)$$

Hence, it is not possible to find the values of p and q from above two intermediate integrals.

So we combine (4) with

$$cx + ep - m_2y = A$$

Thus, we have

$$(m_2 - m_1)\ y = f_1(ax + eq - m_2x) - A$$

$$\Rightarrow \quad ay + eq + m_2x + \phi[(m_2 - m_1)\ y + A]$$

where ϕ is inverse function of f.

$\therefore q = (1/e)\ [-ay + m_2x + \phi\{(m_2 - m_1)\ y + A\}]$

and from (6) $p = (-cx + m_2y + A)/e$.

Substituting in $dz = p\ dx + q\ dy$, we have

$$dz = (1/e)\ [-ay + m_1x + \phi\{(m_2 - m_1)\ y + A\}]\ dy$$

$$\Rightarrow \quad e\ dz = -\ cx\ dx - ay\ dy + m_2(x\ dy + y\ dx) + A\ dx + \phi\{(m_2 - m_1)\ y + A\}\ dy$$

$$\therefore ez = -\ \frac{cx^2}{2} - \frac{ay^2}{2} + m_2xy + Ax + \psi\{(m_2 - m_1)\ y + A\} + B.$$

EXERCISES

1. $2pr + 2qt - 4pq(rt - s^2) = 1$.

Ans. $3z = 3c \pm 2\,(x + a)^{3/2} \pm 2(y + b)^{3/2}$

2. $3s + (rt - s^2) = 2$.

[**Hint:** $\lambda = -1$ or $\lambda = \dfrac{1}{2}$. Intermediate integrals are

$p - y = f(q - 2x),\ p - 2y = F(q + x)$

3. $r + t - (rt - s^2) = 1$.

[**Hint:** $\lambda = \infty$, intermediate integral $p - x = f(q - y)$]

Ans. $z = \dfrac{1}{2}(x^2 + y^2) + ax + by + c$

or $z = \dfrac{1}{2}(x^2 + y^2) + nx + \psi(y + nx)$

4. $7r - 8s - 3t(rt - s^2) = 36$.

Ans. $2x = \alpha, -\beta,\ 2y = \psi'(\beta) - \phi'(\beta)$;

$2x = 3x^2 - 6xy - 7y^2 + \phi(a) - \psi(\beta) + 2\beta y$.

5. $r + 4s + t + (rt - s^2) = 2$. **(Raj., 93)**

Ans. $z = -\dfrac{1}{2}x^2 - \dfrac{1}{2}y^2\ \ 3xy + Ax + \psi(2y + A) + B$.

6. $r = t$ (wave equation). **Ans.** $z = f_1)x + y + f_2(y - x)$

7. $x^2r - 2xs + t + q = 0$. **(Agra, 91; Delhi, Hons. 89, 92, 94)**

Ans. $z = f_1(y + \log x) + xf_2(y + \log x)$

8. $x^2r + 2xys + y^2t = 0$.

(Garhwal, 93; Meerut 90; Rohilkhand, 93)

Ans. $z = y\ f_1(y/x) + f_2(y/x)$

8. $(r - t)\ xy - s\ (x^2 - y^2) = qx - py$.

Ans. $z = f_1(x^2 - y^2) + f_2(y/x)$

9. $2y\ 1 + y^2t = 1$. **Ans.** $yz = y \log y - f(x) + y\ F(x)$.

10. $r = 2y^2$. **Ans.** $z = x^2y^2 + xf(y) + F(y)$.

11. $r - 2s + t = \sin(2x + 3y)$.

Ans. $z = f(x + y) + xF(x + y) - \sin\,(2x + 3y)$.

12. $rx^2 - 3sxy + 2ty^2 + [x + 2qy = x + 2y$. **(IAS, 95)**

Ans. $z = x + y + f(xy) + F(x^2y)$

13. $z(qs - pt) = pq^2$. **(Agra, 98)**

Ans. $z = a \log(y - b) + c$

14. $q^2r - 2pqs + p^2t = pq^2$. **(IAS, 96)**

Ans. $y = e^x f_1(z) + f_2(z)$

15. $rt - s^2 + 1 = 0$, **Ans.** $z = xy + \frac{1}{2}\{\phi(\alpha) = \psi(\beta)\} + \beta y$

where $x = \frac{1}{2}(\alpha, \beta)$, $y = \frac{1}{2}\{\psi'(\beta) - \phi'(\alpha)\}$

16. $3r + 4s + t + (rt - s^2) = 1$.

Ans. $z = ax + by - \frac{1}{2}x^2 + 2xy - \frac{3}{2}y^2 + c$

or $z = 2xy - \frac{1}{2}(x^2 + 3y^2) + nx + \phi(y + mx)$.

7

Linear Partial Differential Equations

7.1 Linear Partial Differential Equation

The Linear Partial Differential Equation of the second order in n independent variables $x_1, x_2, \ldots x_n$, can be written as follows:

$$\sum_{i=1}^{n}\sum_{j=1}^{n} a_{ij} \frac{\partial^2 u}{\partial x_i \partial x_j} + \sum_{i=1}^{n} b_i \frac{\partial u}{\partial x_i} + cu = 0 \qquad \ldots(1)$$

where a_{ij}, b_i and c are constants or functions of $x_1, x_2, \ldots x_n$.

Here let δ_i represent $\frac{\partial}{\partial x_i}$ $\qquad (i = 1, 2, \ldots, n)$

and $\quad \delta_i\ \delta_j$ represent $\frac{\partial^2}{\partial x_i\ \partial x_j}$ $\qquad \begin{pmatrix} i = 1, 2, 3, \ldots, n \\ j = 1, 2, 3, \ldots, n \end{pmatrix}$

Now, let us consider the operator

$$\phi = \sum_{i=1}^{n}\sum_{j=1}^{n} a_{ij}\ \delta_i\ \delta_j \qquad \ldots(2)$$

for all real values of δ_i positive or negative

At a point $x_1, x_2, \ldots, x_n$ we call (1):

(a) *elliptic* if ϕ can be positive for all real values of ∂_i and it reduces to zero only when all ∂_i' are zero.

(b) *hyperbolic* if ϕ can be both positive or negative

and (c) *parabolic* if the determinant Δ vanishes, where

$$\Delta = \begin{vmatrix} a_{11} & a_{12} & \dots & a_{1n} \\ a_{21} & a_{22} & \dots & a_{2n} \\ \dots & \dots & \dots & \dots \\ \dots & \dots & \dots & \dots \\ a_{n1} & a_{n2} & \dots & a_{n3} \end{vmatrix} = 0$$

If a_{ij} are functions of $x_1, x_2, \dots x_n$ the same differential equation can be elliptic, hyperbolic and parabolic at different points.

If a_{ij} are constants, the equation will have the same nature throughout.

Now, let us consider the equation of second order in two independent variables x and y.

$$A \frac{\partial^2 u}{\partial x^2} + B \frac{\partial^2 u}{\partial x \partial y} + C \frac{\partial^2 u}{\partial y^2} + f\left(x, y, u, \frac{\partial u}{\partial x}, \frac{\partial u}{\partial y}\right) = 0 \qquad \dots(3)$$

where A is positive.

Here $f = A\delta_1^2 + B\delta_1\delta_2 + C_2^2$.

The equation (3) is

(a) elliptic if $B^2 - 4\,AC < 0$

(b) hyperbolic if $B^2 - 4AC < 0$

and (c) parabolic if $B^2 - 4AC = 0$.

Notes:

(1) If A, B, C are constants then the nature of the equation (3) will be the same in the whole region i.e., for all values of x and y. The nature will depend on $B^2 - 4AC$,

so that it will be elliptic if $B^2 - 4AC < 0$

it will be hyperbolic if $B^2 - 4AC > 0$

it will be parabolic if $B^2 - 4AC = 0$.

(2) If A, B, C are functions of x and y then the nature of equations (3) will not be same in the whole region i.e., for all values of x and y.

It will be elliptic in the region where $B^2 - 4AC < 0$

It will be hyperbolic in the region where $B^2 - 4AC > 0$

It will be parabolic in the region where $B^2 - 4AC = 0$

Example 1:

Classify the following operators:

(a) $\dfrac{\partial^2 u}{\partial t^2} + \dfrac{\partial^2 u}{\partial x\, \partial t} + \dfrac{\partial^2 u}{\partial x^2}$

(b) $\frac{\partial^2 u}{\partial t^2} - 4\frac{\partial^2 u}{\partial x\, \partial t} + 4\frac{\partial^2 u}{\partial x^2}$

(c) $\frac{\partial^2 u}{\partial t^2} + 4\frac{\partial^2 u}{\partial x\, \partial t} + 4\frac{\partial^2 u}{\partial x^2}$.

Solution:

(a) Here $A = 1, B = 1, C = 1$

$B^2 - 4AC = 1 - 4 = -3 < 0$ ∴ elliptic

(b) Here $A = 1, B = 4, C = 1$

$B^2 - 4AC = 16 - 4 = 12 > 0$ ∴ hyperbolic.

(c) Here $A = 1, B = 4, C = 4$

$B - 4AC = 16 - 16 = 0 = 0$ ∴ parabolic. **Ans.**

Example 2:

Find where the following operators are hyperbolic, parabolic and elliptic.

(a) $\frac{\partial^2 u}{\partial t^2} + t\frac{\partial^2 u}{\partial x\, \partial t} + x\frac{\partial^2 u}{\partial x^2}$

(b) $x^2\frac{\partial^2 u}{\partial t^2} - \frac{\partial^2 u}{\partial x^2} + u$

and (c) $t\frac{\partial^2 u}{\partial t^2} + 2\frac{\partial^2 u}{\partial x\, \partial t} + x\frac{\partial^2 u}{\partial x^2} + \frac{\partial x}{\partial x}$.

Solution:

(a) Here $A = t, B = t, C = x$

∴ $B^2 - 4AC = t^2 - 4x$

∴ The operator is hyperbolic if $t^2 - 4x > 0$

i.e. if $t^2 > 4x$

and parabolic if $t^2 = 4x$

and elliptic if $t^2 < 4x$

(b) Here $A = x^2, B = 0, C = -1$

∴ $B^2 - 4AC = t^2 = 4x^2$

∴ The operator is hyperbolic if $4x^2 > 0$

i.e., if $x > 0$

parabolic if $4x^2 = 0$

i.e., $x = 0.$

Since $4x^2$ cannot be negative, $\therefore$ the function cannot be elliptic.

(c) Here A = t, B = 2, C = x

$B^2 - 4AC = {}^4 - 4tx$

$\therefore$ The operator is hyperbolic if $4 - 4tx > 0$

i.e., $xt < 1$

parabolic if $xt = 1$

and elliptic if $xt > 1.$ **Ans.**

Example 3:

Solve $(1 + q^2)\, r - 2\, pqs + (1 + p^2)t$

$$+ (1 + p^2 + q^2)^{-1\,2} (r! - s^2) = -(1 + p^2 + q^2)^{3\,2}.$$

Solution:

Here $R = 1 + q^2$, $S = -2\, pq$, $T = 1 + p^2$,

$U = (1 + p^2 + q^2)^{-1/2}$, $V = -(1 + p^2 + q^2)^{3/2}$

From λ equation $\lambda^2\, (RT + UV) + US\lambda + U^2 = 0$, we have

$$[(1 + p^2)(1 + q)^2 - (1 + p^2 + q^2)]\, \lambda^2$$

$$- \frac{2pq}{\sqrt{(1 + p^2 + q^2)}}\lambda + \frac{1}{(1 + p^2 + q^2)} = 0$$

$\Rightarrow$ $p^2q^2\, (1 + p^2 + q^2)\, \lambda^2 - 2pq\, \sqrt{(1 + p^2 + q^2)}\, \lambda + 1 = 0$

$\Rightarrow$ $\{pq\, \sqrt{(1 + p^2 + q^2)}\lambda - 1\}^2 = 0$

$$\therefore \quad \lambda = \frac{1}{pq\sqrt{(1 + p^2 + q^2)}}.$$

$\therefore$ Intermediate function is given by

$$U\, dy + \lambda T\, dy + \lambda U\, dp = 0$$

and $$U\, dx + \lambda R dy + \lambda U\, dq = 0$$

$$\Rightarrow \quad (1 + p^2 + q^2)^{-1/2}\, dy + \frac{(1 + p^2)}{pq\sqrt{(1 + p^2 + q^2)}}\, dx$$

$$+ \frac{dp}{pq(1 + p^2 + q^2)} = 0$$

and $$(1 + p^2 + q^2)^{-1/2}\, dx + \frac{(1 + q^2)}{pq\sqrt{(1 + p^2 + q^2)}}\, dx + \frac{dq}{pq(1 + p^2 + q^2)} = 0$$

$$\Rightarrow \quad pq\, dy + (1 + p^2)\, dx + \frac{dp}{\sqrt{(1 + p^2 + q^2)}} = 0 \quad \text{...(1)}$$

and $$pq\, dx + (1 + q^2)\, dy + \frac{dq}{\sqrt{(1 + p^2 + q^2)}} = 0 \quad \text{...(2)}$$

Eliminating dy, we have

$$[(1 + p^2)(1 + q)^2 - p^2q^2]dx + \frac{(1 + q^2)dp}{\sqrt{(1 + p^2 + q^2)}} - \frac{pq\, dq}{\sqrt{(1 + p^2 + q^2)}} = 0$$

$$\Rightarrow \quad dx + \frac{(1 + q^2)\, dp - pq\, dq}{(1 + p^2 + q^2)^{3/2}} = 0$$

$$\Rightarrow \quad dx + \frac{(1 + p^2 + q^2)dp}{(1 + p^2 + q^2)^{3/2}} - \frac{p^2\, dp + pq\, dq}{(1 + p^2 + q^2)^{3/2}} = 0$$

$$\Rightarrow \quad dx + (1 + p^2 + q^2)^{-1/2}\, dp + pd\,(1 + p^2 + q^2)^{-1/2} = 0$$

$$\therefore \quad x + p\,(1 + p^2 + q^2)^{-1/2} = A. \quad \text{...(3)}$$

Similarly eliminating dx from (1) and (2) and simplifying, we have

$$y + q\,(1 + p^2 + q^2)^{-1/2} = B. \quad \text{...(4)}$$

From equation (3) and (4), we have $\frac{x - A}{y - B} = \frac{p}{q}$

$$\therefore\ p = \frac{x - A}{y - B}q. \quad \text{...(5)}$$

Putting in (3), we have

$$x + \frac{x - A}{y - B}q\left\{1 + \left(\frac{x - A}{y - B}\right)^2 q^2 + q^2\right\}^{-1/2} = A$$

$$\Rightarrow \quad (x - A) = \frac{x - A}{y - B}q\left\{1 + \frac{\{(x - A)^2 + (y - B)^2\}q^2}{(y - B)^2}\right\}^{-1/2} = 0$$

$$\Rightarrow \quad \left[1+\frac{\{(x-A)^2+(y-B)^2\}q^2}{(y-B)^2}\right]=\frac{q^2}{(y-B)^2}$$

$$\Rightarrow \quad (y-B)^2 = q^2\,[1-\{(x-A)^2+(ty-B)^2\}]$$

$$\therefore\ q=\frac{y-B}{\sqrt{\left[1-\{(x-A)^2+(y-B)^2\}q^2\right]}}$$

From (5)

$$p=\frac{x-B}{\sqrt{\left[1-\{(x-A)^2+(y-B)^2\}q^2\right]}}$$

Substituting these values in

$dz = p\,dx + q\,dy$, we have

$$dz=\frac{(x-A)\,dx+(y-B)dy}{\sqrt{\left[1-\{(x-A)^2+(y-B)^2\}q^2\right]}}$$

Integrating, $z = [1-\{(x-A)^2+(y-B)^2\}]^{1/2} + C$,

$\therefore (z-C)^2 = 1-(x-A)^2-(y-B)^2$

$\Rightarrow \quad (x-A)^2+(y-B)^2+(z-C)^2 = 1.$ **Ans.**

Example 4:

Classify the equation

$$(1-x^2)\frac{\partial^2 z}{\partial x^2}-2xy\frac{\partial^2 z}{\partial x\,\partial y}+(1-y^2)\frac{\partial^2 z}{\partial y^2}+x\frac{\partial z}{\partial x}+3x^2y\frac{\partial z}{\partial y}-2z=0.$$

Solution:

Let us consider the operator

$$\phi = A\,\delta_1^{\,2}+B\delta_1\delta_2+C\delta_2^{\,2} \text{ where } \delta_1=\frac{\partial}{\partial x},\ \partial_2=\frac{\partial}{\partial y}$$

$\therefore \quad A = 1-x^2,\ B = -2xy,\ C = 1-y^2.$

$\therefore \quad B^2-4AC = 4x^2y^2-4(1-x^2)(1-y^2)$

$\qquad\qquad = 4(-1+x^2+y^2)$

Since A, B, C are function of x and y, the differential equation is hyperbolic in the region

where $B^2-4AC > 0$ i.e. $x^2+y^2 > 1$

parabolic at points on the circle $x^2 + y^2 = 1$ since $B^2 - 4AC = 0$ and elliptic in the region where $x^2 + y^2 < 1$ since $B^2 - 4AC < 0$ **Ans.**

Example 5:

Classify the following equations:

(a) $$\frac{\partial^2 u}{\partial x^2} + \frac{\partial^2 u}{\partial y^2} + \frac{\partial^2 u}{\partial z^2} = 0.$$ *(Laplace's Equation)*

(b) $$\frac{\partial^2 u}{\partial x^2} + \frac{\partial^2 u}{\partial y^2} + \frac{\partial^2 u}{\partial z^2} = \frac{1}{C^2}.\frac{\partial^2 u}{\partial t^2}$$ *(Wave Equation)*

(c) $$\frac{\partial^2 u}{\partial x^2} + \frac{\partial^2 u}{\partial y^2} + \frac{\partial^2 u}{\partial z^2} = \frac{1}{C^2}.\frac{\partial u}{\partial t}.$$ *(Heat Equation)*

Solution:

(a) Here the operator

$\phi = \delta_1^2 + \delta_2^2 + \delta_3^2$

$a_{11} = a_{22} = a_{33} = 1,\ a_{13} = a_{23} = a_{31} = 0$

f is +ve for all real values of $\delta_1, \delta_2, \delta_3$.

$\therefore$ The equation is elliptic.

(b) Here the operator

$$\phi = \partial_1^2 + \partial_2^2 + \partial_3^2 - \frac{1}{C^2}\ \delta_4^2.$$

This can be both positive and negative.

Hence the equation is hyperbolic.

(c) Here $a_{11} = a_{22} = a_{33} = 1,\ a_{44} = 0.$

$a_{13} = a_{23} = a_{31} = a_{14} = a_{24} = a_{34} = 0$ etc.

$$\begin{vmatrix} 1 & 0 & 0 & 0 \\ 0 & 1 & 0 & 0 \\ 0 & 0 & 1 & 0 \\ 0 & 0 & 0 & 0 \end{vmatrix} = 0$$

Therefore the equation is parabolic. **Ans.**

7.2 Homogeneous Linear Equations

An equation of the form

$$x^n \frac{d^n y}{dx^n} + A_1 x^{n-1} \frac{d^{n-1}}{dx^{n-1}} + ... + A_n y = X \qquad ...(1)$$

where $A_1, A_2, ..., A_n$ are constants and X is either a constant or a function of x is called the **Homogeneous Linear Equation.**

7.3 Method of Solution

To solve such equation we generally put $x - e^z$.

$$\therefore \quad z = \log x \text{ and } \frac{dz}{dx} = \frac{1}{x}$$

$$\therefore \quad \frac{dy}{dx} = \frac{dy}{dz}\frac{dz}{dx} = \frac{1}{x}\frac{dy}{dz}$$

$$\Rightarrow \quad \frac{dy}{dz} = x\frac{dy}{dx} \quad \therefore \quad x\frac{d}{dx} = \frac{d}{dz}$$

Now, $x\frac{d}{dx}\left(x^{n-1}\frac{d^{n-1}y}{dx^{n-1}}\right) = x^n\frac{d^ny}{dx^n} + (n-1)x^{n-1}\frac{d^{n-1}y}{dx^{n-1}}$

$$\Rightarrow \quad x^n\frac{d^ny}{dx^n} = \left(x\frac{d}{dx} - n + 1\right)x^{n-1}\frac{d^{n-1}y}{dx^{n-1}}$$

$$= (D - n + 1)\, x^{n-1}\frac{d^{n-1}y}{dx^{n-1}} \qquad ...(2)$$

where $x\frac{d}{dx} \equiv \frac{d}{dz} \equiv D$

$\therefore$ Putting n = 2, 3, ... in (2), we have

$$x^2\frac{d^2y}{dx^2} = (D-1)\, x\frac{dy}{dx} = (D-1)\, Dy.$$

$$x^3\frac{d^3y}{dx^3} = (D-2)\, x^2\frac{d^2y}{dx^2} = (D-2)(D-1)\, Dy.$$

... ...

... ...

$$\therefore \quad x^n\frac{d^ny}{dx^n} = (D-n+1)(D-n+2)\ldots(D-1)\, Dy.$$

Substituting in (1), we have

$$[\{D(D-1)\ldots(D-n+1)\} + A_1\{D(D-1)\ldots(D-n+2)\} + \ldots + A_{n-1}D + A_n]\, y = Z$$

where Z is the function of z into which X is changed

$\Rightarrow \quad f(D)\, y = Z.$ ...(3)

This is an equation with constant coefficients and therefore can be integrated as follows:

To find C.F.: Solve A.E. $f(m) = 0$

(i) If $\alpha_1, \alpha_2, \ldots \alpha_n$ are roots of A.E. (all distinct).

Then C.F. $= c_1 e^{\alpha_1 z} + c_2 e^{\alpha_2 z} + \ldots + c_n e^{\alpha_n z}$

$$= c_1 x^{\alpha_1 z} + c_2 x^{\alpha_2 z} + \ldots + c_n x^{\alpha_n z}.$$

(ii) In case there are r roots alike each equal to α (Real) and the rest all different.

$$\text{Then C.F.} = \left(c_1 + c_2 z + \ldots + c_r z^{r-1}\right) e^{\alpha z} + c_{r+1} e^{\alpha_{r+1} z} + \ldots + c_n e^{\alpha_n z}$$

$$= [c_1 + c_2 \log x + \ldots + c_r (\log x)^{r-1}] x^{\alpha} + c_{r+1} x^{\alpha_{r+1}} + \ldots + c_n e^{\alpha_n}.$$

(iii) In case roots are imaginary say of the form $\alpha \pm i\beta$

Then C.F. corresponding to such root is

$(c_1 \cos \beta z + c_2 \sin \beta) e^{\alpha z}$

$= [c_1 \cos (\beta \log x) + c_2 \sin (\beta \log x)] x^{\alpha}$

$\Rightarrow \quad c_1 e^{az} \cos (\beta z + c_2) = c_1 x^{a} \cos (\beta \log x + c_2).$

(iv) In case the imaginary roots $\alpha \pm i\beta$ occur r times.

Then C.F. corresponding to these roots will be

$$e^{\alpha z} [(c_1 + c_2 z + \ldots + c_r z^{r+1}) \cos \beta z + (c'_1 + c'_2 z + \ldots + c'_r z^{r-1}) \sin \beta z]$$

$$= x^{\alpha}[\{c_1 + c_2 \log x + \ldots + c_r (\log x)^{r-1}\} \cos (\beta \log x)]$$

$$+ \{c'_1 + c'_2 \log x + \ldots + c'_r (\log x)^{r-1}\} \sin (\beta \log x)]$$

To find P.I.: Particular Integral is given by $\dfrac{1}{f(D)} Z$.

Methods

(i) Factorize the operator $\dfrac{1}{f(D)}$

$$\text{then P.I.} = \frac{1}{f(D)} Z = \frac{1}{(D - \alpha_1)(D - \alpha_2) \ldots (D - \alpha_{n-1})} . Z$$

$$= \frac{1}{(D - \alpha_1)(D - \alpha_2) \ldots (D - \alpha_{n-1})} e^{\alpha_n z} \int e^{-\alpha_n z} . Z\, dx$$

Similarly operate with other remaining factors in succession.

(ii) The operator $\dfrac{1}{f(D)}$ may also be separated into P.F.'s

$$\text{Then P.I.} = \frac{1}{f(D)} Z = \frac{c_1}{D-\alpha_1} Z + \frac{c_2}{D-\alpha_2} Z + \ldots + \frac{c_n}{D-\alpha_n} Z$$

$$= c_1 . e^{\alpha_1 z} \int e^{-\alpha_1 z} . Z dz + \ldots + c_n e^{\alpha_n z} \int e^{-\alpha} Z dz$$

Short Methods

(iii) *When Z is of the form $e^{\alpha z}$*

then $\text{P.I.} = \frac{1}{f(D)} e^{\alpha z} = \frac{1}{f(a)} e^{\alpha z}$ provided $f(a) \neq 0$.

(b) *When Z is of the form sin az or cos az*

then $\text{P.I.} = \frac{1}{f(D^2)} \cos az = \frac{1}{f(-a^2)} \cos az$

and $\text{P.I.} = \frac{1}{f(D^2)} \sin az = \frac{1}{f(-a^2)} \sin az$

provided $f(-a^2) \neq 0$.

(iv) *If Z is of the form z^m*

then $\text{P.I.} = \frac{1}{f(D)} z^m$.

Expand $[f(D)]^{-1}$ in ascending powers of D, retaining terms as far D^m and operate each term on z^m.

(v) *If Z is of the form $e^{az}.z$,*

$$\text{P.I.} = \frac{1}{f(D)} e^{az}.z = e^{az} = \frac{1}{f(D+a)} Z$$

(vi) *If Z is of the form $z\,\phi(z)$.*

$$\text{P.I.} = \frac{1}{f(D)} z\phi(z) = z . \frac{1}{f(D)} \phi(z) = \frac{f'(D)}{[f(D)]^2} . Z.$$

Example 1:

Solve $x^2 \frac{d^2y}{dx^2} + x \frac{dy}{dx} + y = 2 \log x.$

(Kanpur, 91; Agra, 92; Gauhati, 95)

Solution:

Putting $x = e^z$ and denoting $\frac{d}{dz}$ by D, the given equation becomes

$[D(D-1) - D + 1]\, y = 2z$

$\Rightarrow \quad (D-1)^2 y = 2z$

A.E. is $(m-1)^2 = 0 \quad \therefore \quad m = 1, 1$

$\therefore$ C.F. $= (c_1 + c_2 z)\, e^z = (c_1 + c_2 \log x)\, x$

and P.I. $= \dfrac{1}{(D-1)} \,.\, 2z = (1-D)^{-2} \,.\, 2z$

$= (1 + 2D) + \ldots) \,.\, 2z = 2z + 4 = 2 \log x + 4.$

Hence, the required solution is

$$y = (c_1 + c_2 \log x)\, x + 2 \log x + 4.$$ **Ans.**

Example 2:

Solve $x^2 \dfrac{d^2y}{dx^2} - 3x \dfrac{dy}{dx} + y = \dfrac{\log x \,.\, \sin(\log x) + 1}{x}.$

(Meerut, T.D.C. 96; Karnatak, 91; Osmania, 98)

Solution:

Putting $x = e^t$, the equation becomes

$[D(D-1) - 3D + 1]y = e^{-z}(z \sin z + 1)$

$\Rightarrow$ $[D^2 - 4D + 1]y = e^{-z} \sin z + e^{-z}$

A.E. is $m^2 - 4m + 1 = 0$ or $m = 2 \pm \sqrt{3}$

$\therefore$ C.F. $= c_1 e^{(2+\sqrt{3})z} + c_2 e^{(2-\sqrt{3})z}$

$$= \left(c_1 e^{\sqrt{3}z} + c_1 e^{-\sqrt{3}z}\right) e^{2z} = \left(c_1 x^{\sqrt{3}} + c_x e^{-\sqrt{3}}\right) x^2$$

P.I. corresponding to e^{-z}

$$= \frac{1}{D^2 - 4D + 1} . e^{-z} = \frac{1}{(-1)^2 - 4(-1) + 1} . e^{-z} = \frac{1}{6} x^{-1}$$

and P.I. corresponding to $e^{-z} z \sin z = \dfrac{1}{D^2 - 4D + 1} . e^{-z}\, z \sin z$

$$= e^{-z} . \frac{1}{(D-1)^2 - 4(D-1) + 1} . z \sin z$$

$$= e^{-z} . \frac{1}{D^2 - 6D + 6} . z \sin z$$

$$= \text{I.P. of } e^{-z} . \frac{1}{D^2 - 6D + 6} . z e^{iz}$$

$$= \text{I.P. of } e^{-z} e^{iz} \frac{6}{(D-i)^2 - 6(D+i) + 6} . z$$

$$= \text{I.P. of } e^{-z} e^{iz} \frac{6}{D^2 + (2i-6)D + (5-6i)} . z$$

$$= \text{I.P. of } e^{-z} e^{iz} \frac{1}{(5-6i)} . \left[1 + \frac{2i-6}{5-6i} D + \frac{1}{5-6i} D^2\right]^{-1} . z$$

$$= \text{I.P. of } e^{-z} e^{iz} \frac{1}{(5-6i)} . \left[1 + \frac{2i-6}{5-6i} D \ldots\right] . z$$

$$= \text{I.P. of } e^{-z} e^{iz} \frac{1}{(5-6i)} . \left[z - \frac{2i-6}{5-6i}\right]$$

$$= \text{I.P. of } e^{-z} e^{iz} \frac{(5+6i)}{5^2+6^2} . \left[z - \frac{(2i-6)(5+6i)}{5^2+6^2}\right]$$

$$= \text{I.P. of } \frac{e^{-z}}{61} (\cos z + i \sin z)(5+6i)\left[z + \frac{1}{61}(26i + 42\right]$$

$$= \frac{e^{-z}}{61} [6z \cos z + 5z \sin z + \frac{42}{61} . 6 \cos z + 5 . \frac{42}{61} \sin z + \frac{26}{61} . (5 \cos z - 6 \sin z)]$$

$$= \frac{x^{-1} \log x}{61} [6 \text{ fcos } (\log x) + 5 \sin (\log x)] + \frac{2x^{-1}}{3721} [\ 191 \cos (\log x) + 27 \sin (\log x)]$$

Hence the solution is

$$y = \left(c_1 x^{\sqrt{3}} + c_2 x^{-\sqrt{3}}\right) x^2 + \frac{1}{6} x^{-1} + \frac{x^{-1} \log x}{61} . [6 \cos (\log x) + 5 \sin (\log x)] + \frac{2x^{-1}}{3721} [\ 191 \cos (\log x) + 27 \sin (\log x)]$$

Example 3:

Solve $x^4 \frac{d^3y}{dx^3} + 2x^3 \frac{d^2y}{dx^2} - x^2 \frac{dy}{dx} + xy = 1$.

Solution:

The given equation may be written as

$$x^3 \frac{d^3y}{dx^3} + 2x^3 \frac{d^2y}{dx^2} - x\frac{dy}{dx} + y = \frac{1}{x}$$

Putting $x = e^z$, the equation becomes

$$[D (D - 1) (D - 2) + 2D (D - 1) - D + 1] y = e^{-z}$$

$$\Rightarrow \quad (D - 1)^2 (D + 1)y = e^{-z}$$

A.E. is $(m - 1)^2 (m + 1) = 0$ or $m = 1, 1, -1$

$\therefore$ C.F. $= (c_1 + c_2z)e^z + c_3e^{-z} = (c_1 + c_2 \log x) . x + c_3 x^{-1}$

and $$\text{P.I.} = \frac{1}{(D-1)^2 (D+1)e^{-z}} = \frac{1}{(D+1)} \frac{1}{(D-1)}.e^{-z}$$

$$= \frac{1}{(D+1)}.\frac{1}{4}.e^{-z} = \frac{1}{4}.e^{-z} \frac{1}{D-1+1}.1$$

$$= \frac{1}{4}ze^{-z} = \frac{1}{4x}\log x.$$

Hence, the required solution is

$$y = (c_1 + c_2 \log x)\, x + \frac{c_3}{x} + \frac{1}{4x}\log x.$$

Example 4:

Solve $(x^2 D^2 - xD + 4)\, y = \cos (\log x) + x \sin (\log x)$.

Solution:

Putting $x = e^z$ and denoting $\frac{d}{dz}$ by D', the given equation becomes

$$[D' (D' - 1) - D' + 4]y = \cos z + e^z \sin z$$

$$[D'^2 - 2D' + 4]\, y = \cos z + e^z \sin z$$

A.E. $\quad m^2 - 2m + 4 = 0$ or $m = 1 \pm \sqrt{3}i$.

$\therefore$ C.F. $= e^z [c_1 \cos \sqrt{3}\, z + c_2 \sin \sqrt{3}\, z]$

$$= x\, [c_1 \cos (\sqrt{3} \log x) + c_2 \sin (\sqrt{3} \log x)]$$

and $$\text{P.I.} = \frac{1}{D'^2 - 2D' + 4}.\cos z + \frac{1}{D'^2 - 2D' + 4} e' \sin z$$

$$= \frac{1}{-1^2 - 2D' + 4}.\cos z + e^z \frac{1}{(D'+1)^2 - 2(D'+1) + 4}. \sin z$$

$$= \frac{1}{3-2D'}\cos z + e^z \frac{1}{D'^2+3}\sin z$$

$$= \frac{(3+2D')}{9-4D'^2}.\cos z + e^z \frac{1}{-1^2+3}\sin z$$

$$= \frac{3\cos z - 2\sin z}{9+4} + \frac{e^2 \sin z}{2}$$

$$= \frac{1}{13}\ [3\cos(\log x) - 2\sin(\log x)] + (x/2)\sin(\log x)$$

Hence, the complete solution is

$$y = x[c_1 \cos(\sqrt{3}\log x) + c_2 \sin(\sqrt{3}\log x)]$$

$$+ \frac{1}{13}\ [3\cos(\log x) - 2\sin(\log x)] + (x/2)\sin(\log x)].\ \textbf{Ans.}$$

Example 5:

Solve $x^2 \frac{d^2y}{dx^2} + 2x\frac{y}{dx} - 20y = (x+1)^2.$ **(Osmania 89)**

Solution:

Putting $x = e^z$, the equation becomes

$[D(D-1) + D - 20]\, y = (e^z + 1)^2$

$\Rightarrow \quad (D+5)(D-1)\, y = e^{2z} + 2e^z + 1$

A.E. is $(m+5)(m-4) = 0 \qquad \therefore m = 4, -5$

C.F. $= c_1e^{4z} + c_2e^{-5z} = c_1x^4 + c_2x^{-5}$

and P.I. $= \frac{1}{(D+5)(D-4)}.e^{2z} + \frac{1}{(D+5)(D-4)}.2e^z$

$$+ \frac{1}{(D-5)(D-4)}.e^0$$

$$= \frac{e^{2z}}{7.(-2)} + \frac{2e^z}{6.(-3)} + \frac{1}{5(-4)} = -\frac{1}{14}x^2 - \frac{1}{9}x - \frac{1}{20}.$$

Hence the complete solution is

$$y = c_1x^4 + c_2x^{-5} - \frac{1}{14}x^2 - \frac{1}{9}x - \frac{1}{20}.$$

Example 6:

Solve $x^3 \frac{d^3y}{dx^3} + 2x^2 \frac{d^2y}{dx^2} + 2y = 10\left(x + \frac{1}{x}\right)$

(Kanpur, 96; Vikram, 95)

Solution:

Putting $x = e^z$, denoting $\frac{d}{dz}$ by D,

the given equation becomes

$$[D(D-1)(D-2) + 2D(D-1) + 2]y = 10(e^z + e^{-z})$$

$\Rightarrow$ $(D^3 - D^2 + 2)y = 10(e^z + e^{-z})$

$\Rightarrow$ $(D+1)(D^2 - 2D + 2)y = 10(e^z + e^{-z})$

A.E. is $(m+1)(m^2 - 2m + 2) = 0$

$\therefore$ $m = -1, 1 \pm i$

$\therefore$ C.F. $= c_1 e^{-z} + c_2 e^z \cos(z + c_3)$

$= c_1 x^{-1} + c_2 \cos(\log x + c_3)$

and $\text{P.I.} = \dfrac{10}{(D+1)(D^2 - 2D - 2)} e^z + \dfrac{10}{(D+1)(D^2 - 2D + 2)} . e^{-z}$

$$= \frac{10e^z}{(1+1)(1-2+2)} + \frac{10}{(1+2+2)} . \frac{1}{D+1} e^{-z}.$$

$$= 5e^z + 2e^{-z} \frac{1}{D-1+1} . 1$$

$$= 5e^z + 2e^{-z}\left(\frac{1}{D}.1\right) = 5e^z + 2ze^{-1} - 5x + 2x^{-1} \log x.$$

Hence, the complete solution is

$y = c_1 x^{-1} + c_2 x \cos(\log x + c_3) + 5x + 2x^{-1} \log x.$ **Ans.**

Example 7:

Solve $(x^2D^2 + 3xD + 1)\, y = \dfrac{1}{(1-x)^2}$

(Rohilkhand, 91, 92; Kanpur, 90)

Solution:

Putting $x = e^z$ and denoting $\frac{d}{dz}$ by D', the equation becomes

$$[D'(D'-1) + 3D' + 1]y = \frac{1}{(1-e^z)^2}$$

$\Rightarrow$ $(D'+1)^2 y = \dfrac{1}{(1-e^z)^2}$

A.E. is $(m+1)^2 = 0$ $\therefore$ $m = 1, -1$

$\therefore$ C.F. – $(c_1 + c_2 z)\, e^{-z} = (c_1 + c_2 \log x) \,.\, x^{-1}$

$$\text{P.I.} = \frac{1}{(D'+1)^2} \cdot \frac{1}{(1-e^z)^2} = \frac{1}{(D'+1)} \frac{1}{(D'+1)} \cdot \frac{1}{(1-e^z)^2}$$

Now, let $\frac{1}{(D'+1)^2} \cdot \frac{1}{(1-e^z)^2} = v$

$$\Rightarrow \quad (D'+1)v = \frac{1}{(1-e^z)^2} \quad \text{or} \quad \frac{dv}{dz} + v = (1-e^z)^{-2}$$

$$\text{I.F.} = e^{\int 1.dz} = e^z$$

$$\therefore v.e^z = \int e^z (1-e^z)^{-2}\, dz = (1-e^z)^{-1}$$

$$\Rightarrow \quad v = \frac{1}{(D'+1)} \cdot \frac{1}{(1-e^z)^2} = e^{-z} (1-e^z)^{-1}$$

$$\therefore \text{P.I.} = \frac{1}{(D'+1)} \,.\, e^{-z} (1-e^z)^{-1}$$

$$= e^{-z} \int e^z \,.\, e^{-z} (1-e^z)^{-1}\, dz = e^{-z} \int \frac{1}{1-e^z} \,.\, dz$$

$$= e^{-z} \int \frac{1}{x(1-x)}\, dx$$

Putting $x = e^z$

$$= e^{-z} \int \left(\frac{1}{x} + \frac{1}{1-x}\right) dx$$

so that $dx = e^z \,.\, dz$

$dz = 1/x\ dx$

$$= e^{-z} [\log x - \log (1-x)] = x^{-1} \log \frac{x}{1-x}$$

Hence, the complete solution is

$$y = (c_1 + c_2 \log x) \frac{1}{x} + \frac{1}{x} \log \frac{x}{1-x}.$$

Ans.

Example 8:

Solve $x^2 \frac{d^2y}{dx^2} + 4x \frac{dy}{dx} + 2y = e^x$.

Solution:

Putting $x - e^x$ the equation becomes

$$[D(D-1) + 4D + 2]\, y = e^{e^z}$$

$\Rightarrow \quad (D+2)(D+1)y = e^{e^z}$

A.E. is $(m+2)(m+1) = 0$ $\quad \therefore \ m = -2, -1.$

$\therefore$ C.F. $= c_1 e^{-2x} + c_2 e^{-z} = c_1 x^{-2} + c_2 x^{-1}$

and $\quad$ P.I. $= \dfrac{1}{(D+2)(D+1)} e^{e^z} = \left(\dfrac{1}{D+1} - \dfrac{1}{D+2}\right) e^{e^z}$

$$= \frac{1}{D+1} e^{e^z} = \frac{1}{D+2} e^{e^z}$$

Now, let $\quad \dfrac{1}{D+1} e^{e^z} = v$

$\Rightarrow \quad (D+1)v = e^{e^z} \quad$ or $\quad \dfrac{dv}{dz} + v = e^{e^z}$

$$\text{I.F.} = e^{\int 1.dz} = e^z$$

$\therefore \ v.e^z = \int e^z . e^{e^z} \, dz = e^{e^z}$

$\therefore \quad v = \dfrac{1}{x} e^x \quad$ or $\quad \dfrac{1}{D+1} e^{e^z} = \dfrac{1}{x} e^x$

Similarly, $\dfrac{1}{D+2} e^{e^z} = \left(\dfrac{1}{x} - \dfrac{1}{x^2}\right) e^x.$

Hence, the solution is $y = c_1 x^{-2} + c_2 x^{-1} + \dfrac{1}{x^2} e^x.$ **Ans.**

7.4 Equations Reducible to Homogeneous Form

An equation of the form

$$(a+bx)^n \frac{d^n y}{dx^n} + P_1 (a+bx)^{n-1} \frac{d^{n-1} y}{dx^{n-1}} + \dots + P_{n-1} (a+bx) \frac{dy}{dx} + P_n y = F(x)$$

where $P_1, P_2, \dots, P_n$ are all constants, can be reduced to the homogeneous linear form by putting $e^z = a + bx$ and then can be solved by the method indicated.

Example 1:

Solve $[(3x+2)^2 D^2 + 3(3x+2) D - 36] y = 3x^2 + 4x + 1.$

Solution:

We have $[(3x+2)^2 D^2 + 3(3x+2) D - 36] y = 3x^2 + 4x + 1.$

Putting $3x + 2 = e^z$ and denoting $\frac{d}{dz}$ by D', the equation reduces to

$$[3^2 D'(D' - 1) + 3.3 D' - 36]y = 3\left(\frac{e^z - 2}{3}\right)^2 + 4\left(\frac{e^z - 2}{3}\right) + 1$$

$\Rightarrow$ $$(D'^2 - 4)\, y = \frac{1}{27}(e^{2z} - 1)$$

A.E. is $m^2 - 4 = 0$ or $m = 2, -2$

$\therefore$ $$\text{C.F.} = c_1 e^{2x} + c_2 e^{-2z}$$
$$= c_2 (3x + 2)^2 + c_2 (3x + 2)^{-2}$$

and $$\text{P.I.} = \frac{1}{27}.\frac{1}{(D'^2 - 4)}.e^{3z} - \frac{1}{27}\frac{1}{D'^2 - 4}.1$$

$$= \frac{1}{27}.e^{2z}\frac{1}{(D' + 2)^2 - 4}1 + \frac{1}{27.4}\left(1 - \frac{D'^2}{4}\right)^{-1}.1$$

$$= \frac{e^{2z}}{27}.\frac{1}{D'^2 + 4D'}.1 + \frac{1}{108}\left(1 + \frac{D'^2}{4}\right)1$$

$$= \frac{e^{2z}}{27}.\frac{1}{4D'}\left(1 + \frac{D'^2}{4}\right)^{-1}1 + \frac{1}{108}$$

$$= \frac{e^{2z}}{27}.\frac{1}{4D'}1 + \frac{1}{108}$$

$$= \frac{1}{108}(ze^{2z} + 1) = \frac{1}{108}[(3x + 2)^2 \log(3x+2) + 1]$$

Hence, the complete solution is

$$y = c_1 (3x + 2)^2 + c_2 (3x + 2)^{-2} + \left(\frac{1}{108}\right)[(3x + 2)^2 \log(3x+2) + 1].$$

Ans.

Example 2:

Solve $2x^2y\frac{d^2y}{dx^2} + 4y^2 = x^2\left(\frac{dy}{dx}\right)^2 + 2xy\frac{dy}{dx}$

after making it homogeneous by the substituting $y = z^2$.

Solution:

We have

$$y = z^2 \qquad \therefore \qquad \frac{dy}{dx} = 2z\frac{dz}{dx}$$

and $$\frac{d^2y}{dx^2} = 2z\frac{d^2z}{dx^2} + 2\left(\frac{dz}{dx}\right)^2.$$

Substituting the values in the given equation, we have

$$2x^2z^2\left[2z\frac{d^2y}{dx^2} + 2\left(\frac{dz}{dx}\right)^2\right] + 4z^2 = x^2\left(2z\frac{dz}{dx}\right)^2 + 2xz^2.2z\frac{dz}{dx}$$

$$\Rightarrow \qquad x^2\frac{d^2z}{dx^2} - x\frac{dz}{dx} + z = 0$$

which is homogeneous equation.

Putting $x = e^v$ and denoting $\frac{d}{dv}$ by D, it becomes

$$[D(D-1) - D + 1]\, z = 0$$

$$\Rightarrow \qquad (D-1)^2 z = 0$$

A.E. is $(m-1)^2 = 0 \quad \Rightarrow m = 1, 1.$

Hence the solution is

$$z = (c_1 + c_2 v)e^v = (c_1 + c_2 \log x)x$$

$$\therefore y = z^2 = (c_1 + c_2 \log x)^2 x^2.$$

Example 3:

Solve $(x+a)^2\frac{d^2y}{dx^2} - 4(x+a)\frac{dy}{dx} + 6y = x.$ **(Meerut, TDC 97 (BP))**

Solution:

Putting $x + a = e^z$ and denoting $\frac{d}{dx}$ by D, the equation becomes

$$[D(D-1) - 4D + 6]\, y = e^z - a$$

$$(D-3)(D-2)\, y = e^z - a$$

A.E. $(m-3)(m-2) = 0$ or $m = 3, 2$

$$\therefore \text{C.F.} = c_1e^{3z} + c_1e^{2z} = c_1(x+a)^3 + c_2(x+a)^2$$

and $$\text{P.I.} = \frac{1}{(D-3)(D-2)}.e^z - \frac{1}{(D-3)(D-2)}.a$$

$$= \frac{2^z}{(1-3)(1-2)} - \frac{1}{6}\left(1 - \frac{5}{6}D + \frac{D^2}{6}\right)^{-1}.a.$$

$$= \frac{1}{3}e^z - \frac{1}{6}a = \frac{1}{2}(x + a) - \frac{1}{6}a.$$

Hence, the complete solution is

$$y = c_1 (x + a)^3 + c_2 (x + a)^3 + \frac{1}{2} (x + a) - \frac{a}{6}.$$ **Ans.**

Example 4:

Solve $(1 + x)^2 \frac{d^2y}{dx^2} + (1 + x) \frac{dy}{dx} + y = 4 \cos \log (1 + x).$

(Meerut, 93, (P))

Solution:

Putting $x + 1 = e^z$ and denoting $\frac{d}{dz}$ by D, the equation becomes

$$[D (D - 1) + D + 1] y = 4 \cos z$$

$\Rightarrow \quad (D^2 + 1) y = 4 \cos z$

A.E. is $m^2 + 1 = 0 \quad \Rightarrow \quad m = \pm i$

$\therefore$ C.F. $= c_1 \cos (z + c_2) = c_1 \cos [\log (x + 1) + c_2]$.

and P.I. $= \frac{1}{D^2 + 1} 4 \cos z = 4 \left(\frac{z}{2} \sin z\right)$

$$= 2 z \sin z = 2 \log (x + 1) \sin \log (x + 1).$$

Hence, the required solution is

$$y = c_1 \cos [\log (x + 1) + c_2] + 2 \log (x + 1) \sin \log (x + 1).$$

EXERCISES

1. $(2x - 1)^2 \frac{d^3y}{dx^3} + (2x - 1) \frac{dy}{dx} - 2y = 0.$

Ans. $y = c_1 (2x - 1) + c_2 (2x - 1) \cosh \left[(\sqrt{3}/2)\log(2x-1)+c_3\right]$

2. $(x + 1)^2 \frac{d^2y}{dx^2} + (x + 1) \frac{dy}{dx} = (2x + 3)(2x + 4)$ **(Rajsthan, 95)**

Ans. $y = c_1 + c_2 \log (1 + x) + x^2 + 8x + [\log (x + 1)]^2$

3. $(3x + 2)^2 \frac{d^2y}{dx^2} + 5(3x + 2) \frac{dy}{dx} - 3y = x^2 + x + 1.$

Ans. $y = c_1 (3x + 2)^{1/3} + c_2 (3x + 2)^{-1} + \frac{1}{405} (3x + 2)^2 - \frac{11}{105} (3x + 2) - \frac{2}{27}$

4. $(x^3D^3 + 6x^2D^2 + 8xD - 8)\, y = x^2$.

Ans. $y = c_1 x + x^{-2} [c_2 \cos (2 \log x) + c_3 \sin (2 \log x)] + \frac{1}{20} x^2$.

5. $x^2 \frac{d^2y}{dx^2} - 3x \frac{dy}{dx} + 5y = x^2 \sin (\log x)$. **(Meerut, 91 (P))**

Ans. $y = x^2 [c_1 \cos (\log x) + c_2 \sin (\log x)] - (x^2/2) \log x \cos (\log x)$

6. $x^3y_3 + 2x^2y_2 + 3xy_1 - 3y = x^2 + x$.

Ans. $y = c_1x + c_2 \cos (\sqrt{3} \log x) + c_3 \sin (\sqrt{3} \log x) + (x/4) \log x + x^2/7$

7. $x^2 \frac{d^2y}{dx^2} + 7x \frac{dy}{dx} + 13y = \log x$. **(Meerut, TDC 97)**

Ans. $y = x^{-3} [c_1 \cos (2 \log x) + c_2 \sin (2 \log x)] + \frac{1}{169} (13 \log x - 6)$

8. $x^2 \frac{d^2y}{dx^2} - x \frac{dy}{dx} - 3y = x^2 \log x$.

Ans. $y = c_1x^3 + (1/x)c_2 - \frac{1}{9} x^2 (3 \log x + 2)$

9. $x^2 \frac{d^2y}{dx^2} - 3x \frac{dy}{dx} + 4y = 2x^2$. **(Gorakhpur, 91; Vikram 93)**

Ans. $y = (c_1 + c_2 \log x)\, x^2 + (\log x)^2 x^2$.

10. $x^3 \frac{d^3y}{dx^3} + 3x^2 \frac{d^2y}{dx^2} + x \frac{dy}{dx} + y = x \log x$.

(Meerut, TDC 94 (BP))

Ans. $y = c_1/x + c_2 \sqrt{x} \cos \left(\frac{1}{2}\sqrt{3} \log x + c_3\right) + \frac{1}{2} x \left(\log x - \frac{3}{2}\right)$

11. $x^2 \frac{d^2y}{dx^2} - x \frac{dy}{dx} + 2y = x \log x$.

(Jiwaji, 90, 92; Madurai, 96)

Ans. $y = xc_1 \cos (\log x + c_2) + x \log x$.

12. $x^2 \frac{d^2y}{dx^2} + 4x \frac{dy}{dx} + 2y = x + \sin x$

Ans. $y = c_1/x + c_2/x^2 + x/6 - (1/x^2) \sin x$.

13. $(x^4D^4 + 6x^3 D^3 + 9x^2 D^2 + 3xD + 1) y = (1 + \log x)^2$

Ans. $y = (c_1 + c_2 \log x) \cos \log x + (c_3 + c_{43} \log x) \sin (\log x) + (\log x)^2 + 2 \log x - 3.$

14. $x^2 \dfrac{d^2y}{dx^2} - 4x \dfrac{dy}{dx} - 4y = x^4.$ **(Meerut, 91, 92, 93; Agra, 91)**

Ans. $y = c_1x^4 + c_2/x + \frac{1}{5}x^4 \log x.$

15. $x^2 \dfrac{d^2y}{dx^2} - 4x \dfrac{dy}{dx} + 6y = x^4.$ **(Meerut, TDC 97(P); Agra, 97)**

Ans. $y = c_1x^2 + c_2x^3 + \frac{1}{2} x^4.$

16. $x^3 \dfrac{d^3y}{dx^3} - x^2 \dfrac{d^2y}{dx^2} + 2x \dfrac{dy}{dx} - 2y = x^3 + 3x.$

(Karnataka, 93; Sagar, 96)

Ans. $y = (c_1 + c_2 \log x) x + c_3x^2 + \frac{1}{4}x^3 - \frac{3}{2} x (\log x)^2$

17. $x^2 \dfrac{d^2y}{dx^2} + 2x \dfrac{dy}{dx} = \log x.$

Ans. $y = c_1 + (1/x) c_2 + \frac{1}{2} (\log x)^2 - (\log x).$

8

Singular Solution

8.1 Singular Solution

Sometimes a solution of differential equation (not of the first degree) can be found without involving any arbitrary constant and this solution cannot be obtained for any particular value of the arbitrary constant in the general solution. This is called **Singular Solution**.

8.2 Discriminant

The discriminant of a quadratic equation

$$ax^2 + bx + c = 0 \quad \text{is} \quad b^2 - 4ac.$$

If $\phi(x, y, c) = 0$ is the solution of the differential equation

$$f(x, y, p) = 0,$$

then p-discriminant is obtained by eliminating p between

$$f(x, y, p) = 0 \quad \text{and} \quad \frac{\partial f}{\partial p} = 0.$$

and c-discriminant is obtained by eliminating c between

$$\phi(x, y, c) = 0 \quad \text{and} \quad \frac{\partial \phi}{\partial c} = 0.$$

16.3 Extraneous Loci

Let f (x, y, c) = 0 be the primitive of the differential equation f(x, y, p) = 0

If $\psi(x, y) = 0$ is its singular solution, then $\psi(x, y) = 0$ is a factor in both p-discriminant and c-discriminant.

Each discriminant may have other factor which correspond to other loci associated with primitive. In general these loci do not satisfy the differential equation, therefore these are sometimes called **Extraneous Loci.**

Tac-Locus: p-discriminant gives equal values of p but these values may belong to two curves of the system that are not consecutive i.e. these curves have different c, s. Locus of such points is called the **Tac-Locus**.

For example consider a family of circles, all of equal radii, whose centres lie on a straight line. The two circles, which are not consecutive, touch at P, i.e., have same values of p but the direction of the tangent at P is not the direction of the line of centres which is the locus of the points of contact of the circles. The lines of centres is called the **Tac-Locus**.

Nodal-Locus: c-discriminant gives equal values of c but these values may belong to the nodes (i.e. double points with distinct tangents) which are also ultimate points of intersection of the consecutive curves. Locus of such points is called the **Nodal-Locus.**

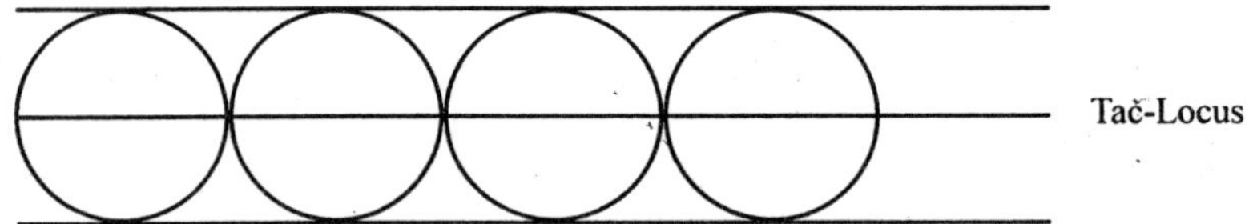

Cusp-Locus: c-discriminant gives equal values of c but these values may belong to the cusp (i.e., double points with coincident tangents) which are also ultimate points of intersection of the consecutive curves. Their locus is called the **Cusp-Locus.**

8.4 Important

The p-discriminat equated to zero may include the envelope (E) as a factor once, the cusp, locus (C) once, and the tac-locus (T) twice and the c-discriminant equated to zero may include envelope (E) once, the cusp-locus (C) thrice and the nodal locus (N) twice.

i.e., p-discriminant $\sim ECT^2 = 0$

and c-discriminant $= EC^3N^2 = 0$

The singular solution is obtained as a common factor from the c and p-discriminants and it must satisfy the differential equation.

Example 1:

Solve the differential equation

$$(px^2 + y^2)(px + y) = (p + 1)^2$$

by reducing it to Clairaut's Form and find its singular solution.

Solution:

We have $(px^2 + y^2)(px + y) = (p + 1)^2$.

Let us substitute $x + y = u$ and $xy = v$

so that $\dfrac{du}{dx} = 1 + \dfrac{dy}{dx} = 1 + p$

and $\dfrac{dv}{dx} = x\dfrac{dy}{dx} + y = xp + y$

$$\therefore \quad \frac{dv}{du} = \frac{\frac{dv}{dx}}{\frac{du}{dx}} = \frac{xp + y}{1 + p}$$

$\Rightarrow \quad P(1 + p) = xp + y$ where $P = dv/du$

$$\Rightarrow \quad p = \frac{y - P}{P - x}.$$

Putting this value of p in the given equation,

$$\left(\frac{y - p}{P - x}.x^2 + y^2\right)\left(\frac{y - p}{P - x}.x + y\right) = \left(\frac{y - p}{P - x} + 1\right)^2$$

$\Rightarrow \quad [P(y^2 - x^2) - (xy^2 - yx^2)](-Px + Py) - (y - x)^2$

$\Rightarrow \quad [P(v + x) - xy] P = 1]$

$$\Rightarrow \quad Pu - v = \frac{1}{P} \quad \Rightarrow \quad v = Pu - \frac{1}{P}$$

which is Clairaut form.

hence the general solution is

$$v = cu - \frac{1}{c}$$

$\Rightarrow \quad c^2u - cv - 1 = 0 \quad \Rightarrow \quad c^2(x + y) - xyc - 1 = 0$

$\therefore$ the c-discriminant (EN^2C^3) is

$$x^2y^2 + 4(x + y) = 0 \qquad \text{...(1)}$$

Also the given differential equation can be written as

$$p^2(x^3 - 1) + p(x^2y + y^2x - 2) + (v^3 - 1) = 0$$

$\therefore$ the p-discriminant (ECT^2) is

$$(x^2y + y^2x - 2)^2 - 4(x^3 - 1)(y^3 - 1) = 0$$

$\Rightarrow \quad [xy(x + y) - 2]^2 - 4(x^3y^3 - x^3 - y^3 + 1) = 0$

$\Rightarrow$ $[x^2y^2 (x + y)^2 - 4xy (x + y) - 4x^3y^3 + 4(x^3 + y^3) = 0$

$\Rightarrow$ $x^2y^2 (x^2 + 2xy + y^2 - 4xy) - 4xy (x + y) + 4(x + y) (x^2 - xy + y^2) = 0$

$\Rightarrow$ $x^2y^2 (x - y)^2 + 4(x + y) (x - y)^2 = 0$

$\Rightarrow$ $(x - y)^2 [x^2y^2 + 4(x + y)] = 0$

From equations (1) and (2) the singular solution is

$$x^2y^2 + 4(x + y) = 0.$$

and $x - y = 0$ is the tac-locus as it occurs twice in p-discriminant.

Ans.

Example 2:

Obtain the singular solution of the equation

$$p^2y^2 \cos^2 \alpha - 2pxy \sin^2 \alpha + y^2 - x^2 \sin^2 \alpha = 0$$

directly from the equation and also from its complete primitive explaining the geometrical significance of the irrelevant factors that present themselves.

Solution:

The given equation is

$$p^2y^2 \cos^2 \alpha - 2pxy \sin^2 \alpha + (y^2 - x^2 \sin^2 \alpha) = 0 \quad ...(1)$$

$\therefore$ the p-discriminant (ECT^2) is

$$4x^2y^2 \sin^2 \alpha - 4y^2 \cos^2 \alpha (y^2 - x^2 \sin^2 \alpha) = 0$$

$\Rightarrow$ $y^2 x^2 \sin^2 \alpha (\sin^2 \alpha + \cos^2 \alpha) - y^4 \cos^2 \alpha = 0$

$\Rightarrow$ $y^2 \cos^2 \alpha (x^2 \tan^2 \alpha - y^2) = 0 \quad ...(2)$

Again from (1) we have

$$p = \frac{2xy \sin^2 \alpha + \sqrt{[4x^2y^2 \sin^4 \alpha - 4y^2 \cos^2 \alpha (y^2 - x^2 \sin^2 \alpha)]}}{2y^2 \cos^2 \alpha}$$

$\Rightarrow$ $yp = x \tan^2 \alpha \pm \sec \alpha \sqrt{[x^2 \tan^2 \alpha - y^2]}$

$\Rightarrow$ $\pm \dfrac{y\,dy - x \tan^2 \alpha\, dx}{2\sqrt{[x^2 \tan^2 \alpha - y^2)]}} = \sec \alpha\, dx$

Integrating, we get $\sqrt{[x^2 \tan^2 \alpha - y^2]} = c - x \sec \alpha$

$\Rightarrow$ $x^2 \tan^2 \alpha - y^2 = (c - x \sec \alpha)^2$

$\Rightarrow$ $c^2 - 2cx \sec \alpha + x^2 (\sec^2 \alpha - \tan^2 \alpha) + y^2 = 0$

$\Rightarrow$ $c^2 - 2cx \sec \alpha + x^2 + y^2 = 0 \quad ...(3)$

which is the general solution (complete primitive) of the given equation (1).

∴ the c-discriminant (EC^3N^2) is

$$4x^2 \sec^2 \alpha - 4(x^2 + y^2) = 0$$

$$\Rightarrow \quad x^2 \tan^2 \alpha - y^2 = 0. \qquad \ldots(4)$$

Clearly the factor $x^2 \tan^2 \alpha - y^2$ occurs only once both in (2) and (4). Hence the singular solution is

$$x^2 \tan^2 \alpha - y^2 = 0$$

$$\Rightarrow \quad y = \pm x \tan \alpha.$$

Clearly y = 0 is the tac-locus as it occurs twice in p-disc.

Hence the given equation (1) represents a family of circles (2) whose envelope is given by

$$y = \pm \tan \alpha.$$ **Ans.**

Example 3:

Solve and test for singular solution $p^3 - 4pxy + 8y^2 = 0$.

Solution:

We have $p^3 - 4pxy + 8y^2 = 0$.

Here put $y = v^2$.

So that $\dfrac{dy}{dx} = 2v\dfrac{dv}{dx}$

$$\Rightarrow \quad p = 2vP \qquad \text{where } P = dv/dx$$

Putting in the given equation, we have

$$8v^3P^3 - 4.2vP.xv^2 + 8v^4 = 0$$

$$\Rightarrow \quad P^3 - xP + v = 0$$

$$\Rightarrow \quad v = Px - P^3$$

which is Clairaut's form

Hence the general solution is $v = c_1x - c_1^3$

$$\Rightarrow \quad v^2 = y = c_1^2(x - c_1^2)^2$$

$$\Rightarrow \quad y = c(x - c)^2 \qquad \ldots(1)$$

where $c_1^2 = c$.

Differentiating (1) partially w.r.t. c, we have

$$0 = (x - c)^2 - 2c(x - c)$$

$$\Rightarrow \quad (x - c)(x - 3c) = 0$$

$$\therefore x = c \quad \text{or} \quad x = 3c$$

Putting c = x in (1), we have y = 0

and putting c = x/3 in (1), we have $y = \frac{x}{3}.\frac{9x^2}{9}$

$\Rightarrow$ $27y - 4x^3 = 0.$

Therefore c-discs (EC^3N^2) is

$$y\ (27y - 4x^3) = 0, \qquad ...(3)$$

Again the given different equation is

$$p^3 - 4pxy + 8y^2 = 0$$

Differentiating (3), partially w.r.t. 'p' we have

$$3p^2 - 4xy = 0. \qquad ...(4)$$

Eliminating p between (3) and (4), we have the p-disc. (ECT^2) as

$$8\left(\frac{xy}{3}\right)^{3/2} - 4xy.2\sqrt{\left(\frac{xy}{3}\right)} + 8y^2 = 2$$

$$\Rightarrow \quad \sqrt{\left(\frac{xy}{3}\right)}.\left(\frac{xy}{3} - xy\right) = -y^2$$

$$\Rightarrow \quad \sqrt{\left(\frac{xy}{3}\right)}.\frac{2}{3}x = y$$

$$\Rightarrow \quad \frac{xy}{3}.\frac{4x^2}{9} = y^2$$

$$\Rightarrow \quad y(27y - 4x^2) = 0 \qquad ...(5)$$

Hence, the singular solutions are

$$y = 0 \text{ and } 27y - 4x^3 = 0$$

Here when c = 0, y = 0 from (1) also.

Hence, y = 0 is a particular integral of the equation. **Ans.**

Example 4:

Obtain the primitive and singular solution of the differential equation

$$4p^2x\ (x - a)\ (x - b) = [3x^2 - 2x\ (a + b) + ab]^2$$

Specify the nature of the loci which are not solutions but which are obtained with the singular solution.

Solution:

The given equation is

$$4p^2x\ (x - a)\ (x - b) + 0.p - [3x^2 - 2x.\ (a + b) + ab]^2 = 0 \qquad ...(1)$$

$\therefore$ the p-disc (ECT^2) is

$$x(x-a)(x-b)[3x^2 - 2x(a+b) + ab]^2 = 0 \quad ...(2)$$

Again from (1), we have

$$p = \frac{dy}{dx} = \pm \frac{3x^2 - 2x(a+b) + ab}{\sqrt{[x(x-a)(x-b)]}}$$

$$\Rightarrow \quad dy = \pm \frac{3x^2 - 2x(a+b) + ab}{2\sqrt{[x^3 - (a+b)x^2 \ abx]}} dx.$$

Integrating, we have $y = \pm \sqrt{[x^2 - (a+b)x^2 + abx]} + c.$

$$\Rightarrow \quad (y-c)^2 = x(x-a)(x-b) \quad ...(3)$$

which is the complete primitive of the given differential equation.

Equation (2) can be written as

$$c^2 - 2cy + [y^2 - x(x-a)(x-b)] = 0$$

$\therefore$ the c-disc. (EC^3N^2) is

$$4y^2 - 4[y^2 - x(x-a)(x-b)] = 0$$

$$\Rightarrow \quad x(x-a)(x-b) = 0 \quad ...(4)$$

Clearly $x(x-a)(x-b)$ are the common factors in (2) and (4) and the differential equation is satisfied by $x = 0$, $x = a$, $x = b$.

Hence, the singular solutions are

$$x = 0, x = a, x = b.$$

Again $3x^2 - 2x(a+b) + ab = 0$ which occurs twice only in p-disc. represents tac-locus. **Ans.**

Example 5:

Examine the equation

$$y^2 - 2pxy + p^2(x^2 - 1) = m^2$$

for singular solution:

Solution:

The given equation can be written as

$$(x^2 - 1)p^2 - 2xyp + (y^2 - 1) = m^2 \quad ...(1)$$

$$\Rightarrow \quad (px - y)^2 = p^2 + m^2$$

$$\Rightarrow \quad y = px \pm \sqrt{(p^2 + m^2)}$$

which is the Clairaut form.

Hence, the general solution is

$$y = cx \pm \sqrt{(c^2 + m^2)}$$

$\Rightarrow \quad (y - cx)^2 = c^2 + m^2$

$\Rightarrow \quad c^2(x^2 - 1) - 2xyc + (y^2 - m^2) = 0 \quad \ldots(2)$

Hence, from equations (1) and (2) both x and c-discriminant are

$4x^2y^2 - 4(x^2 - 1)(y^2 - m^2) = 0$

$\Rightarrow \quad y^2 + m^2x^2 = m^2$

which is the singular solution. **Ans.**

Example 6:

Reduce the equation $xyp^2 - (x^2 + y^2 - 1)p + xy = 0$ to Clairaut's Form by substituting $x^2 = u$ and $y^2 = v$. Hence show that the equation represents a family of conics touching the four sides of a square.

Solution:

Given that $x^2 = u$ and $y^2 = v$

$$\therefore \quad \frac{dv}{du} = \frac{\frac{dv}{dx}}{\frac{du}{dx}} = \frac{2y.\frac{dy}{dx}}{2x}$$

$\Rightarrow \quad P = \frac{y}{x}p \quad$ where $P = \frac{dv}{du}$

$\Rightarrow \quad p = \frac{xP}{y}$

Substituting in the given equation, we have

$$xy.\frac{x^2P^2}{y^2} - (x^2 + y^2 - 1)\frac{xP}{y} + xy = 0$$

$\Rightarrow \quad x^2P^2 - (x^2 + y^2 - 1)P + y^2 = 0$

$\Rightarrow \quad uP^2 - (u + v - 1)P + v = 0$

$\Rightarrow \quad v(-P + 1) - u(P - P^2) + P = 0$

$\Rightarrow \quad v = uP + \frac{P}{P - 1}$

which is Clairaut form.

Hence the general solution is $v = uc + \frac{c}{c - 1}$

$\Rightarrow \quad y^2(c - 1) = x^2c(c - 1) + c$

$\Rightarrow \quad x^2c^2 - (x^2 + y^2 - 1)c + y^2 = 0 \quad \ldots(1)$

$\therefore$ the c-disc. (EN^2C^2) is

$$(x^2 + y^2 - 1)^2 - 4x^2y^2 = 0$$

$$\Rightarrow \quad (x^2 + y^2 - 1 - 2xy)(x^2 + y^2 - 1 + 2xy) = 0$$

$$\Rightarrow \quad [(x - y)^2 - 1][(x + y)^2 - 1] = 0$$

$$\Rightarrow \quad (x - y - 1)(x - y + 1)(x + y - 1)(x + y + 1) = 0. \qquad \ldots(2)$$

Again from the given equation p-disc. (ECT^2) is

$$(x^2 + y^2 - 10^2 - 4x^2y^2 = 0 \qquad \ldots(3)$$

which is the same as the c-disc.

The common factors occurring once in p- and c-disc., give singular solution i.e., envelope of the family of conics given by (1) is

$$x - y - 1 = 0,\ x - y + 1 = 0,\ x + y - 1 = 0,\ x + y + 1 = 0.$$

These four lines clearly form a square.

Hence the given differential equation represents conics (1) touching the four sides of a square. **Ans.**

Example 7:

Reduce the differential equation $(px - y)(x - py) = 2p$ to Clairaut's form by substituting $x^2 = u$ and $y^2 = v$ and find its complete primitive and its singular solution, if any.

Solution:

Given that $x^2 = u$ and $y^2 - v$

$$\therefore \quad \frac{dv}{du} = \frac{\frac{dv}{dx}}{\frac{du}{dx}} = \frac{2y.\frac{dy}{dx}}{2x} = \frac{y}{x}P$$

$$\therefore\ p = \frac{x}{y}P \qquad \text{where } P = \frac{dv}{du}.$$

Substituting in the given equation, we have

$$\left(\frac{x^2}{y}P - y\right)(x - xP) = 2\frac{x}{y}P$$

$$\Rightarrow \quad (x^2P - y^2)(1 - P) = 2P$$

$$\Rightarrow \quad uP - v = \frac{2P}{1 - P}$$

$$\Rightarrow \quad v = Pu - \frac{2P}{1 - P}$$

Which is Clairaut's Form.

Hence, the general solution is $v = c_4 - \frac{2c}{1-c}$

$\Rightarrow \quad y^2 = cx^2 - \frac{2c}{1-c}$

$\Rightarrow \quad y^2 (1 - c) = cx^2 (1 - c) - 2c$

$\Rightarrow \quad c^2x^2 - c (x^2 + y^2 - 2) + y^2 = 0$

$\therefore$ c-disc. (EC^3N^2) is

$(x^2 + y^2 - 2)^2 - 4x^2y^2 = 0$

$\Rightarrow \quad (x^2 + y^2 - 2 - 2xy) (x^2 + y^2 - 2 + 2xy) = 0$

$\Rightarrow \quad [(x - y)^2 - 2] [(x + y)^2 - 2] = 0$

$\Rightarrow \quad (x - y - \sqrt{2}) (x - y + \sqrt{2}) (x + y - \sqrt{2}) (x + y + \sqrt{2}) = 0$ ...(1)

The given equation can be written as

$xyp^2 - p (x^2 - y^2 - 2) + xy = 0$ ...(2)

$\therefore$ p-disc. (ECT^2) is

$(x^2 + y^2 - 2)^2 - 4x^2y^2 = 0.$ ...(3)

which is the same as the c-disc.

The common factors of p and c-disc. occurring once in them give singular solutions.

Hence, the singular solutions are given by

$$(x - y + \sqrt{2}) (x - y - \sqrt{2}) (x + y - \sqrt{2}) (x + y + \sqrt{2}) = 0.$$

Ans.

Example 8:

Reduce the equation $x^2p^2 + py (2x + y) + y^2 = 0$, where $p = dy/dx$ to Clairaut's form by putting $u = y$ and $v = xy$ and find its complete primitive and also its singular solution.

Solution:

We have $u = y$ and $v = xy$

so that $\frac{du}{dx} = \frac{dy}{dx}$ and $\frac{dv}{dx} = x\frac{dy}{dx} + y$

$$\therefore \quad \frac{dv}{du} = \frac{\frac{dv}{dx}}{\frac{du}{dx}} = \frac{x\frac{dy}{dx} + y}{\frac{dy}{dx}} = \frac{xp + y}{p}$$

$$\Rightarrow \quad p\frac{dv}{du} = px + y$$

$$\Rightarrow \quad p = \frac{y}{p - x} \text{ where } P = \frac{dv}{dx}.$$

Putting this value of p in the given equation, we have

$$x^2 \frac{y^2}{(P-x)^2} + \frac{y^2}{P-x}(2x + y) + y^2 = 0$$

$$\Rightarrow \quad x^2 + (2x + y) . (P - x) + (P - x)^2 = 0$$

$$\Rightarrow \quad P^2 - xy + yP = 0$$

$$\Rightarrow \quad xy = yP + P^2 \quad \Rightarrow \quad v = Pu + P^2$$

which is of Clairaut's Form.

Hence the general solution is

$$v = cu + c^2 \qquad \Rightarrow \qquad c^2 + yc - xy = 0$$

$\therefore$ c-discriminant (EC^3N^2) is

$$y^2 + 4xy = 0 \qquad \Rightarrow \qquad y(y + 4x) = 0$$

Also from the given equation, p-discriminant (ECT^2) is

$$y^2 (2x + y)^2 - 4x^2y^2 = 0$$

$$\Rightarrow \quad y^2 (y^2 + 4xy) = 0$$

$$\Rightarrow \quad y.y^2 (y + 4x) = 0.$$

Now, y (y + 4x) occur both in c and p-discriminant and both y = 0 and y + 4x = 0 satisfy the given differential equation.

Hence, y = 0 and y + 4x = 0 are both singular solutions. **Ans.**

Example 9:

Find the complete primitive and the singular solution of the differential equation

$$\sin\left(x\frac{dy}{dx}\right)\cos y = \cos\left(x\frac{dy}{dx}\right)\sin y + \frac{dy}{dx}.$$

Solution:

We have $\sin\left(x\frac{dy}{dx}\right)\cos y = \cos\left(x\frac{dy}{dx}\right)\sin y + \frac{dy}{dx}$

The given equation can be written as

$$\sin(xp)\cos y = \cos(xp)\sin y + p \qquad \text{where } p = \frac{dy}{dx}$$

$\Rightarrow \quad \sin(xp - y) = p$

$\Rightarrow \quad xp - y = \sin^{-1}p$

$\Rightarrow \quad y = xp - \sin^{-1} p$ ...(1)

which is Clairaut's form.

Hence, the general solution is $y = cx - \sin^{-1} c$ ...(2)

from equations (1) and (2) it is clear that p and c-disc. shall be the same.

Differentiating (2), partially w.r.t. 'c' we have

$$0 = x - \frac{1}{\sqrt{(1 - c^2)}}$$

$$\Rightarrow \quad x^2 = \frac{1}{1 - c^2} \quad \text{or} \quad 1 - c^2 = \frac{1}{x^2}$$

$$\Rightarrow \quad c = \frac{\sqrt{(x^2 - 1)}}{x}.$$

Putting in (2), the c-disc., is

$$y = \frac{\sqrt{(x^2 - 1)}}{x}.x - \sin^{-1} \frac{\sqrt{(x^2 - 1)}}{x}$$

$$\Rightarrow \quad y = \sqrt{(x^2 - 1)} - \sin^{-1} \frac{\sqrt{(x^2 - 1)}}{x} \qquad \text{...(3)}$$

p-disc. is also the same.

Hence, the equation (3) is the singular solution. **Ans.**

Example 10:

Obtain the complete primitive and singular solution of the equation $4xp^2 = (3x - a)^2$, explaining the geometrical significance of the irrelevant factors that present themselves.

Solution:

From the given equation, we have

$$p = \frac{dy}{dx} = \pm \frac{3x - a}{2\sqrt{x}} = \pm\left(\frac{3}{2}\sqrt{x} - \frac{a}{2}.\frac{1}{\sqrt{x}}\right)$$

$$\Rightarrow \quad dy = \pm\left(\frac{3}{2}\sqrt{x} - \frac{a}{2\sqrt{x}}\right)dx$$

Integrating, we have

$$y = c \pm (x^{3/2} - a\sqrt{x})$$

$\Rightarrow$ $$(y - c)^2 = x(x - a)^2 \qquad \ldots(1)$$

which is the complete solution of the given equation

$$4xp^2 = (3x - a)^2. \qquad \ldots(2)$$

Equation (2) can be written as

$$4xp^2 + 0.p - (3x - a)^2 = 0$$

$\therefore$ p-discriminant (ECT^2) is

$$0 + 4.4x\,(3x - a)^2 = 0 \quad \text{or} \quad x(3x - a)^2 = 0 \qquad \ldots(3)$$

Also (1) can be written as

$$c^2 - 2yc + y^2 - x(x - a)^2 = 0$$

$\therefore$ c-discriminant (EC^3N^2) is

$$4y^2 - 4.1\,.\,\{y^2 - x(x - a)^2\} = 0$$

$$x\,(x - a)^2 = 0 \qquad \ldots(4)$$

Hence, $x = 0$ which is the common factor in equation (3) and (4) is the singular solution.

The factor $x - a = 0$ which occurs twice in c-disc, but does not occur in p-disc, is the nodal locus.

And the factor $3x - a = 0$ which occurs twice in p-disc, but does not occur in c-disc, is the Tac-locus.

EXERCISES

1. $p^2 + 2px - y = 0$. **(Raj., 91)**

   Ans. $(2x^3 + 3xy + c)^2 + 4\,(x^2 + y^2)^3$, $x^2 + y = 0$ the cusp-locus.

2. $x^3p^2 + x^2yp + a^3 = 0$ **(Raj., 93)**

   Ans. $a^3x + cxy + c^2 = 0$, $x(xy^3 - 4a^3) = 0$ singular solution, $x = 0$ tac-locus.

3. $y^2\,(1 + p^2) = r^2$

   Ans. $y^2 + (x + c)^2 = r^2$, $y^2 - r^2 = 0$ singular solution $y = 0$ tac-locus.

4. $xp^2 - (x - a)^2 = 0$

   Ans. $(y - c)^2 = \frac{4}{9}x\,(x - 3a)^2$, $x = 0$ singular solution; $x - a = 0$, tac-locus; $x - 3a = 0$ nodal locus.

5. $(xp - y)^2 = p^2 - 1$.

   Ans. $(y - cx)^2 = c^2 - 1$, $-y^2 + x^2 = 1$ singular soluton.

6. $(a^2 - x^2)\left[\frac{dy}{dx}\right]^2 + 2xy\frac{dy}{dx} + (b^2 - y^2) = 0.$

Ans. $y = cx \pm \sqrt{[(a^2c^2 + b^2)]}$. $x^2/a^2 + y^2/b^2 = 1.$

7. $xp^2 - 2yp + 4x = 0$

Ans. $c^3x^3 - 2yc + 4 = 0,\ y = \pm 2x.$

8. $x^2 (y - px) = yp^2.$

Ans. $y^2 = x^2c + c^2,\ x^4 + 4y^2 = 0$ singular solution, $x = 0$ tac-locus.

9. $(8p^2 - 27)x = 12p^2y$ **(Raj., 92)**

Ans. $xpx^2 - c(y + c)^2] = 0,\ 4y^3 + 22x^3 = 0.$

10. $(1 + p)^3 = \frac{27}{8a}(x + y)(1 - p)^3$

Ans. $(x - y + c)^3 = a(x + y)^2,\ x + y = 0.$

11. Transform the equation

$$(2x^2 + 1) . \frac{dy}{dx} + (x^2 + 2xy + y^2 + 2)\frac{dy}{dx} + 2y^2 + 1 = 0$$

to Clairaut's form by the substituting $x + y = u$, $xy - 1 = v$ and interpret. Find its singular solution also.

Ans. $v = vP + P^2$ where $P = \frac{dv}{du}$, $(x + y)^2 + 4(xy - 1) = 0.$

12. Reduce the equation $xp^2 - 2yp + x + 2y = 0$ to Clairaut's Form by putting $x^2 = u$ and $y - x = v$. Hence, obtain and interpret the primitive and singular solution of the equation.

Ans. $v = uP + (1/2)P$ where $P = dv/du$,
$2c^2x^2 - 2c(y - x) + 1 = 0,\ y - x = \pm\sqrt{2}c.$

13. Solve and find complete primitive and singular solution of the differential equation

$$3y = 2px - \frac{2p^2}{x}.$$

Ans. $9y^3 + 4c(3y - x^3) + 4c^2 = 0$, singular solution $x^3 - 6y = 0.$

9

Exact Differential Equations (Particular Forms)

9.1 Exact Differential Equations (Kanpur, 91)

Definition : A differential equation of the form

$$f\left(\frac{d^n y}{dx^n}, \frac{d^{n-1}y}{dx^{n-1}}, \ldots, \frac{dy}{dx}\right) = \psi(x) \qquad \ldots(1)$$

is said to be exact if it can be derived by differentiating directly an equation of the next lower order of the form

$$P\left(\frac{d^{n-1}y}{dx^{n-1}}, \frac{d^{n-2}y}{dx^{n-2}}, \ldots, \frac{dy}{dx}, y\right) = \int \phi(x) + c. \qquad \ldots(2)$$

The equation (2) is known as first integral of (1).

9.2 Condition of Exactness for a Linear Diff. Equation of Order n. (Kanpur, 90; Burdwan, 97)

Consider the linear differential equation

$$P_0 \frac{d^n y}{dx^n} + P_1 \frac{d^{n-1}y}{dx^{n-1}} + \ldots + P_n y = \phi(x) \qquad \ldots(1)$$

where P_0, P_1, ..., P_n are functions of x.

Let the equation (1) be obtained by differentiating once the equation

$$P_0 \frac{d^{n-1}y}{dx^{n-1}} + Q_1 \frac{d^{n-2}y}{dx^{n-2}} + \ldots + Q_{n-1} y = \int \phi(x)\,dx + c \qquad \ldots(2)$$

where Q_1, Q_2, ..., Q_{n-1} are functions of x.

Differentiating equation (2), w.r.t. x, we have

$$P_0\frac{d^ny}{dx^n}+(P'_0+Q_1)\frac{d^{n-1}y}{dx^{n-1}}+(Q'_1+Q_2)\frac{d^{n-2}y}{dx^{n-2}}+\ldots$$

$$+(Q'_{n-2}+Q_{n-1})\frac{dy}{dx}+Q'_{n-1}\,y=\phi(x) \qquad \ldots(3)$$

Now, equations (1) and (3) are the same equations. Hence comparing (1) and (3), we have

$$P_1 = P'_0 + Q_1,\ P_2 = Q'_1 + Q_2,\ P_3 = Q'_2 + Q_3$$

$$\ldots \quad \ldots \quad \ldots \quad \ldots \quad \ldots$$

$$P_{n-1} = Q'_{n-2} + Q_{n-1} \text{ and } P_n = Q'_{n-1}$$

$$\therefore Q_1 = P_1 - P'_0,\ Q_2 = Q'_1 = P_1 - P_1' + P_0''$$

$$\ldots \quad \ldots \quad \ldots \quad \ldots \quad \ldots$$

$$Q_{n-1} = P_{n-1}' + P_{n-2}'' - P_{n-3}''' + \ldots + (-1) + P_0^{(-1)}$$

$\therefore$ From $P_n = Q_{n-1}$, we have

$$P_n = P_{n-1}' - P_{n-2}'' + P_{n-3}''' \ldots + (-1)^{n-1} P_0^{(n)}$$

$$\Rightarrow \quad P_n - P_{n-1}' + P_{n-2}'' - P_{n-3}''' \ldots + (-1)^{(n)} P_0^{(n)} = 0$$

which is the condition for the given equation to be exact.

Also the first integral is

$$P_0\frac{d^{n-1}y}{dx^{n-1}}+(P_1-P'_0)\frac{d^{n-2}y}{dx^{n-2}}+(P_1-P_1'+P_0'')\frac{d^{n-3}y}{dx^{n-3}}+\ldots$$

$$\ldots+[P_{n-1}'-P_{n-2}'+\ldots+(-1)^{n-1}P_0^{n-1}]y=\int\phi(x)\,dx+c.$$

Example 1:

Solve $\frac{d^3y}{dx^3}+\cos x\frac{d^2y}{dx^2}-2\sin x\frac{dy}{dx}-y\cos x=\sin 2x.$

Solution:

We have $\frac{d^3y}{dx^3}+\cos x\frac{d^2y}{dx^2}-2\sin x\frac{dy}{dx}-y\cos x=\sin 2x.$

Here $P_0 = 1,\ P_1 = \cos x,\ P_2 = -2\sin x,\ P_3 = -\cos x.$

$\therefore$ The given equation will be exact if

$$P_3 - P_2' + P_1'' - P_0''' = 0$$

i.e., $-\cos x + 2\cos x - \cos x - 0 = 0.$

Hence, the equation of exactness is satisfied Therefore the given equation is exact.

The first integral is

$$P_0 \frac{d^2y}{dx^2} + (P_1 - P_0') \frac{dy}{dx} + (P_2 - P_1' + P_0'')\, y = \int \sin 2x\, dx + c_1.$$

$$\Rightarrow \quad \frac{d^2y}{dx^2} + \cos x \frac{dy}{dx} - \sin x.\, y = -\frac{1}{2} \cos 2x + c_1 \qquad ...(1)$$

for (1), $P_0 = 1$, $P_1 = \cos x$, $P_2 = -\sin x$

and $P_2 - P_1' + P_0'' = -\sin x + \sin x + 0 = 0$

i.e. condition of exactness for (1) is satisfied.

Hence the integral of (1) is

$$P_0 \frac{dy}{dx} + (P_1 - P_0')\, y = -\frac{1}{2} \int \cos 2x\, dx + c_1 + c_2$$

$$\Rightarrow \quad \frac{dy}{dx} + \cos x.y = -\frac{1}{4} \sin 2x\, dx + c_1 x + c_2$$

which is linear, I.F. $= e^{\int \cos x\, dx} = e^{\sin x}.$

Hence the solution is

$$y.e^{\sin x} = -\frac{1}{4} \int \sin 2x.e^{\sin x}\, dx + \int (c_1 x + c_2)\, e^{\sin x}\, dx + c_3$$

$$= -\frac{1}{2} \int \sin x.\cos x\, e^{\sin x}\, dx + \int (c_1 x + c_2)\, e^{\sin x}\, dx + c_3$$

$$= -\frac{1}{2} \int te^t\, dt + \int (c_1 x + c_2)\, e^{\sin x}\, dx + c_3 \qquad \text{where } t = \sin x$$

$$= -\frac{1}{2} \int (te^t - e^t) + \int (c_1 x + c_2)\, e^{\sin x}\, dx + c_3$$

$$= -\frac{1}{2} e^{\sin x} (\sin x - 1) \int (c_1 x + c_2)\, e^{\sin x}\, dx + c_3$$

which is the required solution. **Ans.**

Example 2:

Solve $\sin x \dfrac{d^2y}{dx^2} - \cos x \dfrac{dy}{dx} + 2y \sin x = 0.$

Solution:

We have $\sin x \dfrac{d^2y}{dx^2} - \cos x \dfrac{dy}{dx} + 2y \sin x = 0.$

Here $P_0 = \sin x$, $P_1 = -\cos x$, $P_2 = 2 \sin x$.

Condition of exactness is

$$P_2 - P_1' + P_0'' = 0$$

i.e., $\quad 2\sin x - \sin x - \sin x = 0$ is satisfied.

Hence the given equation is exact.

$\therefore$ The first integral is

$$P_0 \frac{dy}{dx} + (P_1 - P'_0)\, y = c_1$$

$$\Rightarrow \quad \sin x \frac{dy}{dx} + (-\cos x - \cos x)\, y = c_1$$

$$\Rightarrow \quad \frac{dy}{dx} - 2\cot x . y = c_1 \operatorname{cosec} x$$

which is linear.

$$\text{I.F.} = e^{-\int 2 \cot x\, dx} = e^{-2 \log \sin x} = \frac{1}{\sin^2 x} = \operatorname{cosec}^2 x.$$

Hence the solution is

$$y. \operatorname{cosec}^2 x = c_1 \int \operatorname{cosec} x. \operatorname{cosec}^2 x\, dx + c_2$$

$$= c_1 \int \operatorname{cosec}^3 x\, dx + c_2$$

$$= c_1 \left[-\frac{\cot x. \operatorname{cosec} x}{2} + \frac{1}{2} \int \operatorname{cosec} x\, dx \right] + c_2$$

since $\int \operatorname{cosec}^n x\, dx = -\dfrac{\cot x \operatorname{cosec}^{n-2} x}{n-1} + \dfrac{n-2}{n-1} \int \operatorname{cosce}^{n-2} x\, dx$

$$\Rightarrow \quad y \operatorname{cosec}^2 x = -\frac{c_1}{2} \cot x \operatorname{cosec} x + \frac{c_1}{2} \log \tan \frac{x}{2} + c_2$$

$$\Rightarrow \quad y = -\frac{c_1}{2} \cos x + \frac{c_1}{2} \sin^2 x \log \tan \frac{x}{2} + c_2 \sin^2 x$$

which is the required solution. **Ans.**

Example 3:

Solve $(ax - bx^2) \dfrac{d^2y}{dx^2} + 2a \dfrac{dy}{dx} + 2by = x.$ **(Raj., 94)**

Solution:

We have $(ax - bx^2)\dfrac{d^2y}{dx^2} + 2a\dfrac{dy}{dx} + 2by = x$

Here $P_0 = ax - bx^2$, $P_1 = 2a$, $P_2 = 2b$

and $\qquad P_2 - P_1' + P_0'' = 2b - 0 - 2b = 0$

i.e., the condition of exactness is satisfied.

Hence the given equation is exact.

$\therefore$ The first integral is

$$P_0\frac{dy}{dx} + (P_1 - P_0')\,y = \int x\,dx + c_1$$

$$\Rightarrow \quad (ax - bx^2)\frac{dy}{dx} + (2a - a + 2bx)\,y = \frac{x^2}{2} + c_1$$

$$\Rightarrow \quad (ax - bx^2)\frac{dy}{dx} + (a + 2bx)\,y = \frac{1}{2}x^2 + c_1$$

$$\Rightarrow \quad \frac{dy}{dx} + \frac{a + 2bx}{(a - bx)x}\,y = \frac{x}{2(a - bx)} + \frac{c_1}{x(a - bx)}$$

which is linear.

$$\therefore \quad \text{I.F.} = e^{\int \frac{a+2bx}{(a-bx)}dx} = e^{\int\left(\frac{1}{x} + \frac{3b}{a-bx}\right)dx} = e^{\log x - 3\log(a - bx)} = \frac{x}{(a - bx)^3}.$$

Hence the solution is

$$y.\frac{x}{(a - bx)^3} = \frac{1}{2}\int\frac{x^2\,ax}{(a - bx)^4} + c_1\int\frac{dx}{(a - bx)^4} + c_2$$

$$= \frac{1}{2}\int\frac{(a^2/b^2)\sin^4\theta\,(2a/b)\sin\theta\cos\theta\,d\theta}{a^4\cos^8\theta}$$

$$+ \frac{c_1}{3b(a - bx)^3} + c_2$$

Putting $bx = a\sin^2\theta$ so that

$$dx = \frac{2a}{a}\sin\theta\cos\theta\,d\theta$$

$$\sin^2\theta = \frac{bx}{a}, \qquad \cos^2\theta = 1 - \frac{bx}{a}$$

$$= \frac{1}{ab^2}\int\tan^5\theta.\sec^2\theta\,d\theta + \frac{c_1}{3b(a - bx)^3} + c_2$$

$$= \frac{1}{6ab^3} . \tan^6 \theta + \frac{c_1}{3b(a - bx)^3} + c_2$$

$$= \frac{1}{6ab^3} \frac{b^3x^3}{(a - bx)^3} + \frac{c_1}{3b(a - bx)^3} + c_2 \qquad \because \tan^2 \theta = \frac{bx}{a - bx}$$

$$\Rightarrow \quad yx = \frac{x^3}{6a} + \frac{c_1}{3b} + c_2(a - bx)^3$$

which is the required solution. **Ans.**

Example 4:

Solve $x^2 \frac{d^2y}{dx^2} + 3x\frac{dy}{dx} + y = \frac{1}{(1 - x)^2}$

Solution:

Here $P_0 = x^2$, $P_1 = 3x$, $P_2 = 1$

and $\qquad P_2 - P_1' + P_0'' = 1 - 3 + 2 = 0$

i.e., the condition of exactness is satisfied.

Hence the given equation is exact.

∴ The first integral is

$$P_0 \frac{dy}{dx} + (P_1 - P'_0)\, y = \int \frac{1}{(1 - x)^2} dx + c_1$$

$$\Rightarrow \quad x^2 \frac{dy}{dx} + (3x - 2x)\, y = \frac{1}{(1 - x)} + c_1$$

$$\Rightarrow \quad \frac{dy}{dx} + \frac{1}{x}y = \frac{1}{x^2(1 - x)} + \frac{c_1}{x^2}$$

which is linear. I.F. $= e^{\int \frac{1}{x}dx} = x.$

Hence the required solution is

$$yx = yx = \int x\left[\frac{1}{x^2(-x)} + \frac{c_1}{x^2}\right] dx + c_2$$

$$= \int \frac{1}{x(1 - x)} dx + c_1 \int \frac{dx}{x} + c_2$$

$$= \int \left(\frac{1}{x} + \frac{1}{1 - x}\right) dx + c_1 \int \frac{dx}{x} + c_2$$

$$= \log x - \log (1 - x) + c_1 \log x + c_2$$

$$= \log \frac{x}{1-x} + c_1 \log x + c_2$$

which is the required solution. **Ans.**

Example 5:

Solve $(1 + x + x^2) \dfrac{d^3y}{dx^3} + (3 + 6x) \dfrac{d^2y}{dx^2} + 6 \dfrac{dy}{dx} = 0.$

(Gorakhpur, 90; Rohilkhand, 91)

Solution:

We have $(1 + x + x^2) \dfrac{d^3y}{dx^3} + (3 + 6x) \dfrac{d^2y}{dx^2} + 6 \dfrac{dy}{dx} = 0.$

Here $P_0 = 1 + x + x^2$, $P_1 = 3 + 6x$, $P_2 = 6$, $P_3 = 0$

and $\quad P_3 - P_2' + P_1'' - P_0''' = 0 - 0 + 0 - 0 = 0.$

Hence the condition of exactness is satisfied.

$\therefore$ the given equation is exact

$\therefore$ Its first integral is

$$P_0 \frac{d^2y}{dx^2} + (P_1 - P_0') \frac{dy}{dx} + (P_2 - P_1' + P_0'')y = c_1$$

$$\Rightarrow \quad (1 + x + x^2) \frac{d^2y}{dx^2} + [(3 + 6x) - (1 + 2x)] \frac{dy}{dx} + (6 - 6 + 2)y = c_1$$

$$\Rightarrow \quad (1 + x + x^2) \frac{d^2y}{dx^2} + (2 + 4x) \frac{dy}{dx} + 2y = c_1. \qquad \ldots(1)$$

Now, for (1), $P_0 = 1 + x + x^2$, $P_1 = 2 + 4x$, $P_2 = 2$

and $\quad P_2 - P_1' + P_0'' = 2 - 4 + 2 = 0.$

Hence, equation (1) is also exact.

$\therefore$ its first integral is $P_0 \dfrac{dy}{dx} + (P_1 - P_0')\, y = \int c_1\, dx + c_2$

$$\Rightarrow \quad (1 + x + x^2) \frac{dy}{dx} + [2 + 4x - (1 + 2x)]\, y = c_1 x + c_2$$

$$\Rightarrow \quad (1 + x + x^2) \frac{dy}{dx} + (1 + 2x)\, y = c_1 x + c_2$$

$$\Rightarrow \quad \frac{dy}{dx} + \frac{1 + 2x}{1 + x + x^2} y = \frac{c_1 x + c_2}{1 + x + x^2} \qquad \ldots(2)$$

which is linear

$\therefore$ I.F. $= e^{\int \frac{(1+2x)}{(1+x+x^2)} dx} = (1 + x + x^2)$

$\therefore$ The general solution is

$$(1 + x + x^2)\, y = \int (c_1 x + c_2)\, dx + c_3$$

or $(1 + x + x^2)y = \frac{1}{2} c_1 x^2 + c_2 x + c_3$

which is the required solution. **Ans.**

Example 6:

Solve $(x^3 - 4x) \frac{d^3y}{dx^3} + (9x^2 - 12) \frac{d^2y}{dx^2} + 18x \frac{dy}{dx} + 6y = 0.$

Solution:

We have $(x^3 - 4x) \frac{d^3y}{dx^3} + (9x^2 - 12) \frac{d^2y}{dx^2} + 18x \frac{dy}{dx} + 6y = 0.$

Here $P_0 = x^3 - 4x$, $P_1 = 9x^2 - 12$, $P_2 = 18x$, $P_3 = 6$.

and $\qquad P_3 - P_2' + P_1'' - P_0''' = 6 - 18 + 18 - 6 = 0.$

Hence the given equation is exact.

The first integral is

$$P_0 \frac{d^2y}{dx^2} + (P_1 - P_0') \frac{dy}{dx} + (P_2 - P_1' + P_0'')\, y = c_1$$

$$\Rightarrow \quad (x^3 - 4x) \frac{d^2y}{dx^2} + [(9x^2 - 12) - (3x^2 - 4)] \frac{dy}{dx}$$
$$+ [18x - 18x + 6x]\, y = c_1$$

$$\Rightarrow \quad (x^3 - 4x) \frac{d^2y}{dx^2} + (6x^2 - 8) \frac{dy}{dx} + 6xy = c_1. \qquad ...(1)$$

Now from (1),

$P_0 = x^3 - 4x$, $P_1 = 6x^2 - 8$, $P_2 = 6x$

and $\qquad P_2 - P_1' - P_0'' = 6x - 12x + 6x = 0$

Hence (1) is also exact.

$\therefore$ its first integral is

$$P_0 \frac{dy}{dx} + (P_1 - P'_0)\, y = \int c_1\, dx + c_2$$

$$\Rightarrow \quad (x^3 - 4x) \frac{dy}{dx} + [(6x^2 - 8) - (3x^2 - 4)]\, y = c_1 x + c_2$$

$$\Rightarrow \quad (x^3 - 4x)\frac{dy}{dx} + (3x^2 - 4)\,y = c_1x + c_2$$

$$\Rightarrow \quad \frac{dy}{dx} + \frac{2x^2 - 4}{x^3 - 4x}\,y = \frac{c_1x + c_2}{x^3 - 4x} \qquad ...(2)$$

which is linear. I.F. $e^{\int \frac{3x^2 - 4}{x^3 - 4x}dx} = (x^3 - 4x)$

∴ The solution is

$$y.(x^3 - 4x) = \int (c_1x + c_2)\,dx + c_3$$

$$\Rightarrow \quad y(x^3 - 4x) = c_1\frac{x^2}{2} + c_2x + c_3$$

which is the required solution. **Ans.**

9.3 Integrating Factors

It may be noticed that sometimes an equation becomes exact after it has been multiplied by a suitable factor called the integrating factor.

Case I. In case the coefficients P_0, P_1, ... P_n etc. are of the type $(A_0x^s + A_1x^t + ...)$ etc., then integrating factor is of the type x^m.

In such cases multiply all terms by x^m and then applying the condition of integrability the particular value of m is obtained.

Case II. In case the coefficients P_0, P_1,..., etc. are trigonometrical functions, the integrating factor is obtained by trial method with is also a trigonometrical factor.

Example 1:

Solve $\frac{d^2y}{dx^2} + 2\sin x,\frac{dy}{dx} + 3y = \tan^2 x \sec x.$

Solution:

This equation is not exact. Multiplying it by cos x, we have

$$\cos x\frac{d^2y}{dx^2} + 2\sin x.\frac{dy}{dx} + 3\cos x.y = \tan^2 x \qquad ...(1)$$

for which $P_2 - P_1' + P_0'' = 0$.

Hence, equation (1) is exact equation.

$\therefore$ Its first integral is

$$P_0 \frac{dy}{dx} + (P_1 - P_0')y = \int \tan^2 x \, dx + c_1$$

$$\cos x \frac{dy}{dx} + (2 \sin x + \sin x)\, y = \int (\sec^2 x - 1) dx + c_1$$

$$\Rightarrow \quad \frac{dy}{dx} + 3y \tan x = \sec x \tan x - x \sec x + c_1 \sec x.$$

which is linear.

$$\text{I.F.} = e^{3\int \tan x \, dx} \; e^{-3 \log \cos x} = \frac{1}{\cos^3 x} = \sec^3 x$$

Hence the solution is

$$y \sec^3 x = \int \sec^4 x \tan x \, dx - \int x \sec^4 x \, dx + c_1 \int \sec^4 x \, dx + cx_2$$

$$= \int (1 + \tan^2 x) \tan x \sec^2 x \, dx - \int x(1 + \tan^2 x) \sec^2 x \, dx$$

$$+ c_1 \int 1 + \tan^2 x \; c \sec^2 x \, dx + c_2$$

$$= \int \tan x \sec^2 x \, dx + \int \tan^3 x \sec^2 x \, dx - \int x \sec^2 x \, dx$$

$$- \int x \tan^2 x \sec^2 x \, dx + c_1 \int \sec^2 x \, dx$$

$$+ c_1 \int x \tan^2 x \sec^2 x \, dx + c_2$$

$$= \frac{1}{2} \tan^2 x + \frac{1}{4} \tan^4 x - x \tan x + \int \tan x \, dx$$

$$-(x/3) \tan^3 x + \frac{1}{3} \int \tan^3 x \, dx + c_1 \tan x$$

$$+(c_1/3) \tan^3 x + c_2.$$

$$= \frac{1}{2} \tan^2 x + \frac{1}{4} \tan^4 x - x \tan x + \log \sec x - (x/3) \tan^3 x$$

$$+ \frac{1}{3} \int (\sec^2 x - 1) \tan x \, dx + c_1 \tan x + (c_1/3) \tan^3 x + c_2$$

$$= \frac{1}{2} \tan^2 x + \frac{1}{4} \tan^4 x - x \tan x + \log \sec x - (x/3) \tan^2 x$$

$$+ \frac{1}{3} \cdot \frac{1}{2} \tan^2 x - \frac{1}{3} \log \sec x + c_1 \tan x + (c_1/3) \tan^3 x + c_2$$

$$= \frac{2}{3}\tan^2 x - \frac{1}{4}\tan^4 x - x\tan x + \frac{2}{3}\log\sec x$$

$$(x/3)\tan^3 x + c_1 \tan x + (c_1/3)\tan^3 x + c_2$$

which is the required solution. **Ans.**

Example 2:

Solve $\sin^2 x \dfrac{d^2y}{dx^2} = 2y.$

Solution:

The given equation can be written as

$$\frac{d^2y}{dx^2} - 2\,\text{cosec}^2 x.y = 0$$

which is not exact.

Multiplying be cot x, we have

$$\cot x \frac{d^2y}{dx^2} - 2\cot x\,\text{cosec}^2 x.y = 0 \qquad ...(1)$$

For (1), $P_0 = \cot x$, $P_1 = 0$, $P_2 = -2\cot x\,\text{cosec}^2 x$.

$\therefore \quad P_2 - P_1' + P_0'' = -2\cot x.\,\text{cosec}^2 x - 0 + \cot x\,\text{cosec}^2 x = 0$

Thus (1) is an exact equation.

$\therefore$ Its first integral is $P_0 \dfrac{dy}{dx} + (P_1 - P_0')\,y = c_1$

$$\Rightarrow \quad \cot x.\frac{dy}{dx} + (0 + \text{cosec}^2 x).\,y = c_1$$

$$\Rightarrow \quad \frac{dy}{dx} + \frac{\text{cosec}^2 x}{\cot x}y = \frac{c_1}{\cot x}$$

which is linear.

$$\text{I.F.} = e^{\int \frac{\text{cosec}^2 x}{\cot x}dx} = e^{-\log\cot x} = \frac{1}{\cot x} = \tan x.$$

Hence the solution is

$$y.\tan x = c_1 \int \tan^2 x\,dx + c_2 = c_1 \int (\sec^2 x - 1)\,dx + c_2$$

$$= c_1(\tan x - x) + c_2$$

which is the required solution. **Ans.**

Example 3:

Solve $\sqrt{x}\,\dfrac{d^2y}{dx^2} + 2x\dfrac{dy}{dx} + 3y = x.$ **(Gauhati, 96; Bombay, 93; Raj., 91)**

Solution:

We have $\sqrt{x}\,\dfrac{d^2y}{dx^2} + 2x\dfrac{dy}{dx} + 3y = x$

Here the condition of integrability

$$P_2 - P_1' + P_0'' = 0 \text{ is not satisfied.}$$

Let it become exact after multiplying by x^m.

$\therefore$ Multiplying the given equation by x^m, we have

$$x^{m+1/2}\frac{d^2y}{dx^2} + 2x^{m+1}\frac{dy}{dx} + 3x^m y = x^{m+1} \qquad ...(1)$$

for (1) $P_0 = x^{m+1/2}$, $F_1 = 3x^{m+1}$, $P_2 = 3x^m$,

$\therefore$ (1) is exact if

$$P_2 - P_1' + P_0'' = 0$$

$$\Rightarrow \quad 3x - 2(m+1)\,x^m + \left(m + \frac{1}{2}\right)\left(m - \frac{1}{2}\right) x^{m-3/2} = 0$$

$$\Rightarrow \quad (2m-1)\left[-x^m + \frac{1}{4}(2m+1)\,x^{m-3/2}\right] = 0$$

$$\therefore \quad m = \frac{1}{2}$$

i.e., equation (1) is exact if $m = \dfrac{1}{2}$ $\quad\therefore$ I.F $= \sqrt{x}$.

Putting $m = \dfrac{1}{2}$, the exact equation is

$$x\frac{d^2y}{dx^2} + 2x^{3/2}\frac{dy}{dx}\ 3x^{1/2}\,y = x^{3/2} \qquad ...(2)$$

$\therefore$ Its first integral is $\quad P_0\dfrac{d^2y}{dx^2} + (P_1 - P_0')\,y = \int x^{3/2}\,dx + c_1$

$$\Rightarrow \quad x\frac{d^2y}{dx^2} + (2x^{3/2} - 1)\,y = \frac{2}{5}x^{3/2} + c_1$$

$$\frac{dy}{dx} + \left(2x^{1/2} - \frac{1}{x}\right)y = \frac{2}{5}x^{3/2} + \frac{c_1}{x}$$

which is linear, I.F. $= e^{\int\left(2x^{1/2}-\frac{1}{2}\right)dx} = e^{\frac{4}{3}x^{3/2}-\log x} . = \frac{1}{x}e^{\frac{4}{3}x^{3/2}}$

Hence the complete solution is

$$y.\frac{1}{x}e^{\frac{4}{3}x^{3/2}} = \frac{2}{3}\int x^{3/2}.\frac{1}{x}e^{\frac{4}{3}x^{3/2}}\,dx + c_1\int\frac{1}{x}.\frac{1}{x}e^{\frac{4}{3}x^{3/2}}\,dx + c_2$$

$$= \frac{2}{5}\int x^{1/2}e^{\frac{4}{3}x^{3/2}}\,dx + c_1\int\frac{1}{x^2}e^{\frac{4}{3}x^{3/2}}\,dx + c_2$$

$$= \frac{1}{5}e^{\frac{4}{3}x^{3/2}} + c_1\int\frac{1}{x^2}e^{\frac{4}{3}x^{3/2}}\,dx + c_2$$

which is the required solution. **Ans.**

Example 4:

$$x^5\frac{d^2y}{dx^2} + 3x^2\frac{dy}{dx} + (3-6x)\,x^2y = x^4 + 2x - 5.$$

Solution:

We have $x^5\dfrac{d^2y}{dx^2} + 3x^2\dfrac{dy}{dx} + (3-6x)\,x^2y = x^4 + 2x - 5.$

Here the conditions of integrability is not satisfied.

$\therefore$ Let the equation become exact after it has been multiplied by x^m.

Multiplying the given equation by x^m, we have

$$x^{m+5}\frac{d^2y}{dx^2} + 3x^{m+3}\frac{dy}{dx} + (3-6x)\,x^{m+2}\,y = x^{m+4} + 2x^{m+1} - 5x^m \quad ...(1)$$

For (1), $P_0 = x^{m+5}$, $P_1 = 3x^{m+3}$, $P_2 = (3-6x)\,x^{m+2}$

(1) is exact if

$P_2 - P_1' + P_0'' = 0$

$\Rightarrow \quad (3-6x)x^{m+2} - 3\,(m+3)x^{m+2} + (m+5)\,(m+4)\,x^{m+3} = 0$

$\Rightarrow \quad (m^2 + 9m + 14)\,x^{m+3} - 3\,(m+2)\,x^{m+2} = 0$

$\Rightarrow \quad (m+2)\,[(m+7)\,x^{m+3} - 3x^{m+2}] = 0$

$\therefore \quad m = -2$

i.e. x^2 is the I.F.

Putting m = – 2 in (1), we have

$$x^3\frac{d^2y}{dx^2}+3x\frac{dy}{dx}+(3-6x)\,y=x^2+\frac{2}{x}-\frac{5}{x^2}$$

which is the exact equation.

∴ Its first integral is

$$P_0\frac{dy}{dx}+(P_1-P_0')\,y=\int\left(x^2+\frac{2}{x}-\frac{5}{x^2}\right)dx+c_1$$

$$\Rightarrow\qquad x^3\frac{dy}{dx}+(3x-3x^2)\,y=\frac{x^3}{3}+2\log x\,\frac{5}{x}+c_1$$

$$\Rightarrow\qquad \frac{dy}{dx}+\frac{3(1-x)}{x^2}\,y=\frac{1}{3}+\frac{2}{x^3}\log x+\frac{5}{x^4}+\frac{c_1}{x^3}$$

which is linear

$$\text{I.F.}=e^{\int\frac{3(1-x)}{x^2}dx}=e^{3\int\left(\frac{1}{x^2}-\frac{1}{x}\right)dx}=e^{3\left(-\frac{1}{x}\log x\right)}=\frac{1}{x^3}e^{-3/x}$$

Hence the solution is

$$\frac{1}{x^3}e^{-3/x}=\left(\frac{1}{3}+\frac{2}{x^3}\log x+\frac{5}{x^4}+\frac{c_1}{x^3}\right).\frac{1}{x^3}e^{-3/x}\,dx$$

which is the required solution. **Ans.**

9.4 Non-Linear Equation

Sometimes the equations which are not linear can be solved by the trial method. This method can also be used in linear equations.

Example 1:

Solve $x^2y\frac{d^2y}{dx^2}+\left(x\frac{dy}{dx}-y\right)^2-3y^2=0.$

(Kanpur, 97, 98; Raj. 91, 93)

Solution:

The given equation can be written as

$$du=x^2y\frac{d^2y}{dx^2}+x^2\left(\frac{dy}{dx}\right)^2-2xy\frac{dy}{dx}-2y^2=0. \qquad ...(1)$$

∴ $u = c_1$ is first integral of (1)

Now, $x^2y\frac{d^2y}{dx^2}$ is obtained from differentiation of $x^2y\frac{dy}{dx}$

$\therefore$ Let $u_1 = x^2y\dfrac{dy}{dx}$

so that $du_1 = x^2y\dfrac{d^2y}{dx^2} + x^2\left(\dfrac{dy}{dx}\right)^2 + 2xy\dfrac{dy}{dx}$

so that $du - du_1 = 4xy\dfrac{dy}{dx} - 2y^2.$

Again let $u_2 = -2xy^2$, so that $du_2 = -4xy\dfrac{dy}{dx} - 2y^2$

$\therefore du - du_1 - du_2 = 0 \quad \Rightarrow \quad u = u_1 + u_2 = c_1.$

$\therefore$ The first integral is

$$x^2y\frac{dy}{dx} - 2xy^2 = c_1. \qquad ...(2)$$

Now, putting $y^2 = y$ so that $2y\dfrac{dy}{dx} = \dfrac{dv}{dx}$

the equation (2) reduces to

$$\frac{1}{2}x^2\frac{dv}{dx} - 2xv = c_1$$

$$\frac{dv}{dx} - \frac{4}{x}v = \frac{2c_1}{x^2}$$

which is linear equation.

$$\text{I.F.} = {}^{-4\int 1/x\,dx} = e^{-4\log x} = \frac{1}{x^4}.$$

Hence the solution is

$$v.\frac{1}{x^4} = 2c_1\int\frac{1}{x^2}\cdot\frac{1}{x^4}\,dx + c_2$$

$$y^2.\frac{1}{x^4} = 2c_1\left(-\frac{1}{5x^5}\right) + x_2$$

$\Rightarrow \quad xy^2 = a_1 + a_2x^5.$

which is the required solution. **Ans.**

Example 2:

Solve $x^2y\dfrac{d^2y}{dx^2} + \left(x\dfrac{dy}{dx} - y\right)^2 = 0,$ **(Kanpur, 97)**

Solution:

The given equation can be written as

$$du = x^2 y \frac{d^2y}{dx^2} + x^2 \left(\frac{dy}{dx}\right)^2 - 2xy\frac{dy}{dx} + y^2 = 0. \qquad ...(1)$$

Let $u_1 = x^2 y \dfrac{dy}{dx}$

so that $du_1 = x^2 y \dfrac{d^2y}{dx^2} + x^2 \left(\dfrac{dy}{dx}\right)^2 + 2xy\dfrac{dy}{dx}$

$\therefore du - du_1 = -4xy \dfrac{dy}{dx} + y^2.$

Again let $u_2 = -2xy^2$ so that $du_2 = -4xy \dfrac{dy}{dx} - 2y^2$

$\therefore du - du_1 - du_2 = -3y^2.$

Hence the equation is not exact.

So dividing the given equation (1) by x^2, we have

$$dv = y \frac{d^2y}{dx^2} + \left(\frac{dy}{dx}\right)^2 - \frac{1}{x^2}\left(2xy\frac{dy}{dx} - y^2\right) = 0.$$

Now, let $v_1 = y\dfrac{dy}{dx}$, so that $dv_1 = y\dfrac{d^2y}{dx^2} + \left(\dfrac{dy}{dx}\right)^2$

$\therefore dv + dv_1 = -\dfrac{1}{x^2}\left(2xy\dfrac{dy}{dx} - y^2\right) = -\dfrac{2y}{x}\dfrac{dy}{dx} + \dfrac{y^2}{x^2}$

Again let $v_2 = -\dfrac{y^2}{x}$ so that $dv_2 = -\dfrac{2y}{x}\dfrac{dy}{dx} + \dfrac{y^2}{x^2}$

$\therefore \quad dv - dv_1 - dv_2 = 0 \quad \Rightarrow \quad dv = dv_1 + dv_2$

$\therefore \quad v = v_1 + v_2 = c_1$

$$\Rightarrow \quad y\frac{dy}{dx} - \frac{y^2}{x} = c_1. \qquad ...(2)$$

Now, let $\dfrac{1}{2}y^2 = \omega$ so that $y\dfrac{dy}{dx} = \dfrac{d\omega}{dx}$

from equation (2), we have $y\dfrac{d\omega}{dx} - \dfrac{2}{x}\omega = c_1$

which is linear $\therefore$ I.F. $= e^{\int -(2/x)dx} = \dfrac{1}{x^2}.$

Hence the solution is $\omega\dfrac{1}{x^2} = \int c_1 . \dfrac{1}{x^2}dx + c_2.$

$$\Rightarrow \quad \omega = x^2\left(-\frac{c_1}{x}+c_2\right) \Rightarrow \frac{y^2}{2} = x(-c_1 + c_2x)$$

$$\Rightarrow \quad y^2 = 2x(-c_1 + c_1x).$$

Example 3:

Show that the equation

$$y + 3x\frac{dy}{dx} + 2y\left(\frac{dy}{dx}\right)^2 + \left(x^2 + 2y^2\frac{dy}{dx}\right)\frac{d^2y}{dx^2} = 0$$

is exact and find its integral. **(Meerut, 90)**

Solution:

The given equation can be written as

$$du = x^2\frac{d^2y}{dx^2} + 2y^2\frac{dy}{dx}\frac{d^2y}{dx^2} + 2x\left(\frac{dy}{dx}\right)^3 + 3x\frac{dy}{dx} + y = 0. \quad ...(1)$$

Now, let $u_1 = x^2\dfrac{dy}{dx}$ so that $du_1 = x^2\dfrac{d^2y}{dx^2} + 2x\dfrac{dy}{dx}$

$$\therefore\ du - du_1 = 2y^2\frac{dy}{dx}\frac{d^2y}{dx^2} + 2y\left(\frac{dy}{dx}\right) + x\frac{dy}{dx} + y = 0. \quad ...(2)$$

Again let $u_2 = y^2\left(\dfrac{dy}{dx}\right)^2$

So that $\quad du_2 = 2y^2\dfrac{dy}{dx}\dfrac{d^2y}{dx^2} + 2y\left(\dfrac{dy}{dx}\right)^3$

$$\therefore\ du - du_1 - du_2 = x\frac{dy}{dx} + y. \quad ...(3)$$

Again let $u_3 = xy$ so that $du_3 = x\dfrac{dy}{dx} + y$

$$\therefore\ du - du_1 - du_2 - du_3 = 0$$

$$\Rightarrow \quad d\left[x^2\frac{dy}{dx} + y^2\left(\frac{dy}{dx}\right)^2 + xy\right] = 0.$$

Hence, the given equation is exact and its first integral is

$$x^2\frac{dy}{dx} + y^2\left(\frac{dy}{dx}\right)^2 + xy = c_1.$$

Ans.

9.5 An Equation Which does not Contain x Directly

Let the equation of this form be

$$f\left(\frac{d^n y}{dx^n}, \frac{d^{n-1}y}{dx^{n-1}}, \ldots, y\right) = 0$$

Putting $\frac{dy}{dx} = p.$

so that $\frac{d^2y}{dx^2} = \frac{dp}{dx} = \frac{dp}{dy}\cdot\frac{dy}{dx} = p\frac{dp}{dy}$

$$\frac{d^3y}{dx^3} = \frac{d}{dx}\left(p\frac{dp}{dy}\right) = \frac{d}{dy}\left(p\frac{dp}{dy}\right)\cdot\frac{dy}{dx} = p^2\frac{d^2p}{dy^2} + p\left(\frac{dp}{dy}\right)^2, \ldots \text{etc.}$$

The given equation becomes

$$f\left(\frac{d^{n-1}p}{dy^{n-1}}, \ldots, p, y\right) = 0$$

which is of $(n-1)^{th}$ order between x and y and may be solved.

Let $p = f(y)$.

$\therefore$ the required solution is $\int \frac{dy}{f(y)} = x + c.$ **Ans.**

Example 1:

Solve $y\frac{d^2y}{dx^2} - \left(\frac{dy}{dx}\right)^2 = y^2 \log y.$ **(Kanpur, 98)**

Solution:

The given equation does not contain x directly.

$\therefore$ let $\frac{dy}{dx} = p$ so that $\frac{d^2y}{dx^2} = \frac{dp}{dx} = p\frac{dp}{dy}$

$\therefore$ the given equation becomes

$$yp\ yp\frac{dp}{dy} - p^2 = y^2 \log y$$

$$\Rightarrow \quad p\frac{dp}{dy} - \frac{1}{y}p^2 = y \log y.$$

Putting $p^2 = v$ so that

$p\frac{dp}{dy} - \frac{1}{2}\frac{dv}{dx}$, we have

$$\frac{1}{2}\frac{dv}{dx} - \frac{1}{y}v = y \log y \Rightarrow \frac{dv}{dx} - \frac{1}{2}v = 2y \log y$$

which is linear equation.

$\therefore$ $$\text{I.F.} = e^{-2\int 1/y\, dy} = e^{-2\log y} = \frac{1}{y^2}$$

$\therefore$ $$v.\frac{1}{y^2} = \int \frac{1}{y^2}\, 2y \log y\, dy + c_1$$

$\Rightarrow$ $$p^2.\frac{1}{y^2} = (\log y)^2 + c_1$$

$\Rightarrow$ $$p = y\sqrt{\{c_1 + (\log y)^2\}}$$

$\Rightarrow$ $$\frac{dy}{\sqrt{\{c_1 + (\log y)^2\}}} = dy \Rightarrow \frac{du}{\sqrt{(c_1 + u^2)}} = dx$$ where $u = \log y$

$\therefore$ $$\log [u + \sqrt{(c_1 + u^2)}] = x + \log c_2$$

$\Rightarrow$ $$u = \sqrt{(c_1 + u^2)} = c_2 e^x$$

$\Rightarrow$ $$\log y + \sqrt{\{c_1 + (\log y)^2\}} = c_2 e^x$$

$\Rightarrow$ $$c_1 + (\log y)^2 = (c_2 e^x - \log y)^2$$

$\Rightarrow$ $$c_1 = c_2^2 e^{2x} - 2c_2 e^x \log y$$

$\Rightarrow$ $$\log y = k_1 e^x + k_2 e^{-x}$$

which is the required solution. **Ans.**

9.6 An Equation Which does not Contain y Directly

Let the equation of this form be

$$f\left(\frac{d^n y}{dx^n}, \frac{d^{n-1}y}{dx^{n-1}}, \ldots, \frac{dy}{dx}, x\right) = 0. \quad \ldots(1)$$

Putting $\frac{dy}{dx} = p$, so that $\frac{d^2y}{dx^2} = \frac{dp}{dx}, \frac{d^3y}{dx^3} = \frac{d^2p}{dx^2}, \ldots$etc.

the equation (1) becomes

$$f\left(\frac{d^{n-1}y}{dx^{n-1}}, \frac{d^{n-2}y}{dx^{n-2}}, \ldots, p, x\right) = 0$$

which is of (n – 1)th order between x and p and may be solved for p.

If $\quad p\frac{dy}{dx} = F(x)$

the $\quad p \int F(x)\, dx + c$

is the solution of the given equation (1).

Note: If the given equation (not containing y directly) is of the form

$$\left(\frac{d^n y}{dx^n}, \frac{d^{n-1}y}{dx^{n-1}}, \ldots, \frac{d^m y}{dx^m}, n\right) = 0$$

$m < n$

i.e., the equation of nth order in which lowest derivative is of order m, the substitution is

$$\frac{d^m y}{dx^m} = q, \text{ so that } \frac{d^{m-1}y}{dx^{m-1}} = \frac{dq}{dx} \ldots \text{ etc.}$$

Thus, the new equation obtained will be of order $(n - m)$. **Ans.**

Example 1:

Solve $(1 + x^2)\dfrac{d^2y}{dx^2} + 1 + \left(\dfrac{dy}{dx}\right)^2 = 0.$

Solution:

given equation does not contain y-directly.

$\therefore$ Putting $\dfrac{dy}{dx} = p$ so that $\dfrac{d^2y}{dx^2} = \dfrac{dp}{dx}$

the given equation reduces to

$$(1 + x^2)\frac{dp}{dx} + 1 + p^2 = 0 \Rightarrow \frac{dp}{1 + p^2} + \frac{dx}{1 + x^2} = 0.$$

Integrating, $\tan^{-1} p + \tan^{-1}x = \tan^{-1} c_1$

$$\Rightarrow \quad \tan^{-1}\frac{p + x}{1 - px} = \tan^{-1} c_1 \Rightarrow \frac{p + x}{1 - px} = c_1$$

$$\Rightarrow \quad p + x = c_1(1 - px)$$

$$\Rightarrow \quad p = \frac{dy}{dx} = \frac{c_1 - x}{1 + c_1 x} = \frac{1}{c_1}\left(\frac{1 + c_1^2}{1 + c_1 x} - 1\right).$$

Integrating, $y = \dfrac{\left(1 + c_1^2\right)}{c_1^2}\log(1 + c_1 x) - \dfrac{1}{c_1}x + c_2$

which is the required solution. **Ans.**

Example 2:

$$2x\frac{d^3y}{dx^3}\cdot\frac{d^2y}{dx^2} = \left(\frac{d^2y}{dx^2}\right)^2 - a^2.$$

Solution:

The given equation does not contain y directly.

$\therefore$ Let $\dfrac{d^2y}{dx^2} = q \left(\text{Since } \dfrac{d^2y}{dx^2} \text{ is the lowest derivative} \right)$

so that $\dfrac{d^3y}{dx^3} = \dfrac{dq}{dx}$.

$\therefore$ the given equation becomes

$$2x\frac{dq}{dx}q = q^2 - a^2 \quad \Rightarrow \quad \frac{2q\,dq}{q^2 - a^2} = \frac{dx}{x}$$

Integrating, we have $\log(q^2 - a^2) = \log x + \log c_1$

$\Rightarrow \quad q^2 - a^2 = c_1x$

$\Rightarrow \quad q = \dfrac{dq}{dx} = \sqrt{(a^2 + a_1x)}$

$\Rightarrow \quad dp = \sqrt{(a^2 + c_1x)}\, dx$

Integrating, $p = \dfrac{2}{3c_1}(a^2 + c_1x)^{3/2} + c_2$

$\Rightarrow \quad \dfrac{dy}{dx} = \dfrac{2}{3c_1}(a^2 + c_1x)^{3/2} + c_2$

Integrating, we get $y = \dfrac{2}{3c_1}\cdot\dfrac{2}{5c_1}(a^2 + c_1x)^{5/2} + c_2x + c_3$

$\Rightarrow \quad y = \dfrac{4}{15c_1^2}(a^2 + 3c_1x)^{5/2} + c_2x + c_3$

which is the required solution. **Ans.**

Example 3:

Solve $y(1 - \log y)\dfrac{d^2y}{dx^2} + (1 + \log y)\left[\dfrac{dy}{dx}\right]^2 = 0.$

(Raj., 90; Kanpur, 99, 92)

Solution:

We have $y(1 - \log y)\dfrac{d^2y}{dx^2} + (1 + \log y)\left[\dfrac{dy}{dx}\right]^2 = 0.$

The given equation does not contain x directly.

$\therefore$ Putting $\dfrac{dy}{dx} = p$, so that $\dfrac{d^2y}{dx^2} = \dfrac{dp}{dx} = p\dfrac{dp}{dy}$.

The given equation becomes

$$y\,(1 - \log y)\; p\frac{dp}{dy} + (1 + \log y)\,p^2 = 0$$

$$\Rightarrow \quad \frac{dp}{dy} + y\frac{(1 + \log y)}{(1 - \log y)}\,dy = 0$$

$$\therefore \quad \log p = -\int \frac{1 + \log y}{y(1 - \log y)}\,dy + c_1 \qquad \text{where } \log y = t$$

$$= -\int \frac{1 + t}{1 - t}\,dt + c_1 \qquad \text{so that } \frac{1}{y}\,y = dt.$$

$$\therefore \quad \log p = \int\left(1 + \frac{2}{t - 1}\right) dt + c_1$$

$$= t + 2\log(t - 1) + c_1$$

$$= \log y + 2\log(\log y - 1) + c_1$$

$$\Rightarrow \quad p = \frac{dy}{dx} = cy\,(\log y - 1)^2$$

$$\Rightarrow \quad \frac{dy}{y(\log y - 1)} = cdx$$

Integrating $-\dfrac{1}{\log y - 1} = cx + c_2$

$$\Rightarrow \quad 1 - \log y = \frac{1}{cx + c_2}$$

which is the required solution. **Ans.**

Example 4:

Solve $\dfrac{d^2y}{dx^2} + \dfrac{dy}{dx} + \left(\dfrac{dy}{dx}\right)^3 = 0.$

Solution:

We have $\dfrac{d^2y}{dx^2} + \dfrac{dy}{dx} + \left(\dfrac{dy}{dx}\right)^3 = 0.$

The given equation does not contain y directly.

$\therefore$ let $\dfrac{dy}{dx} = p$ so that $\dfrac{d^2y}{dx^2} = \dfrac{dp}{dx}$

Putting in the given equation, we have

$$\frac{dp}{dx} + p + p^2 = 0$$

$$\Rightarrow \quad \frac{dp}{p(1+p^2)} + dx = 0$$

$$\Rightarrow \quad \left(\frac{1}{p} - \frac{p}{1+p^2}\right) dp + dx = 0.$$

Integrating, $\log p - \frac{1}{2} \log(1 + p^2) + x = \log c_1$

$$\Rightarrow \quad \frac{p}{\sqrt{(1+p^2)}} = c_1 e^{-x}$$

$$\Rightarrow \quad p^2 + c_1^2 e^{-2x}.\ (1 + p^2)$$

$$\Rightarrow \quad p^2(1 - c_1^2 e^{-2x}) = c_2^2 e^{-2x}$$

$$\Rightarrow \quad p = \frac{dy}{dx} = \frac{c_1 e^{-x}}{\sqrt{(1 - c_1^2 e^{-2x})}}.$$

Integrating, $y = -\sin^{-1}(c_1 e^{-x}) + c_2$,
which is the required solution. **Ans.**

9.7 An Equation of the Form $\frac{d^n y}{dx^n} = f(x)$.

The equation can be integrated directly.

Integrating, we have

$$\frac{d^{n-1}y}{dx^{n-1}} = \int f(x)\, dx + c_1.$$

Integrating again and again we shall get required solution.

9.8 An Equation of the Form $\frac{d^2y}{dx^2} = f(y)$.

Multiplying both sides of the given equation by $2\frac{dy}{dx}$, we have

$$2\frac{dy}{dx}\frac{d^2y}{dx^2} = 2\frac{dy}{dx} f(y).$$

Integrating w.r.t., 'x' we have

$$\left(\frac{dy}{dx}\right)^2 = 2\int f(y).\frac{dy}{dx}\, dx + c_1 = \int f(y)\, dy + c_1$$

which we may integrate further after taking square root of both sides.

Example 1:

Solve $\frac{d^2y}{dx^2} = sec^2\, y\, tan\, y.$

Solution:

Multiplying both sides of the given equation by $2\frac{dy}{dx}$, we have

$$2\frac{dy}{dx}\frac{d^2y}{dx^2} = 2\sec^2 y \tan y\frac{dy}{dx}.$$

Integrating, w.r.t., 'x', we have

$$\left(\frac{dy}{dx}\right)^2 = 2\int \sec^2 y \tan y\, dy + c_1$$

$$= \tan^2 y + c_1 = \frac{\sin^2 y + c_1 \cos^2 y}{\cos^2 y}$$

$$\Rightarrow \quad \cos y \frac{dy}{dx} \sqrt{\{c_1 - (c_1 - 1)\sin^2 y\}}$$

$$\Rightarrow \quad \frac{\cos y\, dy}{\sqrt{\{c_1 - (c_1 - 1)\sin^2 y\}}} = dx.$$

Integrating, we have $\frac{1}{\sqrt{(c_1 - 1)}} \sin^{-1}\left[\sqrt{\left(\frac{c_1 - 1}{c_1}\right)} \sin y\right] = x + c_2$

$$\Rightarrow \quad \sin^{-1}\left[\sqrt{\left(\frac{c_1 - 1}{c_1}\right)} \sin y\right] = \sqrt{(c_1 - 1)}\,(x + c_2)$$

$$\Rightarrow \quad \sqrt{\left(\frac{c_1 - 1}{c_1}\right)} \sin y \sin\left[\sqrt{(c_1 - 1)}\,(x + c_2)\right]$$

which is the required of the form. **Ans.**

9.9 An Equation of the Form

$$f\left(\frac{d^ny}{dx^{n-1}}, \frac{d^{n-2}y}{dx^{n-2}}, x\right) = 0$$

In such equation put $\frac{d^{n-2}y}{dx^{n-2}} = q.$

$\therefore$ the given equation reduces to

$$f\left(\frac{d^2q}{dx^2}, q, x\right) = 0$$

which can be solved for q. i.e., let $q = \frac{d^{n-2}y}{dx^{n-2}} = \phi(x)$

which can be solved or y by the method of successive integration.

Example 1:

Solve $\frac{d^2y}{dx^2} = x^2 \sin x.$

Solution:

We have $\frac{d^2y}{dx^2} = x^2 \sin x.$

Integrating the given equation, we have

$$\frac{dy}{dx} = x^2 (-\cos x) + \int 2x \cos x \, dx + c_1$$

$$= -x^2 \cos x + 2x \sin x - 2 \int \sin x \, dx + c_1$$

$$= -x^2 \cos x + 2x \sin x + 2 \cos x + c_1$$

Integrating again, we have

$$y = -x^2 \sin x + 2 \int x \sin x \, dx + 2 \int x \sin x \, dx + 2 \sin x + c_1x + c_2$$

$$= -x^2 \sin x + 4x (-\cos x) + 4 \int \cos x \, dx + 2 \sin x + c_1x + c_2$$

$$= -x^2 \sin x - 4x \cos x + 6 \sin x + c_1x + c_2$$

which is the required solution. **Ans.**

9.10 Equations in Which Order of the Differential Coefficients Differ by Unity

The equation do such type can be written as

$$f\left(\frac{d^ny}{dx^n}, \frac{d^{n-1}y}{dx^{n-1}}, x\right) = 0$$

Putting $\frac{d^{n-1}y}{dx^{n-1}} = q$ so that $\frac{d^ny}{dx^n} = \frac{dq}{dx}$

The given equation reduces to

$$f\left(\frac{dq}{dx}, q, x\right) = 0.$$

from which q and ultimately y can be found out.

Example 1:

Solve $a\dfrac{d^2y}{dx^2} = \left[1 + \left(\dfrac{dy}{dx}\right)^2\right]^{1/2}$.

Solution:

We have $a\dfrac{d^2y}{dx^2} = \left[1 + \left(\dfrac{dy}{dx}\right)^2\right]^{1/2}$

$\dfrac{dy}{dx} = p$ so that $\dfrac{d^2y}{dx^2} = \dfrac{dp}{dx}$.

Putting in the given equation, we have

$$a\frac{dp}{dx} = \left[1 + p^2\right]^{1/2} \Rightarrow \frac{dp}{\sqrt{(1 + p^2)}} = \frac{dx}{a}$$

Integrating, we get $\sin h^{-1} p = \dfrac{x}{a} + c_1$

$$\therefore\ p = \sin h\left(\frac{x}{a} + c_1\right) \Rightarrow dy = \sin h\left(\frac{x}{a} + c_1\right) dx$$

$$y = a \cos h\left(\frac{x}{a} + c_1\right) + c_2$$

which is the required solution. **Ans.**

Example 2:

Solve $\dfrac{d^4y}{dx^4} - a^2\dfrac{d^2y}{dx^2} = 0$.

Solution:

We have $\dfrac{d^4y}{dx^4} - a^2\dfrac{d^2y}{dx^2} = 0$.

Putting, $\dfrac{d^2y}{dx^2} = q$ so that $\dfrac{d^4y}{dx^4} - \dfrac{d^2q}{dx^2}$

the equation becomes $\dfrac{d^2q}{dx^2} - a^2q = 0$

A.E. is $m^2 - a^2 = 0 \quad \therefore \quad m = \pm a$.

$\therefore q = c_1e^{ax} + c_2e^{-ax}$

$$\Rightarrow \quad \frac{d^2y}{dx^2} = c_1e^{ax} + c_2e^{-ax}$$

$$\therefore \quad \frac{dy}{dx} = \frac{c_1}{a}e^{ax} - \frac{c_2}{a}e^{-ax} + c_3$$

and $y = \frac{c_1}{a^2}e^{ax} + \frac{c_2}{a^2}e^{-ax} + c_1x + c_4$

which is the required solution. **Ans.**

EXERCISES

1. $x^4\frac{d^2y}{dx^2} + x^2(x-1)\frac{dy}{dx} + xy = x^3 - 4$

Ans. $ye^{1/x} = \int\left(1 + \frac{c_1}{x}\right)e^{1/x}\,dx - 2e^{1/x}\left(\frac{1}{x} - 1\right) + c_2$

2. $x^2\frac{d^2y}{dx^2} + (4x^2 - 3x)\frac{dy}{dx} + (2x-3)y = 0.$

Ans. $xy + a + be^{3/x}$

3. $2x^2(x+1)\frac{d^2y}{dx^2} + x(7x+3)\frac{dy}{dx} - 3y = x^2.$

Ans. $5(x+1)\,y = c_1x + c_2x^{3/2} + 5x^2/7.$

4. $\frac{d^2y}{dx^2} = \sec^2 y \tan y \tan y$, *given that* $y = 0$ *and* $\frac{dy}{dx} = 1$, *when* $x = 0$.

Ans. $y = \sin^{-1} x$

5. $y\frac{d^2y}{dx^2} + 1 = \left(\frac{dy}{dx}\right)^2$ **(Meerut, 93(P)**

Ans. $c_1y + \sin h\,(c_1x + c_2)$

6. $(1+y^2)\frac{d^2y}{dx^2} = 2\frac{dy}{dx}\left[1 + y\frac{dx}{dx}\right]$

Ans. $-2y^2 + 2c_1y - c_1^2\log(c_1 + 2y) = 8x + c_2$

7. $\frac{d^2y}{dx^2} + \frac{dy}{dx}\left[\frac{dx}{dx}\right]^3 = 0$ **Ans.** $y = c_1 - \sin^{-1}(c_2e^{-x})$

8. $x\frac{d^2y}{dx^2} + x\left[\frac{dx}{dx}\right]^2 - \frac{dy}{dx} = 0.$ **Ans.** $y = \log(x^2 + 2c_1e^{2x}) + c_2$

9. $2\dfrac{d^2y}{dx^2} - \left(\dfrac{dy}{dx}\right)^2 + 4 = 0$ **Ans.** $y = 2x - 2 \log (1 - c_1 e^{2x}) + c_2$

10. $(1 - x^2)\dfrac{d^2y}{dx^2} - x\dfrac{dy}{dx} = 2.$ **Ans.** $y = c_1 \sin^{-1} x + (\sin^{-1} x)^2 + c_2$

11. $\dfrac{d^3y}{dx^3} = \sin^2 x$ **Ans.** $p = \dfrac{1}{12}x^3 + \dfrac{1}{16}\sin 2x + \dfrac{1}{2}c_1x^2 + c_2x + c_3$

12. $\dfrac{d^3y}{dx^3} = xe^x$ **Ans.** $y = e^x (x - 3) + \dfrac{1}{2}c_1x^2 + c_2x + c_3.$

13. $\dfrac{d^2y}{dx^2} + \dfrac{a^2}{y}\ 0.$ **Ans.** $\sqrt{2}ax + c_2 = \int \dfrac{1}{\sqrt{(c_1 - \log y)}}$

14. $\dfrac{d^2y}{dx^2} = e^y$ **Ans.** $\dfrac{\sqrt{(2e^y + c^2)} - c}{\sqrt{(2e^y + c^2)} + c} = c_1e^{ax}.$

15. $\dfrac{d^3y}{dx^3}\cdot\dfrac{d^2y}{dx^2} = 2.$ **Ans.** $y = \dfrac{1}{60}(4x = c_1)^{5/2} + c_2x + c_3.$

16. $\dfrac{d^3y}{dx^3} = \log x.$ **Ans.** $36y = 6x^2 \log x - 11x^3 + c_1x^2 + c_2x + c_3.$

17. $\dfrac{d^3y}{dx^3} + (x^2 + x + 3)\dfrac{d^2y}{dx^2} + (4x + 2)\dfrac{dy}{dx} + 2y = 0.$

Ans. $y.xe^{\frac{x}{2}(x+2)} = \int\left(c_1 + \dfrac{c_2}{x}\right)e^{\frac{x}{2}(x+2)}\,dx + c_3.$

18. $(3^2 - x)\dfrac{d^3y}{dx^3} + (8x^2 - 3)\dfrac{d^2y}{dx^2} + 14x\dfrac{dy}{dx} + 4y = \dfrac{2}{x^3}.$

(Raj., 93; Gorakhpur, 94)

Ans. $yx\sqrt{(x^2 - 1)} = \sec^{-1} x + c_1\sqrt{(x^3 - 1)} + c_2 \log [x + \sqrt{(x^2 - 1)} + c_3$

19. $\dfrac{d^2y}{dx^2} - \cot x\dfrac{dy}{dx} + 2y = \cos x.$

Ans. $y \operatorname{cosec}^2 x = c_2 + 1/2 \log \tan x/2 - c_1 (1/2 \operatorname{cosec} x \cot x + 1/2 \log \tan x/2)$

20. $\dfrac{1}{4}\left(\dfrac{d^3y}{dx^3}\right)^2 + x\dfrac{d^3y}{dx^3} - \dfrac{d^2y}{dx^2} = 0.$

Ans. $y = 1/3\, c_1x^3 + 1/2\, c_1^2 x^2 + c_2x + c_3$

21. $\frac{d^2y}{dx^2} - \frac{a^2}{x(a-x^2)} \cdot \frac{dy}{dx} = \frac{x^2}{a(a^2-x^2)}$.

Ans. $y = -c_1(a^2 - x^2)^{1/2} - (1/2a)\,x^2 + c_2$.

22. $\left(\frac{dy}{dx}\right)^2 - y\frac{d^2y}{dx^2} = n\left\{\left(\frac{dv}{dx}\right)^2 + a^2\left(\frac{d^2y}{dx^2}\right)^2\right\}^{1/2}$

(Kanpur, 97, 91; Sagar, 94)

Ans. $cy = -n\,(1 + a^2c^2)^{1/2} + c_2e^{cx}$

23. $\sin^3 y \frac{d^2y}{dx^2} = \cos y$.

Ans. $\sin\left(\sqrt{k_1}\,x + k_2\right) + \sqrt{\left(\frac{k_1}{k_1 - 1}\right)}\cos y = 0$

24. $\frac{d^ny}{dx^n} - \frac{d^{n-2}}{dx^{n-2}} = 8\cos 3x$.

Ans. $y = c_1\cos x + c_2\sin x - \frac{1}{3^{n-1}}\cos\left\{3x - \frac{\pi}{2}(n-2)\right\}$

25. $\left(\frac{d^ny}{dx^n}\right)^2 = 4\frac{d^{n-1}y}{dx^{n-1}}$

$y = \frac{2x^{n+1}}{(n+1)} + 2c_1\frac{x^n}{n!} + c_1^2\frac{x^{n-1}}{(n-1)}\; c_2x^{n-2} + c_3\,x^{n-3}$

$+ \ldots + c_{n-1}\,x + c_n$.

26. $yy_2 + (y_1^2 + a^2y_2^2)^{1\,2} = y_1^2$.

Ans. $cy + \sqrt{(1 + a^2c^2)} = c_1e^{cx}$

27. $2y\frac{d^3y}{dx^3} + 2\left(y + \frac{dy}{dx}\right)\frac{d^2y}{dx^2} + 2\left(\frac{dy}{dx}\right)^2 = 2$.

Ans. $y = x^2 + c_1 + c_2x + c_3e^{-3}$

28. $(1 + x^2)\frac{d^2y}{dx^2} + 3x\frac{dy}{dx} + y = 0$.

Ans. $y\sqrt{(1 + x^2)} = c_2 + c_1\log[x + \sqrt{(1 + x^2)}$

29. $x\frac{d^3y}{dx^3} + (x^2 - 3)\frac{d^2y}{dx^2} + 4x\frac{dy}{dx} + 2y = 0$ **(Meerut, 97)**

Ans. $y.\frac{e^{\frac{1}{2}x^2}}{x^5} = c_1\int\frac{1}{x^5}e^{\frac{1}{2}x^2}\,dx + c_2\int\frac{.1}{x^6}e^{\frac{1}{2}x^2}\,dx + c_3$

30. $\frac{d^2y}{dx^2} + 2e^x \frac{dy}{dx} + 2e^x y = x^3$

Ans. $ye^{2e^x} = \frac{1}{3}\int x^3 e^{2e^x} dx + \int c_1 e^{2e^x} dx + c_2$

31. $\frac{d^2y}{dx^2} + \frac{dy}{dx} = e^x$ **Ans.** $y = \frac{1}{2} e^x + c_1 + c_2 e^{-x}$

32. $\frac{d^2y}{dx^2} + 2 \sin x \frac{dy}{dx} + 2y \cos x = 0$

Ans. $y = e^{2\cos x} \int c_1 e^{-2\cos x} dx + c_2 e^{2\cos x}$

33. Show that the equation

$$(x - 4x) \frac{d^3y}{dx^3} + (9x^2 - 4) \frac{d^2y}{dx^2} + 18x \frac{dy}{dx} + 6y = 6$$

is exact ans solve it.

Ans. $y(x^2 - 4)^2 = x^4 - 12x^2 + c_1(1/2\, x^{3/2} - 4x \log x) + c_2(x^2 + 4) + c_3 x.$

34. Show that the equation

$$\left(y^2 + 2x^2 \frac{dy}{dx}\right) \frac{d^2y}{dx^2} + 2(x + y) \left(\frac{dy}{dx}\right)^2 + x \frac{dy}{dx} + y = 0$$

is exact and find its first integral.

Ans. $y^2 \frac{dy}{dx} + x^2 \left(\frac{dy}{dx}\right)^2 + xy = c.$

10

Numerical Integration

10.1 Introduction

The method given in the preceding chapters are obtaining solution in finite form only apply to certain spacial types of deferential equations. If an equations does not belong to one of these special types. It is usually not clear how one should proceed in an attempt to obtain a solution exactly methods are available for finding to an any desired degree of accurately the numerical solution of any ordinary differential equations. In this chapter, it is intended to discuss there methods for first order differential equations.

10.2 Simpson's Rule

Simpson's Rule is one of the formulae which enables us to evaluate approximately an area, or a definite integral, even when the analytic relation between y and x is not known. Here we should know the values of y for values x sufficiently close to one another.

Simpson's Rule States that Approximate Value of

$$\int_a^b y dx \text{ is}$$

$$\frac{1}{3}h\,[y_1 + y_{2n+1} + 2\,(y_3 + y_5 + \ldots + y_{2n-1}) + 4\,(y_2 + y_4 + \ldots + y_{2n})]$$

when n is any positive number, $h = (b - a)/2n$ and y_r is the value of y corresponding to the value $a + (r - 1)\,h$ of x.

To prove this, let the functional relation between y and x be $y = f(x)$.

Let, P_1, P_2,..., be the points on the curve $y = f(x)$ whose x coordinates are a, $a + h$, $a + 2h$, ... Also let us suppose that the curve (parabola) whose equation referred to (a + h, 0) as origin is of the form $y = A + Bx + Cx^2$ and passes throgh P_1, P_2, P_3,....., P_n.

Obviously the parabola will be sufficiently close approximation to the curve y = f (x) between the points P_1 and P_2.

Now, the area A_1 between the curve P_1, P_2, P_3 the axis of x and the ordinates of P_1 and P_3 is

$$A_1 = \int_{-h}^{h} \left(A + Bx + Cx^2\right) dx = \left(Ax + \frac{Bx^2}{2} + \frac{Cx^3}{3}\right)_{-h}^{h}$$

$$= 2Ah + \frac{2C}{3} h^3 . \qquad ..(1)$$

Now, the abscissae of P_1, P_2, P_3 with (a + h, 0) as origin are – h, 0 and h, therefore, we have

$$y_1 = A - Bh + Ch^2,\ y_2 = A, \qquad y_3 = A + Bh + Ch^2$$

$$\therefore A\ y_2 \text{ and } C = \frac{y_1 + y_3 - 2y_2}{2h^2}$$

∴ From (1), we have

$$A_1 = 2y_2h + \frac{2}{3}h^3 . \frac{y_1 + y_3 - 2y_2}{2h^2} \Rightarrow A_1 = \frac{h}{3}[y_1 + y_3 + 4y_2].$$

Similarly, the area between the parabolic are P_3, P_4, P_5 the axis of x and the ordinates of P_3, P_5 is

$$A_2 = \frac{h}{3}[y_3 + y_5 + 4y_4].$$

and so on.

Adding, the required area is

$$(h/3)[(y_1 + y_3 + 4y_2 + (y_3 + y_5 + 4y_4) + ... + (y_{2n-1} + y_{2n+1} + 4y_{2n}]$$

$$= (h/3)[y_1 + y_{2n+1} + 2(y_3 + y_5 + ... + y_{2n-1}) + 4(y_2 + y_4 + ... + y_{2n})].$$

Example 1:

Obtain an approximate value of log 2, by calculating the definite integral $\int_1^2 \frac{dx}{x}$ *by Simpson's Rule, using eleven ordinates.*

Solution :

We have $\int_1^2 \frac{dx}{x} = (\log x)_1^2 = \log 2$

Here a = 1, b = 2, n = 5, $h = \frac{b-a}{2n} = \frac{2-1}{10} = \frac{1}{10}$

$$y = \frac{1}{x}, \; y_1 = \frac{1}{1} = 1, \; y_2 = \frac{1}{1.1} = .90909$$

$$y_3 = \frac{1}{1.2} = .8333, \quad y_4 = \frac{1}{1.3} = .76923$$

$$y_5 = \frac{1}{1.4} = .71422. \quad y_6 = \frac{1}{1.5} = .66667$$

$$y_7 = \frac{1}{1.6} = .62500, \quad y_8 = \frac{2}{1.7} = .58824$$

$$y_9 = \frac{2}{1.8} = .55556, \quad y_{10} = \frac{1}{1.9} = .52632.$$

$$y_{11} = \frac{1}{2} = .50000$$

$\therefore$ By Simpson's Rule $\int_1^2 \frac{dx}{x}$.

$$= \frac{1}{3} \cdot \frac{1}{10} [y_1 + y_{11} + 2 (y_3 + y_5 + y_7 + y_9) + 4 (y_2 + y_4 + y_6 + y_8 + y_{10})]$$

$$= \frac{1}{30} [1 + 50000 + 2 (.83333 + .71429 + .62500 + .55556) + 4 (.90909 + .76923 + .66667 + .58824 + .52632)]$$

$$= \frac{1}{30} [20.79456] = .69315.$$ **Ans.**

Example 2:

Verify Simpson's Rule by finding an approximate value of $\int_0^4 e^x dx$ *and compare it with the exact value, having given that*

$$e^0 = 1, \; e = 2.72, \; e^2 = 7.39, \; e^3 = 20.09, \; e^4 = 54.60.$$

Solution:

Here a = 0, b = 4, n = 2, $h = \frac{b-a}{2n} = 1$.

By Simpson's Rule, an approximate value is

$$(h/3) [y_1 + y_5 + 2y_3 + 4 (y_2 + y_4)]$$

$$= \frac{1}{3} [1 + 54.60 + 2 \times (7.39) + 4 \times (2.72 + 20.09)] = 53.97$$

while the correct value is

$$\int_0^4 e^x dx = \left(e^x\right)_0^4 = e^4 - 1 = 54.60 - 1 = 53.60.$$ **Ans.**

10.3 Numerical Approximation Direct Form the Differential Equation

Generally the integration becomes complicated and impracticable and so we have to use other methods. Below we consider the problem geometrically.

Let the differential equation be $\frac{dy}{dx} = f(x, y)$ where $f(x, y)$ is determinate for all values of x and y.

Now, the given differential equation determines a family of curves. If the co-ordinates of P are (a, b), on one of the curves then

$$\left(\frac{dy}{dx}\right) \text{ at } (a, b) = f(a, b).$$

Also if abscissa of Q is (a + h) then y (co-ordinate of

Q) = BQ = BC + CO = b + PC tan QPC

Let the tangent at P to the curve meet the ordinate BQ in D.

Since h is small, tan QPC = tan DPC (approximately)

= b + hf (a, b).

Hence y = b + hf (a, b).

Again if E is the middle-point of PD the co-ordinates of E are $\left\{a + \frac{h}{2}, b + \frac{h}{2} f(a, b)\right\}$ and the gradient of the curve which passes through

E is $f\left\{a + \frac{h}{2}, b + \frac{h}{2} f(a, b)\right\}$.

Now, if we take the chord PQ to be parallel to this tangent of the other curve, then

y = BQ = BC + CQ = b + PC tan QPC

= b + hf {a + h/2, b + (1/2) hf (a, b)}.

Example 1:

Find the value of x = 1.2 for the differential equation

$$\frac{dy}{dx} = 2x - \frac{y}{x} \text{ given that } y = 2 \text{ when } x = 1.$$

Solution:

Here a = 1, b = 2 and h = 1.2 – 1 = 0.2

$$f(x, y) = 2x - \frac{y}{x},\ f(1, 2) = 2 - \frac{2}{1} = 0$$

$$f\{a + h/2,\ b + h/2\ f(a, b)\} = f(1 + 0.1,\ 2 + 0.1 \times 0)$$

$$= 2 \times 1.1. - \frac{2}{1.2} = 2.2 - 1.818 = .382$$

$\therefore y = 2 + 0.2 \times .382 = 2.076.$ **Ans.**

Example 2:

Find y when x = 2.7 for differential equation

$$\frac{dy}{dx} = (x^2 - y)^{1/3} - 1 \text{ given that } y = 4 \text{ when } x = 2.3.$$

Solution:

Here a = 2.3, b = 4, h = 2.7 – 2.3 = .4

$$f(x, y) = (x^2 - y)^{1/3} - 1$$

$f(2.3, 4) = (5.29 - 4)^{1/3} - 1 = (1.29)^{1/3} - 1 = .086$ approximately.

$$f\left\{a + \frac{h}{2},\ b + \frac{h}{2} f(a, b)\right\}$$

$= f(2.3 + 0.2,\ 4 + 0.2 \times .086) = f(2.5,\ 4.0172) = (6.25,\ 4.0172)^{1/3} - 1$

$= (2.2328)^{1/3} - 1 = 0.306$ approximately.

$\therefore y = 4 + 0.4 \times 0.306 = 4.122$ approximately. **Ans.**

EXERCISES

1. Calculate $\int_2^{10} \frac{dx}{1+x}$ by Simpson's Rule, using nine ordinates.

   [Ans. 1.299]

2. Use Simpson's Rule, taking five ordinates, to find an approximation to two decimal places to the value of the integral.

   $$\int_2^1 \sqrt{\left(x - \frac{1}{x}\right)}\, dx.$$

   [Ans. 84]

3. Find y when x = 1.2 for differential equation

   $\frac{dy}{dx} = 2 - \frac{y}{x}$; given that y = when x = 1. **[Ans.** 2.036]

4. Find y when x = .03 for the differential equation

   $\frac{dy}{dx} = x + y^2$ given that y = 0 when x = 0.

11

Wave, Heat, Laplace and Diffusion Equations

11.1 One-Dimensional Wave Equation

(Meerut, 81; Poona, 70)

Let us consider small transverse vibrations of an elastic string, which is stretched to length I and then fixed at the end points. Suppose that, the string is distorted and then at a certain instant, say $t = 0$ it is released and allowed to vibrate. Here the problem is to find its deflection $u(x, t)$ at any point x and at any time $t > 0$.

For this let us make the following assumptions :

1. The motion of the string is a small transverse vibration in a vertical plane i.e., each particle of the string moves strictly vertically, and the deflection and the slope at any point of the string are small in absolute value.
2. The tension caused by stretching the string before fixing it at the end points is so large that the action of the gravitational force on the string can be neglected.
3. The string is homogeneous i.e., the mass of the string per unit length is constant. The string is perfectly elastic and does not offer any resistance to bending.

Consider the forces acting on a small portion PQ of the string. Since the string does not offer resistance to bending (according to our assumption), the tensions T_1 and T_2 and P and Q respectively are tangential to be curve of the string.

Since there is no motion in horizontal direction.

$\therefore T_1 \cos\alpha = T_2 \cos\beta = T$ (constant). ...(1)

If δs is the length of PQ then its mass is $\rho\ \delta s$, where r is the mass per unit length of the string.

Resultant vertical force acting on PQ is $T_2 \sin\beta - T_1 \sin\alpha$ (negative sign appears because the vertical component of T_1 at P is directed downwards).

By Newton's Second Law of Motion we have

$$T_2 \sin\beta - T_1 \sin\alpha = (\rho\ \sigma s)\ \frac{\partial^2 u}{\partial t^2},$$

(since $\frac{\partial^2 u}{\partial t^2}$ is the acceleration of PQ, vertically upwards)

$$\Rightarrow \quad \frac{T_2 \sin\beta}{T_2 \cos\beta} - \frac{T_1 \sin\alpha}{T_1 \cos\alpha} = \frac{\rho\ \delta s}{T}\frac{\partial^2 u}{\partial t^2} \quad \text{(by using (1))}$$

$$\Rightarrow \quad \tan\beta - \tan\alpha = \frac{\rho\ \delta x}{T}\frac{\partial^2 u}{\partial t^2}$$

[since the slope of the curve of the string is small, we may be replace δs by δx]

$$\Rightarrow \quad \left(\frac{\partial u}{\partial x}\right)_{x+\partial x} - \left(\frac{\partial u}{\partial x}\right)_x = \frac{\rho\ \delta x}{T}\frac{\partial^2 u}{\partial t^2}$$

[since $\tan\alpha$ and $\tan\beta$ are the slopes of curve of the string at x and $x + \delta x$]

$$= \quad \frac{\left(\frac{\partial u}{\partial x}\right)_{x+\partial x} - \left(\frac{\partial u}{\partial x}\right)_x}{\delta x} = \frac{\rho}{T}\frac{\partial^2 u}{\partial t^2}$$

$$\Rightarrow \quad \frac{u_x\,(x+\delta x.t) - u_x\,(x,t)}{\delta x} = \frac{\rho}{T}\frac{\partial^2 u}{\partial t^2}$$

Now, as $\delta x \to 0$, we have

$$\frac{\partial^2 u}{\partial x^2} = \frac{\rho}{T}\frac{\partial^2 u}{\partial t^2}$$

which may be written as $\frac{\partial^2 u}{\partial t^2} = c^2\frac{\partial^2 u}{\partial x^2}$, where $c^2 = \frac{T}{\rho}$.

This known as one **Dimensional Wave Equation.**

Notes:

1. Above partial **derivatives** are used because u depends on 'x' and also on 't'.

2. For the physical constant $\frac{T}{\rho}$ the notation c^2 has been chosen to indicate that this constant is positive.

11.2 Two Dimensional Wave Equations

Let us consider the motion of a stretched membrane. The membrane is stretched and then fixed along its entire boundary in the xy-plane.

Let us make the following assumptions :

1. The membrane, is homogeneous, i.e., the mass of the membrane, per unit area is constant. The membrane is perfectly flexible and is so that it does not offer any resistance to bending.
2. The tension T per unit length caused by stretching the membrane is the same at all points and in all directions and does not change during motion.
3. The deflection u (x, y, t) of the membrane during the motion is small compared with the since of the membrane, and all angles of inclination are small.

Consider the forces acting on a small portion of the membrane. Since the deflection of the membrane and the angles of inclination are small, the sides of the portion are approximately equal to dx and dy. If T is the tension (force) per unit length then the force acting on the edges of the portion are $T\delta x$ and $T\delta y$ approximately. These forces are tangential to the membrane (since according to our assumption the membrane is perfectly flexible.

The horizontal components of the forces at one pair of opposite edges are

$$T\delta y \cos \text{ and } T\,\delta y \cos \beta$$

which are approximately equal since α and β are small.

i.e., $\cos \alpha$ and $\cos \beta$ are close to 1.

Hence, the horizontal components of the forces at opposite edges are approximately equal. Therefore, the motion of the particles of the membrane in horizontal direction will be negligibly small. Therefore we may assume that each, particle of the membrane moves vertically.

Now, the resultant of the vertical components of the forces, along the edges parallel to the yu-plane are

$$T\delta y \sin \beta - T\,\delta y \sin \alpha$$

(minus sign with the second component appears because it is directed downwards).

Since α and β are small, therefore this resultant is equivalent to

$$T\,\delta y (\tan \beta - \tan \alpha) = T\,\delta y\,(u_x\,(x + \delta x, y_1) - u_x\,(x, y_2)]$$

where the subscripts x denote partial derivatives and y_1 and y_2 are values between y and y + δy.

Similarly, the resultant of the vertical components of the forces acting on the other two edges of the portion is

$$T\delta x\ [u_y\ (x_1, y + \delta y) - u_y\ (x_3, y)]$$

where the subscripts denote partial derivatives and x_1 and x_2 are values between x and x + δx.

Hence, the resultant of the vertical components of the forces acting on the portion under consideration is

$T\delta y\ [u_x\ (x + \delta x, y_1) - u_x\ (x, y_2)] + T\delta x\ [u_y\ (x_1, y + dy) - u_y\ (x_2, y)]$ which by Newton's Second Law of motion is equal to

$$\rho\ \delta x\ \delta y.\frac{\partial^2 u}{\partial t^2}$$

where ρ is the mass per unit area of the membrane and δx δy the area of the portion under consideration and $\frac{\partial^2 u}{\partial t^2}$ the acceleration,

Therefore

$$\rho\delta x.\delta y\frac{\partial^2 u}{\partial t^2} = T\delta y\ [u_x\ (x + \delta x, y_1) - u_x\ (x, y_2)]$$
$$+ T\delta x\ [u_y\ (x_1, y + \delta y) - u_y\ (x_2, y)]$$

$$\Rightarrow \frac{\partial^2 u}{\partial t^2} = \frac{T}{r}\left[\frac{ux\left(x+\delta x, y_1\right)-u_x\left(x, y_2\right)}{\delta x}+\frac{u_y\left(x_1, y+\delta y\right)-u_y\left(x_2, y\right)}{\delta y}\right]$$

Now $\delta x \to 0$

and $\delta y \to 0$

we have $\frac{\partial^2 u}{\partial t^2} = \frac{T}{\rho}\left(\frac{\partial^2 u}{\partial x^2}+\frac{\partial^2 u}{\partial y^2}\right)$

$$\Rightarrow \frac{\partial^2 u}{\partial t^2} = c^2\left(\frac{\partial^2 u}{\partial x^2}+\frac{\partial^2 u}{\partial y^2}\right) \quad ...(1)$$

where $c^2 = \frac{T}{\rho}$

This equation is called *Two Dimensional Wave Equation.*

This equation (1), may also be written as

$$\frac{\partial^2 u}{\partial t^2} = c^2\nabla^2 u$$

where ∇^2 is the Laplacian operator for two-dimension.

11.3 Heat Equation **(Meerut, 90)**

Thermal energy is transferred from warmer to colder regions interior to a solid body by conduction. It is convenient to refer to that transfer as the flow of heat, as if heat were a fluid or gas which diffuses through the body from regions of high concentration into regions of low concentration of that fluid.

Physical experiments show that the ratio of flow is proportional to the gradient of the temperature. Therefore velocity $\vec{v}$ = of heat flow in a body is given by $\vec{v} = -k$ grad U where U (x, y, z, t) is the temperature, t is the time, and K is a positive constant, called the *Thermal Conductivity* of the body.

Now, let us consider a region R in the body and S be its boundary surface.

Therefore, the amount of heat leaving R per unit of time is

$$\iint_S v_m \, dA$$

where $v_M = \vec{v}.\vec{m}$ is the component of $\vec{v}$ in the direction of the outer unit normal vector $\hat{\vec{M}}$ of S.

But from "*Gauss's Divergent Theorem* we know that, *the normal surface integral of a function $\vec{F}$ over the boundary of a closed region is equal to the volume integral of div $\vec{F}$ taken throughout the region."*

$$\therefore \iint_S v_M \, dA = \iint_S \vec{v}.\hat{\vec{M}} \quad \iiint_R \text{div } (\vec{v}) \, dv$$

$$= -K \iiint_R \text{div (grad U) dx dy da}$$

$$= -K \iiint_R \nabla^2 U \, dx \, dy \, dz. \qquad ...(1)$$

Also the total amount of heat H in R is

$$H = \iiint_R \sigma\rho U \, dx \, dy \, dz.$$

where σ is the specific heat of the material of the body, and ρ is its density.

$\therefore$ the time rate of decrease of H is

$$-\frac{\partial H}{\partial t} = -\iiint_R \sigma\rho \frac{\partial U}{\partial t} \, dx \, dy \, dz$$

and this must be equal to the amount of heat leaving R per unit of time, given by (1)

$$\therefore -\iiint_R \sigma\rho \frac{\partial U}{\partial t} \, dx \, dy \, dz = -K - \iiint_R \nabla^2 U \, dx \, dy \, dz$$

$$\Rightarrow -\iiint_R \left(\sigma\rho\frac{\partial U}{\partial t} - K\nabla^2 U\right) dx\, dy\, dz = 0.$$

Since this holds for any region R in the body, the integrand (if continuous) must be zero everywhere, i.e.,

$$\sigma\rho\frac{\partial U}{\partial t} - K\nabla^2 U = 0 \Rightarrow \frac{\partial U}{\partial t} = \frac{K}{\sigma\rho}\nabla^2 U$$

$$\Rightarrow \qquad \frac{\partial U}{\partial t} = c^2\nabla^2 U, \text{ where } c^2 = \frac{K}{\sigma\rho}.$$

This partial differential equation is called the **Heat Equation.**

(Meerut, 90)

If the temperature within a body are independent of z, i.e., if there is no flow of heat in the z direction, the heat equation reduces to the equation for two **Dimensional Flow** parallel to the xy plane

$$\frac{\partial U}{\partial t} = c^2\left(\frac{\partial^2 U}{\partial x^2} + \frac{\partial^2 U}{\partial y^2}\right).$$

For the dimensional flow parallel to the x-axis, the equation becomes

$$\frac{\partial U}{\partial t} = c^2\frac{\partial^2 U}{\partial x^2}.$$

11.4 Laplace's Equation

We obtained the heat equation as

$$\frac{\partial U}{\partial t} = c^2\nabla^2 U$$

when temperature are in steady state, i.e. when U does not vary with time, then the heat equation becomes $\nabla^2 U = 0$.

i.e.,
$$\frac{\partial^2 U}{\partial x^2} = \frac{\partial^2 U}{\partial y^2} + \frac{\partial^2 U}{\partial z^2} = 0.$$

This is known as **Laplace Equation.**

11.5 Laplace's Equation in Terms of Spherical Coordinates

If (x, y, z) are the Cartesian coordinates and (r, θ, ϕ) the spherical coordinates of a point P, then

$$x = r\sin\theta\cos\phi,\ y = r\sin\phi,\ z = r\cos\theta$$

$$\therefore r^2 = x^2 + z^2,\ \tan\phi = \frac{y}{x} \text{ i.e. } \phi = \tan^{-1}\frac{y}{x}$$

and $\tan\theta = \dfrac{\sqrt{(x^2+y^2)}}{z}$ i.e. $\theta = \tan^{-1} = \dfrac{\sqrt{(x^2+y^2)}}{z}$

$$\therefore \frac{\partial r}{\partial x} = \frac{x}{r} = \sin\theta\cos\phi,\ \frac{\partial r}{\partial y} = \frac{y}{r} = \sin\theta\sin\phi,\ \frac{\partial r}{\partial z} = \frac{z}{r} = \cos\theta$$

$$\frac{\partial\theta}{\partial x} = \frac{\cos\theta\cos\phi}{r},\ \frac{\partial\theta}{\partial y} = \frac{\cos\theta\sin\phi}{r},\ \frac{\partial\theta}{\partial z} = -\frac{\sin\theta}{r}$$

$$\frac{\partial\phi}{\partial x} = -\frac{\sin\phi}{r\sin\theta},\ \frac{\partial\phi}{\partial y} = \frac{\cos\phi}{r\sin\theta},\ \frac{\partial\phi}{\partial z} = 0$$

Then $\dfrac{\partial U}{\partial x} = \dfrac{\partial U}{\partial r}.\dfrac{dr}{\partial x} + \dfrac{\partial U}{\partial\theta}.\dfrac{\partial\theta}{\partial x} + \dfrac{\partial U}{\partial\phi}.\dfrac{\partial\phi}{\partial x}$

$$= \frac{\partial U}{\partial r}.\sin\theta\cos\phi + \frac{\partial U}{\partial\theta}.\frac{\cos\theta\cos\phi}{r} + \frac{\partial U}{\partial f}.\left(-\frac{\sin\phi}{r\sin\theta}\right)$$

$$\therefore \frac{\partial}{\partial x} \equiv \sin\theta\cos\phi\frac{\partial}{\partial r} + \frac{\cos\theta\cos\phi}{r}\frac{\partial}{\partial\theta} - \frac{\sin\theta}{r\sin\theta}\frac{\partial}{\partial\phi}$$

$$\therefore \frac{\partial^2 U}{\partial x^2} = \frac{\partial}{\partial x}\left(\frac{\partial U}{\partial x}\right)$$

$$= \left(\sin\theta\cos\phi\frac{\partial}{\partial r} + \frac{\cos\theta\cos\phi}{r}\frac{\partial}{\partial\theta} - \frac{\sin\phi}{r\sin\theta}\frac{\partial}{\partial\phi}\right)$$

$$\times\left(\sin\theta\cos\phi\frac{\partial}{\partial r} + \frac{\cos\theta\cos\phi}{r}\frac{\partial U}{\partial\theta} - \frac{\sin\phi}{r\sin\theta}\frac{\partial U}{\partial\phi}\right)$$

$$= \sin^2\theta\cos^2\phi\frac{\partial^2 U}{\partial r^2} + 2\frac{\sin\theta\cos\theta\cos^2\phi}{r}\frac{\partial^2 U}{\partial r\,\partial\theta}$$

$$-2\frac{\sin\theta\cos\theta\cos^2\phi}{r^2}\frac{\partial U}{\partial\theta} - 2\frac{\sin\phi\cos\phi}{r}\frac{\partial^2 U}{\partial r\,\partial\phi}$$

$$+\frac{\sin\phi\cos\phi}{r^2}\frac{\partial U}{.\partial\phi} + \frac{\cos^2\theta\cos^2\phi}{r}\frac{\partial U}{\partial r} + \frac{\cos^2\theta\cos^2\phi}{r^2}\frac{\partial^2 U}{\partial\theta^2}$$

$$-2\frac{\cos\theta\sin\phi\cos\phi}{r^2\sin\theta}\frac{\partial^2 U}{\partial\theta\,\partial\phi} + \frac{\cos^2\theta\sin\phi\cos\phi}{r^2\sin^2\theta}\frac{\partial U}{\partial\phi}$$

$$+\frac{\sin^2}{r}\frac{\partial U}{\partial r}+\frac{\cos\theta\sin^2\phi}{r^2\sin\theta}\frac{\partial U}{\partial\theta}+\frac{\sin^2\phi}{r^2\sin^2\theta}\frac{\partial^2 U}{\partial\phi^2}+\frac{\sin\phi\cos\phi}{r^2\sin^2\theta}\frac{\partial U}{\partial\phi} \quad ...(1)$$

Again $\frac{\partial U}{\partial y}=\frac{\partial U}{\partial r}.\frac{\partial r}{\partial y}+\frac{\partial U}{\partial\theta}.\frac{\partial\theta}{\partial y}+\frac{\partial U}{\partial\phi}.\frac{\partial\phi}{\partial y}$

$$=\frac{\partial U}{\partial r}.\sin\theta\sin\phi+\frac{\partial U}{\partial\phi}.\frac{\cos\theta\sin\phi}{r}+\frac{\partial U}{\partial\phi}.\frac{\cos\phi}{r\sin\theta}$$

$$\therefore\ \frac{\partial}{\partial y}\equiv\sin\theta\sin\phi\frac{\partial}{\partial r}+\frac{\cos\theta\sin\phi}{r}\frac{\partial}{\partial\theta}+\frac{\cos\phi}{r\sin\theta}\frac{\partial}{\partial\phi}$$

$$\therefore\ \frac{\partial^2 U}{\partial y^2}=\left(\sin\theta\sin\phi\frac{\partial}{\partial r}+\frac{\cos\theta\sin\phi}{r}\frac{\partial}{\partial\theta}+\frac{\cos\phi}{r\sin\theta}\frac{\partial}{\partial\phi}\right)$$

$$\times\left(\sin\theta\sin\phi\frac{\partial U}{\partial r}+\frac{\cos\theta\sin\phi}{r}\frac{\partial U}{\partial\theta}+\frac{\cos\phi}{r\sin\theta}\frac{\partial U}{\partial\phi}\right)$$

$$=\sin^2\theta\sin^2\phi\frac{\partial^2 U}{\partial r^2}+2\frac{\sin\theta\cos\theta\sin^2\phi}{r}\frac{\partial^2 U}{\partial r\,\partial\theta}$$

$$-\frac{\sin\phi\cos\phi}{r^2}\frac{\partial U}{\partial\phi}+\frac{\cos^2\theta\sin^2\phi}{r}\frac{\partial U}{\partial r}$$

$$+\frac{\cos^2\theta\sin^2\phi}{r^2}\frac{\partial^2 u}{\partial\theta^2}+2\frac{\cos\theta\sin\phi}{r^2\sin\theta}\frac{\partial^2 U}{\partial\theta\,\partial\phi}$$

$$\frac{\cos^2\theta\sin\phi\cos\phi}{r^2\sin^2\theta}\frac{\partial U}{\partial\phi}+\frac{\cos^2\phi}{r}\frac{\partial U}{\partial r}$$

$$+\frac{\cos\theta\cos^2\phi}{r^2\sin\theta}\frac{\partial U}{\partial\theta}+\frac{\cos^2\phi}{r^2\sin^2\theta}\frac{\partial^2 U}{\partial\phi^2}-\frac{\sin\phi\cos\phi}{r^2\sin^2\theta}\frac{\partial U}{\partial\phi} \quad ...(2)$$

Also $\frac{\partial U}{\partial z}=\frac{\partial U}{\partial r}.\frac{\partial U}{\partial\theta}.\frac{\partial\theta}{\partial z}+\frac{\partial U}{\partial\phi}.\frac{\partial\phi}{\partial z}$

$$=\frac{\partial U}{\partial r}.\cos\theta+\frac{\partial U}{\partial\theta}\left(-\frac{\sin\theta}{r}\right)$$

$$\therefore\ \frac{\partial}{\partial z}\equiv\cos\theta\frac{\partial}{\partial r}-\frac{\sin\theta}{r}\frac{\partial}{\partial\theta}$$

$$\therefore\ \frac{\partial^2 U}{\partial z^2} = \left(\cos\theta\,\frac{\partial}{\partial r} - \frac{\sin\theta}{r}\,\frac{\theta}{\partial\theta}\right)\left(\cos\theta\,\frac{\partial U}{\partial r} - \frac{\sin\theta}{r}\,\frac{\partial U}{\partial\theta}\right)$$

$$= \cos^2\theta\,\frac{\partial^2 U}{\partial z^2} - 2\,\frac{\sin\theta\cos}{r}\,\frac{\partial^2 U}{\partial r\,\partial\theta} + 2\,\frac{\sin\theta\cos\theta}{r^2}\,\frac{\partial U}{\partial\theta}$$

$$+ \frac{\sin^2\theta}{r}\,\frac{\partial U}{\partial r} + \frac{\sin^2\theta}{r^2}\,\frac{\partial^2 U}{\partial\theta^2} \qquad ...(3)$$

Adding equations (1), (2) and (3), we have

$$\frac{\partial^2 U}{\partial x^2} + \frac{\partial^2 U}{\partial y^2} + \frac{\partial^2 U}{\partial z^2} = 0 \text{ is transformed to}$$

$$\frac{\partial^2 U}{\partial r^2} + \frac{2}{r}\,\frac{\partial U}{\partial r} + \frac{1}{r^2}\,\frac{\partial^2 U}{\partial\theta^2} + \frac{\cot\theta}{r^2}\,\frac{\partial U}{\partial\theta} + \frac{1}{r^2\sin^2\theta}\,\frac{\partial^2 U}{\partial\phi^2} = 0,$$

which is the **Laplace Differential Equation in Spherical Co-ordinates.** Which may also be written as

$$\nabla^2.U\ \frac{1}{r^2}\,\frac{\partial}{\partial r}\left(r^2\,\frac{\partial U}{\partial r}\right) + \frac{1}{r^2\sin\theta}\,\frac{\partial}{\partial\theta}\left(\sin\theta\,\frac{\partial U}{\partial\theta}\right) + \frac{1}{r^2\sin^2\theta}\,\frac{\partial^2 U}{\partial\phi^2} = 0$$

11.6 Laplace's Equation in Terms of Cylindrical Co-Ordinates

If (x, y, z) are the Cartesian coordinates of the point P whose cylindrical co-ordinates are (r, θ, z), then

$$x = r\cos\theta,\ y = r\sin\theta,\ z = z$$

$$r^2 = x^2 + y^2,\quad \theta\ \tan^{-1}\frac{y}{x}$$

$$\frac{\partial r}{\partial x} = \frac{y}{r} = \cos\theta,\ \frac{\partial r}{\partial y} = \frac{y}{r} = \sin\theta$$

$$\frac{\partial\theta}{\partial x} = \frac{1}{1+\frac{y^2}{x^2}}.\left(-\frac{y}{x^2}\right) = -\frac{\sin\theta}{r} \quad \text{and} \quad \frac{\partial\theta}{\partial y} = \frac{\cos\theta}{r}$$

Then $\dfrac{\partial U}{\partial x} = \dfrac{\partial U}{\partial r}.\dfrac{\partial r}{\partial x} + \dfrac{\partial U}{\partial\theta}.\dfrac{\partial\theta}{\partial x} + \dfrac{\partial U}{\partial z}.\dfrac{\partial z}{\partial x}$

$$= \frac{\partial U}{\partial r}.\cos\theta + \frac{\partial U}{\partial\theta}\left(-\frac{\sin\theta}{r}\right)$$

$$\therefore \frac{\partial}{\partial x} \equiv \cos\theta \frac{\partial}{\partial r} - \frac{\sin\theta}{r}\frac{\partial}{\partial \theta}$$

$$\therefore \quad \frac{\partial^2 U}{\partial x^2} = \frac{\partial}{\partial x}\left(\frac{\partial U}{\partial x}\right) = \left(\cos\theta \frac{\partial}{\partial r} - \frac{\sin\theta}{r}\frac{\partial}{\partial \theta}\right)\left(\cos\partial\frac{\partial U}{\partial r} - \frac{\sin\theta}{r}\frac{\partial U}{\partial \theta}\right)$$

$$= \cos^2\theta \frac{\partial^2 U}{\partial r^2} - 2\frac{\sin\theta\cos\theta}{r}\frac{\partial^2 U}{\partial r\,\partial\theta}$$

$$+2\frac{\sin\theta\cos\theta}{r^2}\frac{\partial U}{\partial \theta} + \frac{\sin^2\theta}{r}\frac{\partial U}{\partial r} + \frac{\sin^2\theta}{r^2}\frac{\partial^2 U}{\partial \theta^2} \qquad \text{...(1)}$$

Similarly,

$$\frac{\partial^2 U}{\partial y^2} = \sin^2\theta\frac{\partial^2 U}{\partial r^2} + \frac{\sin\theta\cos\theta}{r}\frac{\partial^2 U}{\partial r\,\partial\theta} - 2\frac{\sin\theta\cos\theta}{r^2}\frac{\partial U}{\partial \theta}$$

$$+\frac{\cos^2\theta}{r}\frac{\partial U}{\partial r} + \frac{\cos^2\theta}{r^2}\frac{\partial^2 U}{\partial \theta^2} \qquad \text{...(2)}$$

and $\frac{\partial^2 U}{\partial z^2} = \frac{\partial^2 U}{\partial z^2}$.

Adding (1), (2) and (3), we have

$$\nabla^2 U - \frac{\partial^2 U}{\partial x^2} + \frac{\partial^2 U}{\partial y^2} + \frac{\partial^2 U}{\partial z^2} = 0 \text{ is transformed to}$$

$$\frac{\partial^2 U}{\partial r^2} + \frac{1}{x}\frac{\partial U}{\partial r} + \frac{1}{r^2}\frac{\partial^2 U}{\partial \theta^2} + \frac{\partial^2 U}{\partial z^2} = 0$$

which is the Laplace's differential equation in cylindrical co-ordinates. This may also the written as

$$\frac{1}{r}\frac{\partial}{\partial r}\left(r\frac{\partial U}{\partial r}\right) + \frac{1}{r^2}\frac{\partial^2 U}{\partial \theta^2} + \frac{\partial^2 U}{\partial z^2} = 0.$$

11.7 Diffusion Equation

Let us consider the flow of electricity in a long insulated cable. Suppose the flow to be one dimensional the current i and be voltage E at any point in the cable can be completely specified by one coordinate x and a time variable t.

Consider the fall of potential in a linear element of length dx situated at the point x, we have

$$-\delta E = iE\ \delta x + L\ \delta x\ \frac{\partial i}{\partial t}, \qquad ...(1)$$

where R and L are resistance and inductance per unit length respectively.

If there is a capacitance to earth of C per unit length and conductance G per unit length, then

$$-\delta i = GE\ \delta x + C\ \delta x\ \frac{\partial E}{\partial t} \qquad ...(ii)$$

These equation (i) and (ii) are equivalent to the differential equations

$$\frac{\partial E}{\partial x} + Ri + L\frac{\partial i}{\partial t} = 0$$

$$\frac{\partial i}{\partial x} + GE + C\frac{\partial E}{\partial t} = 0. \qquad ...(iv)$$

Differentiating (iii) w.r.t. x and (iv) w.r.t. 't' we have

$$\frac{\partial^2 E}{\partial x^2} + R\frac{\partial i}{\partial x} + L\frac{\partial^2 i}{\partial x\,\partial t} = 0 \qquad ...(v)$$

and $$\frac{\partial^2 i}{\partial x\,\partial t} + G\frac{\partial E}{dt} + C\frac{\partial^2 E}{\partial t^2} = 0. \qquad ...(vi)$$

Eliminating $\frac{\partial i}{\partial x}$ and $\frac{\partial^2 i}{\partial x\,\partial t}$ from (iv), (v) and (vi), we have

$$\frac{\partial^2 E}{\partial x^2} = CL\frac{\partial^2 E}{\partial t^2} + (CR + GL)\frac{\partial E}{\partial t} + RGE \qquad ...(vii)$$

Similarly differentiating (iii) w.r.t. t and (iv) w.r.t. x, we have

$$\frac{\partial^2 E}{\partial t\,\partial x} + R\frac{\partial i}{\partial t} + L\frac{\partial^2 i}{\partial t^2} = 0$$

and $$\frac{\partial^2 i}{\partial x^2} + G\frac{\partial E}{\partial x} + C\frac{\partial^2 E}{\partial t\,\partial x} = 0. \qquad ...(ix)$$

Eliminating $\frac{\partial E}{\partial x}$ and $\frac{\partial^2 E}{\partial x\,\partial t}$ from equations (iii), (viii) and (ix), we have.

$$\frac{\partial^2 i}{\partial x^2} = CL\frac{\partial^2 i}{\partial t^2} + (CR + GL)\frac{\partial i}{\partial t} + RGi. \qquad ...(x)$$

From (vii) and (x), we find that E and i both satisfy the second-order partial differential equation

$$\frac{\partial^2 \phi}{\partial x^2} = CL\frac{\partial^2 \phi}{\partial t^2} + (CR + GL)\frac{\partial \phi}{\partial t} + RG\phi.$$

Equation (xi) is called the **Telegraphy Equation.**

If the leakage to ground is small, then G and L may be taken to be zero.

From equation (xi), we have $\frac{\partial^2 \phi}{\partial x^2} = CR\frac{\partial \phi}{\partial t}$

$$\Rightarrow \qquad \frac{\partial^2 \phi}{\partial x^2} = \frac{1}{k}\frac{\partial \phi}{\partial t} \qquad ...(xii)$$

where $k = \frac{1}{CR}$ is constant.

Equation (xii) is known as **One Dimensional Diffusion Equation.**

In three dimensions the diffusion equation is $\nabla^2 \phi = \frac{1}{k}\frac{\partial \phi}{\partial t}$.

Notes:

(i) The equation (xii) is also sometimes called the **telegraphic equation.**

(ii) The above equation (diffusion equation) is satisfied by the temperature when there is no generation or absorption of heat in the solid, as a result of chemical reaction and as the solid is not under going radioactive decay or is not absorbing radioaction. In this case $k = \frac{k'}{\sigma\rho}$ where k' is thermal conductivity of the body σ the specific head of the material of the body and ρ is its density.

EXERCISES

1. A string is tightly stretched between its two ends and given a small transverse vibration. Obtain the differential equation of its motion. **(Meerut, 91)**

2. Establish the heat flow equation

$$\frac{\partial v}{\partial t} = h^2 \nabla^2 v$$

where h^2 represents the diffusively of the substance and v stands for the temperature. **(Meerut, 90)**

12

Boundary Value Problems

12.1 A Boundary Value Problem

Many of the problems in Physics are governed by partial differential equations. When the partial differential equation is satisfied within a certain regio, and in addition certain conditions (known as boundary conditions) are to be satisfied on the boundary of the region, the problem is said to be a Boundary Value Problem.

12.2 Solution by Separation of Variables

One of the most well known methods of solving linear partial differential equations is the product method (or the method of separation of variables).

Consider the second order partial differential equation

$$Rr + Ss + Tt + Pp + Qq + Zz = F. \quad \text{...(1)}$$

Now, let the solution of (1) be given by

$$z = X(x).Y(y) \quad \text{...(2)}$$

where X (x) and Y (y) are functions of x and y alone respectively

Substituting equation (2) in (1), and simplifying we have

$$\frac{1}{C}.\{f(D).X\} = \frac{1}{Y}\{g(D')Y\} \quad \text{...(3)}$$

where f (D) and g (D') are functions of

$D \equiv \frac{\partial}{\partial x}$ and $D' \equiv \frac{\partial}{\partial y}$ respectively.

Now, the L.H.S. of (3) is a function of x alone and R.H.S. is a function of y alone, and the two can not be equal unless each is equal to a constant value λ.

Thus, we have

$$f(D).X = \lambda X \text{ and } g(D') = \lambda Y. \qquad ...(4)$$

Thus, the solution of (1) reduces to the solution of a pair of ordinary differential equations given by (4).

12.3 Solution of One Dimensional Wave Equation

(Meerut, 97 (S), 95, 97, 98; Kanpur, 97)

In the last chapter we have seen that the vibration of an elastic string are governed by the onė dimensional wave equation

$$\frac{\partial^2 u}{\partial t^2} = c^2 \frac{\partial^2 u}{\partial y^2} \qquad ...(A)$$

where u (x, t) is the deflection of the string stretched between the fixed points (0, 0) and (*l*, 0). Let f(x) be the initial deflection and g(x) the initial velocity of the string. Therefore, the function u(x, t) is required to satisfy (A) and all the boundary and initial conditions, i.e.

Boundary condition $u(0, t) = 0$

and $u(l, t) = 0 \qquad ...(B)$

and initial conditions $u(x, 0) = f(x) \qquad ...(C)$

and $$\left(\frac{\partial u}{\partial t}\right)_{t=0} = g(x). \qquad ...(D)$$

Now, our problem is to find a solution of (A) satisfying the conditions (B), (C) and (D).

(I) Solution by Separation of Variables

It is clear from (A) that u (x, t) is a function of x and t. Therefore suppose that (A) has a solution of the form

$$u(x, t) = F(X)\,T(t) = FT \text{ (say)} \qquad ...(1)$$

where F is a function of x only and T that of t only.

Thus, substituting in (A), we have

$$F\frac{d^2T}{dt^2} = c^2 T \frac{d^2F}{dx^2}$$

$$\Rightarrow \qquad \frac{1}{F}\frac{d^2F}{dx^2} = \frac{1}{c^2T}\frac{d^2T}{dt^2}.$$

Now, the left-hand side is a function of the independent variable x, while the right hand side is a function of the independent variable t. The two cannot be equal to each other, unless both reduce to a constant value.

Hence, $\frac{1}{F}\frac{d^2F}{dx^2} = \frac{1}{c^2T}\frac{d^2T}{dt^2} = 0 \Rightarrow k^2 \Rightarrow -k^2$

Therefore in the three cases, we have

$$\frac{d^2F}{dx^2} = 0, \text{ and } \frac{d^2T}{dt^2} = 0$$

$$\frac{d^2F}{dx^2} = k^2F = 0, \frac{d^2T}{dt^2} - k^2c^2T = 0$$

$$\frac{d^2F}{dx^2} = k^2F = 0, \frac{d^2T}{dt^2} - k^2c^2T = 0.$$

The general solution in these three cases are

(i) $F = Ax + B, T = Ct + D$

(ii) $F = Ae^{kx} + Be^{-kx}, \; T = Ce^{ket} + De^{-ket}$.

(iii) $F = A \cos kx + B \sin kx$, $T = C \cos ket + D \sin ket$.

Now from boundary conditions (B) and (i), we have

$$u(0, t) = F(0)\, T(t) = 0$$

and $$u(l, t) = F(1)\, T(t) = 0$$

which gives either $T(t) = 0$

$\Rightarrow$ $F(0) = 0$ and $F(l) = 0$.

But $T(t) \neq 0$ otherwise from (1), we have $u(x, t) = 0$

$\therefore$ $F(0) = 0$ and $F(l) = 0$.

Therefore (i) $F(0) = B = 0$

and $F(l) = Al + B = 0$

giving $A = B = 0$. Hence $F(x) = 0$ and therefore $u(x, t) = 0$.

Also from (ii) $F(0) = A + B = 0$

and $F(l) = Ae^{kl} + Be^{-kl} = 0$,

giving $A = B = 0$, so that $F(x) = 0$ and therefore $u(x, t) = 0$.

Hence the solutions (i) and (ii) do not constitute the solution of the wave equation (A).

The third solution (iii) is periodic (in time).

Hence we investigate a solution of the type.

$$u(x, t) = (A \cos kx + B \sin kx)(C \cos ket + D \sin ket) \qquad ...(2)$$

which will satisfy the different equation (A).

Let us now adjust the constants so that the solution satisfies the boundary conditions (B) also

$\therefore$ u (0, t) = (A + 0) (C cos ket + D sin ket) = 0. Hence A = 0

and $\therefore$ u (l, t) = (0 + B sin kl) (C cos ket + D sin ket) = 0.

This will be satisfied if kl = nπ

$$\Rightarrow \qquad k = \frac{n\pi}{l}$$

where n = 1, 2, 3,... (i.e. a positive integer).

Hence, a solution of the equation (A) satisfying the two boundary conditions (B) is

$$u_n (x, t) = \left(C_n \cos \frac{n\pi ct}{l} + D_n \sin \frac{n\pi ct}{l}\right) \sin \frac{n\pi x}{l} \qquad ...(3)$$

(Kanpur, 97)

Now, from equation (3) and initial conditions (C) and (D), we have

$$u (x, 0) = C_n \sin \frac{n\pi x}{l} = f(x)$$

and $$\left(\frac{\partial u}{\partial t}\right)_{t=0} = \left[-\frac{n\pi c}{l} C_n \sin \frac{n\pi ct}{l} + \frac{n\pi c}{l} D_n \cos \frac{n\pi ct}{l}\right]_{t=0} \sin \frac{n\pi x}{l}$$

$$= \frac{n\pi c}{l} D_n \sin \frac{n\pi x}{l} = g(x).$$

Evidently these will not be satisfied if we take a single term as our solution.

Since the equations (A) is linear and homogeneous therefore the sum of any number of different solutions of (A) will still be a solution of equation (A).

Thus, instead of (3), the solution may be taken as

$$u (x, t) = \sum_{n=1}^{\infty} \left(C_n \cos \frac{n\pi ct}{l} + D_n \sin \frac{n\pi ct}{l}\right) \sin \frac{n\pi x}{l} \qquad ...(4)$$

This will also satisfy boundary conditions (B).

Now, from equation (4) and the initial conditions (C) and (D), we have

$$u (x, 0) = \sum_{n=1}^{\infty} C_n \sin \frac{n\pi x}{l} = f(x) \qquad ...(5)$$

and $$\left(\frac{\partial u}{\partial t}\right)_{t=0} = \sum_{n=1}^{\infty} \frac{n\pi c}{l} D_n \sin \frac{n\pi x}{l} = g(x).$$

The left hand sides can be considered as the Fourier sine expansions of the right hand sides. Hence

$$C_n = \frac{2}{l} \int_0^l f(x) \sin \frac{n\pi x}{l} dx \qquad ...(6)$$

and $$\frac{n\pi c}{l} D_n = \frac{2}{l} \int_0^l g(x) \sin \frac{n\pi x}{l} dx. \qquad ...(7)$$

These values of C_n and D_n completely satisfy the solution (4).

Thus, u (x, t) given by (4) with the coefficients (6) and (7) is the solution of the wave equation (A) that satisfies the conditions (B), (C) and (D).

Particular Case: If the initial velocity g (x) is zero

i.e., $$\left(\frac{\partial u}{\partial t}\right)_{t=0} = g(x) = 0 \text{ therefore } D_n = 0$$

and (4) reduces to the form

$$u(x, t) = \sum_{n=1}^{\infty} C_n \cos l_n t \sin \frac{n\pi x}{l}, \text{ where } \lambda_n = \frac{cn\pi}{l}$$

$$= \frac{1}{2} \sum_{n=1}^{\infty} C_n \left[\sin \frac{n\pi}{l}(x - ct) + \sin \frac{n\pi}{l}(x + ct)\right]$$

$$= \frac{1}{2} \sum_{n=1}^{\infty} C_n \sin \frac{n\pi}{l}(x - ct) + \frac{1}{2} \sum_{n=1}^{\infty} C_n \sin \frac{n\pi}{l}(x + ct)$$

The two series $\frac{1}{2} \sum_{n=1}^{\infty} C_n \sin \frac{n\pi}{l}(x - ct)$ and $\sum_{n=1}^{\infty} C_n \sin \frac{n\pi}{l}(x + ct)$ are obtained by substituting (x – ct) and (x + ct) respectively for the variable x in the Fourier Sine Series (5) for f (x).

Thus, we may take

$$u(x, t) = \frac{1}{2} [F(x - ct) + F(x + ct)] \qquad ...(8)$$

as the solution of wave equation (A) where F is the odd periodic extension of f with the period $2l$.

Note: The functions given by (3) are called the **Eigen Functions,** or **Characteristic Function** and the values $\lambda_n = \frac{cn\pi}{l}$ are called the **Eigen Values or Characteristic Values,** of the vibrating string. The set $\lambda_1, \lambda_2, \ldots,$ is called **Spectrum.**

(II) D'Alembert's Solution **(Meerut, 95)**

The solution of the wave equation (A) in the form (8), can be obtained by transforming it [i.e. (A)] in a suitable way.

Let us introduce the new independent variables v and w given by

$$\left.\begin{aligned} \text{by} \qquad & v = x + ct \\ \text{and} \qquad & w = x - ct \end{aligned}\right\} \qquad \text{...(1)}$$

$$\therefore \qquad \frac{\partial v}{\partial x} = 1 \text{ and } \frac{\partial w}{\partial x} = 1 .$$

Therefore, $\dfrac{\partial u}{\partial x} = \dfrac{\partial u}{\partial v} \cdot \dfrac{\partial v}{\partial x} + \dfrac{\partial u}{\partial w} \cdot \dfrac{\partial w}{\partial x} = \dfrac{\partial u}{\partial v} + \dfrac{\partial u}{\partial w}$

$$\frac{\partial}{\partial x} \equiv \frac{\partial}{\partial v} + \frac{\partial}{\partial w}$$

Now, $\dfrac{\partial^2 u}{\partial x^2} = \dfrac{\partial}{\partial x}\left(\dfrac{\partial u}{\partial x}\right) = \left(\dfrac{\partial}{\partial v} + \dfrac{\partial}{\partial w}\right)\left(\dfrac{\partial u}{\partial v} + \dfrac{\partial u}{\partial w}\right)$

$$\therefore \quad \frac{\partial^2 u}{\partial x^2} = \frac{\partial^2 u}{\partial v^2} + 2\frac{\partial^2 u}{\partial x\, \partial w} + \frac{\partial^2 u}{\partial w^2} . \qquad \text{...(2)}$$

Again $\dfrac{\partial v}{\partial t} = c$ and $\dfrac{\partial w}{\partial t} = -c$.

$$\therefore \quad \frac{\partial u}{\partial t} = \frac{\partial u}{\partial v} \cdot \frac{\partial v}{\partial t} + \frac{\partial u}{\partial w} \cdot \frac{\partial w}{\partial t} = c\left(\frac{\partial u}{\partial v} - \frac{\partial u}{\partial w}\right)$$

Therefore $\qquad \dfrac{\partial^2 u}{\partial t^2} = c^2\left(\dfrac{\partial}{dv} - \dfrac{\partial}{dw}\right)\left(\dfrac{\partial u}{\partial v} - \dfrac{\partial u}{\partial w}\right)$

$$= c^2\left(\frac{\partial^2 u}{\partial v^2} - 2\frac{\partial^2 u}{\partial v\, \partial w} + \frac{\partial^2 u}{\partial w^2}\right) . \qquad \text{...(3)}$$

Substituting from (2) and (3) in (A), we have

$$c^2\left(\frac{\partial^2 u}{\partial v^2} - 2\frac{\partial^2 u}{\partial w\, \partial w} + \frac{\partial^2 u}{\partial w^2}\right) = c^2\left(\frac{\partial^2 u}{\partial v^2} + 2\frac{\partial^2 u}{\partial v\, \partial w} + \frac{\partial^2 u}{\partial w^2}\right)$$

$$\Rightarrow \qquad \frac{\partial^2 u}{\partial u\, \partial w} = 0 .$$

Integrating, w.r.t. 'w' we have $\dfrac{\partial u}{\partial v} = F(v)$

where F (v) is an arbitrary function of v.

Integrating this w.r.t. 'v' we have

$$u = \phi (v) + \psi (w) \quad \text{where } \phi\, F (v)\, dv = \phi (v)$$

and y (w) is an arbitrary function of w.

$$\therefore \quad u (x, t) = \phi (x + ct) + \psi (x - ct). \qquad ...(4)$$

This is known as **D' Alembert's Solution** of the wave equation (A).

Now $\quad \dfrac{\partial u}{\partial t} = c\phi' (x + ct) - c\psi' (x - ct).$

$\therefore$ using the initial conditions (C) and (D), we have

$$u (x, 0) = \phi (x) + \psi (x) = f (x), \qquad ...(5)$$

and $\quad \left(\dfrac{\partial u}{\partial t}\right)_{t=0} = c\phi' (x) - c\psi' (x) = 0. \qquad ...(6)$

Taking the particular case when g (x) = 0.

From (7), we have

$$\psi' (x) = f' (x).$$

$$\therefore \quad \psi (x) = f (x) + A.$$

$\therefore$ from (5), we have

$$2\phi(x) + A = f (x).$$

$$\therefore \quad \phi(x) = \frac{1}{2} [f (x) - A]$$

and $\quad \phi(x) = \phi(x) + A = \dfrac{1}{2} [f (x) + A].$

Hence, from equation (4), we have

$$u (x, t) = \frac{1}{2} [f (x + ct) + f (x - ct)]. \qquad ...(7)$$

Now, using boundary condition (B) and (C), we have

$$u (0, t) = \frac{1}{2} [f (ct) + f (-ct)] = 0.$$

$$u (l, t) = \frac{1}{2} [f (l + ct) + f (l - ct) = 0,$$

which shows that the function f must be odd and have period $2l$.

Thus, equation (7) is the solution of wage equation (A) satisfying the boundary conditions (B) (C) and (D).

12.4 Solution of Two Dimensional Wave Equation

(Meerut, 91 (P), 97; Kanpur, 95)

In the last chapter, we have seen that the oscillations of a perfectly flexible membrane stretched to a uniform tension T are governed by the two dimensional wave equation

$$\frac{\partial^2 u}{\partial t^2} = c^2\left(\frac{\partial^2 u}{\partial x^2} + \frac{\partial^2 u}{\partial y^2}\right) \quad ...(A)$$

where $c^2 = \frac{T}{\rho}$

where (x, y, t) is the deflection of the membrane.

Let f (x, y) be the initial deflection and g (x, y) be the initial velocity of the membrane. Therefore, the function u (x, y, t) is required to satisfy (A) and all the boundary and initial conditions i.e.

Boundary conditions
$$\left.\begin{aligned} u(0, y, t) &= 0 \\ u(a, y, t) &= 0 \\ u(x, 0, t) &= 0 \\ u(x, b, t) &= 0 \end{aligned}\right\} \quad ...(B)$$
and

for all t.

and initial conditions $u(x, y, 0) = f(x, y)$ ...(C)

$$\left(\frac{\partial u}{\partial t}\right)_{t=0} = g(x, y) \quad ...(D)$$

Now, our problem is to find a solution of (A) satisfying the conditions (B), (C) and (D).

It is clear from (A) that u (x, y, t) is a function of x, y and t. Therefore suppose that (A) have a solution of the form

$$u(x, y, t) = F(x)\,Y(y)\,T(t) = FYT \text{ (say)} \quad ...(1)$$

where F is a function of x only, Y is that of y only and T is that of t only.

Thus, substituting in (A), we have

$$FY\frac{d^2T}{dt^2} = c^2\left(YT\frac{d^2T}{dx^2} + FT\frac{d^2Y}{dy^2}\right)$$

$$\Rightarrow \quad \frac{1}{c^2T}\frac{d^2T}{dt^2} = \frac{1}{F}\frac{d^2F}{dx^2} + \frac{1}{Y}\frac{d^2Y}{dy^2}.$$

Now, the left hand side is a function of the independent variable, t, while the right hand side is a function of independent x and y. The two cannot be equal to each other, unless both reduce to a constant value. Again the two terms

$$\frac{1}{F}\frac{d^2F}{dx^2} \text{ and } \frac{1}{Y}\frac{d^2Y}{dy^2}$$

on the right hand side are functions of x and y respectively and their sum cannot be equal to a constant unless each of these is constant. Thus we have three possibilities.

(i) $\dfrac{1}{c^2T}\dfrac{d^2T}{dt^2} = 0, \dfrac{1}{F}\dfrac{d^2F}{dx^2} = 0, \dfrac{1}{Y}\dfrac{d^2Y}{dy^2} = 0.$

(ii) $\dfrac{1}{c^2T}\dfrac{d^2T}{dt^2} = k^2, \dfrac{1}{F}\dfrac{d^2F}{dx^2} = k_1^2, \dfrac{1}{Y}\dfrac{d^2Y}{dy^2} = k_2^2$

where $k^2 = k_1^2 + k_2^2$

and (iii) $\dfrac{1}{c^2T}\dfrac{d^2T}{dt^2} = -k^2, \dfrac{1}{F}\dfrac{d^2F}{dx^2} = -k_1^2, \dfrac{1}{Y}\dfrac{d^2Y}{dy^2} = -k_2^2$

where $k^2 = k^2 = k_1^2 + k_2^2$.

The general solution in these three cases are

(a) $F = A_1x + B_1 \quad Y = A_2y + B_2 \quad T = A_2t + B_2$

(b) $F = A_1e^{k_1x} + B_1e^{-k_1x}, Y = A_2e^{k_2y} + B_2e^{-k_2y},$

$T = A_2e^{ckt} + B_3e^{-ckt}$

and (c) $F = A_1 \cos k_1x + B_1 \sin k_1x$, $Y = A_2 \cos k_2y + B_2 \sin k_2y$,

$T = A_2 \cos ckt + B_2 \sin ckt$.

From the boundary conditions (B) and (1), we have

$$\left.\begin{aligned} u(0, y, t) &= F(0)\,Y(y)\,Y(t) = 0 \\ u(a, y, t) &= F(a)\,Y(y)\,T(t) = 0 \end{aligned}\right\} \text{ from which it follows that } F(0) = 0,\ F(a) = 0.$$

$$\left.\begin{aligned} u(x, 0, t) &= F(x)\,Y(0)\,T(t) = 0 \\ u(x, b, t) &= F(x)\,Y(b)\,T(t) = 0 \end{aligned}\right\} \text{ from which it follows that } F(0) = 0,\ F(a) = 0.$$

Hence, the solution (a) and (b) do not constitute the solution of the wave equation (A). (I his can be shown as in the last article).

The third solution (c) is periodic in time.

Hence, we investigate a solution of the type

$$u(x, y, t) = (A_1 \cos k_1x + B_1 \sin k_1x)(A_1 \cos k_2y + B_2 \sin k_2y)$$
$$(A_3 \cos ckt + B_3 \sin ckt) \quad ...(2)$$

which will certainly satisfy the differential equation (A).

Let us, now adjust the constant so that he solution satisfies the boundary conditions (B) also.

$$u(0, y, t) = (A_1 + 0)(A_2 \cos k_2y + B_2 \sin k_2y)(A_3 \cos ckt + B_3 \sin ckt) = 0$$

$$\therefore A_1 = 0$$

$$u(a, y, t) = (0 + B_1 \sin k_1a)(A_2 \cos k_2y + B_2 \sin k_2y)(A_2 \cos ckt + B_2 \sin ckt) = 0$$

$\therefore \sin k_1a = 0 \quad \Rightarrow k_1a = m\pi$

$$\Rightarrow \qquad k_1 = \frac{m\pi}{a} .(m = 1, 2, 3,...).$$

Similarly, using the other two boundary conditions, we have

$$A_2 = 0 \text{ and } k_2 = \frac{n\pi}{b}, (n = 1, 2, 3,...).$$

$\therefore$ from (2), we have

$$u_{mn}(x, y, t) = (A_{mn} \cos ck_{mn}t + B_{mn} \sin ck_{mn}t) \sin \frac{m\pi x}{a} \sin \frac{n\pi y}{b} \quad ...(3)$$

$$\text{where } k^2 = k^2_{mn} = \pi^2 \left(\frac{m^2}{a^2} + \frac{n^2}{b^2}\right).$$

Since the equation (A) is linear and homogeneous therefore the sum of any number of different solutions of (A) will still be a solution of equation (A).

Thus instead of (3) an appropriate solution of (A) is taken as

$$u(x, y, t) = \sum_{n=1}^{\infty} \sum_{m=1}^{\infty} (A_{mn} \cos ck_{mn}t + B_{mn} \sin ck_{mn}t) \sin \frac{m\pi x}{a} \sin \frac{n\pi y}{b} \quad ...(4)$$

$$k^2 = k^2_{mn} = k^3 \left(\frac{m^2}{a^2} + \frac{n^2}{b^2}\right).$$

This will also satisfy the boundary conditions (B).

Now, from equation (4) and the initial conditions (C) and (D), we have

$$u(x, y, 0) = \sum_{m=1}^{\infty} \sum_{n=1}^{\infty} A_{mn} \sin \frac{m\pi x}{a} \sin \frac{n\pi y}{b} = f(x, y).$$

The series is called the double Fourier sine series of (x, y)

i.e. $$A_{mn} = \frac{2}{a} \cdot \frac{2}{b} \cdot \int_{x=0}^{a} \int_{y=0}^{b} f(x, y) \sin \frac{m\pi x}{a} \sin \frac{n\pi y}{b} dx\, dy$$

$$\Rightarrow \quad A_{mn} = \frac{4}{ab} \cdot \int_{x=0}^{a} \int_{y=0}^{b} B_{mn} \sin \frac{m\pi x}{b} \sin \frac{n\pi y}{b} dx\, dy \quad \text{...(5)}$$

and $$\left(\frac{\partial u}{\partial t}\right)_{t=0} = \sum_{m=1}^{\infty} \sum_{n=1}^{\infty} ck_{mn} B_{mn} \sin \frac{m\pi x}{b} \sin \frac{n\pi y}{b} = g(x, y)$$

i.e., $$ck_{mn} B_{mn} = \frac{4}{ab} \int_{x=0}^{a} \int_{y=0}^{b} g(x, y) \sin \frac{m\pi x}{a} \sin \frac{n\pi y}{b} dx\, dy$$

$$\Rightarrow \quad B_{mn} = \frac{4}{ab\, ck_{mn}} \int_{x=0}^{a} \int_{y=0}^{b} g(x, y) \sin \frac{m\pi y}{a} \sin \frac{n\pi y}{b} dx\, dy \quad \text{...(6)}$$

Thus, the solution of the two dimensional wave equation (A) is given by (4) with coefficients (5) and (6), satisfying all the conditions, (B, (C) and (D).

Particular Case **(Meerut, 91 (R)**

If the initial velocity g (x, y) = 0

i.e., $$\left(\frac{\partial u}{\partial t}\right)_{t=0} = g(x, y) = 0$$

then $$B_{mn} = 0$$

and therefore the solution of the wave equation (A) is given by

$$u(x, y, t) = \sum_{m=1}^{\infty} \sum_{n=1}^{\infty} A_{mn} \cos ck_{mnt} \sin \frac{m\pi x}{a} \sin \frac{n\pi y}{b}$$

where $$A_{mn} = \frac{4}{ab} \int_{x=0}^{a} \int_{y=0}^{b} f(x, y) \sin \frac{m\pi x}{a} \sin \frac{n\pi y}{b} dx\, dy$$

and $$k^2 = \pi^2 \left(\frac{m^2}{a^2} + \frac{n^2}{b^2}\right)$$

Note: $k^2 = k_{mn}^2 = \pi \left(\frac{m^2}{a^2} + \frac{n^2}{b^2}\right)$

The frequencies of possible oscillation are given by this equation. These quantities are known as Eigen Values (or Characteristic Values) of the vibrating membrane.

12.5 Vibration of a Circular Membrane

(Poona, 90; Meerut, 97, 96R)

The vibration of a circular membrane are governed by the two dimensional wave equation

$$\frac{\partial^2 u}{\partial t^2} = c^2 \left(\frac{\partial^2 u}{\partial x^2} + \frac{\partial^2 u}{\partial y^2} \right)$$

changing it into polar by the substitution.

$x = r \cos \theta$ and $y = r \sin \theta$, we have

$$\frac{\partial^2 u}{\partial t^2} = c^2 \left(\frac{\partial^2 u}{\partial r^2} + \frac{1}{r} \frac{\partial u}{\partial r} + \frac{1}{r^2} \frac{\partial^2 u}{\partial \theta^2} \right). \quad \text{...(A)}$$

Let $f(r, \theta)$ be the initial displacement and $g(r, \theta)$ the initial velocity of the membrane. Therefore, the function $u(r, \theta, t)$ is required to satisfy (A) and all the boundary and initial conditions, i.e.

Boundary conditions $u(a, \theta, t) = 0$ ...(B)

$(-\pi \le \theta \le \pi, t \ge 0)$

Initial conditions

$$u(r, \theta, 0) = f(r\ \theta) \quad \text{...(C)}$$

and $\left(\frac{\partial u}{\partial t} \right)_{t=0} = g(r, \theta) \quad 0 \le r \le a, -\pi \le \theta \le \pi$

Let us suppose

$$u(r, \theta, t) = R(r) F(\theta) T(t)$$
$$= RFT \text{ (say).} \quad \text{...(1)}$$

Substituting in (A), we have

$$RF \frac{d^2T}{dt^2} = c^2 \left[FT \frac{d^2R}{dr^2} + \frac{1}{r}.FT \frac{dR}{dr} + \frac{1}{r^2}.RT \frac{d^2F}{d\theta^2} \right]$$

$$\Rightarrow \quad \frac{1}{c^2T} \frac{d^2T}{dt^2} = \frac{1}{R} \frac{d^2R}{dr^2} + \frac{1}{rR} \frac{dR}{dr^2} + \frac{1}{r^2}.\frac{1}{F} \frac{d^2F}{d\theta^2}$$

(1) satisfies (A), (B) if

$$\frac{1}{c^2T}\frac{d^2T}{dt^2}=\frac{1}{R}\frac{d^2R}{dr^2}+\frac{1}{rR}\frac{dR}{dr}+\frac{1}{r^2F}\frac{d^2F}{d\theta^2}=-k^2 \qquad ...(2)$$

where $-k^2$ is any constant. We separate variable again in equations (2) and write $\frac{1}{F}\frac{d^2F}{d\theta^2}=-m^2$.

Thus, the ordinary differential equations for R, F and T are

$$\frac{d^2R}{dr^2}+\frac{1}{R}\frac{dR}{dr}+\left(k^2-\frac{m^2}{r^2}\right)R=0 \qquad ...(3)$$

$$\frac{d^2F}{d\theta^2}+m^2F=0 \qquad ...(4)$$

$$\frac{d^2T}{dt^2}+c^2k^2T=0. \qquad ...(5)$$

Equation (4) has solution of the form

$$F = Ae^{\pm im\theta} \quad m = 0, 1, 2, \ldots\ldots$$

Introducing the new independent variable $s = kr$ so that $\frac{1}{r}=\frac{k}{s}$, we have

$$\frac{dR}{dr}=\frac{dR}{ds}.\frac{ds}{dr}=k\,\frac{dR}{ds}$$

$$\Rightarrow \qquad \frac{d^2R}{dr^2}=k^2\,\frac{d^2R}{ds^2}$$

$$\frac{d^2R}{ds^2}+\frac{1}{s}\frac{dR}{ds}+\left(1-\frac{m^2}{s^2}\right)R=0$$

which is Bessel's equation, whose general solution is

$$R = c_1 J_m (s) + c_2 Y_m (s)$$

$$\Rightarrow \qquad R = c_1 J_m (kr) + c_2 Y_m (kr).$$

Since the deflection of the membrane is always finite while Y_m becomes infinite as r approaches zero, we cannot use Y_m and must choose $c_2 = 0$.

Thus, using the boundary condition (B).

$$u (a, \theta, t) = R (a) F (\theta) T (t) = 0.$$

$\therefore$ R (a) = 0 otherwise if f (θ) = 0, $\Rightarrow$ T (t) = 0, then u = 0.

$$R (a) = c_1 J_m (ka) = 0$$

$$\therefore \qquad J_m (ka) = 0. \qquad ...(6)$$

Let $k_{m1}, k_{m2}, \ldots$ be the positive roots of (6).

The corresponding general solution of (5) is

$$T = a_{mn} \cos ck_{mn} t + b_{mn} \sin ck_{mn} t$$

$$\Rightarrow \quad T = a'_{mn} e \pm ik_{mn} ct$$

Thus, we get the general solution as

$$u(r, \theta, t) = \sum_{m=1}^{\infty} \sum_{n=1}^{\infty} \left(A_{mn} \cos ck_{mn} t + B_{mn} \sin ck_{mn} t\right) e^{\pm im\theta} J_m(k_{mn} r)$$

$$\Rightarrow \quad u(r, \theta, t) = \sum_{m=1}^{\infty} \sum_{n=1}^{\infty} \left\{A_{mn} J_m\left(k_{mn} r\right)\right\} e^{\left(\pm im\theta \pm ik_{mn}ct\right)}$$

satisfying the boundary condition (B), where

A_{mn}, B_{mn} are constants.

Considering the solution of the wave equation (A) which are symmetrical radially i.e., when the solution is independent of θ, we get the general solution as

$$u(r, t) = \sum_{n=1}^{\infty} (A_n \cos ck_n t + B_n \sin ck_n t) J_0(k_n r) \quad \text{...(7)}$$

$$\Rightarrow \quad u(r, t) = \sum_{n=1}^{\infty} A_n J_0(k_n r). e^{\pm ick_n t}$$

where $k_1, k_2, \ldots$ are the positive roots of the equation

$$j_0(ka) = 0.$$

Now, from equation (7) and the initial condition (C) when $t = 0$, we have

$$u(r, 0) = \sum_{n=1}^{\infty} A_n J_0(k_a r) = f(r)$$

$f(r, 0)$ becomes $f(r)$ when independent of θ.

Hence, A_n must be the coefficients of the Fourier Bessel series which represent $f(r)$ in terms of $J_0(k_n r)$, i.e.

$$A_n = \frac{2}{a^2 J_1^2(k_n a)} \int_b^a rf(r) J_0(k_a r)\, dr \quad \text{...(8)}$$

$r = 1, 2, \ldots$

Also from (D),

$$\left(\frac{\partial u}{\partial t}\right)_{t=0} = \sum_{n=1}^{\infty} ck_n B_n J_n(k_n r) = g(r)$$

$g(r, \theta)$ becomes $g(r)$ when independent of θ.

$\therefore$ From Fourier Bessel series,

$$ck_n\ B_n = \frac{2}{a^2\ J_1^2\,(k_n\ a)} \int_0^a rg\,(r)\ J_0\,(k_n\ r)\ dr, \quad n = 1, 2...$$

$$\Rightarrow \quad B_n = \frac{2}{ck_n\ a^2\ I_1^2(k_n\ a)} \int_0^a rg\,(r)\ J_0\,(k_n\ r)\ dr \qquad ...(9)$$

Hence, equation (7) is the solution of the wave equation with the coefficients given by the equations (8) and (9), which is radially symmetric.

Particular Case **(Meerut, 97)**

If $$\left(\frac{\partial u}{\partial t}\right)_{t=0} = g\,(r) = 0,$$

then $B = 0.$

and hence the solution of the wave equation (A) is given by

$$u\ (r,\ t) = \sum_{n=1}^{\infty} A_n\, J_0\,(k_n\ r)\ \cos\,(c\ k_n\ 1) \qquad ...(10)$$

where A_n is given by the equation (8).

Example 1:

Find the deflection u (x, y, t) of the square membrane with a = b = 1 and c = 1, if the initial velocity is zero and the initial deflection if f(x, y) = A sin πx sin 2xy. **(Meerut, 91 (P), 90)**

Solution:

Here the vibration of the square membrane are governed by two dimensional wave equation

$$\frac{\partial^2 u}{\partial t^2} = c^2 \left(\frac{\partial^2 u}{\partial x^2} + \frac{\partial^2 u}{\partial y^2}\right)$$

Here the boundary conditions are

$$u\ (0,\ y,\ t) = 0$$
$$u\ (1,\ y,\ t) = 0$$
$$u\ (x,\ 0,\ t) = 0$$
$$u\ (x,\ 1,\ t) = 0$$

and the initial conditions are

$$u\ (x,\ y,\ 0) = f\ (x,\ y) = A\ \sin\ \pi x\ \sin\ 2py$$

and $$\left(\frac{\partial u}{\partial t}\right)_{t=0} = 0.$$

Proceeding as in (Particular case), we have

$$u(x, y, t) = \sum_{n=1}^{\infty} \sum_{m=1}^{\infty} A_{mn} \cos K_{mn} t \sin m\pi x \sin n\pi y. \quad ...(1)$$

Since $c = 1, a = b = 1$

$$K_{mn}^2 = \pi^2 (m^2 + n^2)$$

where, $A_{mn} = 4 \int_0^1 \int_0^1 f(x, y) \sin m\pi x. \sin n\pi y \, dx \, dy$

$$= 4A \int_{x=0}^{1} \int_{y=0}^{1} \sin \pi x \sin m\pi x. \sin 2\pi y \sin n\pi y \, dx \, dy.$$

Obviously $A_{m1} = A_{m3} = A_{m5} ... = 0$ each.

But $A_{m2} = 4A \int_{x=0}^{1} \int_{y=0}^{1} \sin \pi x \sin m\pi x. \sin^2 2\pi y \, dx \, dy$

$$= 2A \int_{x=0}^{1} \int_{y=0}^{1} \sin \pi x \sin m\pi x (1 - \cos 4\pi y) \, dx \, dy$$

$$= 2A \int_{x=0}^{1} \sin \pi x \sin m\pi x \left(y - \frac{1}{4\pi} \sin 4\pi y \right)_0^1 dx$$

$$= 2A \int_{x=0}^{1} \sin \pi x \sin m\pi x \, dx$$

Again obviously $A_{22} = A_{22} = ... = 0$ (each).

But $A_{12} = 2A \int_0^1 \sin^2 \pi x \, dx = A \int_0^1 (1 - \cos 2\pi x) \, dx$

$$= A \left(1 - \frac{1}{2\pi} \sin 2\pi x \right)_0^1 = A.$$

$\therefore$ from (1), $u(x, y, t) = A_{12} \cos K_{12} t \sin \pi x \sin 2\pi y$

$$= A \cos \sqrt{5\pi t} \sin \pi x \sin 2\pi y.$$

Since all other coefficients are zero $K_{12}^2 = \pi^2 (1^2 + 2^3)$.

$\Rightarrow \quad K_{12} = \sqrt{5\pi}$. **Ans.**

Example 2:

A string is stretched between the fixed points (0, 0) and (l, 0) and released at rest from the initial deflection given by

$$f(x) \begin{cases} \frac{2k}{l} x & \text{when } 0 < x < \frac{l}{2} \\ \frac{2k}{l}(l - x) & \text{when } \frac{l}{2} < x < l \end{cases}$$

Find the deflection of the string at any time t.

Solution:

Proceeding as in (particular case) the deflection of the string is given by

$$u(x, t) = \sum_{n=1}^{\infty} c_n \cos\frac{n\pi ct}{l} \sin\frac{n\pi x}{l}$$

where

$$c_n = \frac{2}{l}\int_0^l f(x)\sin\frac{n\pi x}{l}\,dx$$

$$= \frac{2}{l}\cdot\left[\int_0^{l/2}\frac{2k}{l}x\sin\frac{n\pi x}{l}\,dx + \int_{l/2}^{l}\frac{2k}{l}(l-x)\sin\frac{n\pi x}{l}\,dx\right]$$

$$= \frac{2}{l}\cdot\frac{2k}{l}\left[\int_0^{l/2}x.\sin\frac{n\pi x}{l}\,dx + \int_{l/2}^{l}(l-x)\sin\frac{n\pi x}{l}\,dx\right]$$

$$= \frac{4k}{l^2}\cdot\left[\left\{-x\,\frac{l}{n\pi}\cos\frac{n\pi x}{l}\right\}_0^{l/2} + \frac{l}{n\pi}\int_0^{l/2}\cos\frac{n\pi x}{l}\,dx\right.$$

$$\left. + \left\{-(l-x)\,\frac{l}{n\pi}\cos\frac{n\pi x}{l}\right\}_{l/2}^{l} \cos\frac{n\pi x}{l}\,dx\right]$$

$$= \frac{4k}{l^2}\cdot\left[-\frac{l^2}{2n\pi}\cos\frac{n\pi}{2} + \frac{1}{n\pi}\cdot\frac{1}{n\pi}\left(\sin\frac{n\pi x}{l}\right)_0^{l/2}\right.$$

$$\left. -0 + \frac{l}{2}\cdot\frac{l}{n\pi}\cos\frac{n\pi}{2} - \frac{l}{n\pi}\cdot\frac{l}{n\pi}\left(\sin\frac{n\pi x}{l}\right)_{l/2}^{l/2}\right]$$

$$= \frac{4k}{l^2}\cdot\left[\frac{l^2}{n^2\pi^2}\sin\frac{n\pi}{2} - \frac{l^2}{n^2\pi^2}\left(\sin n\pi - \sin\frac{n\pi}{2}\right)\right]$$

$$= \frac{4k}{l^2}\cdot\left[\frac{2l^2}{n^2\pi^2}\sin\frac{n\pi}{2}\right] = \frac{8k}{n^2\pi^2}.\sin\frac{n\pi}{2}$$

$$\therefore u(x, t) = \sum_{n=1}^{\infty}\frac{8k}{n^2\pi^2}\sin\frac{n\pi}{2}\cos\frac{n\pi ct}{t}\sin\frac{n\pi x}{l}$$

$$= \frac{8k}{\pi^2}\left[\frac{1}{1^2}\sin\frac{\pi}{l}x\cos\frac{\pi c}{l}t - \frac{1}{3^2}\sin\frac{3\pi}{l}x\cos\frac{3\pi c}{l}t + \ldots\right].$$ **Ans.**

Example 3:

The points of trisection of a string are pulled aside through a distance b on opposite sides of the position of equilibrium and the string is released from rest. Derive an expression for the displacement of the string at any subsequent time and show that the mid point of the string always remains at rest. **(Meerut, 97 (S), 99)**

Solution:

Let OA be the equilibrium position of the string of length $l = 3a$ and L, M its points of trisection which are pulled through a distance b on opposite sides and then released.

Here the displacement of the string will be the solution of the one dimensional wave equation

$$\frac{\partial^2 u}{\partial t^2} = c^2 \frac{\partial^2 u}{\partial x^2}.$$

Its initial deflection is given as follows.

$$f(x) = \begin{cases} b\,\dfrac{x}{a} & \text{when } 0 \le x \le a \\ \dfrac{b(3a-2x)}{a} & \text{when } a \le x \le 2a \\ \dfrac{b(x-3a)}{a} & \text{when } 2a \le x \le 3a \end{cases}$$

equation of the line OL' is

$$y = \frac{b-0}{a-0}(x-0)$$

$$y = \frac{b}{a}x.$$

equation of L' M '

$$y - b = \frac{-b-b}{2a-a}(x-a)$$

$$y = \frac{b(3a-2x)}{a}$$

equation of M 'A

$$y - (-b) = \frac{0-(-b)}{3a-2a}(x-2a)$$

$$y = \frac{b(x-3a)}{a}.$$

and initial velocity $g(x) = 0$.

Now, proceeding similarly as in (particular case) we have deflection

$$(u, x, t) = \sum_{n=1}^{\infty} c_n \cos\frac{cn\pi}{3a}t \sin\frac{n\pi x}{3a}$$

$$\text{where } c_n = \frac{2}{3a}\int_0^{2a} f(x)\sin\frac{n\pi x}{3a}\,dx$$

$$= \frac{2}{3a}\left[\frac{b}{a}\int_0^a x\sin\frac{n\pi x}{3a}\,dx + \frac{b}{a}\int_a^{2a}(3a-2x)\frac{n\pi x}{3a}\,dx + \frac{b}{a}\int_{2a}^{3a}(x-3a)\sin\frac{n\pi x}{3a}\,dx\right]$$

Integrating by parts

$$c_n = \frac{2b}{3a^2}\left[\left(-x.\frac{3a}{n\pi}\cos\frac{n\pi x}{3a}\right)_0^a + \frac{3a}{n\pi}\int_0^a \cos\frac{n\pi x}{3a}\,dx\right.$$

$$+\left\{-(3a-2x)\frac{3a}{n\pi}\cos\frac{n\pi x}{3a}\right\}_a^{2a} - \frac{6a}{n\pi}\int_a^{2a}\cos\frac{n\pi x}{3a}\,dx$$

$$\left.+\left\{-(x-3a)\frac{3a}{n\pi}\cos\frac{n\pi x}{3a}\right\}_{2a}^{2a} - \frac{3a}{n\pi}\int_{2a}^{3a}\cos\frac{n\pi x}{3a}\,dx\right]$$

$$= \frac{2b}{3a^2}\left[-\frac{3a^2}{n\pi}\cos\frac{n\pi}{3} + \frac{9a^2}{n^2\pi^2}\left(\sin\frac{n\pi x}{3a}\right)_a^a\right.$$

$$+\frac{3a^2}{n\pi}\cos\frac{2n\pi}{3} + \frac{3a^2}{n\pi}\cos\frac{n\pi}{3} - \frac{18a^2}{n^2\pi^2}\left(\sin\frac{n\pi x}{3a}\right)_0^{2a}$$

$$\left.-\frac{3a^2}{n\pi}\cos\frac{2n\pi}{3} + \frac{9a^2}{n^2\pi^2}\left(\sin\frac{n\pi x}{3a}\right)_{2a}^{3a}\right]$$

$$= \frac{2b}{3a^2}\left[\frac{9a^2}{n^2\pi^2}\sin\frac{n\pi}{3} - \frac{18a^2}{n^2\pi^2}\left(\sin\frac{2n\pi}{3} - \sin\frac{n\pi}{3}\right)\right.$$

$$\left.+\frac{9a^2}{n^2\pi^2}\left(\sin n\pi - \sin\frac{2n\pi}{3}\right)\right]$$

$$= \frac{2b}{3a^2}.\frac{27a^2}{n^2\pi^2}\left[\sin\frac{n\pi}{3} - \sin\frac{2n\pi}{3}\right]$$

$$= \frac{18b}{n^2\pi^2}\left[\sin\frac{n\pi}{3} - \sin\left(n\pi - \frac{n\pi}{3}\right)\right]$$

$$= \frac{18b}{n^2\pi^2}\left\{1+(-1)^n\right\}\sin\frac{n\pi}{3}$$

$\therefore c_n = 0$ when n is odd

and $\quad c_n = \dfrac{18b}{n^2\pi^2}2\sin\dfrac{n\pi}{3}$ when n is even say n = 2m.

$$\therefore u(x, t) = \frac{86b}{\pi^2}\sum_{m=1}^{\infty}\frac{1}{(2m)^2}.\sin\frac{2m\pi}{3}.\sin\frac{2m\pi x}{3a}.\cos\frac{2m\pi ct}{3a}$$

$$= \frac{9b}{\pi^2} \sum_{m=1}^{\infty} \frac{1}{m^2} \sin \frac{2m\pi}{3} \sin \frac{2m\pi}{3a} x . \cos \frac{2m\pi ct}{3a}$$ **Ans.**

The displacement of the mid-point of the string is obtained by putting

$x = \frac{3a}{2}$ in the above expression. Since when

$$x = \frac{3a}{2}, \sin \frac{2m\pi}{3a} x = \sin m\pi = 0.$$

$\therefore u (x, t) = 0$

i.e., the mid-point of the string always remains at rest.

Example 4:

Find the deflection u (x, t) of the following string (length) $l = \pi$, ends fixed, and $c^2 = 1$) corresponding to zero initial velocity and initial deflection

$$f(x) = k (\sin x - \sin 2x).$$

Solution:

Deflection u (x, t) of the string is given by

$$u (x, t) = \sum_{n=1}^{\infty} c_n \cos nct \sin nx \qquad \text{since } l = \pi$$

$$= \sum_{n=1}^{\infty} c_n \cos nt \sin nx \qquad \text{as } c = 1$$

where $$c_n = \frac{2}{\pi} \int_0^{\pi} f(x) \sin nx \, dx$$

$$= \frac{2}{\pi} \int_0^{\pi} k (\sin x - \sin 2x) \sin nx \, dx$$

$$= \frac{2k}{\pi} \int_0^{\pi} \sin x \sin nx \, dx - \frac{2k}{\pi} \int_0^{\pi} \sin 2x \sin nx \, dx$$

It is obvious that $c_3 = c_4 = \ldots = 0$ (each)

and $$c_1 = \frac{2k}{\pi} \int_0^{\pi} \sin^2 x \, dx - \frac{2k}{\pi} \int_0^{\pi} \sin 2x \sin x \, dx$$

$$= \frac{2k}{\pi} \int_0^{\pi} \frac{1}{2}(1 - \cos 2x) \, dx - \frac{k}{\pi} \int_0^{\pi} (\cos x - \cos 3x) \, dx$$

$$= \frac{k}{\pi}, \pi = k.$$

Also $c_2 = \frac{2k}{\pi}\int_0^{\pi} \sin x \sin 2x\, dx - \frac{2k}{\pi}\int_0^{\pi} \sin^2 2x\, dx$

$$= \frac{k}{\pi}\int_0^{\pi} (\cos x - \cos 3x)\, dx - \frac{k}{\pi}\int_0^{\pi} (1-\cos 4x)\, dx$$

$$= 0 - \frac{k}{\pi}.\pi = -k.$$

Hence, $u(x, t) = c_1 \cos t \sin x + c_2 \cos 2t \sin 2x$

$= k(\cot t \sin x - \cos 2t \sin 2x).$ **Ans.**

Example 5:

A string is stretched between the fixed points (0, 0) and (1, 0) and released at rest from the position $u = A \sin \pi x$. Find the formula for its subsequent displacement $u(x, t)$.

Solution:

Here the vibration of the string are governed by the one dimensional wave equation.

$$\frac{\partial^2 u}{\partial t^2} = c^2 \frac{\partial^2 u}{\partial x^2}$$

Boundary conditions are $u(0, t) = 0$

and $u(1, t) = 0$

and the initial conditions $u(x, 0) = A \sin \pi x$

and $\left(\frac{\partial u}{\partial t}\right)_{t=0} = 0.$

Hence, proceeding as in (particular case), the formula for subsequent displacements $u(x, t)$ of the string is

$$u(x, t) = \sum_{n=1}^{\infty} c_n \cos(nc\pi t) \sin n\pi x$$

where $c_n = 2\int_0^1 A \sin \pi x . \sin n\pi x\, dx.$

Here all the coefficients $c_2, c_3 \ldots$ etc. are zero, each, except c_1, which is given by

$$c_1 = A\int_0^1 2\sin^2 \pi x\, dx = A\int_0^1 (1-\cos 2\pi x)\, dx$$

$$= A.$$

Hence, $u(x, t) = c_1 \cos c\pi t \sin \pi x$

$= A \cos c\pi t \sin \pi x.$ **Ans.**

12.6 Solution of One-dimensional Heat Equation

The heat flow in a body of homogeneous material is governed by the heat equation

$$\frac{\partial u}{\partial t} = c^2 \nabla^2 u \qquad c^2 = \frac{k}{\sigma\rho}$$

$$= c^2 \left(\frac{\partial^2 u}{\partial x^2} + \frac{\partial^2 u}{\partial y^2} + \frac{\partial^2 u}{\partial z^2} \right)$$

where $u(x, y, z, t)$ is the temperature in the body, k is the thermal conductivity, σ is the specific heat ρ is the density of the material of the body.

If we consider the heat flow in a long thin bar or wire of constant cross section and homogeneous material which is along x-axis and is perfectly insulated laterally, so that the heat flows in the x-direction only, u depends on x and t only and therefore the heat equation becomes

$$\frac{\partial u}{\partial t} = c^2 \frac{\partial^2 u}{\partial x^2} \qquad \text{...(A)}$$

which is known as one dimensional heat equation.

Let the solution of the equation (A) be of the form

$$u(x, t) = F(x)\, T(t) = FT \text{ (say)}$$

where F is a function of x only and T that of t only.

Substituting in (1), we have

$$F\frac{dT}{dt} = c^2 T \frac{d^2F}{dx^2}$$

$$\Rightarrow \quad \frac{1}{F}\frac{d^2F}{dx^2} = \frac{1}{c^2 T}\frac{dT}{dt}$$

Since left hand side is a function of independent variable x, while right hand side is a function of independent variable t, the two cannot be equal to each other unless both reduce to a constant value.

$$\therefore \quad \frac{1}{F}\frac{d^2F}{dx^2} = \frac{1}{c^2 T}\frac{dT}{dt} = 0 \Rightarrow k^2. \Rightarrow -k^2$$

and hence in the three cases, we have

(i) $\dfrac{d^2F}{dx^2} = 0$ and $\dfrac{dT}{dt} = 0$

(ii) $\frac{d^2F}{dx^2} - k^2F = 0$ and $\frac{dT}{dt} = k^2c^2T$

(iii) $\frac{d^2F}{dx^2} + k^2F = 0$ and $\frac{dT}{dt} + k^2c^2T = 0$.

The general solutions in these cases are

(a) $F = Ax + B, \quad T = C$

(b) $F = Ae^{kx} + Be^{kx}, \ T = Ce^{k^2c^2t}$

(c) $F = A \cos kx + B \sin kx, \ T = Ce^{k^2c^2t}$

Case I: Both Ends at Zero Temperature

To determine the solution of the equation (A); where l is the length of the rod whose ends are kept at temperature zero and f (x) is the initial temperature in the bar. **(Meerut, 99 (S), 90, 91, 91 (P))**

The boundary conditions are

$$u(0, t) = 0 \quad ...(1)$$

and $$u(l, t) = 0 \quad ...(2)$$

for all t.

and the initial condition is

$$u(x, 0) = f(x) \quad ...(3)$$

$0 < x < l$

0 — l — x

Now from boundary conditions (1), (2), we have

$$u(0, t) = F(0)\, T(t) = 0$$

and $$u(l, t) = F(l)\, T(t) = 0.$$

$\therefore \quad F(0) = 0$ and $F(l) = 0$.

[since if we take T (t) = 0 then u, (x, t) = 0].

Thus the solutions (a) and (b) do not constitute the solution as they given A = B = 0 i.e. F = 0 and hence u (x, t) = 0.

Thus from (c), we have

$F(0) = A = 0$ and $F(l) = 0 + B \sin kl = 0$.

Now $B \neq 0$ otherwise F = 0 and hence u (x, t) = 0.

$\therefore \quad \sin kl = 0$

$\Rightarrow \quad kl = n\pi$

$\Rightarrow \quad k = \frac{n\pi}{l}, \qquad n = 1, 2, 3,$

Hence, for some value of n

$$u_n(x, t) = B_n \sin \frac{n\pi}{l} x . e^{\left(-n^2\pi^2c^2/l^2\right)t}$$

are solution of (A) satisfying equations (1) and (2).

Therefore, for each value of n, we may take the solution as

$$u(x, t) = \sum_{n=1}^{\infty} u_n(x, t)$$

$$= \sum_{n=1}^{\infty} B_n \sin \frac{\pi n}{l} x . e^{-n^2\pi^2c^2t/l^2} \quad ...(5)$$

from this and (3), we have

$$u(x, 0) = \sum_{n=1}^{\infty} B_n \sin \frac{\pi n}{l} x = f(x).$$

Hence, equation (2) satisfies the initial condition (1) if the coefficient B_n is so chosen that u (x, 0) becomes the Fourier Sine series of f (x).

i.e., $$B_n = \frac{2}{l} \int_0^l f(x) \sin \frac{n\pi}{l} x \, dx. \quad ...(6)$$

Thus, the series (5) with the coefficient (6) is the solution of the one dimensional heat equation.

Case II: One Face at Temperature u_0

To obtain the temperature function of the wire as t increase if it is at uniform temperature v_0 through the wire and the end x = 0 is kept at zero temperature and the end x = 1 is kept at a constant temperature u_0.

$$u(0, t) = 0 \quad ...(7)$$

$$u(l, t) = u_0 \quad ...(8)$$

for all t.

and initial condition is

$$u(x, 0) = v_0. \quad ...(9)$$

In this case, the solution given by (b) is inadmissible, as in this case the temperature uy = FT increases infinitely with time. The solution (c) by itself is inadequate to given the complete solution, since in this case the temperature will tend to zero as $t \to \infty$. Hence our solution must be a combination of solution of the types (a) and (c). So we can write

$$u(x, t) = u_s(x) + u_t(x, t) \quad ...(10)$$

where $u_s(x)$ is the temperature distribution after a long period of time when the rod has reached a steady state of temperature distribution, $u_t(x, t)$

indicates the transient effects which die down with the passage of time. These two must be the solution of the types (a) and (c) respectively.

It is obvious that when the end x = 0 is maintained at temperature u = 0 and the end x = l at u = u_0, ultimately there will be uniform gradation of temperature.

$$\therefore \quad u_s(x) = \frac{u_0}{l}x.$$

$\therefore$ from (10), we have $u(x, t) = \dfrac{u_0}{l}x + u_t(x, t)$. ...(11)

Now, with the help of (7), (8) and (9), we get boundary and initial conditions for $u_t(x, t)$ as follows

$$u(0, t) = u_t(0, t) = 0. \quad ...(12)$$

$$u(l, t) = u_0 + u_t(1, t) = u_0.$$

$$\therefore \quad u_t(l, t) = 0 \quad ...(13)$$

and

$$u(x, 0) = \frac{u_0}{l}x + u_t(x, 0) = v_0.$$

$$\therefore \quad u_t(x, 0) = v_0 - \frac{u_0}{l}x. \quad ...(14)$$

$\therefore$ Let us take

$$u_t(x, t) = (A' \cos kx + B' \sin kx)\ e^{-k^2c^2t} \quad ...(15)$$

Now, from equation s (12) and (15), we have

$$u_t(0, t) = A'\,e\,e^{-k^2c^2t} = 0 \qquad \therefore A' = 0$$

Then (15) and (13), gives

$$u_t(l, t) = B' \sin kl = 0$$

$\therefore \sin kl = 0 \Rightarrow kl = n\pi$

$$\Rightarrow \quad k = n\pi/ \qquad (n = 1, 2, ...)$$

$\therefore$ a solution for $u_t(x, t)$ is

$$B_n \sin \frac{n\pi}{l}x.e^{-\left(\frac{n\pi c}{l}\right)^2 t} \qquad (n = 1, 2,...).$$

Now, adding the solution for different n, the general solution may be written as

$$u_t(x, t) = \sum_{n=1}^{\infty} B_n \frac{n\pi}{l}x.e^{-\left(\frac{n\pi c}{l}\right)^2 t} \quad ...(16)$$

From this and equation (14), we have

$$u_t(x, 0) = v_0 - \frac{u_0}{l}x = \sum_{n=1}^{\infty} B_n \sin\frac{n\pi}{l}x$$

$\therefore$ R.H.S. is the Fourier sine expansion of the function $v_0 - \frac{u_0}{l}x$

$$\therefore\ B_n = \frac{2}{l}\int_0^l \left(v_0 - \frac{u_0}{l}x\right)\sin\frac{n\pi}{l}x\,dx$$

Integrating by parts, we have

$$= \frac{2}{l}\left[\left\{\left(v_0 - \frac{u_0}{l}x\right)\left(-\cos\frac{n\pi}{l}x\right)\frac{l}{n\pi}\right\}_0^l - \frac{u_0}{l}\int_0^l \frac{l}{n\pi}\cos\frac{n\pi x}{l}\,dx\right]$$

$$= \frac{2}{l}\left[-(v_0 - u_0)\frac{l}{n\pi}\cos n\pi + \frac{l}{n\pi}v_0 - \frac{u_0}{n\pi}.\frac{l}{n\pi}\left(\sin\frac{n\pi x}{l}\right)_0^l\right]$$

$$= \frac{2}{l}\left[-\frac{l}{n\pi}(v_0 - u_0)(-1)^n + \frac{l}{n\pi}v_0 - \frac{lu_0}{n^2\pi^2}(\sin n\pi - 0)\right]$$

$$= \frac{2}{n\pi}\left[v_0 - (-1)^n(v_0 - u_0)\right]$$

$\therefore$ from (11) and (16)

$$u(x, t) = \frac{u_0}{l}x + \frac{2}{\pi}\sum_{n=1}^{\infty}\frac{1}{n}\left[v_0 - (-1)^n(v_0 - u_0)\right].e^{-(n\pi c/l)^2 t}\sin\frac{n\pi}{l}x \quad ...(17)$$

which is the required solution.

Particular Case: If the initial temperature of wire is zero i.e. $v_0 = 0$,

$$\text{then } u(x, t) = \frac{u_0}{l}\left[x - \frac{2l}{\pi}\sum_{n=1}^{\infty}\frac{1}{n}(-1)^n.e^{-(n\pi c/l)^2 t}\sin\frac{n\pi}{l}\right] \quad ...(18)$$

Case III: Case of Infinite Bar

To find solution of the heat equation (A) in the case of a bar which extends to infinity on both sides and insulated laterally. The initial temperature of the bar is f (x). **(Meerut, 96 R)**

In this case there is no boundary condition but only the initial condition

$$u(x, 0) = f(x). \quad ...(19)$$

$(-\infty < x < \infty)$

In this case the solution (a) and (b) do not constitute our solution. Hence taking the third solution (c).

$$F = A\cos px + B\sin px,\quad T = ce^{-c^2p^2t}$$

[Taking the constant as $-p^2$ instead of $-k^2$]

Hence, $u(x, t) = FT = (C \cos px + D \sin px)\, e^{-c^2p^2t}$ ...(20)

Here f (x)is not periodic, it is natural to use Fourier integrals in the present case instead of Fourier series.

Since C and D in (20) are periodic, we may consider them as functions of p and write C = C (p), D = D (p).

Since the heat equation is linear and homogeneous

$$\therefore u(x, t) = \int_0^{\infty} u(x, t, p)\, dp$$

$$= \int_0^{\infty} \{C(p) \cos px + D(p) \sin px\}\, e^{-c^2p^2t} \quad ...(21)$$

Now, equation (21) is the solution of heat equation (A) provided this integral exists and can be differentiated twice w.r.t 'x' and once w.r.t. 't'.

Now, from equations (21) and (19), we have

$$u(x, 0) = \int_0^{\infty} [C(p) \cos px + D(p) \sin px]\, dp = f(x)$$

$$\therefore C(p) = \frac{1}{\pi} \int_{-\infty}^{\infty} f(v) \cos pv\, dv$$

and $$D(p) = \frac{1}{\pi} \int_{-\infty}^{\infty} f(v) \sin pv\, dv$$

$$\therefore u(x, 0) = \frac{1}{\pi} \int_0^{\infty} \left[\int_{-\infty}^{\infty} f(v) \cos(px - pv)\, dv \right] dp$$

Thus, equation (21) becomes

$$u(x, t) = \frac{1}{\pi} \left[\int_{-\infty}^{\infty} f(v) \cos(px - pv).e^{c^2p^2t}\, dv \right] dp$$

$$= \frac{1}{\pi} \left[\int_{-\infty}^{\infty} f(v) \left[\int_0^{\infty} e^{-c^2p^2t} \cos(x - v)\, p.dp \right] dv \right] \quad ...(22)$$

assuming that we may invert the order of integration.

Now, $$\int_0^{\infty} e^{-c^2p^2t} \cos(x - v)\, p\, dp$$

$$= \frac{1}{c\sqrt{t}} \int_0^{\infty} e^{-s^2} \cos \frac{(x - v)}{2c\sqrt{t}} 2s.ds. \quad \text{Putting } cp\sqrt{t} = s$$

so that $dp = \dfrac{1}{c\sqrt{t}} ds$.

$$= \frac{1}{c\sqrt{t}} \cdot \frac{\sqrt{\pi}}{2} e^{-\left(\frac{x-v}{2c\sqrt{t}}\right)^2} = \frac{\sqrt{\pi}}{2c\sqrt{t}} e^{-(x-v)^2/4c^2t}$$

Since $\int_0^{\infty} e^{-s^2} \cos 2bs \, dx = \frac{\sqrt{\pi e}^{-b^2}}{2}$

$\therefore$ from (22)

$$u(x, t) = \frac{1}{\pi} \int_{-\infty}^{\infty} f(v) \frac{\sqrt{\pi}}{2c\sqrt{t}} e^{-(x-v)^2/4c^2t} \, dv$$

$$= \frac{1}{2c\sqrt{\pi t}} \int_{-\infty}^{\infty} f(v).e^{-(x-v)^2/4c^2t} \, dv$$

$$= \frac{1}{\sqrt{\pi}} \int_{-\infty}^{\infty} f\left(x + 2c\omega\sqrt{t}\right).e^{-\omega^2} \, d\omega$$

Putting $\frac{(v-x)}{2c\sqrt{t}} = \omega$ so that $dx = -\ 2c\sqrt{t}\, d\omega$.

This function is the required solution.

Case IV: Insulated Faces

To find the temperature u (x, t) in a bar of length l which is perfectly insulated, also at the ends at x = 0 and x = 1; assuming that the initial temperature of the bar u (x, 0) = f (x). **(Meerut, 98)**

The flux of heat across the faces $x = 0$ and $x = l$ is proportional to $\frac{\partial u}{\partial x}$ at that end, since these end are insulated.

$\therefore$ Boundary conditions are

$$u_x(0, t) = 0 \quad \text{...(24)}$$

$$u_x(l, t) = 0. \quad \text{...(25)}$$

Initial condition is

$$u_x(x, 0) = f(x)$$

$$0 < x < l.$$

In this case, the solution given by (b) is inadmissible as in this the temperature u = FT increases indefinitely with time. The solution (c) by itself is inadequate to given the complete solution, since in this case the temperature will tend to zero as $t \to \infty$.

If the solution given by (a), then $u = FT = A'x + B'$.

But from this and (24), $u_x(0, t)\ A' = 0$.

$$u = B' \text{ (constant)} \quad \text{...(27)}$$

Again, if the solution of the heat equation (A) is given by (c) then

$$u(x, t) = (C' \cos kx + D' \sin kx)\ e^{-k^2c^2t} \qquad ...(28)$$

from this and (24), we have

$$u_x(0, t) = D'k\,e^{-k^2c^2t} = 0 \qquad \therefore\ D' = 0$$

and then from (28) and (25), we have

$$u_x(l, t) = \sin kl = \theta$$

$$\Rightarrow \qquad kl = n\pi \qquad (n = 1, 2,...)$$

$$\Rightarrow \qquad k = \frac{n\pi}{l}.$$

$\therefore$ for each value of n, we have a solution of (A) of the type

$$u_x(x, t) = A_n \cos\frac{n\pi}{l}x.e\left(\frac{n\pi C}{l}\right)^2 t$$

Since we know that, if each of the n functions satisfies a linear homogeneous differential equation, then every linear combination of those functions also satisfies the differential equation.

Hence, the complete solution of linear homogeneous differential equation (A), may be written as

$$u(x, t) = B' + \sum_{n=1}^{\infty} A_n \cos\frac{n\pi}{l}x.e^{-\left(\frac{n\pi c}{l}\right)^2 t}$$

from this and (26), we have

$$(u, x) = f(x) = B' + \sum_{n=1}^{\infty} A_n \cos\frac{n\pi x}{l} \qquad ...(29)$$

Multiplying both sides of (29) by cos $\frac{n\pi x}{l}$ and then integrating between 0 to l, we have

$$A_n \frac{2}{l}\int_0^l f(x) \cos\frac{n\pi x}{l} dx.$$

Again, integrating both sides of (29) between the limits 0 to l, we have

$$B_n \frac{1}{l}\int_0^l f(x)\ dx$$

$$= \frac{1}{2}.\frac{2}{l}\int_0^l f(x).\cos\frac{n\pi}{l}\ 0\ dx$$

$$= \frac{1}{2}A_0,$$

Hence, the complete solution of (A) is given by

$$u(x, t) = \frac{1}{2}A_0 + \sum_{n=1}^{\infty} A_0 \cos\frac{n\pi x}{l}.e^{-(n\pi c/l)^2 t} \qquad ...(30)$$

where $$A_n = \frac{2}{l}\int_0^l f(x) \cos\frac{n\pi x}{l} dx. \qquad ...(31)$$

Equation (30) with coefficient (31) determine the temperature in the bar.

Example 1:

Find that temperature u (x, t) in a bar of silver (length 10 cm, constant cross-section of area 1 cm², density 10.6 gm/cm³, thermal conductivity 1.04 cal/cm. deg sec, specific heat 0.056 cal/gm deg.) which is perfectly insulated laterally, whose ends are kept at temperature 0°C and whose initial temperature (in °C) is f (x).

where

$$(a)\ f(x) = \begin{cases} x & \text{if } 0 < x < 5 \\ 10 - x & \text{if } 5 < x < 10 \end{cases}$$

(b) $f(x) = x(10 - x)$.

Solution:

From case I, we know that the temperature in a bar is given by

$$u(x, t) = \sum_{n=1}^{\infty} B_n \sin\frac{n\pi}{l}x.e^{-\frac{n^2\pi^2c^2t}{l}} \quad \text{where } c^2 = \frac{k}{\sigma\rho}$$

(k is thermal conductivity, σ the specific heat and ρ is the density of the material of the bar)

and $$B_n = \frac{2}{l}\int_0^l f(x) \sin\frac{n\pi x}{l} dx.$$

In the present case l = 10 cm.

and $$c^2 = \frac{1.04}{(0.056).(10.6)} = 1.752$$

$$(a)\ B_n = \frac{1}{10}\int_0^{10} f(x) \sin\frac{n\pi x}{10} dx$$

$$= \frac{1}{5}\left[\int_0^5 x \sin\frac{n\pi}{10}x.dx + \int_5^{10} (10 - x) \sin\frac{n\pi}{10} x\, dx\right.$$

Integrating by parts, and simplifying as

$$= \frac{40}{n^2\pi^2}\sin\frac{n\pi}{2}.$$

Thus, $\quad B_n = 0$ when n is odd

$$\frac{40}{n^2\pi^2} \quad (n = 1, 5, 9, ...)$$

and $$= -\frac{40}{n^2\pi^2} \quad (n = 3, 7, 11, ...)$$

$$\therefore \quad u(x, t) = \frac{40}{\pi^2}\left[\sin\frac{\pi x}{10} e^{-0.01752\pi^2 t} - \frac{1}{3^2}.\sin\frac{3\pi x}{10}.e^{-0.01752\left(3^2\pi^2\right)t} + ...\right]$$ **Ans.**

(b) $$B_n = \frac{2}{10}\int_0^{10} x\left(10-x\right)\sin\frac{n\pi}{10}x\,dx.$$

Integrating by parts taking x (10 – x) as first function

$$= \frac{1}{5}\left[\left\{-x\left(10-x\right).\frac{10}{n\pi}\cos\frac{n\pi}{10}x\right\}_0^{10} + \int_0^{10}\left(10-2x\right).\frac{10}{n\pi}\cos\frac{n\pi}{10}x.dx\right]$$

$$= \frac{1}{5}.\frac{20}{n\pi}\int_0^{10}\left(5-x\right)\cos\frac{n\pi}{10}x\,dx$$

$$= \frac{4}{n\pi}\left[\left\{\left(5-x\right)\frac{10}{n\pi}\sin\frac{n\pi}{10}x\right\}_0^{10} + \int_0^{10}\frac{10}{n\pi}\sin\frac{n\pi}{10}x.dx\right]$$

$$= \frac{40}{n^2\pi^2}\int_0^{10}\sin\frac{n\pi}{10}x.dx = \frac{40}{n^2\pi^2}\frac{10}{n\pi}\left(-\cos\frac{n\pi}{10}x\right)_0^{10}$$

$$= \frac{400}{n^3\pi^3}\,(1 - \cos n\pi).$$

$\therefore B_n = 0$ if n is even and $B_n = \dfrac{800}{n^3\pi^3}$ if n is odd.

Hence, $$u(x, t) = \sum_{n=1}^{\infty} B_n \sin\frac{n\pi x}{10}.e^{-\frac{1.752}{100}n^2\pi^2 t}$$

$$= \frac{800}{\pi^2}\left[\frac{1}{1^3}\sin\frac{px}{10}.e^{-0.01752\pi^2 t} + \frac{1}{3^3}\sin\frac{3\pi x}{10}.e^{-001752(3\pi)^2 t} + ...\right].$$

Ans.

Example 2:

Find the temperature u(x, t) in a slab whose ends x = 0 and x = l are kept at temperature zero and whose initial temperature f(x) = sin $\frac{\pi x}{l}$.

Solution:

Here $B_n = \frac{2}{l}.\int_0^l \sin\frac{\pi x}{l}.\frac{n\pi x}{l}dx$

It is obvious that $B_2 = B_3 = ... = 0$ (each).

But $\quad B_1 = \frac{2}{l}.\int_0^l \sin^2\frac{\pi x}{l}\, d:$

$$= \frac{1}{l}.\int_0^l\left(1 - \cos\frac{n\pi x}{l}\right)dx = 1.$$

$$\therefore \quad u(x, t) = \sum_{n-1}^{\infty} B_n e - c^2n^2\pi^2 t/l^2 \sin\frac{n\pi x}{l}.e^{-\pi^2c^2t/l^2}.\sin\frac{\pi x}{l}. \qquad \textbf{Ans.}$$

Example 3:

Solve the equation

$$\frac{\partial^2 v}{\partial x^2} = \frac{\partial v}{\partial t}$$

given that v = 0, when t = ∞, and when x = 0, and when x = 1.

(Meerut, 91 (P))

Solution:

Given Differential equation is

$$\frac{\partial^2 v}{\partial x^2} = \frac{\partial v}{\partial t} \qquad ...(A)$$

and the given conditions are

$$v(0, t) = 0 \qquad ...(B)$$

$$v(l, t) = 0 \qquad ...(C)$$

and $\quad v(x, \infty) = 0 \qquad ...(D)$

Now, the problem is to find the solution of (A) satisfying (B) (C) and (D).

By the method of separation of variables

let $\quad v = v(x, t) = FT$

where F is a function of x only, and T that of t only.

Substituting in (A), we have

$$T\frac{\partial^2 F}{\partial x^2} = F\frac{dT}{dt}$$

$$\Rightarrow \quad \frac{1}{F}\frac{\partial^2 F}{\partial x^2} = \frac{1}{T}\cdot\frac{dT}{dt} = 0 \text{ or } k^2 \text{ or } -k^2.$$

[Since the two sides are functions of different independent variables].

and hence in three cases, we have

(i) $\quad \dfrac{d^2F}{dx^2} = 0 \quad$ and $\quad \dfrac{dT}{dt} = 0$

(ii) $\quad \dfrac{d^2F}{dx^2} - k^2F = 0 \quad$ and $\quad \dfrac{dT}{dt} = k^2T$

(iii) $\quad \dfrac{d^2F}{dx^2} + k^2F = 0 \quad$ and $\quad \dfrac{dT}{dt} + k^2T = 0$

The general solution in these three cases are

(a) $\quad F = Ax + B \qquad T = C$

(b) $\quad F = Ae^{kx} + Be^{-kx} \qquad T = Ce^{k^2t}$

and (c) $\quad F = A\cos kx + B\sin kx \quad T = Ce^{-k^2t}.$

Now, since the condition (D) implies that as $t \to \infty$, $v(x, t) \to 0$, which is possible only in case (C).

Hence, (a) and (b) do not constitute the solution.

$\therefore$ We can take

$$v(x, t) = (A\cos kx + B\sin kx)\cdot Ce^{-k^2t}$$

$$= (A'\cos kx + B'\sin kx)\cdot e^{-k^2t} \qquad \text{...(2)}$$

from (B) and (2), we have

$$v(0, t) = A'e^{-k^2t} = 0 \qquad \therefore A' = 0$$

$\therefore$ from (C) and (2), we have

$$v(l, t) = (B'\sin kt)\cdot e^{-k^2t} = 0$$

Now, since B' ≠ 0 otherwise v(x, t) = 0.

$\therefore \quad \sin kl = 0 \quad \text{or} \quad kl = n\pi$

$\therefore \quad k = \frac{n\pi}{l}, \; n = 1, 2, \ldots$

∴ for some value of n, the solution may be taken as

$$v_n(x, t) = B_n \sin \frac{n\pi x}{l} . e^{-\frac{n^2\pi^2 x}{l^2}}$$

Hence, for each value of n, the solution is

$$v(x, t) = \sum_{n=1}^{\infty} v_n(x, t)$$

$$= \sum_{n=1}^{\infty} B_n \sin \frac{n\pi x}{l} . e^{-\frac{n^2\pi^2 x}{l^2}}$$ **Ans.**

Example 4:

Solve $\frac{\partial u}{\partial t} - c^2 \frac{\partial^2 u}{\partial x^2} = 0,$ *for* $0 < x < \pi, \; t > 0$

$u_x(0, t) = 0$

$u_x(\pi, t) = 0$

and $u(x, 0) = \sin x.$

Solution:

We have

$$u(x, t) = \frac{1}{2} A_0 + \sum_{n-1}^{\infty} A_n \cos \frac{n\pi x}{\pi} . e^{-(n\pi c/\pi)^2 t}$$

$$= \frac{1}{2} A_0 + \sum_{n-1}^{\infty} A_n \cos nx, e^{-n^2 c^2 t}$$

where $A_n = \frac{2}{\pi} \int_0^{\pi} \sin x . \cos nx \, dx$

$$= \frac{1}{\pi} \int_0^{\pi} [\sin(n+1)x - \sin(n-1)x] dx$$

$$= \frac{1}{\pi} \left[-\frac{\cos(n+1)x}{n+1} + \frac{\cos(n-1)x}{n-1} \right]_0^{\pi}$$

Example 5:

Find the temperature u(x, t) in a bar of length l, which is perfectly insulated laterally whose ends are kept at temperatures zero and in trial temperature is

$$f(x) = \begin{cases} x & \text{when } 0 < x < l/2 \\ l - x & \text{when } l/2 < x < l \end{cases}$$ **(Meerut, 89)**

Solution:

$$\text{Here } B_n = \frac{2}{l}\int_0^{l/2} f(x)\sin\frac{\pi n}{l}x\,dx$$

$$= \frac{2}{l}\left[\int_0^{l/2} f(x)\sin\frac{\pi n}{l}x\,dx + \int_{l/2}^{l}(l-x)\sin\frac{\pi n}{l}x\,dx\right]$$

Integrating by parts, we have

$$= \frac{2}{l}\left[\left(-x.\frac{l}{n\pi}\cos\frac{\pi n}{l}x\right)_0^{l/2} + \frac{l}{n\pi}\int_0^{l/2}\cos\frac{\pi n}{l}x\,dx\right.$$

$$\left. + \left\{(l-x)\left(-\frac{l}{n\pi}\right)\cos\frac{n\pi x}{l}\right\}_{l/2}^{l} - \frac{l}{n\pi}\int_{l/2}^{l}\cos\frac{n\pi x}{l}dx\right]$$

$$= \frac{2}{l}\left[-\frac{l^2}{2n\pi}\cos\frac{n\pi}{2} + \frac{l^2}{n^2\pi^2}\left(\sin\frac{n\pi}{l}x\right)_0^{l/2}\right.$$

$$\left. + \frac{l}{2}.\frac{l}{n\pi}\cos\frac{n\pi}{2} - \frac{l^2}{n^2\pi^2}\left(\sin\frac{n\pi}{l}x\right)_{l/2}^{l}\right]$$

$$= \frac{2}{l}\left[\frac{2l^2}{n^2\pi^2}\sin\frac{n\pi}{2}\right] = \frac{4l}{n^2\pi^2}\sin\frac{n\pi}{2}.$$

Thus, $B_n = 0$ when n is given

and $B_n = \frac{4l}{n^2\pi^2}$ (n = 1, 5, 9, ...)

and $B_n = -\frac{4l}{n^2\pi^2}$ (n = 3, 7, 11, ...)

$$\therefore\ u(x, t) = \sum_{n=1}^{\infty} B_n \sin\frac{\pi n}{l}x.e^{-\frac{n^2\pi^2c^2}{l^2}t}$$

$$= \frac{4l}{\pi^2}.\left[\frac{1}{l^2}\sin\frac{n\pi}{l}.e^{-\frac{c^2\pi^2}{l^2}t} - \frac{1}{3^2}\sin\frac{2\pi x}{l}.e^{-\frac{3^2\pi^2c^2}{l^2}t} + \ldots\right]$$ **Ans.**

Example 6:

Find the temperature in a rod (length 1 and conductivity k), which is at a uniform temperature of 50°C. Suddenly at t = 0, the end x = 0 is cooled to 0°C by an application of ice, and the end x = l is heated to 100°C by an application of stream, and these two temperatures are maintained at ends. Furthermore, the rod is insulated along its length so that no transfer of heat can occure from the sides.

Solution:

In this case the temperature u (x, t) in the rod satisfies the one dimensional heat equation

$$c^2\frac{\partial^2 u}{\partial x^2} = \frac{\partial u}{\partial t} \quad \ldots(1)$$

$$c^2 = \frac{k}{\rho\sigma}$$

k = thermal conductivity,

ρ = density,

σ = specific heat

and boundary conditions

$$u(0, t) = 0 \quad \ldots(2)$$

$$u(1, t) = 100 \quad \ldots(3)$$

and the initial condition

$$u(x, 0) = 50 \quad \ldots(4)$$

$0°C \frac{50°C}{100°C} x$

Substituting in (1), we have

$$\frac{1}{F}\frac{d^2F}{dx^2} = \frac{1}{c^2T}\frac{dT}{dt}.$$

Since the left and right hand sides of this equation being respectively functions of x and t can be equal when they reduce to constant. Taking the constant equal to λ^2, 0 and $-\lambda^2$ respectively we get the following differential equations.

(i) $\dfrac{d^2F}{dx^2} - \lambda^2 F = 0$ $\qquad \dfrac{dT}{dt} - c^2\lambda^2 T = 0$

(ii) $\dfrac{d^2F}{dx^2} = 0,$ $\qquad \dfrac{dT}{dt} = 0$

and (iii) $\dfrac{d^2F}{dx^2} + \lambda^2F = 0,$ $\qquad \dfrac{dT}{dt} + c^2\lambda^2T = 0.$

their general solutions are given by

(a) $F = Ae^{\lambda x} + Be^{-\lambda x}, \quad T = Ce^{c^2\lambda^2 t}$

(b) $F = Ae + B, \qquad T = C$

and (c) $F = A\cos\lambda x + B\sin\lambda x, \; T = Ce^{-c^2\lambda^2 t}.$

The solution given by (i) is inadmissible as in this case the temperature u = FT increases indefinitely with time. The solution (iii) by itself is inadequate to give complete solution, since in this case the temperature will tend to zero as t ¹ ¥. Hence our solution must be a combination of solution of the types (ii) and (iii). So we can write

$$u(x, t) = u_s(x) + u_t(x, t) \qquad \text{...(6)}$$

where $u_s(x)$ is the temperature distribution after a long period of time when the rod has reached a steady state of temperature distribution, $u_s(x, t)$ indicates the transient effects which die down with the passage of time. These $u_s(x)$ and $u_t(x, t)$ must be the solutions of the types (ii) and (iii) respectively.

It is clear that, when the end x = 0 is maintained at temperature u = 0 and the end x = l at u = 100, ultimately there will be uniform gradation of temperature.

$$\therefore u_s(x) = \frac{100}{l} x.$$

$$\therefore u(x, t) = \frac{100}{l} x + u, (x, t). \qquad \text{...(7)}$$

Now, with the help of equations (2), (3) and (4) we see that the boundary and initial conditions for $u_t(x, t)$ are boundary conditions

$$u(0, t) = u_t(0, t) = 0, \qquad \text{...(8)}$$

$$u(l, t) = 100 + u_t(l, t) = 100.$$

$$\therefore \qquad u_t(l, t) = 0$$

and initial condition

$$u(x, 0) = \frac{100}{l}x + u_t(x, 0) = 50$$

$$\therefore \quad u_t(x, 0) = 50 - \frac{100}{l}x. \qquad \ldots(10)$$

∴ Let us take

$$u_t(x, t) = (A' \cos \lambda x + B' \sin \lambda x)e^{-c^2\lambda^2 t} \qquad \ldots(11)$$

Now, from (8) and (11), we have

$$u_t(0, t) = A'e^{-c^2\lambda^2 t} = 0, \quad \therefore \quad A' = 0.$$

Then (9) and (11) gives

$$u_t(l, t) = B' \sin \lambda l e^{-c^2\lambda^2 t} = 0$$

$$\Rightarrow \qquad \sin \lambda l = 0$$

$$\Rightarrow \qquad \lambda l = np$$

$$\therefore \qquad \lambda = \frac{n\pi}{l} \qquad (n = 1, 2, 3, \ldots)$$

∴ a solution of $u_t(x, t)$ is

$$B_n \sin \frac{n\pi}{l}x.e^{-c^2\lambda^2\pi^2 t/l^2}.$$

Now, adding the solution for different n, the general solution may be written as

$$u_t(x, t) = \sum_{n=1}^{\infty} B_n \sin\frac{n\pi}{l}x.e^{-c^2\lambda^2\pi^2 t/l^2} \qquad \ldots(12)$$

from this and (10), we have

$$u_t(x, 0) = 50 - \frac{100}{l}x = \sum_{n=1}^{\infty} B_n \sin\frac{n\pi}{l}x$$

∴ R.H.S. is the Fourier sine expansion of the function

$$50 - \frac{100}{l}x$$

$$\therefore \quad B_n = \frac{2}{l}\int_0^l \left(50 - \frac{100}{l}x\right) \sin \frac{n\pi}{l}x \, dx \qquad \text{(Integrating by parts)}$$

$$= \frac{2}{l}\left[\left\{-\left(50 - \frac{100}{l}x\right).\frac{l}{n\pi}\cos\frac{n\pi x}{l}\right\}_0^l - \frac{100}{l}\int_0^l \frac{l}{n\pi}\cos\frac{n\pi x}{l}\,dx\right]$$

$$= \frac{2}{l}\left[\frac{50l}{n\pi}\cos n\pi + 50\frac{l}{n\pi} - \frac{100}{l}.\frac{l}{n\pi}.\frac{l}{n\pi}\left(\sin\frac{l}{n\pi}x\right)_0^l\right]$$

$$= \frac{100}{n\pi}(\cos n\pi + 1)$$

$\therefore$ $B_n = 0$ where n is odd

and $B_n = \dfrac{200}{n\pi}$ where n is even

So from (12)

$$u_t(x, t) = \frac{200}{\pi}\left[\frac{1}{2}\sin\frac{2\pi x}{l}.e^{-2^2c^2\pi^2t/l^2} + \frac{1}{4}\sin\frac{4\pi x}{l}e^{-4^2c^2\pi^2t/l^2} +..,\right]$$

$$= \frac{100}{\pi}\left[e^{-2^2c^2\pi^2t/l^2}\sin\frac{2\pi x}{l} + \frac{1}{2}e^{-4^2c^2\pi^2t/l^2}\sin\frac{4\pi x}{l} +..\right]$$

Hence, from equation (7) the complete solution is

$$u(x, t) = \frac{100}{l}x + \frac{100}{\pi}\left[e^{(2\pi c/l)^2}\sin\frac{2\pi x}{l} + \frac{1}{2}e^{-(4\pi c/l)^2 t}\sin\frac{4\pi x}{l} + ...\right]$$

Ans.

Example 7:

Find the temperature u (x, t) in a slab whose ends x = 0 and x = l are kept at temperature zero and whose initial temperature f (x) is given by

$$f(x) = \begin{cases} A & \text{when } 0 < x < \dfrac{l}{2} \\ 0 & \text{when } \dfrac{l}{2} < x < l \end{cases}$$ **(Meerut, 91)**

Solution:

From cos E.I., we have

$$u(x, t) = \sum_{n=1}^{\infty} B_n \sin\frac{n\pi x}{l}.e^{-\frac{c^2\pi^2n^2t}{l^2}}$$

where $B_n = \dfrac{2}{l}\int_0^l f(x)\sin\dfrac{n\pi x}{l}\,dx$

$$= \frac{2}{l}\left[\int_0^{l/2} A \sin\frac{n\pi x}{l}\, dx + \int_{l/2}^{l} 0.\sin\frac{n\pi}{l}x\, dx\right]$$

$$= \frac{2}{l}\frac{Al}{n\pi}\left(-\cos\frac{n\pi x}{l}\right)_0^{l/2} = \frac{2A}{n\pi}\left(1-\cos\frac{n\pi}{2}\right)$$

$$= \frac{4A}{n\pi}\sin^2\frac{n\pi}{4}$$

$$\therefore u(x, t) = \frac{4A}{\pi}\sum_{n=1}^{\infty}\frac{1}{n}\sin^2(n\pi/4).e^{-\frac{c^2n^2\pi^2t}{l^2}}\sin\frac{n\pi x}{l}. \qquad \textbf{Ans.}$$

Example 8:

Determine the solution of one dimensional heat equation $\frac{\partial\theta}{\partial t} = a^2\frac{\partial^2\theta}{\partial x^2}$ *under the boundary conditions*

$\theta(0, t) = 0,\ \theta(l, t) = 0 \qquad t > 0$

and the initial condition

$\theta(x, 0) = x \qquad 0 < x < l$

l being the length of the bar.

Solution:

Proceeding similarly as in case I, we have

$$\theta(x, t) = \sum_{n=1}^{\infty} B_n \sin\frac{n\pi}{l}x.e.^{-n^2\pi^2a^2t/l^2}$$

$$\text{where } B_n = \frac{2}{l}\int_0^l f(x).\sin\frac{n\pi}{l}x\, dx$$

$$= \frac{2}{l}\int_0^l x.\sin\frac{n\pi}{l}x\, dx \qquad \text{as } f(x) = x \text{ here.}$$

Integrating by parts

$$= \frac{2}{l}\left[\left(-x\frac{l}{n\pi}\cos\frac{n\pi}{l}x\right)_0^l + \frac{l}{n\pi}\int_0^l \cos\frac{n\pi}{l}x\, dx\right]$$

$$= \frac{2}{l}\left[-\frac{l^2}{n\pi}\cos n\pi + \frac{l}{n\pi}.\frac{l}{n\pi}\left(\sin\frac{n\pi}{l}x\right)_0^l\right]$$

$$= -\frac{2l}{n\pi}\cos n\pi .$$

$$B_n = \frac{2l}{n\pi} \quad \text{for odd values of N}$$

and $$B_n = -\frac{2l}{n\pi} \quad \text{for even values of n.}$$

$$\therefore \quad \theta(x, t) = \sum_{n=1}^{\infty} B_n \sin\frac{n\pi}{l} x . e^{-\frac{n^2\pi^2a^2r}{l^2}}$$

$$= \frac{2l}{\pi}\left[\frac{1}{l}\sin\frac{\pi}{l}x . e^{-\frac{x^2a^2t}{l^2}} - \frac{l}{2}\sin\frac{2\pi}{l}x . e^{-\frac{2^2\pi^2a^2t}{l^2}}\right.$$

$$\left. + \frac{1}{3}.\sin\frac{2\pi}{l}x . e^{-\frac{3^2\pi^2a^2t}{l^2}} - \frac{1}{4}\sin\frac{4\pi}{l}x . e^{-\frac{4^2\pi^2a^2t}{l^2}} - \ldots\right]$$ **Ans.**

12.7 Solution of Two Dimensional Laplace's Equation

Let us consider the steady state temperature distribution in a rectangular sheet of metal. Hence the temperature is everywhere independent of time and therefore the temperature distribution within the plate is obtained by solving the differential equation

$$\frac{\partial^2 u}{\partial x^2} + \frac{\partial^2 u}{\partial y^2} = 0. \qquad \ldots(A)$$

This is the Laplace's Equation is Two-Dimension.

Case I. *To find the steady state temperature distribution in a thin plae bounded by the lines x = 0; x = 1; y = 0 and y = ∞ assuming that heat cannot escape from either surface; the sides x = 0; x = 1 being kept at temperature zero. The lower edge y = 0 is kept at temperature f(x) and the edge y = ∞ at temperature zero.* **(Meerut, 99 (S), 91)**

Solution:

Here the temperature distribution within the plate is obtained by solving (A), since the heat cannot escape the either surface, so temperature is independent of z.

Boundary conditions are

$$u(0, y) = 0 \qquad \ldots(B)$$

$$u(l, y) = 0 \qquad \ldots(C)$$

$$u(x, 0) = f(x) \quad \text{...(D)}$$

and $$u(x, \infty) = 0 \quad \text{...(E)}$$

Let us consider the solution of (A) in the form

$$u(x, y) = F(x)\, G(y) = FG \text{ (say)},$$

where F and G are functions of x and y respectively.

Substituting in (A), we have

$$G\frac{d^2F}{dx^2} + F\frac{d^2G}{dy^2} = 0$$

$$\frac{1}{F}\frac{d^2F}{dx^2} = -\frac{1}{G}\frac{d^2G}{dy^2}.$$

Since left hand side is the function of independent variable x only and right hand side is the function of independent variable y only the two cannot be equal unless search reduces to a constant.

$$\therefore \quad \text{Let } \frac{1}{F}\frac{d^2F}{dx^2} = -\frac{1}{G}\frac{d^2G}{dy^2} = -k^2.$$

The negative constant has been chosen to suit the boundary conditions.

Therefore the corresponding differential equations are

$$\frac{d^2F}{dx^2} + k^2F = 0 \text{ and } \frac{d^2G}{dy^2} - k^2G = 0$$

whose general solution are

$$F = A\cos kx + B\sin kx \text{ and } G = Ce^{ky} + De^{-ky},$$

$$\therefore u(x, y) = FG$$

$$= (A\cos kx + B\sin kx)(Ce^{ky} + De^{-ky}) \quad \text{...(2)}$$

Since from (E), $u(x, \infty) = 0 \therefore C = 0$ otherwise $u \to \infty$ as $y \to \infty$ and hence $u(x, y) = (A'\cos kx + B'\sin kx)\, e^{-ky}$. ...(3)

Now, from this and (B)

$$u(0, y) = A'e^{-ky} = 0 \quad \therefore \quad A' = 0$$

and then from (3) and (C), we have

$$u(l, y) = B'\sin kl \,.\, e^{-ky} = 0$$

$$\therefore \quad \sin kl = 0$$

$$kl = n\pi$$

$$k = \frac{n\pi}{l} \quad (n = 1, 2, 3, ...)$$

Thus, for each value of n we have the solution like

$$u_n(x, y) = B_n \sin \frac{n\pi}{l} x \cdot e^{-n\pi y/t}, \qquad (n = 1, 2, ...)$$

Hence, for different values of n, the solution may be taken as

$$u(x, y) = \sum_{n=1}^{\infty} u_n(x, y)$$

$$\Rightarrow \quad u(x, y) = \sum_{n=1}^{\infty} B_n \sin \frac{n\pi}{l} x \cdot e^{-n\pi y/t} \qquad ...(4)$$

from this and (D), we have

$$u(x, 0) = \sum_{n=1}^{\infty} B_n \sin \frac{n\pi}{l} x = f(x).$$

The left hand side is the Fourier Sine series for f(x).

$$\therefore \quad B_n = \frac{2}{l} \int_0^l f(x) \sin \frac{n\pi}{l} x \, dx \qquad ...(5)$$

Hence, equation (4) with the coefficient (5) is te solution of the Laplace equation satisfying all the boundary conditions (B), (C), (D) and (E).

Case II. *To find the steady state temperature distribution in a thin rectangular plate bounded by the lines x = 0, x = 1, y = 0, y = h, the edges x = 0, x = 1, y = 0 are kept at temperature zero while the edges y = h is kept at temperature f(x).* **(Meerut, 90, 98)**

Solution:

Here the boundary conditions are

$$u(0, y) = 0 \qquad ...(B)$$

$$u(l, y) = 0 \qquad ...(C)$$

$$u(x, 0) = 0 \qquad ...(D)$$

and $$u(x, h) = f(x) \qquad ...(E)$$

Proceeding similarly as in case I, we have

$$u(x, y) = FG = (A \cos kx + B \sin kx)(Ce^{ky} + De^{-ky}) \qquad ...(1)$$

$\therefore$ from (1) and (B), we have

$$u(0, y) = A(Ce^{ky} + De^{-ky}) = 0$$

$\therefore$ $A = 0$

Then from (1) and (C), we have

$$u(l, y) = B \sin kl \, (Ce^{ky} + De^{-ky}) = 0.$$

$\therefore \quad \sin kl = 0$

$\Rightarrow \quad kl = n\pi$

$\Rightarrow \quad k = n\pi/l \quad (n = 1, 2, 3, ...)$

$\therefore$ for each value of n we have the solution like

$$u_n(x, y) = (C_n e^{(k\pi y)/l} + D_n e^{-(n\pi y/l)} \sin \frac{n\pi x}{l} \qquad (n = 1, 2, 3, ...)$$

Hence for different values of n the solution may be taken as,

$$u(x, y) = \sum_{n=1}^{\infty} u_n(x, y)$$

$$\Rightarrow \quad u(x, y) = \sum_{n=1}^{\infty} (C_n e^{(n\pi y)/l} + D_n e^{-(n\pi y/l)} \sin \frac{n\pi}{l} x. \qquad ...(2)$$

Now, from (2) and (D), we have

$$u(x, 0) = C_n + D_n = 0$$

$$\Rightarrow \quad D_n = -C_n.$$

$$\therefore u(x, y) = \sum_{n=1}^{\infty} C_n e^{(n\pi y)/l} + e^{-(n\pi y/l)} \sin \frac{n\pi}{l} x$$

$$u(x, y) = \sum_{n=1}^{\infty} C'_n \sin h \frac{n\pi y}{l} \sin \frac{n\pi}{l} x \qquad ...(3)$$

where $C'_n = 2C_n$

from (3) and (E), we have

$$u(x, h) = \sum_{n=1}^{\infty} C'_n \sin h \frac{n\pi y}{l} \sin \frac{n\pi}{l} x = f(x).$$

$\therefore$ the L.H.S. is the Fourier Sine series for f(x).

$$\therefore C'_n \sin h \frac{n\pi y}{l} = \frac{2}{l} \int_0^l f(x) \frac{n\pi y}{l} dx.$$

$$\Rightarrow \quad C'_n = \frac{2}{l \sin h \frac{n\pi y}{l}} \int_0^l f(x) \sin \frac{n\pi y}{l} dx. \qquad ...(4)$$

Hence, equation (3) with coefficient (4) is the required solution satisfying the boundary conditions (B), (C), (D) and (E).

Case III. *One side insulated : To obtain the steady state temperature in a rectangular plate of length a and width b, the sides of which are kept at temperature zero. the lower end is kept at temperature f(x) and the upper edge is kept insulated.* **(Meerut, 99, 90)**

Solution:

Here the boundary conditions are

$u(0, y) = 0$ ...(B)

$u(a, y) = 0$ ...(C)

$u(x, 0) = f(x)$ ...(D)

and $\left(\frac{\partial u}{\partial y}\right)_{y=b} = 0$ ...(E)

Suppose the solution of (A) tobe as

$u(x, y) = FG$ ...(1)

where F is a function of x only
and G that of y only.

Proceeding similarly as in case I we have the two different equations as

$$\frac{d^2F}{dx^2} + k^2F = 0 \quad \text{and} \quad \frac{d^2G}{dy^2} - k^2G = 0$$

whose general solutions are

$F = A \cos kx + B \sin kx$ and $G = C \cos h\, ky + D \sin h\, ky$

$\therefore$ We can write

$\therefore$ $u(x, y) = FG$

$u(x, y) = (A \cos kx + B \sin kx)(C \cos h\, ky + D \sin h\, ky)$ ...(2)

Applying (B), we have

$u(0, y) = A(C \cos h\, ky + D \sin h\, ky) = 0. \quad \therefore \; A = 0$

and then applying (C), we have

$u(a, y) - B \sin ka\,(C \cos h\, ky + D \sin h\, ky) = 0$

$\therefore$ $\sin ka = 0$

$\Rightarrow$ $ka = n\pi$

$\Rightarrow$ $k = \frac{n\pi}{a}$ $(n = 1, 2, 3, ...)$

$\therefore$ for each value of n we have the solution like

$$u_n(x, y) = \left(C_n \cos h \frac{n\pi}{a} + y\, D_n \sin h \frac{n\pi}{a} y\right) \sin \frac{n\pi}{a} x.$$

$(n = 1, 2, ...)$

$\therefore$ for each value of n, we can write

$$u(x, y) = \sum_{n=1}^{\infty}\left[C_n \cos h \frac{n\pi y}{a} + D_n \sin h \frac{n\pi}{a} y\right] \sin \frac{n\pi}{a} x \quad ...(3)$$

aplying (D), we have

$$u(x,\ 0) = \sum_{n-1}^{\infty} C_n \sin \frac{n\pi}{a} x = f(x).$$

$$\therefore \quad C_n = \frac{2}{a}\int_0^a f(x) \sin \frac{n\pi}{a} x\, dx. \qquad \ldots(4)$$

Applying (E) and (3), we have

$$\left(\frac{\partial u}{\partial y}\right)_{y=b} = \sum_{n-1}^{\infty} \frac{n\pi}{a}\left[C_n \sin h \frac{n\pi}{a} b + D_n \cos h \frac{n\pi}{a} b\right] \sin \frac{n\pi}{a} x = 0.$$

This will be satisfied for every value of x, if

$$C_n \sin h \frac{n\pi}{a} b + D_n \cos h \frac{n\pi}{a} b = 0 \qquad \ldots(5)$$

$$D_n = -\, C_n \tan h \frac{n\pi b}{a}.$$

Hence, equation (3) with coefficients given by (4) and (5) is the solution satisfying the boundary conditions.

12.8 Solution of Two Dimensional Heat Equation

Example 1:

Obtain the steady state temperature distribution in a rectangular metal plate of length a and width b, the sides of which are kept at temperature 0°C, the lower edge is kept at 100°C and the upper edge kept insulated.

Solution:

Here the temperature u(x, y) satisfies the differential equation

$$\frac{\partial^2 u}{\partial x^2} + \frac{d^2 u}{dy^2} = 0$$

and the boundary conditions are

(i) u (0, y) = 0 (ii) u (a, y) = 0

(iii) u (x, 0) = 100 (iv) $\left(\frac{\partial u}{\partial y}\right)_{y=b} = 0$.

We have

$$u(x,\ y) = \sum_{n-1}^{\infty}\left[C_n \cos h \frac{n\pi}{a} y + D_n \sin h \frac{n\pi}{a} y\right] \sin \frac{n\pi}{a} x$$

where $C_n = \frac{2}{a}\int_0^a f(x) \sin \frac{n\pi}{a} x = \frac{2}{a}\int_0^a 100 \sin 0\, x\, dx.$

$$= \frac{200}{n\pi}(1 - \cos n\pi)$$

and $\quad D_n = -C_n \tan h \frac{n\pi}{a} b$

$$\therefore \; u(x, y) = \frac{200}{\pi} \sum_{n-1}^{\infty} n\left(\cos h \frac{n\pi}{a} y - \sin h \tan h \frac{n\pi b}{a}\right).$$

$$(1 - \cos n\pi) \sin \frac{n\pi}{a} x$$

which is the required expression. **Ans.**

Example 2:

A thin rectangular plane whose surface is impervious to heat flow has at $t = 0$ an arbitrary distribution of temperature $f(x, y)$. Its four edges $x = 0$, $x = a$, $y = 0$, $y = b$ are kept at zero temperature. Determine the temperature at a point of the plate as t increases.

(Meerut, 91 (P))

Solution:

The solution of the two-dimensional heat equation

$$\frac{\partial u}{\partial t} = c^2\left(\frac{\partial^2 u}{\partial x^2} + \frac{\partial^2 u}{\partial y^2}\right) \qquad \text{...(A)}$$

$$c^2 = \frac{k}{\sigma\rho}$$

will determine the temperature of any point of plate at any time t.

Here the temperature of the four faces of the plate are kept at zero temperature and the initial temperature of the plate if f (x, y). Therefore, here the function u (x, y, t) is required to satisfy (A) and the following boundary and initial conditions:

$$\text{Boundary conditions} \quad \left.\begin{aligned} u(0, y, t) &= 0 \\ u(a, y, t) &= 0 \\ u(x, 0, t) &= 0 \\ u(x, b, t) &= 0 \end{aligned}\right] \qquad \text{...(B)}$$

for all t.

and initial condition $\quad u(x, y, 0) = f(x, y)$. ...(C)

Let us suppose that (A) has a solution of the form

$$u(x, y, t) = F(x)\,Y(y)\,T(t)$$
$$= FYT \text{ (say)} \qquad \ldots(1)$$

where F is a function of x only, Y that of y only and T that of t only.

Substituting in (A) and simplifying, we have

$$\frac{1}{C^2 T}\frac{dT}{dt} = \frac{1}{F}\frac{d^2F}{dx^2} + \frac{1}{Y}\frac{d^2Y}{dy^2}$$

Now, the left hand side is a function of the independent variable t, while the right hand side is the sum of two functions of independent variables x and y. Thus, change in one variable x, y or t will not affect the other two functions. Therefore in order that (1) may satisfy (A), we have three possibilities.

(i) $\quad \dfrac{1}{F}\dfrac{d^2F}{dx^2} = 0, \quad \dfrac{1}{Y}\dfrac{d^2Y}{dy^2} = 0, \quad \dfrac{1}{C^2T}\dfrac{dT}{dt} = 0$

(ii) $\quad \dfrac{1}{F}\dfrac{d^2F}{dx^2} = k_1^2, \quad \dfrac{1}{Y}\dfrac{d^2Y}{dy^2} = k_2^2, \quad \dfrac{1}{c^2T}\dfrac{dT}{dt} = k^2$

where $k^2 = k_1^2 + k_2^2$

and (iii) $\quad \dfrac{1}{F}\dfrac{d^2F}{dx^2} = -k_1^2, \quad \dfrac{1}{Y}\dfrac{d^2Y}{dy^2} = -k_2^2, \quad \dfrac{1}{C^2T}\dfrac{dT}{dt} = -k^2$

where $k^2 = k_1^2 + k_2^2$.

The solutions of the differential equations (i) and (ii) do not constitute our solutions. It can be shown easily as in other articles.

The general solutions of the differential equation given by (iii) are

$$F = A_1 \cos k_1x + B_1 \sin k_1x, \quad Y = A_2 \cos k_2y + B_2 \sin k_2y$$

and $\quad T = Ae^{-c^2k^2t}$.

Hence we investigate solution of the type

$$u(x, y, t) = (A_1 \cos k_1x + B_1 \sin k_1x). - c^2k^2t \qquad \ldots(1)$$
$$(A'_2 \cos k_2y + B'_2 \sin k_1x)$$

which will certainly satisfy (A).

Let us now adjust the constants so that the solution satisfies the boundary conditions (B) also.

Now, $u(0, y, t) = A_1 (A'_2 \cos k_2y + B'_2 \sin k_2y)\, e^{-c^2k^2t} = 0$

$\therefore \quad A_1 = 0$

and then $u(a, y, t) = B_1 \sin k_1 a (A'_2 \cos k_2 y + B'_2 \sin k_2 y) e^{-c^2 k^2 t} = 0.$

$$\therefore \quad \sin k_1 a = 0$$

$$\Rightarrow \quad k_1 a = mp$$

$$\Rightarrow \quad k_1 = \frac{m\pi}{a} \qquad (m = 1, 2, 3, ...)$$

Similarly, using the other two boundary conditions, we have

$$(n = 1, 2, 3, ...)$$

Thus, for different values of m and n, we get the solution of the form

$$A_{mn} e^{-c^2 k^2 mnt} \sin \frac{m\pi}{a} x \sin \frac{n\pi}{b} y k^2 = k^2_{mn} = \pi^2 \left(\frac{m^2}{a^2} + \frac{n^2}{b^2}\right)$$

Hence for all values of m and n, we get the solution of (A), as

$$u(x, y, t) = \sum_{m-1}^{\infty} \sum_{n-1}^{\infty} A_{mn} e^{-c^2 k^2 mnt} \sin \frac{m\pi}{a} x \sin \frac{n\pi}{b} y \qquad ...(3)$$

which satisfy the boundary conditions (B).

Now, from equation (3) and the initial condition (C), we have

$$u(x, y, 0) = \sum_{m-1}^{\infty} \sum_{n-1}^{\infty} A_{mn} \sin \frac{m\pi}{a} x \sin \frac{n\pi}{b} y = f(x, y)$$

The series is called the double Fourier series of f(x, y)

$$\therefore \quad A_{mn} = \frac{2}{a}, \frac{2}{b} \int_{x=0}^{a} \int_{y=0}^{b} f(x, y) \sin \frac{m\pi}{a} x \sin \frac{n\pi}{b} y \, dx \, dy. \qquad ...(4)$$

Thus, the solution of heat equation (A) satisfying the boundary conditions (B) and initial conditions (C) is given by (3) with the coefficient given by equation (4) and

$$k^2_{mn} = k^2 \left(\frac{m^2}{a^2} + \frac{n^2}{b^2}\right). \qquad \begin{array}{l} m = 1, 2, ... \\ n = 1, 2, ... \end{array}$$

Thus, function u(x, y, t) determine the subsequent temperature of the plate as t increases.

Example 3:

A rectangular plate bounded by the lines $x = 0$, $y = 0$, $x = a$, $y = b$ has an initial distribution temperature given by

$$V = A \sin \frac{\pi x}{a} \sin \frac{\pi y}{b}.$$

The edges are kept at zero temperature and the plane faces are impervious to heat. Find the temperature at any point and time.

Solution:

Here the temperature at any point and time are given by the two-dimensional heat equation

$$\frac{\partial u}{\partial t} = c^2\left(\frac{\partial^2 u}{\partial x^2} + \frac{\partial^2 u}{\partial y^2}\right)$$

We have

$$u(x, y, t) = \sum_{m-1}^{\infty}\sum_{n-1}^{\infty} A_{mn} e^{-c^2 k^2_{mn} t} \sin\frac{m\pi}{a}x \sin\frac{n\pi}{b}y$$

where $\quad k^2{}_{mn} = \pi^2\left(\frac{m^2}{a^2} + \frac{n^2}{b^2}\right)$

and $\quad A_{mn} = \frac{4}{ab}\int_{x=0}^{a}\int_{y=0}^{b} f(x, y)\sin\frac{m\pi}{a}x \sin\frac{n\pi}{b}y\, dx\, dy$

$$= \frac{4A}{ab}\int_{x=0}^{a} \sin\frac{\pi x}{a}\sin\frac{\pi y}{b}\sin\frac{m\pi}{a}x\sin\frac{n\pi}{b}y\,dx\,dy$$

$$= \frac{4A}{ab}\int_{x=0}^{a} \sin\frac{\pi x}{a}\sin\frac{m\pi}{a}x\left[\int_{y=0}^{b}\sin\frac{n\pi}{b}\sin\frac{n\pi}{b}y\,dy\right]dx$$

Now, $\quad \int_{y=0}^{b}\sin\frac{\pi y}{b}\sin\frac{n\pi}{b}y\,dy = 0 \qquad$ for $n = 2, 3, \ldots$

but for $n = 1$.

$$= \int_{y=0}^{b}\left(\sin\frac{\pi y}{a}\right)^2 dy = \frac{1}{2}\int_0^b\left(1 - \cos\frac{2\pi y}{b}\right)dy = \frac{b}{a}$$

$$\therefore\ A_{m1} = \frac{4A}{ab}\int_0^a \frac{b}{2}\sin\frac{\pi x}{a}\sin\frac{m\pi x}{a}dx$$

$$= \frac{2A}{a}\int_0^a \sin\frac{\pi x}{a}\sin\frac{m\pi x}{a}dx$$

which is equal to zero for $m > 1$

for $m = 1$.

$$A_{11} = \frac{2A}{a}\int_{x=0}^{a}\sin^2\frac{\pi x}{a}dx\ \frac{A}{a}\int_0^a\left(1 - \cos\frac{2\pi x}{a}\right)dx$$

and $k^2_{11} = \pi^2\left(\frac{1}{a^2}+\frac{1}{b^2}\right)$

$$\therefore\quad u(x, y, t) = \sum_{m-1}^{\infty}\sum_{n-1}^{\infty} A_{mn} e^{-c^2k^2_{mn}t} \sin\frac{m\pi}{a}x \sin\frac{n\pi}{b}y$$

$$= A_{11}e^{-c^2k^2_{11}t}\sin\frac{\pi n}{a}\sin\frac{\pi y}{b} \text{ Since } A_{mn} = 0$$

for all values of $m > 1$, $n > 1$.

$$= Ae^{-c^2\pi^2\left(\frac{1}{a^2}+\frac{1}{b^2}\right)t}\sin\frac{x\pi}{a}\sin\frac{\pi y}{b}$$

which given the temperature of the plate at any point and time.

12.9 Solution of Two-Dimensional Laplace's Equation Given in the Cylindrical Coordinates

As discussed in the Laplace's Differential Equation in cylindrical coordinates is written as

$$\frac{\partial^2 u}{\partial r^2}+\frac{1}{r}\frac{\partial u}{\partial r}+\frac{1}{r^2}\frac{\partial^2 y}{\partial \theta^2}+\frac{\partial^2 u}{\partial z^2} = 0.$$

If we assume that u is independent of the coordinate z, then the above equation may be written as

$$\frac{\partial^2 u}{\partial r^2}+\frac{1}{r}\frac{\partial u}{\partial r}+\frac{1}{r^2}\frac{\partial^2 u}{\partial \theta^2} = 0. \qquad ...(1)$$

It is obvious that u is a function of r and θ only.

$\therefore$ Let us suppose $u = FR$ ...(2)

where F is a function of θ only and R that of r only. Substituting in (1), we have

$$F\frac{d^2R}{dr^2}+\frac{1}{r}F\frac{dR}{dr}+\frac{1}{r^2}R\frac{d^2F}{d\theta^2} = 0$$

$$\Rightarrow\quad \frac{1}{R}\left[r^2\frac{d^2R}{dr^2}+r\frac{dR}{dr}\right] = -\frac{1}{F}\frac{d^2F}{d\theta^2}.$$

Now, any change in θ in R.H.S. produces no change in L.H.S. and any change in r in L.H.S. produces no change in R.H.S. Thus the two sides can not be equal unless both reduce to a constant.

$$\therefore\quad \text{let } \frac{1}{R}\left[r^2\frac{d^2R}{dr^2}+r\frac{dR}{dr}\right] = -\frac{1}{F}\frac{d^2F}{d\theta^2} = n^2.$$

Thus, we get the two differential equations as

$$\frac{d^2F}{d\theta^2} + n^2F = 0 \qquad ...(3)$$

and $$r^2\frac{d^2R}{dr^2} + r\frac{dR}{dr} - n^2 R = 0 \qquad ...(4)$$

whose general solution when $n \neq 0$, are

$$F = A \cos n\theta + B \sin n\theta. \qquad ...(5)$$

$$R^* = Cr^n + Dr^{-n} \qquad ...(6)$$

where A, B, C and D are arbitrary constants.

If $n = 0$, then the equations (3) and (4), become

$$\frac{d^2F}{d\theta^2} = 0$$

and $$r^2\frac{d^2R}{dr^2} + r\frac{dR}{dr} = 0$$

whose general solutions are

$$F = A_0\theta + B_0$$

and** $R = C_0 \log r + D_0$. ...(7)

The solutions of Laplace's equation in cylindrical coordinates when u is independent of z are called circular harmonics and n is called the degree of the harmonic.

* Equation (4) is homogeneous linear equation with varibale coefficient.

$\therefore$ Put $r = e^z$

$$\frac{dr}{dz} = e^z = r \quad \frac{dR}{dr} = \frac{dR}{dz}\cdot\frac{dz}{dr}$$

$$r\frac{dR}{dr} = \frac{dR}{dz} \quad \therefore r\frac{d}{dr} \equiv \frac{d}{dz}$$

$$r\frac{d}{dr}\left(r\frac{dR}{dr}\right) = \frac{d}{dz}\left(\frac{dR}{dz}\right)$$

$$\Rightarrow \quad r^2\frac{d^2R}{dr^2} + r\frac{dR}{dr} = \frac{d^2R}{dz^2}$$

$$\therefore \quad \text{from (4)} \ \frac{d^2R}{dz^2} - n^2R = 0$$

$$\therefore \quad R = Ce^{pz} + De^{-pz} = Cr^n + Dr^{-n}$$

**When $n = 0$, (4) becomes

$$\frac{d^2R}{dz^2} = 0$$

$$\therefore \quad R = C_0z + D_0 = C_0 \log r + D_0.$$

$\therefore$ Circular harmonics of degree zero are

$$u_0 = (A_0\theta + B_0)(C_0 \log r + D_0).$$

and circular harmonics of degree n are

$$u_n = (A_n \cos n\theta + B_n \sin n\theta)(C_nr^n + D_nr^{-n})$$

Hence, the general single-valued solution of Laplace's Equation for all possible values of n is

$$u = a_0 \log r + \sum_{n-0}^{\infty}(A_n \cos n\theta + B_n \sin n\theta)(C_n r^n + D_n r^{-n}) + k.$$

12.10 Solution of the Laplace's Equation Given in Cylindrical Coordinates

The Laplace's Equation in cylindrical coordinates in written as

$$\frac{\partial^2 u}{\partial r^2} + \frac{1}{r}\frac{\partial u}{\partial r} + \frac{1}{r^2}\frac{\partial^2 u}{\partial \theta^2} + \frac{\partial^2 u}{\partial z^2} = 0. \qquad ...(1)$$

It is obvious that u is a function of r, θ and z.

∴ let us suppose

$$u(r, \theta, z) = F_1(\theta) R(r) F_2(z)$$
$$= F_1 R F_2 \text{ (say)} \qquad ...(2)$$

where F_1 is a function of θ only, R that of r only and F_2 that of z only. Substituting in (1) and on simplification, we have

$$\frac{1}{R}\frac{d^2R}{dr^2} + \frac{1}{rR}\frac{dR}{dr} + \frac{1}{r^2F_1}\frac{d^2F_1}{d\theta^2} + \frac{1}{F^2}\frac{d^2F_2}{dz^2} = 0$$

$$\Rightarrow \quad \frac{1}{R}\frac{d^2R}{dr^2} + \frac{1}{rR}\frac{dR}{dr} + \frac{1}{r^2F_1}\frac{d^2F_1}{d\theta^2} = -\frac{1}{F^2}\frac{d^2F_2}{dz^2}. \qquad ...(3)$$

The R.H.S. of the above equation is independent of r and q and the L.H.S. is independent of z. Hence the two cannot be equal unless both reduce to a constant.

$$\therefore \quad \frac{1}{R}\frac{d^2R}{dr^2} + \frac{1}{rR}\frac{dR}{dr} + \frac{1}{r^2F_1}\frac{d^2F_1}{d\theta^2} = -\frac{1}{F^2}\frac{d^2F_2}{dz^2} = -k^2$$

from which, we have

$$\frac{1}{R}\frac{d^2R}{dr^2} + \frac{1}{rR}\frac{dR}{dr} + \frac{1}{r^2F_1}\frac{d^2F_1}{d\theta^2} = -k^2 \qquad ...(4)$$

and
$$\frac{d^2F_2}{dz^2} - k^2F_2 = 0 \qquad ...(5)$$

from (4) we have

$$\frac{r^2}{R}\frac{d^2R}{dr^2} + \frac{r}{R}\frac{dR}{dr} + k^2r^2 = -\frac{1}{F_1}\frac{d^2F_1}{d\theta^2}$$

∴ with the same argument as above, we have

$$\frac{r^2}{R}\frac{d^2R}{dr^2} + \frac{r}{R}\frac{dR}{dr} + k^2r^2 = -\frac{1}{F_1}\frac{d^2F_1}{d\theta^2} = m^2$$

from which we have

$$\frac{d^2F_1}{d\theta^2} + m^2F_1 = 0 \qquad \ldots(6)$$

and $$\frac{r^2}{R}\frac{d^2R}{dr^2} + \frac{r}{R}\frac{dR}{dr} + k^2r^2 - m^2 = 0$$

$\Rightarrow$ $$r^2\frac{d^2R}{dr^2} + r\frac{dR}{dr} + (k^2r^2 - m^2)R = 0. \qquad \ldots(7)$$

Putting $kr = x$ so that $\frac{dr}{dx} = \frac{1}{k}$

and $$\frac{dR}{dr} = \frac{dR}{dx}\cdot\frac{dx}{dr} = k\frac{dR}{dx}$$

$\therefore$ $$\frac{d}{dr} = k\frac{d}{dx}$$

and $$\frac{d}{dr}\left(\frac{dR}{dr}\right) = k^2\frac{d}{dx}\left(k\frac{dR}{dx}\right)$$

$\therefore$ $$\frac{d^2R}{dx^2} = k^2\frac{d^2R}{dx^2}$$

In equation (7), we have

$$\frac{x^2}{k^2}.k^2\frac{d^2R}{dx^2} + \frac{x}{k}k\frac{dR}{dx} + (x^2 - m^2)R = 0$$

$\Rightarrow$ $$\frac{d^2R}{dx^2} + \frac{1}{x}\frac{dR}{dx} + \left(1 - \frac{m^2}{x^2}\right)R = 0 \qquad \ldots(8)$$

Now, general solution of (5) and (6) are

$$F_2 = A_1e^{kz} + B_1e^{-kz}$$

and $$F_1 = (A_2 \cos m\theta + B_2 \sin m\theta).$$

Equation (8) is Bessel's equation, therefore its general solution may be written as

$$R = A_3J_m(x) + B_3J_{-m}(x) \text{ when m is fraction}$$

$$= A_3J_m(kr) + B_3j_{=m}(kr)$$

and $$R = A_3J_m(kr) + B_3Y_m(kr) \text{ when m is an integer.}$$

Hence, the general solution of (1) may be written as

$$u(r, \theta, z) = (A_1e^{kz} + B_1e^{-kz})(A_2 \cos m\theta + B_2 \sin m\theta)\{A_3J_m(kr) + B_3J_{-m}(kr)\}$$

when m is not an integer and k = 1, 2, 3, ...

and $$u(r, \theta, z) = (A_1e^{kz} + B_1e^{-kz})(A_2 \cos m\theta + B_2 \sin m\theta)\{A_3J_m(kr) + B_3J_m(kr)\}$$

when m is an integer and k = 1, 2, 3, ...

12.11 Solution of Laplace's Equation Given in Spherical Co-ordinates

The Laplace's Equation in spherical co-ordinates may be written as

$$r^2\frac{\partial^2 u}{\partial r^2} + 2r\frac{\partial u}{\partial r} + \frac{1}{\sin\theta}\frac{\partial}{\partial\theta}\left(\sin\theta\frac{\partial u}{\partial\theta}\right) + \frac{1}{\sin^2\theta}\frac{\partial^2 u}{\partial\phi^2} = 0 \qquad \text{...(1)}$$

It is obvious that v is a function of r, θ and ϕ

$\therefore$ let us suppose

$$u(r, \theta, \phi) = R(r)\,F_1(\theta),\,F_2(\phi)$$
$$= RF_1\,F_2 \text{ (say)}, \qquad \text{...(2)}$$

where R is a function of r, only, F_1 that of θ only and F_2 that of ϕ only. Substituting in (1), we have

$$r^2F_1F_2\frac{d^2R}{dr^2} + 2r\,.\,F_1F_2\frac{dR}{dr} + \frac{1}{\sin\theta}\frac{d}{d\theta}\left(\sin\theta\,.\,RF_2\frac{dF_1}{d\theta}\right) + \frac{1}{\sin^2\theta}.\,RF_1\frac{d^2F_2}{d\phi^2} = 0$$

$$\Rightarrow \quad r^2F_1F_2\frac{d^2R}{dr^2} + 2rF_1F_2\frac{dR}{dr} + \frac{RF_2}{\sin^2\theta}\frac{d}{d\theta}\left(\sin\theta\frac{dF_1}{d\theta}\right) + \frac{1}{\sin^2\theta}.\,RF_1\frac{d^2F_2}{d\phi^2} = 0$$

$$\Rightarrow \quad \left[\frac{r^2}{R}\frac{d^2R}{dr^2} + \frac{2r}{R}\frac{dR}{dr} + \frac{1}{F_1\sin\theta}\frac{d}{d\theta}\left(\sin\theta\frac{dF_1}{d\theta}\right)\right]\sin^2\theta$$
$$= -\frac{1}{F_1}\frac{d^2F_2}{d\phi^2} = m^2. \qquad \text{...(3)}$$

Since the R.H.S. is independent of r and q and the L.H.S. is independent of f, hence the two cannot be equal unless both reduce to a constant.

From (3), we have

$$\frac{d^2F_2}{d\phi^2} + m\,F_2 = 0 \text{ whose solution of } F_2 = Ce^{\pm\, im\phi} \qquad \ldots(4)$$

and $$\frac{r^2}{R}\frac{d^2R}{dr^2} + \frac{2r}{R}\frac{dR}{dr} + \frac{1}{F_1 \sin\theta}\frac{d}{d\theta}\left(\sin\theta\,\frac{dF_1}{d\theta}\right) = \frac{m^2}{\sin^2\theta}$$

$$\Rightarrow \qquad \frac{r^2}{R}\frac{d^2R}{dr^2} + \frac{2r}{R}\frac{dR}{dr} = -\frac{1}{F_1 \sin\theta}\frac{d}{d\theta}\left(\sin\theta\,\frac{dF_1}{d\theta}\right) = \frac{m^2}{\sin^2\theta}$$

$$= k \text{ (Const.)}$$

With the similar argument as before if we take

$$k = n\,(n+1)$$

then the two differential equations obtained are

$$\frac{r^2}{R}\frac{d^2R}{dr^2} + \frac{2r}{R}\frac{dR}{dr} - n(n+1)\,R = 0$$

$$\Rightarrow \qquad r^2\frac{d^2R}{dr^2} + 2r\frac{dR}{dr} - n(n+1)\,R = 0 \qquad \ldots(5)$$

and $$\frac{1}{\sin\theta}\frac{d}{d\theta}\left(\sin\theta\,\frac{dF_1}{d\theta}\right) + \left\{n(n+1)\,R - \frac{m^2}{\sin^2\theta}\right\}F_1 = 0 \qquad \ldots(6)$$

Equation (5) is Homogeneous Linear Equation with variable coefficient.

Substituting $r = e^x$,

so that $\dfrac{dr}{dz} = e^x = r.$

$$\therefore \quad \frac{dR}{dr} = \frac{dR}{dz}\cdot\frac{dz}{dr} = \frac{1}{r}\frac{dR}{dz}.$$

$$\therefore \quad r\frac{d}{dr} = \frac{d}{dz}$$

Let D denote $\dfrac{d}{dz}$

then $$r\frac{d}{dr}\left(r\,\frac{dR}{dr}\right) = r^2\,\frac{d^2R}{dr^2} + r\frac{dR}{dr}.$$

$$\therefore \quad r^2\frac{d^2R}{dr^2} = r\frac{d}{dr}\left(r\frac{d}{dr} - 1\right)R$$

$$= D\,(D-1)\,R.$$

Therefore, equation (5) becomes

$$\{D(D-1)+2D-n(n+1)\}R = 0$$

$$\Rightarrow \quad (D^2 + D - n^2 - n)R = 0$$

$$\Rightarrow \quad (D-n)(D+n^2+n)R = 0$$

$\therefore$ solution of (5) is

$$R = A'e^{nx} - B'e^{-(n+1)z}$$

$$\Rightarrow \quad R = A'r^n + B'r^{-(n+1)}. \qquad ...(7)$$

Now, substituting $\mu = \cos\theta$,

so that $$\frac{dF_1}{d\theta} = -\frac{dF_1}{d\mu}\sin\theta$$

in (6), we have

$$\frac{1}{\sin\theta}\cdot\frac{d}{d\theta}\left(-\sin^2\theta\,\frac{dF_1}{d\mu}\right) + \left\{n(n+1) - \frac{m^2}{(1-\mu^2)}\right\}F_1 = 0$$

$$\Rightarrow \quad (1-\mu^2)\frac{d^2F_1}{d\mu^2} - 2\mu\frac{dF_1}{d\mu} + \left\{n(n+1) - \frac{m^2}{(1-\mu^2)}\right\}F_1 = 0 \qquad ...(8)$$

which is associated Legendra equation whose solution is

$$F_1 = AP^m{}_n(\mu) = AP^m{}_n(\cos\theta).$$

Hence, the solution of (1) is given by

$$u(r,\theta,\phi) = \sum_{n=0}^{\infty}\sum_{m<n=0}^{\infty}\left(A_{m,n}r^n + \frac{B_{m,n}}{r^{n+1}}\right)P^m{}_n(\cos\theta)\,.\,e^{\pm im\phi}$$

12.12 Solution of Three Dimensional Wave Equation in the Spherical Polar Coordinates.

The three dimensional wave equation in spherical polar co-ordinates (r, θ, ϕ) may be written as

$$\frac{\partial^2 u}{\partial r^2} + \frac{2}{r}\frac{\partial u}{\partial r} + \frac{1}{r^2\sin\theta}\frac{\partial}{\partial\theta}\left(\sin\theta\,\frac{\partial u}{\partial\theta}\right) + \frac{1}{r^2\sin^2\theta}\frac{\partial^2 u}{\partial\phi^2}$$

$$= \frac{1}{c^2}\frac{\partial^2 u}{\partial t^2} \qquad ...(1)$$

It is obvious that u is a function of r, θ, ϕ and t.

$\therefore$ Let us assume

$$u(r,\theta,\phi,t) = R(r)\,F_1(\theta)\,F_2(\phi)\,T(t)$$

$$= RF_1F_2T \text{ (say)}. \qquad ...(2)$$

where R is a function of r only, F_1 that of θ only, F_2 that of φ only and T that of t only.

$$F_1F_2\,T\frac{d^2R}{dr^2}+\frac{2F_1.F_2T}{r}\frac{dR}{dr}+\frac{RF_2T}{r^2\sin\theta}.\frac{d}{d\theta}\left(\sin\theta\frac{dF_1}{d\theta}\right)+\frac{RF_1T}{r^2\sin^2\theta}.\frac{d^2F_2}{d\phi^2}$$

$$=\frac{RF_1F_2}{c^2}.\frac{d^2T}{dt^2}$$

$$\Rightarrow\ \frac{1}{R}\frac{d^2R}{dr^2}+\frac{2}{Rr}\frac{dR}{dr}+\frac{1}{r^2F_1\sin\theta}\frac{\partial}{\partial\theta}\left(\sin\theta\frac{dF_1}{d\theta}\right)+\frac{1}{F_2r^2\sin^2\theta}\frac{d^2F_2}{\partial\phi^2}$$

$$=\frac{1}{c^2T}\frac{d^2T}{dt^2}=-k^2.$$

Since the R.H.S. is independent of r, θ and φ and the L.H.S. is independent of t. Hence, the two cannot be equal unless both reduce to a constant.

From (3), we have

$$\frac{d^2T}{dt^2}+c^2k^2T=0 \text{ whose solution is } T=c_1e^{\pm iket}$$

$$\text{and}\ \left[\frac{1}{R}\frac{d^2R}{dr^2}+\frac{2}{Rr}\frac{dR}{dr}+\frac{1}{r^2F_1\sin\theta}\frac{\partial}{\partial\theta}\left(\sin\theta\frac{dF_1}{d\theta}\right)+k^2\right]r^2\sin^2$$

$$=-\frac{1}{F_2}\frac{d^2F_2}{d\phi^2}=m^2 \qquad \ldots(4)$$

with the same argument as above.

Which give

$$\frac{d^2F_2}{\partial\phi^2}+m^2F_2=0 \text{ whose solution is } F_2=C_2e^{\pm im\phi}$$

$$\text{and}\ \frac{r^2}{R}\frac{d^2R}{dr^2}+\frac{2r}{R}\frac{dR}{dr}+\frac{1}{F_1\sin\theta}\frac{d}{d\theta}\left(\sin\theta\frac{dF_1}{d\theta}\right)+k^2r^2=\frac{m^2}{\sin^2\theta}$$

$$\Rightarrow\ \frac{r^2}{R}\frac{d^2R}{dr^2}+\frac{2r}{R}\frac{dR}{dr}+k^2r^2=-\frac{1}{F_1\sin\theta}\frac{d}{d\theta}\left(\sin\theta\frac{dF_1}{d\theta}\right)+\frac{m^2}{\sin^2\theta}$$

$$\Rightarrow\ \frac{r^2}{R}\frac{d^2R}{dr^2}+\frac{2r}{R}\frac{dR}{dr}+k^2r^2=-\frac{1}{F_1}\frac{d^2F_1}{d\theta^2}-\frac{\cot\theta}{F_1}\frac{dF_1}{d\theta}+\frac{m^2}{\sin^2\theta}$$

$$=n(n+1)\ (\text{constant}) \qquad \ldots(4)$$

with the same argument as above

From equation (4), we have two differential equations

$$\frac{d^2R}{dr^2} + \frac{2r}{R}\frac{dR}{dr} + \left\{k^2 - \frac{n(n+1)}{r^2}\right\} R = 0 \qquad \ldots(5)$$

and $$\frac{d^2F_1}{d\theta^2} + \cot\theta \,.\, \frac{dF_1}{d\theta} + \left[-\frac{m^2}{\sin^2\theta} + n(n+1)\right] F_1 = 0 \qquad \ldots(6)$$

Now, putting $R = r^{-1/2}\,\psi(r)$ in (5), we have

$$\frac{d^2\psi}{dr^2} + \frac{1}{r}\frac{d\psi}{dr} + \left\{k^2 - \frac{(n+1/2)^2}{r^2}\right\}\psi = 0. \qquad \ldots(7)$$

$\therefore$ If $n + \frac{1}{2}$ is neither zero nor an integer, then

$$\psi(r) = A\,J_{n+1/2}(kr) + B\,J_{-(n+1/2)}(kr)$$

where A and B are constants and $J_n(z)$ denote the Bessel function of the first kind of order v and argument z.

$\therefore R = r^{-1/2}\,[A'J_{n+1/2}(kr) + B'J_{-(n+1/2)}(kr)$.

Again substituting $m = \cos\theta$ in (6).

so that $$\frac{dF_1}{d\theta} = \frac{dF_1}{d\mu}.\frac{d\mu}{d\theta} = -\frac{dF_1}{d\mu}.\sin\theta$$

and $$\frac{d^2F_1}{d\theta^2} = \frac{d}{d\theta}\left(-\frac{dF_1}{d\mu}\sin\theta\right)$$

$$= -\frac{d^2F_1}{d\mu^2}.\frac{d\mu}{d\theta}\sin\theta - \frac{dF_1}{d\mu}\cos\theta$$

$$= \frac{d^2F_1}{d\mu^2}\sin^2\theta - \frac{dF_1}{d\mu}\cos\theta$$

$$= (1-\mu^2)\frac{d^2F_1}{d\mu^2} - \mu\frac{dF_1}{d\mu}$$

from (6), we have

$$(1-\mu^2)\frac{d^2F_1}{d\mu^2} - 2\mu\frac{dF_1}{d\mu} + \left\{n(n+1) - \frac{m^2}{1-\mu^2}\right\}F_1 = 0 \qquad \ldots(8)$$

which is associated Legendre equation, whose solution is

$F_1 = A'P_n^{\,m}(\mu) = A''P_n^{\,m}(\cos\theta)$ where A" is a constant.

Hence, the complete solution of (1) is

$$u\ (r, \theta, \phi, t) = r^{-1/2} \sum_{m,n,k} [A_{m,n,k}\ J_{n+1/2}(kr)]$$

$$= B_{m,\ n,\ k}\ J_{-(n+1/2)}\ (kr)\ P_n^{\ m}\ (\cos\theta)\ .\ e^{\pm im\phi\ \pm\ iket}$$

Example 1:

The surface S of a sphere of radius a is kept at a fixed distribution of electric potential u = f(0). Find the potential u at all points in space which is assumed to be free of further charge.

Solution:

Since the potential on S is independent of f so is the potential in space.

Thus, $\dfrac{\partial^2 u}{\partial \phi^2} = 0$

and the Laplace's equation in spherical coordinates reduces to the form

$$\frac{\partial^2 u}{\partial r^2} + \frac{2}{r}\frac{\partial u}{\partial r} + \frac{1}{r^2}\frac{\partial^2 u}{\partial \theta^2} + \frac{\cot\theta}{r^2}\frac{\partial u}{\partial \theta} = 0. \qquad \ldots(1)$$

Now, we are required to find the solution of (1) outside and inside the sphere.

Let the solution of (1) be of the form

$$u\ = R\ (r)\ F\ (\theta) = RF\ (\text{say}), \qquad \ldots(2)$$

where R is a function of r only and F that of q only.

Substituting in (1) and on simplification, we have

$$\frac{r^2}{R}\frac{d^2R}{dr^2} + \frac{2r}{R}\frac{dR}{dr} = -\left(\frac{1}{F}\frac{d^2F}{d\theta^2} + \frac{\cot\theta}{F}\frac{dF}{d\theta}\right) = k\ (\text{cosnt.}) \qquad \ldots(3)$$

Since the left hand side is independent of q and right hand side is independent of r, hence the two cannot be equal unless both reduce to a constant.

If $\qquad k = n\ (n + 1)$

then (3) gives two differential equations as

$$r^2\frac{d^2R}{dr^2} + 2r\frac{dR}{dr} - n\ (n + 1)\ R = 0 \qquad \ldots(4)$$

and $$\frac{d^2F}{d\theta^2} + \cos\theta\frac{dF}{d\theta} + n(n + 1)\ F = 0\ . \qquad \ldots(5)$$

The solution of equation (4) as discussed

$$R = Ar^n + Br^{-(n+1)}.$$

Now, substituting u = cos θ

so that $\frac{dF}{d\theta} = \frac{dF}{du}.\frac{du}{d\theta} = -\sin\theta \frac{dF}{du}.$

In equation (5), we have

$$\frac{d}{d\theta}\left(-\sin\theta \frac{dF}{du}\right) + \cot\theta\left(-\sin\theta \frac{dF}{du}\right) + n(n+1)F = 0$$

$$\Rightarrow \quad \sin^2\theta \frac{d^2F}{du^2} - 2\cos\theta . \frac{dF}{du} + n(n+1)F = 0$$

$$\Rightarrow \quad (1-u^2)\frac{d^2F}{du^2} - 2u\frac{dF}{du} + n(n+1)F = 0$$

which is Legender differential equation

$\therefore$ $F = cP_n(U) = cP_n(\cos\theta)$

for different values of n.

Hence, $u = \sum_{n=0}^{\infty}\left(A_n r^n + \frac{B_n}{r^{h+1}}\right) P_n(\cos\theta).$ ...(7)

(Meerut, 97)

which is the solution of equation (1).

Outside the Sphere: Since the potential at r = ∞ is zero i.e., finite, therefore r^n must be absent from the solution (7) for the potential outside the sphere.

Thus, for r > a where a is the radius of the sphere the potential

$$u = \sum_{n=0}^{\infty}\frac{B_n}{r^{n+1}} P_n(\cos\theta) \qquad ...(8)$$

Now, since on the surface of sphere the potential is f(θ)

$\therefore$ $u = f(\theta)$ where $r = a$,

$$f(\theta) = \sum_{n=0}^{\infty}\frac{B_n}{r^{an+1}} P_n(u) \qquad u = \cos q$$

$$a^{n+1} f(\theta) = \sum_{n=1}^{\infty} B_n \; P_n(u).$$

Multiplying both sides by $P_n(u)$ and then integrating between the limits – 1 to + 1, we have

$$B_n \int_{-1}^{+1} P_n^2 (u)\, du = a^{n+1} \int_{-1}^{+1} f(\theta)\ P_n(u)\, du$$

$$\Rightarrow \qquad B_n \frac{2}{2n+1} = a^{n+1} \int_{-1}^{+1} f(\theta)\ P_n(u)\, du \qquad u = \cos\theta$$

$$\Rightarrow \qquad B_n = \frac{2n+1}{2} a^{n+1} \int_{-1}^{+1} f(\theta)\ P_n(\cos\theta) \sin\theta\, d\theta. \qquad ...(9)$$

Hence, the potential outside the sphere is given by equation (8) with coefficient (9).

Inside the Sphere: Since the potential at r = 0 is finite, therefore $\frac{1}{r^{n+1}}$ must be absent from the solution (7) for the potential inside the sphere,

$$\therefore \text{ for } r < a, \text{ the potential } u = \sum_{n=0}^{\infty} A_n\ r^n\ P_n(\cos\theta). \qquad ...(10)$$

Since on the surface of the sphere potential is f(θ).

$$\therefore \qquad u = f(\theta) \qquad \text{where } r = a$$

$$\therefore \qquad f(\theta) = \sum_{n=0}^{\infty} A_n a^n\ P_n(\cos\theta).$$

Proceeding similarly as above

$$A_n = \frac{(2n+1)}{2a^n} \int_0^{\pi} f(\theta)\ (\cos\theta) \sin\theta\, d\theta. \qquad ...(11)$$

Hence, the potential inside the sphere is given by equation (19) with the coefficient (11).

Example 2:

A square plate has its faces and its edges x = 0 and x = π (0 < y < p) insulated. Its edges y = 0 and y =p are kept at temperatures zero and f(x) respectively. Derive this formula for its steady temperatures

$$u(x, y) = \frac{1}{2\pi} a_0 y + \sum_{n=1}^{\infty} a_n \frac{\sin h\ ny}{\sin h\ n\pi} \cos nx$$

$$\textit{where} \qquad a_n = \frac{2}{a} \int_0^{\pi} f(x) \cos nx\, dx \qquad (n = 0, 1, 2, ...)$$

Solution:

Steady temperature of the plate will be governed by the equation

$$\frac{\partial^2 u}{\partial x^2} + \frac{\partial^2 u}{\partial y^2} = 0. \qquad ...(1)$$

Here the boundary conditions are

$$u_x(0, y) = 0 \quad ...(2)$$

$$u_x(\pi, y) = 0 \quad ...(3)$$

$$u(x, 0) = 0 \quad ...(4)$$

and $$u(x, p) = f(x) \quad ...(5)$$

Let $$u(x, y) = F(x)\, G(y)$$

$$= FG \text{ (say)} \quad ...(6)$$

Proceeding we get the general solution in two cases as

(a) $F = Ax + B, \qquad G = Cy + D$

(b) $F = A \cos \lambda x + B \sin \lambda x$, $G = C \cos h\, \lambda y + D \sin h\, \lambda y$.

If we take the solution given by (a) then we have

$$u(x, y) = (Ax + B)(Cy + D) \quad ...(7)$$

$\therefore$ using (4), we have $D = 0$

and using equation (3), we have $A = 0$

$\therefore$ we may take the solution of equation (1), as

$$u(x, y) = A'y. \quad ...(8)$$

Again if we take solution given by (b), we have

$$u(x, y) = (A \cos \lambda x + B \sin \lambda x)(C \cos h\, \lambda y + D \sin h\, \lambda y). \quad ...(9)$$

Using (4), we have $C = 0$

and using (2), we have $B = 0$.

$\therefore$ we may take

$$u(x, y) = B' \cos \lambda x \sin h\, \lambda y$$

Using (3), we have $\sin \lambda\pi = 0$

$\Rightarrow \qquad \lambda p = np \qquad (n = 1, 2, 3,...)$

$\Rightarrow \qquad \lambda = n$

$\therefore$ for each value of n, we take the solution of (1) as

$$u_n(x, y) = A_n \cos nx \sin h\, ny..$$

Hence for each value of n, we may write

$$u(x, y) = A'y + \sum_{n=1}^{\infty} A_n \sinh ny \cos nx \quad ...(10)$$

$\therefore$ using equation (5), we have

Multiplying both sides of (11) by cos nx and then integrating between the limits 0 to π, we have

$$A_n \sin h\, n\pi = \frac{2}{\pi} \int_0^{\pi} f(x) \cos nx\, dx$$

$$\Rightarrow \qquad A_n = \frac{a_n}{\sin h\, n\pi} \text{ where } a_n = \frac{2}{\pi}\int_0^{\pi} f(x)$$

Again integration equation (11), between the limits 0 to π, we have

$$A'\,\pi^2 = \int_0^{\pi} f(x)\,dx$$

$$\therefore \qquad A' = \frac{1}{2\pi}\cdot\frac{2}{\pi}\int_0^{\pi} f(x) \cos nx\,dx = \frac{1}{2\pi}\,a_0.$$

Hence, the formula for steady temperature in the given square plate satisfying all the conditions is

$$u(x, y) = \frac{1}{2\pi}\,a_0 y + \sum_{n=1}^{\infty} a_n \frac{\sin h\, ny}{\sin h\, n\pi} \cos nx$$

where $\quad a_n = \frac{1}{2\pi}\int_0^{p} f(x) \cos nx\,dx$

$(n = 0, 1, 2, ...)$ **Hence Proved.**

Example 3:

In a semi-circular plate of radius a with bounding diameter at 0°C and surface at 100°C, prove that the temperature distribution is given by

$$u = \frac{100}{\pi}\sum \frac{r^{2n-1}}{2n-1}\,\frac{\sin(2n-1)\,\theta}{a^{2n-1}}.$$

Solution:

Here the solution of the two dimensional Laplace's Equation

$$\frac{\partial^2 u}{\partial r^2} + \frac{1}{r}\frac{\partial u}{\partial r} + \frac{1}{r^2}\frac{\partial^2 u}{\partial \theta^2} = 0 \qquad ...(1)$$

will give the temperature distribution on the plate.

Proceeding similarly we have

$$u = a_0 \log r + \sum_{n=1}^{\infty}(A_n \cos n\theta + B_n \sin n\theta)\,(C_n r^n + D_n r^{-n}) + k.$$

Since temperature is finite at $r = 0$.

$\therefore a_0 = D_n = 0$ and since at $r = 0$, $u = 0$ $\qquad \therefore k = 0.$

$$\therefore \; u = \sum_{n=1}^{\infty}(A'_n \cos n\theta + B'_n \sin n\theta)\, r^n. \qquad ...(2)$$

Let $u = u(a)$ at $r = a$, is the radius of the sphere.

$$\therefore u(a) = \sum_{n=1}^{\infty}(A'_n \cos n\theta + B'_n \text{ is n } n\theta)\, n^n. \quad ...(3)$$

$$\therefore A'_n = \frac{2}{\pi}\int_0^{\pi} \frac{u(a)}{a^n} \cos n\theta \, d\theta.$$

Since $u(a) = 100$ for $\pi > \theta > 0$.

$$\frac{200}{\pi a^n}\int_0^{\pi} \cos n\theta \, d\theta = 0$$

and
$$B'_n = \frac{2}{\pi}\int_0^{\pi} \frac{u(a)}{a^n} \sin n\theta \, d\theta = \frac{200}{\pi a^n}\left[-\frac{\cos n\theta}{n}\right]_0^{\pi}$$

$$= \frac{200}{\pi n a^n}(1 - \cos n\pi)$$

$$\therefore \quad B'_n = 0 \text{ for n even}$$

and
$$B'_n = \frac{400}{\pi n a^n} \text{ for n odd.}$$

Hence,
$$u = \frac{400}{\pi}\sum_{m=1}^{\infty} \frac{r^{2m-1}}{(2m-1)\, a^{2m-1}} \sin (2m-1)\,\theta$$

$$= \frac{400}{\pi}\sum_{m=1}^{\infty} \frac{r^{2n-1}}{(2m-1)\, a^{2m-1}} \sin(2n-1)\,\theta. \quad \textbf{Hence Proved.}$$

Example 4:

A rigid sphere of radius a is placed in a stream of fluid whose velocity in the undisturbed state is V. Determine the velocity of the fluid at any point of the disturbed stream.

Solution:

Let the origin be taken at the centre of the sphere and the z-axis is the direction of the velocity q.

The velocity of the fluid is given by vector

$$q = -\text{ grad } \phi \quad ...(1)$$

where ϕ is the velocity potential satisfying the following conditions

(i) $\nabla^2\phi = 0$ ...(2)

(ii) $\dfrac{\partial\phi}{\partial r} = 0$ on $r = a$

and (iii) $\phi \to -Vr\cos\theta = -V\, r\, P_1(\cos\theta)$ as $r \to \infty$.

$$\phi = \sum_{n-0}^{\infty}\left(A_n r^n + \frac{B_n}{r^{n+1}}\right) P_n (\cos\theta) \qquad ...(3)$$

$$\therefore \frac{\partial\phi}{\partial r} = \sum\left\{A_n n r^{n-1} - \frac{(n+1) B_n}{r^{n+2}}\right\} P_n (\cos\theta).$$

$\therefore$ from condition (2), we have

$$nA_n a^{n-1} - \frac{(n+1) B_n}{r^{n+2}} = 0$$

$$\Rightarrow \quad B_n = B_n = \frac{n}{n+1} A_n . a^{2n+1} \qquad ...(4)$$

and from condition (3), we have

As $r \to \infty, \phi \to \sum_{n-0}^{\infty} A_n r^n P_n(\cos\theta) = -Vr P_1(\cos\theta)$.

Comparing $A_1 = -V, A_2 = ... = 0$ (each)

$\therefore$ From (iii), $B_2 + B_3 = ... = 0$ (each)

and $\quad B_1 = -\left(\frac{n}{n+1} V.a^{2n+1}\right)_{n=1} = \frac{Va^3}{2}$

Hence, from equation (3), $f = -V\left(r + \frac{a^3}{2r^2}\right). \cos\theta. \qquad ...(5)$

The velocity of the fluid is given by (1) and (5).

The components of the velocity are given by

$$q_r = -\frac{\partial\phi}{\partial r} = V\left(1 - \frac{a^3}{r^3}\right)\cos\theta$$

and $$q_r = -\frac{1}{r}\frac{\partial\phi}{\partial r} = V\left(1 - \frac{a^3}{r^3}\right)\sin\theta. \qquad \textbf{Ans.}$$

Example 5:

Solve $\frac{\partial^2 u}{\partial x^2} + \frac{\partial^2 u}{\partial y^2}$ *for* $0 < x < \pi, 0 < y < \pi$

$u(x, 0) = x^2$

$u(x, \pi) = 0$

$$\left[\frac{\partial u}{\partial x}\right]_{(0, y)} = \left[\frac{\partial u}{\partial x}\right]_{(\pi, y)} = 0.$$

Solution:

Given

$$\frac{\partial^2 u}{\partial x^2} + \frac{\partial^2 u}{\partial y^2} = 0 \quad \text{...(1)}$$

$$u(x, 0) = x^2 \quad \text{...(2)}$$

$$u(x, \pi) = 0 \quad \text{...(3)}$$

$$u_x(0, y) = 0 \quad \text{...(4)}$$

$$u_x(\pi, y) = 0 \quad \text{...(5)}$$

Let the solution of (1) be taken as

$$u(x, y) = F(x)\, G(y)$$

$$= FG \text{ (say)} \quad \text{...(6)}$$

where F is a function of x only and G that of y only.

Substituting in (1) and simplifying, we have

$$\frac{1}{F}\frac{d^2F}{dx^2} = -\frac{1}{G}\frac{d^2G}{dy^2}.$$

Since change in x in L.H.S. does not alter the R.H.S. and similarly change in y in R.H.S. does not alter the L.H.S., therefore the two cannot be equal unless both reduce to a constant.

$$\therefore \quad \frac{1}{F}\frac{d^2F}{dx^2} = -\frac{1}{G}\frac{d^2F}{dy^2} = 0 \text{ or } -\lambda^2$$

The corresponding differential equations obtained are

(i) $\dfrac{d^2F}{dx^2} = 0, \quad \dfrac{d^2G}{dy^2} = 0.$

(ii) $\dfrac{d^2F}{dx^2} + \lambda^2 F = 0, \quad \dfrac{d^2G}{dy^2} - \lambda^2 G = 0.$

The general solutions in the two cases are

(a) $F = Ax + B, \qquad G = Cy + D$

(b) $F = A\cos\lambda x + B\sin\lambda x,\ G = C\cos h\,\lambda y + D\sin h\,\lambda y.$

Taking the solutions given by (a), we have

$$u(x, y) = (Ax + B)(Cy + D). \quad \text{...(6)}$$

From (6) and (3), we have

$$u(x, \pi) = (Ax + B)(C\pi + D) = 0$$

$\therefore \quad D = -C\pi$

$\therefore \quad u(x, y) = (A'x + B')(y - \pi).$

Then using (4),

$$u_x(0, y) = A'(y - \pi) = 0 \quad \therefore \quad A' = 0$$

$\therefore$ One solution of (1) may be taken as

$$u(x, y) = B'(y - p). \qquad ...(7)$$

Again taking the solution given by (b), we have

$$u(x, y) = (A \cos \lambda x + B \sin \lambda x) . (C \cos h\, \lambda y + D \sin h\, \lambda y) \quad ...(8)$$

From this and (3), we have

$$u(x, \pi) = (A \cos \lambda x + B \sin \lambda x)(C \cos h\, \lambda\pi + D \sin h\, \lambda\pi) = 0$$

$$\therefore \quad D = -\frac{C \cos h\, \lambda\pi}{\sin h\, \lambda\pi}$$

$\therefore$ from (1), we have

$$u(x, y) = (A' \cos \lambda x + C' \sin \lambda x)\left(\cos h\, \lambda y - \frac{\cos h\, \lambda\pi}{\sin h\, \lambda\pi}. \sin h\, \lambda y\right)$$

$$\Rightarrow \quad u(x, y) = (A' \cos \lambda x + C' \sin \lambda x)\frac{\sin h\,(\pi - y)}{\sin h\, \lambda\pi} \qquad ...(9)$$

Now, from equations (4) and (9), we have C' = 0

and then using (5), we have

$$u_x(\pi, y) = -A' \lambda \sin h\, \lambda\pi \frac{\sin h\,(\pi - y)}{\sin h\, \lambda\pi} = 0$$

$\Rightarrow \quad \sin h\, \lambda p = 0$

$\Rightarrow \quad \lambda\pi = n\pi$

$\therefore \quad \lambda = n \qquad (n = 1, 2, 3, ...)$

$\therefore$ for each value of n, we have the solution of (1) as

$$u_x(x, y) = A_n \cos nx, \frac{\sin h\,(\pi - y)}{\sin h\, \lambda\pi}.$$

Hence, the complete solution of (1) may be written as

$$u(x, y) = B'(y - \pi) + \sum_{n=1}^{\infty} A_n \cos nx \frac{\sin h\,(\pi - y)}{\sin h\, \lambda\pi} \qquad ...(10)$$

using (1), we have

$$u(x, 0) = x^2 = -B'\pi + \sum_{n=1}^{\infty} A_n \cos nx \qquad ...(11)$$

Multiplying both sides by cos nx and then integrating between the limits 0 to π, we have

$$A_n = \frac{2}{\pi}\int_0^{\pi} x^2 \cos nx\, dx$$

Integrating by parts

$$A_n = \frac{2}{\pi}\left[\left(\frac{x^2}{n}\sin nx\right)_0^{\pi} - \frac{2}{\pi}\int_0^{\pi} x \sin nx\, dx\right]$$

$$= -\frac{4}{n\pi}\left[\left\{-\frac{x}{\pi}\cos nx\right\}_0^{\pi} - \frac{1}{n} + \int_0^{\pi} \cos nx\, dx\right]$$

$$= -\frac{4}{n\pi}\cos n\pi = \frac{4}{n^2}(-1)^n.$$

Again integrating both side of (11), between the limits 0 to π, we have

$$-B'\pi^2 = \int_0^{\pi} x^2\, dx = \frac{1}{3}\pi^3$$

$$\therefore \quad B' = -\frac{\pi}{3}$$

Hence, the complete solution of (1) is

$$u(x, y) = \frac{\pi}{3}(\pi - y) + 4\sum_{n=1}^{\infty}\frac{(-1)^n \sinh n(\pi - y)\cos nx}{n^2 \sinh n\pi} \qquad \textbf{Ans.}$$

12.13 Solution of the Diffusion Equation. (By the Method of the Separation of Variables)

Case I. Let us consider one-dimensional diffusion equation

$$\frac{\partial^2 \phi}{\partial x^2} = \frac{1}{k}\frac{\partial \phi}{\partial t}. \qquad \text{...(1)}$$

It is obvious that $\phi(x, t)$ is a function of x and t. Therefore suppose (1) has a solution of the form

$$\phi(x, t) = F(x)\, T(t)$$
$$= FT \text{ say} \qquad \text{...(2)}$$

where F is a function of x only and T that of t only.

Thus, substituting in equation (1), we have

$$T\frac{d^2F}{dx^2} = \frac{1}{k}F\frac{dT}{dt}$$

$$\Rightarrow \qquad \frac{1}{F}\frac{d^2F}{dx^2} = \frac{1}{kT}\frac{dT}{dt}$$

Now, the left hand side is a function of the independent variable x, while the right hand side is a function of the independent variable t. The two cannot be equal to each other, unless both reduce to a constant value.

$$\therefore \qquad \text{We have } \frac{d^2F}{dx^2} = \lambda F$$

and
$$\frac{dT}{dt} = k\lambda T.$$

The general solutions are

$$F = A\cos(nx + \epsilon)$$

and
$$T = Be^{-kn^2t}$$

where we have taken $-n^2$ for λ.

$\therefore$ The solution for all values of n is

$$\phi_n(x, t) = C_n \cos(nx + \epsilon_n)\, e^{-kn^2t}$$

where C_n is constant.

Summing over all values of n, we have

$$\phi(x, t) = \sum_{n=0}^{\infty} C_n \cos(nx + \epsilon_n)\, e^{-kn^2t}$$

This solution has the property that

$$\phi \to 0 \text{ as } t \to \infty.$$

and
$$\phi(x, 0) = \sum_{n=0}^{\infty} C_n \cos(nx + \epsilon_n)$$

Case II. If we consider two dimensional diffusion equation.

(Meerut, 93 (P))

$$\frac{\partial^2\phi}{\partial x^2} + \frac{\partial^2\phi}{\partial y^2} = \frac{1}{k}\frac{\partial\phi}{\partial t} \qquad ...(1)$$

then $\quad f(x, y, t) = X(x)\, Y(y)\, T(t).$

Substituting in (i), we have

$$\frac{1}{X}\frac{d^2X}{dx^2} + \frac{1}{Y}\frac{d^2Y}{dy^2} = \frac{1}{kT}\frac{dT}{dt}$$

$\therefore$ we may take

$$\frac{dT}{dt} = -n^2k\,T$$

$$\frac{d^2X}{dx^2} = -l^2X$$

and $$\frac{d^2Y}{dy^2} = -m^2Y$$

where $n^2 = l^2 + m^2$.

The general solutions are

$$T = Ae^{-(l^2+m^2)kt}$$

$$X = B\cos(lx + \epsilon_l)$$

$$X = B\cos(my + \epsilon_m)$$

$$\therefore\quad \phi(x, y, t) = C_{lm}\cos(lx + \epsilon_l)\cos(my + \epsilon_m)\,e^{-(l^2+m^2)kt}$$

Summing over all values of n, we have the solution of (1), as

$$\phi(x, y, t) = \sum_{l=0}^{\infty}\sum_{m=0}^{\infty} C_{lm}\cos(lx + \epsilon_l)\cos(my + \epsilon_m)\,e^{-(l^2+m^2)kt}$$

3.14 Solution of the Diffusion Equation

Example 1:

$$\frac{\partial^2 u}{\partial r^2} + \frac{1}{r}\frac{\partial u}{\partial r} + \frac{1}{r^2}\frac{\partial^2 u}{\partial \phi^2} + \frac{\partial^2 u}{\partial z^2} = \frac{1}{k}\frac{\partial u}{\partial t} \qquad ...(1)$$

Solution:

Let us assume the solution of (1), of the form

$$u(r, \phi, z, t) = R(r)\,F(\phi)\,Z(z)\,T(t) \qquad ...(2)$$

where $R(r)$, $F(\phi)$, $Z(z)$, $T(t)$ are function of r, ϕ, z and t alone respectively.

Substituting from (2) in (1), we have

$$FZT\frac{d^2R}{dr^2} + \frac{FZT}{r^2}\frac{dR}{dr} + \frac{RZT}{r}\frac{d^2F}{d\phi^2} + RFT\frac{d^2Z}{dz^2} = \frac{RFZ}{k}\frac{dT}{dt}$$

$$\Rightarrow\quad \frac{1}{R}\left(\frac{d^2R}{dr^2} + \frac{1}{r}\frac{dR}{dr}\right) + \frac{1}{r^2}\cdot\frac{1}{F}\frac{d^2F}{d\phi^2} + \frac{1}{Z}\frac{d^2Z}{dz^2} = \frac{1}{KT}\frac{dT}{dt} = -\lambda^2 \text{ (say)}$$

which gives $\frac{dT}{dt} + \lambda^2 kT = 0$ ...(3)

whose solution is $T(t) = C_1 e^{-k\lambda^2 t}$

and $\frac{1}{R}\left(\frac{d^2R}{dr^2} + \frac{1}{r}\frac{dR}{dr}\right) + \frac{1}{r^2}\cdot\frac{1}{F}\frac{d^2F}{d\phi^2}\lambda^2 + \frac{1}{Z}\frac{d^2Z}{dz^2} = -m^2$ (say)

which gives $\frac{d^2Z}{dz^2} = m^2 Z$

whose solution is $Z(z) = C_2 e^{\pm mz}$

and $\frac{r^2}{R}\left(\frac{d^2R}{dr^2} + \frac{1}{r}\frac{dR}{dr}\right) + (\lambda^2 + m^2)\, r^2 = -\frac{1}{F}\frac{d^2F}{d\phi^2} = n^2$ (say)

which gives $\frac{d^2F}{d\phi^2} + n^2 F = 0$

whose solution is $F(\phi) = C_2 e^{\pm in\phi}$

and $r^2\frac{d^2R}{dr^2} + r\frac{dR}{dr} + \{\lambda^2 + m^2).\, r^2 - n^2\}\, R = 0$

$$\Rightarrow \quad \frac{d^2R}{dr^2} + \frac{1}{r}\frac{dR}{dr} + \left\{(\lambda^2 + m^2) - \frac{\pi^2}{r^2}\right\} R = 0$$

where solution is

$$R(r) = C_4 J_n\, \{\sqrt{(\lambda^2 + m^2)}.r\} + C_4 J_{-n}\{\sqrt{(\lambda^2 + m^2)}.r\}$$

Hence, the complete solution of equation (1) is

$$u(r, \phi, z, t) = \sum_{\lambda,m,n} A_{lmn} J_n\, \{l^2 + m)^2.r\}.\, e^{\pm mz - k\lambda^2 t \pm in\phi}$$

3.15 Solution of the Diffusion Equation

Example 1:

$$\frac{\partial^2 u}{\partial r^2} + \frac{2}{e}\frac{\partial u}{\partial r} + \frac{1}{r^2 \sin\theta}\cdot\frac{\partial}{\partial\theta}\left\{\sin\theta\,\frac{\partial u}{\partial\theta}\right\} + \frac{1}{r^2\sin^2\theta}\cdot\frac{\partial^2 u}{\partial\phi^2} = \frac{1}{k}\frac{\partial u}{\partial t}. \quad ...(1)$$

(Meerut, 91 (P))

Solution:

Clearly, the solution of the equation (1) is a function of r, θ, φ and t.

Let us assume the solution of (1) of the form

$$u(r, \theta, \phi, t) = R(r)\, F_1(\theta)\, F_2(\phi)\, T(t) \qquad ...(2)$$

where R(r), $F_1(\theta)$, $F_2(\phi)$ and T(t) are functions of r, θ, φ and t alone respectively.

Substituting from (2) in (1), we have

$$F_1F_2T\frac{d^2R}{dr^2} + \frac{2F_1F_2T}{r}\frac{dR}{dr} + \frac{RF_2T}{r^2 \sin\theta}\cdot\frac{\partial}{\partial\theta}\left(\sin\theta.\frac{dF_1}{d\theta}\right)$$

$$+\frac{RF_1T}{r^2\sin^2\theta}\cdot\frac{d^2F_2}{d\phi^2} = \frac{RF_1F_2}{k}\frac{dT}{dt}$$

$$\Rightarrow \quad \frac{1}{R}\left(\frac{d^2R}{dr^2} + \frac{2}{r}\frac{dR}{dr}\right) + \frac{1}{F_1r^2\sin\theta}\frac{d}{d\theta}\left(\sin\theta.\frac{dF_1}{d\theta}\right)$$

$$\frac{1}{F_1r^2\sin\theta}\cdot\frac{d^2F_2}{d\phi^2} = \frac{1}{KT}\frac{dT}{dt} = -\lambda^2 \text{ (say)}$$

which gives $\dfrac{dT}{dt} + \lambda^2kT = 0$

where solution is $T(t) = C_2 e^{-\lambda^2kt}$

and $\left[\dfrac{1}{R}\left(\dfrac{d^2R}{dr^2} + \dfrac{2dR}{rdr}\right) + \dfrac{1}{F_1r^2\sin\theta\, d\theta}\dfrac{d}{}\left(\sin\theta.\dfrac{dF_1}{d\theta}\right) + \lambda^2\right] r^2\sin^2\theta$

$$= -\frac{1}{F_2}\frac{d^2F_2}{d\phi^2}$$

$$= m^2 \text{ (say)}$$

which gives $\dfrac{d^2F_2}{d\phi^2} + m^2 F_2 = 0$

where solution is

$$F_2(f) = C_2e^{\pm im\phi}$$

and $\dfrac{1}{R}\left(\dfrac{d^2R}{dr^2} + \dfrac{2}{r}\dfrac{dR}{dr}\right) + \dfrac{1}{F_1r^2\sin\theta}\cdot\dfrac{d}{d\theta}\left(\sin\theta\,\dfrac{dF_1}{d\theta}\right) + \lambda^2 = \dfrac{m^2}{r^2\sin^2\theta}$

$$\Rightarrow \quad \frac{r^2}{R}\left(\frac{d^2R}{dr^2}+\frac{2}{r}\frac{dR}{dr}\right)+\lambda^2r^2 = -\frac{1}{F_1r^2\sin\theta}\cdot\frac{d}{d\theta}\left(\sin\theta\frac{dF_1}{d\theta}\right)+\frac{m^2}{\sin^2\theta}$$

$$= n(n+1) \text{ say}$$

which gives $\dfrac{d^2R}{dr^2}+\dfrac{2dR}{rdr}+\dfrac{2dR}{rdr}+\left\{\lambda^2-\dfrac{n(n+1)}{r^2}\right\}R=0$ ...(3)

and $\quad -\dfrac{1}{F_1\sin\theta}\cdot\left(\sin\theta.\dfrac{d^2F_1}{d\theta^2}+\cos\theta.\dfrac{dF_1}{d\theta}\right)+\dfrac{m^2}{\sin^2\theta}=n(n+1)$

$$\Rightarrow \quad \frac{d^2F_1}{d\theta^2}+\cot\theta.\frac{dF_1}{d\theta}+\left\{-\frac{m^2}{\sin^2\theta}+n(n+1)\right\}F_1=0. \qquad ...(4)$$

Putting $R=(\lambda r)^{-1/2}\phi(r)$ in (3)

and proceeding as before the solution of (3) and (4) are

$$R(r)=(\lambda r)^{-1/2}(A'J_{n+1/2}(\lambda r)+B'J_{-\left(n+\frac{1}{2}\right)}(\lambda r)]$$

and $\quad F_1(\theta)=A^n P^m{}_n(\cos\theta)$

where A', B', A' are constants.

Hence, the required solution of equation (1) is

$$(r,\theta,t)=\sum_{\lambda,m,n}(\lambda r)^{-1/2}\{A_{mnl}J_{(n+1/2)}(\lambda r)$$

$$+B_{mn\lambda}J_{-(n+1.2.(\lambda r)}(\lambda r)\}P^m{}_n(\cos\theta)e^{\pm im\phi-\lambda^2kt}$$

Example 2:

Find the temperature in a sphere of radius a when its surface is maintained at zero temperature and its initial temperature is f(r, θ).

Solution:

Here the temperature is governed by the three dimensional heat equation in polar coordinates in dependent of ϕ.

$$\therefore \frac{\partial^2u}{\partial r^2}+\frac{2}{r}\frac{\partial u}{\partial r}+\frac{1}{r^2\sin\theta}\frac{\partial}{\partial\theta}\left(\sin\theta\frac{\partial u}{\partial\theta}\right)=\frac{1}{k}\frac{\partial u}{\partial t} \qquad ...(i)$$

Now, the solution of (1) is a function or r, θ and t

∴ Let us assume

$$u(r,\theta,t)=R(r)F_1(q)T(t) \qquad ...(ii)$$

where R(r), F_1(q) and T(t) are functions of r, θ and t alone respectively. substituting from (ii) in (i), we have

$$\frac{1}{R}\left(\frac{d^2R}{dr^2}+\frac{2}{r}\frac{dR}{dr}\right)+\frac{1}{F_1 r^2 \sin\theta}\cdot\frac{d}{d\theta}\left(\sin\theta\frac{dF_1}{d\theta}\right)$$

$$=\frac{1}{kT}\frac{dT}{dt}$$

$$= -\lambda^2 \text{ (say)}$$

∴ we have $\frac{dT}{dt} = -\lambda^2 kT$

whose solution is

$$T(t) = C_1 e^{-\lambda^2 kt}$$

and $$\frac{r^2}{R}\left(\frac{d^2R}{dr^2}+\frac{2}{r}\frac{dR}{dr}\right)+\lambda^2 r^2 = -\frac{1}{F_1\sin\theta}\frac{d}{d\theta}\left(\sin\theta.\frac{dF_1}{d\theta}\right)$$

$$= n(n+1) \text{ say.}$$

⇒ We have

$$\frac{d^2R}{dr^2}+\frac{2}{r}\frac{dR}{dr}+\left(\lambda^2-\frac{n(n+1)}{r^2}\right)R = 0 \qquad \text{...(iii)}$$

and $$-\frac{1}{F_1\sin\theta}\cdot\frac{d}{d\theta}\left(\sin\theta.\frac{dF_1}{d\theta}\right) = n(n+1)$$

⇒ $$\frac{d^2F_1}{d\theta^2}+\cot\theta\frac{dF_1}{d\theta}+n(n+1)F_1 = 0 \qquad \text{...(iv)}$$

Putting $R = (\lambda r)^{-1/2}\psi(r)$ is (iii), we have

$$\frac{d^2\psi}{dr^2}+\frac{1}{r}\frac{d\psi}{dr}+\left\{\lambda^2-\frac{\left(n+\frac{1}{2}\right)^2}{r^2}\right\}\psi = 0.$$

∴ If $n+\frac{1}{2}$ is neither zero nor an integer, then

$$\psi(r) = AJ_{n+1/2}(\lambda r) + BJ_{-(n+1/2)}(\lambda r)$$

where A and B are constants.

∴ $$R(r) = (\lambda r)^{-1/2}\left[AJ_{n+1/2}(\lambda r) + BJ_{-(n+1/2)}(\lambda r)\right]$$

Again substituting $\mu = \cos\theta$ is (iv), so that

$$\frac{dF_1}{d\theta} = -\frac{dF_1}{d\mu}\sin\theta = -\frac{dF_1}{d\mu}\sqrt{(1-\mu^2)}$$

$$\frac{d^2F_1}{d\theta^2} = \frac{d^2F_1}{d\mu^2}.\sin^2\theta - \frac{dF_1}{d\mu}\cos\theta$$

$$= (1+\mu^2)\frac{d^2F_1}{d\mu^2} - \mu\frac{dF_1}{d\mu}$$

we have

$$(1+\mu^2)\frac{d^2F_1}{d\mu^2} - 2\mu\frac{dF_1}{d\mu} + n(n+1)F_1 = 0$$

which is Legendre's equation, whose solution is given by

$$F_1(\theta) = A'P_n(\mu) = A'P_n(\cos\theta)$$

where A' is a constant.

∴ the solution of (i) may be taken as

$$u(r,\theta,t) = \sum_{n,\lambda} A_{n,\lambda}\ (\lambda r)^{-1/2} J_{n+1/2}(\lambda r)\, P_n(\cos\theta).\, e^{-k\lambda^2 t} \qquad \text{...(v)}$$

Since the temperature of the surface of the sphere is zero.

∴ $u(a, \theta, t) = 0$

∴ from (v), we have

$J_{n+1/2}(\lambda a) = 0$

whose roots are $l = \xi_1, \xi_2, \ldots \xi m_1 \ldots$

$$\therefore u(a,\theta,t) = \sum_{n,m} A_{n,m}\ (\xi_m r)^{-1/2} J_{n+1/2}(\xi_m r)\, P_n(\cos\theta) \times e^{=k\xi^2 nt} \qquad \text{...(iv)}$$

Also the initial temperature of the sphere is $f(r, \theta)$.

∴ $u(r, \theta, t) = f(r, \theta) = \sum A_{n,m}\, (\xi_m r)^{-1/2} J_{n+1/2}(\xi_m r)\, P_n(\cos\theta)$

∴ form the theory of Bessel functions and Legendre polynomial.

$$\therefore A_{n,m} = \frac{(2n+1)\,\xi_m^{1/2}}{a^2[J'_{n+1/2}(\xi_m a)]^2}\int_0^a r^{3/2}\ J_{n+1/2}(\xi x_m)\, dr \times \int_0^a P_n(\mu).\ f(r,\theta)\, d\mu. \qquad \text{...(vii)}$$

Hence, the required temperature is given by (vi) and (vii). **Ans.**

Example 3:

Obtain the solution u(r, t) of the three dimensional diffusion equation in the infinite cylinder $0 \le r \le a$, when the initial temperature $u(r, 0) = f(r)$ and the surface $r = a$ is maintained at zero temperature

(Meerut, 97(S), 90)

Solution:

Here the temperature is governed by the heat equation in cylindrical coordinates in dependent of z and ϕ.

$$\therefore \qquad \frac{\partial^2 u}{\partial r^2} + \frac{1}{r}\frac{\partial u}{\partial r} = \frac{1}{k}\frac{\partial u}{\partial t} \qquad ...(1)$$

$$0 < r < a,\ t > 0.$$

and the boundary conditions are

$$u(a, t) = 0 \qquad ...(2)$$

$$t > 0$$

and $$u(r, 0) = f(r) \qquad ...(3)$$

$$(0 \le r \le a).$$

Clearly the solution of (1) is of the form

$$u(r, t) = R(r)\ T(t) \qquad ...(4)$$

where R(r) and T(t) are functions of r and t alone respectively.

Substituting from (iv) in (i), we have

$$\frac{1}{R}\left(\frac{d^2R}{dr^2} + \frac{1}{r}\frac{dR}{dr}\right) = \frac{1}{KT}\frac{dT}{dt} = \lambda^2 \text{ (say)}$$

$$\therefore \qquad \frac{dT}{dt} = -\lambda^2 kT \qquad ...(5)$$

and $$\frac{d^2R}{dr^2} + \frac{1}{r}\frac{dR}{dr} + \lambda^2 R = 0 \qquad ...(6)$$

$$\lambda^2 kt$$

Solution of equation (5) is $\quad T(t) = C_1 e^{-\lambda^2 kt}$

Putting $\lambda r = \mu$ in (5); so that

$$\frac{dR}{dr} = \lambda\frac{dR}{d\mu} \text{ and } \frac{d^2R}{dr^2} = \lambda^2\frac{d^2R}{d\mu^2},$$

we have

$$\frac{d^2R}{d\mu^2} + \frac{1}{\mu}\frac{dR}{d\mu} + R = 0$$

which is Bessel's Equation of zeroth order whose solution is

$$R(r) = C_2J_0(\mu) = C_2J_0(\lambda r).$$

Hence solution of (1) is

$$u(r, t) = \sum_n A_\lambda J_0(\lambda r).e^{-\lambda^2 kt} \qquad ...(7)$$

Using the condition (2), we have $J_0(\lambda a) = 0$.

Let $x_1, x_2, ...x_n, ...$ be the roots of this equation

i.e., $\lambda = \xi_1, \xi_2, ..., \xi_n, ...$

$$\therefore \quad u(r, t) = \sum_n A_n J_0(\xi_n r).e^{-kt\xi_n^2} \qquad ...(8)$$

Using the condition (3), we have

$$u(r, 0) = f(r) = \sum_n A_n J_0(\xi_n r)$$

which is Fourier Bessel Series.

$$\therefore \quad A_n = \frac{2}{a^2 J_1^2(\xi_n a)} \int_0^a v' \ f(v) \ J_0(\xi_n v) \ dv.$$

Hence, the required solution of equation (1) is given by

$$u(r, t) \ \frac{2}{a^2} \sum_n \frac{J_0(\xi_n r)}{J_1^2(\xi_n a)}. \ e^{-kt\xi^2_n} \int_0^a vf(v) \ J_0(\xi_n v) \ dv. \qquad \textbf{Ans.}$$

Example 4:

Find the temperature u (x, t) in a slab whose ends x = 0 and x = l are kept at temperature zero and whose initial temperature f (x) is given by

$$f(x) = \begin{cases} A & \text{when } 0 < x < \frac{l}{2} \\ 0 & \text{when } \frac{l}{2} < x < l \end{cases} \qquad \textbf{(Meerut, 91)}$$

Solution:

From cos e I, we have

$$u(x, t) = \sum_{n=1}^{\infty} B_n \sin \frac{n\pi x}{l}.e - \frac{c^2\pi^2 n^2 t}{l^2}$$

where $B_n = \frac{2}{l}\int_0^l f(x) \sin \frac{n\pi x}{l} dx$

$$= \frac{2}{l}\left[\int_0^{l/2} A \sin \frac{n\pi x}{l} dx + \int_{l/2}^{l} 0.\sin \frac{n\pi}{l} x\, dx\right]$$

$$= \frac{2}{l}\frac{Al}{n\pi}\left(-\cos\frac{n\pi x}{l}\right)_0^{l/2} = \frac{2A}{n\pi}\left(1-\cos\frac{n\pi}{2}\right)$$

$$= \frac{4A}{n\pi}\sin^2\frac{n\pi}{4}$$

$$\therefore u(x, t) = \frac{4A}{\pi}\sum_{n=1}^{\infty}\frac{1}{n}\sin^2(n\pi/4).e^{-\frac{c^2n^2\pi^2t}{l^2}} \sin\frac{n\pi x}{l}. \qquad \textbf{Ans.}$$

EXERCISES

1. The faces $x = 0$, $x = a$ of an infinite slab are maintained at zero temperature. The initial distribution of temperature in the slab is described by the equation $u = f(x)$, $(0 \le x \le a)$. Determine the temperature at a subsequent time t. **(Meerut, 97, 93)**

   Ans. $u(x, t) = \frac{2}{a}\sum_{n=1}^{\infty} e^{-n^2\pi^2 t/a^2} \sin\left(\frac{n\pi x}{a}\right). \int_0^a f(v) \sin\left(\frac{n\pi v}{a}\right) dv.$

2. Solve $\frac{\partial u}{\partial t} - k\frac{\partial^2 u}{\partial t^2} = 0$, for $0 < x < \pi$, $t > 0$

   $u(0, t) = u(\pi, t) = 0$ and $u(x, 0) = \sin^5 x$.

   Ans. $u(x, t) = \frac{3}{4}e^{-kt}. \sin x - \frac{1}{4}e^{-2kt} \sin 2x$

3. Make use of the method of separation of variables to solve

   $\frac{\partial u}{\partial t} = c^2\frac{\partial^2 u}{\partial x^2}$ $0 \le x \le 1$, $t > 0$,

   $u(0, t) = 2$, $u(1, t) = 3$, $t \ge 0$,

   and $u(x, 0) = x(1 - x)$, $0 < x < 1$. **(Meerut, 94)**

   Ans. $u(x, t) = x + 2 +$

   $$\sum_{n=1}^{\infty}\left(\frac{6\cos n\pi - 4}{n\pi} - \frac{4\cos n\pi - 4}{n^3\pi^3}\right)e^{-n^2\pi^2c^2x}.\sin n\pi x.$$

4. Solve $\frac{\partial^2 z}{\partial x^2} = \frac{1}{K}\frac{\partial z}{\partial x}$. by separation of variables. **(Meerut, 97)**

5. Find the steady state temperature in a bar that is insulated at the end x = 0 and held at constant temperature 0°C at the end x = 1

(Meerut, 92)

6. Show that the solution of the equation

$$\frac{\partial^2 \theta}{\partial x^2} = \frac{\partial \theta}{\partial t}$$

satisfying the conditions

(i) $\theta \to$ at $t \to \infty$

(ii) $\theta = 0$ when $x = \pm a$ or all values of $t > 0$,

(iii) $\theta = x$ when $t = 0$ and $-a < x < a$ is

$$\theta = \frac{2a}{p}\sum_{n=1}^{\infty}\frac{(-1)^{n-1}}{n}\sin\left(\frac{n\pi x}{a}\right)\exp.\left(-\frac{n^2\pi^2 t}{a^2}\right)$$ **(Meerut, 93(P)**

7. Find the deflection u(x, t) of the vibrating string (length $l = \pi$, ends fixed and c =) corresponding to zero initial velocity and initial-deflection f(x), given by

Ans. (i) $u(x, t) = \frac{84}{\pi}\left(\cos t \sin x \frac{1}{3^3}\cos 3t \sin 3x + \frac{1}{5^5}\cos 5t \sin 5x + \ldots\right)$

8. Solve $\frac{\partial^2 u}{\partial t^2} = c^2\frac{\partial^2 u}{\partial x^2}$.

where $u(0, t) = 0$ $u(1, t) = 0$

and $u(x, 0) = A \sin \pi x$; $\left(\frac{\partial u}{\partial t}\right)_{t=0} = 0.$ **(Meerut, 90)**

Ans. $u(x, t) = A \cos c\pi t.\sin \pi x.$

9. A straight is tracheated between the fixed points (0, 0) and (1, 0) and released from rest from the position $u(x, 0) = A \sin \sin 2\pi x$. Find the displacements u(x, t). **(Meerut, 94)**

Ans. $u(x, t) = A \sin 2x \cos 2\pi ct.$

10. By separating the variables, show that one dimensional wave equation:

$$\frac{\partial^2 z}{\partial x^2} = \frac{1}{c^2}\frac{\partial^2 z}{\partial t^2}$$

has solutions of the form A exp. ($\pm$ inx $\pm$ inct),
where A and n are constants. **(Meerut, 92 (P))**

11. Obtain a solution of Laplace's equation is Cartesian coordinates:

$$\frac{\partial^2 u}{\partial x^2} + \frac{\partial^2 u}{\partial y^2} + \frac{\partial^2 u}{\partial z^2} = 0$$

by the method of separation of variables. **(Meerut, 91)**

12. Show that the two-dimensional Laplace's Equation in polar coordinates has the solutions of the form:

$$(Ar^n + Br^{-n})\, e^{\pm in\theta}$$

where A, B and n are constants. **(Meerut, 92 (P))**

13. Solve the one dimensional diffusion equation

$$\frac{\partial^2 \theta}{\partial x^2} = \frac{1}{k} \cdot \frac{\partial \theta}{\partial t}$$

in the region $0 \le x \le p$, $t \ge 0$, then

(a) θ remain finite at $t \to \infty$.

(b) $\theta = 0$, if $x = 0$ or π, for all values of t.

(c) At $t = 0$, $\begin{cases} \theta = x, & 0 \le x \le \pi/2 \\ \theta = \pi - x, & \frac{1}{2}\pi \le x \le \pi. \end{cases}$ **(Meerut, 94)**

14. Solve $\dfrac{\partial^2 y}{\partial x^2} = \dfrac{1}{C^2}\dfrac{\partial^2 y}{\partial t^2}$

under the boundary conditions:

$y = 0$, for $x = 0$ and for all values of t,

$\dfrac{\partial y}{\partial t} = 0$, for $t = 0$ and for all values of x,

$y = 0$ for $x = \pi$ and for all values of t,

$y = hx/d$, for $0 \le x \le d$ and $t = 0$

and $y = h(\pi - d)$, for $d \le x \le p$ and $t = 0$. **(Meerut, 98)**

15. Derive the solution of

$$\frac{\partial^2 V}{\partial r^2} + \frac{1}{r}\frac{\partial V}{\partial r} + \frac{\partial^2 V}{\partial z^2} = 0$$

for the region $r \leq 0$, $z \geq 0$, with boundary condition

$V \to 0$ as $z \to \infty$ and $r \to \infty$

$V = f(r)$ an $z = 0$, $r \geq 0$ **(Meerut, 97)**

Ans. $V(r, z) = \int_0^\infty \xi . f(\xi) e^{-\xi z} J_0(\xi r)\, d\xi$

16. Discuss the solution of transverse vibrations of a thin membrane bounded by circle of radius a in xy plane described by the function z(x, y, t) satisfying the wave equation

$$\frac{\partial^2 z}{dr^2} + \frac{1}{r}\frac{\partial z}{\partial r} + \frac{1}{r^2}\frac{\partial^2 z}{\partial \theta^2} = \frac{1}{c^2}\cdot\frac{\partial^2 z}{\partial t^2}$$

with conditions,

$z = 0$ on $r = a$

$z = f(r)$, $\frac{\partial z}{\partial t} = 0$, at $t = 0$ **(Meerut, 97)**

13

Linear Differential Equations (Matrix Method)

13.1 Differential Equation

(Meerut, B.Sc. Hons. 90, 92)

Definition: *An equation relating the values of a function y and one or more of its* **differences** Δy, $\Delta^2 y$... **for each value of some set of** *number A for which each of these functions is defined is called a differential equation over the set A.*

A differential equation is therefore a relation involving differences. If we consider functions y defined for all real numbers x, the following equations are examples of differential equation over the set of real numbers.

$$\Delta y (x) + 2y (x) = 0 \quad ...(1)$$

$$\Delta^2 y (x) + 2\Delta y (x) + y (x) = 2 \quad ...(2)$$

$$\Delta^2 y (x) - xy (x) = 3x + 8 \quad ...(3)$$

$$y (x). \Delta^3 y (x) = 6 \quad ...(4)$$

$$[\Delta y (x)]^2 + [y (x)]^2 = -4 \quad ...(5)$$

Remark:

We shall use the symbol h rather the x to denote a number in the domain of the functions related by the differential equation and write y_k *rather than y(h) for the value of y at h.*

All difference operations are taken with a difference interval equal to 1. Now, equation (1) to (5) are written as follows:

$$\Delta y_h + 2y_h = 0 \quad ...(6)$$

$$\Delta^2 y_h + 2\Delta y_h + y_h = 2 \quad ...(7)$$

$$\Delta^2 y_h - hy_h = 3h + 8 \quad ...(8)$$

$$y_h\, D^3 y_h = 6 \quad ...(9)$$

$$(\Delta y_h)^2 + (y_h)^2 = -4 \quad ...(10)$$

A differential equation written in this form will be understood as being defined over some set of consecutive non-negative integers this set may start with any integer and may be finite or infinite.

In each case we must specify the range of values of the integer h.

13.2 To Write a Differential Equation as a Relation Among the Value of y

We know the relationship between the operator E and Δ of the type

$$\Delta y_h = Ey_h - y_h \quad ...(1)$$

$$\Delta^2 y_h = E^2 y_h - 2Ey_h + y_h \quad ...(2)$$

$$\Delta^2 y_h = E^3 y_h - 3E^2 y_h + 3Ey_h - y_h \quad ...(3)$$

and so on.

We also know that $E^n\, y(x) = y(x + nk)$

$\therefore \quad E^n\, y_h = y(h + nk) = y_{h+n}$ if $k = 1$.

Therefore, under (1), (2), (3) and (4) the equation (6) and (10) of

$$y_{h+1} + y_h = 0 \quad ...(5)$$

$$y_{h+2} = 2 \quad ...(6)$$

$$y_{h+2} - 2y_{h+1} + (1 - h)\, y_h = 3h + 8 \quad ...(7)$$

$$y_h\, y_{h+2} - 3y_h\, y_{h+2} + 3y_h\, y_{h+1} - y_h^{\,2} = 6 \quad ...(8)$$

$$(y_{h-1} - y_h)^2 + y_h^{\,2} = -4. \quad ...(9)$$

13.3. Linear Differential Equation (Definition)

(Meerut, M.Sc 99; B.Sc Hons, 97)

A differential equation over the set of h values 0, 1, 2.. is said to be linear if it can be expressed in the form

$$g_0(h)\, y_{h+n} + g_1(h)\, y_{h+n-1} + ... + g_{n-1}(h)\, y_{h+1} + g_n(h)\, y_h = f(h) \quad ...(1)$$

where $g_0, g_1, g_2, ..., g_n$ and f are functions of h (but not of yh) defined for $h = 0, 1, 2, ...$

13.4 Linear Differential Equation with Constant Coefficients

(Meerut, M.Sc. 90)

Definition: If the coefficients $g_0, g_1, g_2, ..., g_n$ in eqn. (1) are each constant functions then differential equation (1) is called Linear differential equation with constant coefficients. If both g_0 and g_n are non-zero constants, then it is a linear differential equation with constant **coefficients and of order n.**

Consider the following examples

$$3 - y_h = 12 \quad ...(a)$$

$$5y_{h+2} + 2y_{h+1} + y_h = 2^h \quad ...(b)$$

$$4y_{h+2} - y_h = h \quad ...(c)$$

The above are examples of different equations with constant coefficients and are of order 1, 2 and 3 respectively.

When the leading coefficient or the coefficient of y_{h+n} is not 1, it is somethings useful to divide the equation throughout by this leading coefficient.

If (1) is of order g_0 is different from zero and we may divide by g_0. Thus (1) is transformed as

$$y_{h+n} + \frac{g_1(h)}{g_0(h)} y_{h+n-1} + ... + \frac{g_{n-1}(h)}{g_0(h)} y_{h+1} + \frac{g_n(h)}{g_0(h)} y_h = \frac{f(h)}{g_0(h)}. \quad ...(2)$$

If g_0 g_1, g_2,...,g_n are all constants

then $$\frac{g_1(h)}{g_0(h)}, \frac{g_2(h)}{g_0(h)}, ..., \frac{g_n(h)}{g_0(h)}$$

are all constants. Let these constants by $b_1, b_2, ..., b_n$, respectively ($b_n \neq 0$). $\therefore$ equation (2) becomes

$$y_{h+n} + b_1 y_{h+n-1} + ... + b_n y_n = R(h) \quad ...(3)$$

where R(h) is a function of h.

Definition: The equation (3) for which R.H.S.

$$R(h) = 0, \text{ i.e.,}$$

$$y_{h+n} + b_1 y_{h+n-1} + ... + b_n y_n = 0 \quad ...(4)$$

is said to be linear homogeneous differential equation of (3).

(Meerut, M.Sc. 89)

13.5 Order of a Linear Differential Equation

(Meerut M.Sc. 90, B.Sc. Hons., 91)

Definition: *A linear equation over a set A is of order n over A if when written in the form (1) both g_0 and g_n are different from zero at each point of A.*

A linear differential equation is of order n if the difference between the largest and smallest arguments for the function y involved in the equation is n. (Note that the difference interval is 1).

Examples:

$y_{h+2} + 6y_{h+1} - 8y_h - 2y_h = 2h$ order is 2 ...(1)

$y_{h+2} + 4y_{h+1} = 3h$ order is 1 ...(2)

$y_{h+2} + y_h = 4h$ order is 2 ...(3)

$Ay_{h+2} + 2y_{h+1} - 7y_h = 0$ order is 2 ...(4)

(if A does not contain 0)

$4^h\, y_{h+3} - 3^h\, y_{h+2} + 2^h\, y_{h+1} - y_h = 7$ order is 3. ...(5)

Remark:

If, when written in the form (1) either g_0 or g_n is 0 at some point of A, then definition of is not applicable. Thus the order of (4) is undefined if A does contain 0.

13.6 Solution of a Differential Equation

Definition: *A function y is a* **solution** *of a differential equation over a set A if the values of y make the differential equation a true statement for every point of A.*

In other words a function is a solution of a differential equation if it satisfy the differential equation.

A General Solution of a differential equation of order n is a solution which involves n arbitrary periodic constants.

A Particular Solution is a solution obtained from the general solution by assigning particular values to periodic constants.

13.7 Solution of the Equation

$y_{h+1} = Ay_h + C.$

(Meerut, M.Sc., 96, 98; B.Sc. Hons. 90, 92)

The linear first order differential equation has the form

$f_0(h)\, y_{h+1} + f_1(h)\, y_h = g(h),\ h = 0, 1,...$ ...(1)

over the indicated set of h-values. The functions $f_0(h)$ and $f_1(h)$ according to the definition are never zero so if they are constant functions, they are non-zero constants.

Dividing (1) by $f_0(h)$ we have

$$y_{h+1} = -\frac{f_1(h)}{f_0(h)} y_h + \frac{g(h)}{f_0(h)}.$$

If we now suppose that f_0 and f_1 as well as g, are constant functions, we can write this equation in the form

$$y_{h+1} = Ay_h + C, \qquad h = 0, 1, 2, \ldots \qquad \ldots(3)$$

where A and C are constant and $A \neq 0$.

Now we find the solution of (3) with y_0 prescribed.

Put $h = 0$ in (3). Then $y_1 = Ay_0 + C$.

Now put $h = 1$ in (3). Then

$$\begin{aligned} y_2 &= Ay_1 + C \\ &= A(Ay_0 + C) + C \\ &= A^2 y_0 + C(1 + A). \end{aligned}$$

Again put $h = 2$ in (3), Then

$$\begin{aligned} y_3 &= Ay_2 + C \\ &= A[A^2y_0 + C(1 + A)] = C \\ &= A^3y_0 + C(1 + A + A^2) \end{aligned}$$

and so on. In general we have

$$y_h = A^h y_0 + C(1 + A + A^2 + \ldots + A^{h-1}) \qquad \ldots(1)$$

The series within bracket on the right hand side is a geometric series with first term 1 and common ratio A.

$$\therefore 1 + A + A^2 + \ldots + A^{h-1} = \frac{1 - A^h}{1 - A} \quad \text{if} \quad A \neq 1$$

$$= h \quad \text{if} \quad A = 1$$

and hence we may write

$$y_h = \begin{cases} A^h y_0 + \dfrac{C(1 - A^h)}{(1 - A)}, & \text{if } A \neq 1 \\ y_0 + Ch & \text{if } A = 1 \end{cases} \qquad \ldots(4)$$

Remark:

We have also included the case when $h = 0$ for the formula reduces to y_0 when $h = 0$.

Theorem:

The function y given by (4) is a solution, and the only solution of the differential equation (3) with y_0 prescribed. **(Meerut, B.Sc. Hons. 92)**

Proof:

That equation (4) givens a solution of (1) can be seen from direct substituting.

When $A \neq 1$ $y_{h+1} = A^{h+1} y_0 + C \dfrac{1 + A^{h+1}}{1 - A}$

substituting in (3), we have

$$\text{R.H.S.} = Ay_h + C \quad A\left\{A^h y_0 + C\frac{1 - A^h}{1 - A}\right\} + C$$

$$= A^{h+1} y_0 + C\left\{1 + \frac{A - A^{h+1}}{1 - A}\right\}$$

$$= A^{h+1} y_0 + C\,\frac{A - A^{h+1}}{1 - A} = y_{h+1} = \text{L.H.S.}$$

Hence, for y given by (4), $A \neq 1$, (3) reduces to an identity.

If $\quad A = 1.$

Then $\quad y_h = y_0 + Ch$

on substituting it in (3) we have

$$\text{R.H.S.} = Ay_h + C = A(y_0 + Ch) + C$$

$$= y_0 + C(1 + h) \quad A = 1$$

$$= y_{h+1} = \text{L.H.S.}$$

Hence, y given by (4) is the solution of equation (3).

That this is the only solution for which y_0 is prescribed follows immediately from theorem (1).

It is to be noted that our assumption that the differential equation is defined over the set of h-values 0, 1, 2,... is merely for convenience. The result does not depend upon the assumption The following theorem give the more general result.

Theorem:

The linear differential equations

$$y_{h+1} = Ay_h + C,\ h = a,\ a + 1,\ a + 2 \qquad \text{...(1)}$$

taken over the indicated set of h-values (which may or may not continue indefinitely) has infinitely many solutions. If y is a solution, there is a constant D such that

$$y_h = \begin{cases} DA^{h-a} + C\dfrac{1-A^{h-a}}{1-A} \text{ if } A \neq 1,\ h = a, a+1, a+2,\ldots \\ D + C(h-a) \text{ if } A = 1. \end{cases} \qquad \ldots(2)$$

If a single value of y is prescribed for one of the h-values a, a + 1, a + 2,... then a unique solution of (1) is determined. In particular, if y_0 is prescribed, then the unique solution of (1) is given by (2) with $C = y_a$.

Example 1:

Solve

$$y_{h \cdot 1} = -y_h + 2,\ h = 0, 1, 2,\ldots$$

(Meerut, M.Sc 91, 97)

Solution:

Here A = – 1 and C = 2, therefore the solution is given by

$$y_h = A^h y_0 + C\,\frac{1-A^h}{1-A}$$

$$= (-1)^h y_0 + 2.\ \frac{1-(-1)^h}{1+1}$$

$$= (-1)^h y_0 + 1 - (-1)^h,\ h = 0, 1, 2,\ldots$$

If h = 0 or an even integer, then $(-1)^h = 1$, and if h is an odd integer then $(-1)^h = -1$.

∴ we obtain the following sequence of values

$$y_0, -y_0 + 2, y_0, -y_0 + 2,\ldots$$ **Ans.**

Example 2:

Solve the differential equation

$$2y_{h+1} - y_h = 4,\ h = 0, 1, 2,\ldots$$

with the initial condition $y_0 = 3$.

(Meerut, M.Sc. B.Sc Hons. 90)

Solution:

The given equation in the standard form is

$$y_{h+1} = \frac{1}{2}\,y_h + 2.$$

Hence, $A = \frac{1}{2}$ and C = 2. Therefore the solution is given by

$$y_h = A^h y_0 = C.\ \frac{1-A^h}{1-A}$$

$$= 3\left(-\frac{1}{2}\right)^h + 2.\frac{1-\left(-\frac{1}{2}\right)^h}{1-\frac{1}{2}} = 4 - \left(\frac{1}{2}\right)^h,$$

h = 0, 1, 2,...

Thus, we obtained the following sequence of values

3, 7/2, 15/4, 31/8,...

Example 3:

Solve $(E - a)\, y_h = 0,\ h = 0, 1, 2,...$

or $y_{h+1} = a y_h$

a is a constant. **(Meerut, M.Sc. 98)**

Solution:

Comparing the given differential equation

$$y_{h+1} = a y_h$$

with the differential equation

$$y_{h+1} = A y_h + C$$

We have $A = a,\ C = 0.$

Hence, the solution is given by

$$y_h = A^h y_0 + C\,\frac{1 - A^h}{1 - A} = a^h\, v_0.$$

$\therefore$ The sequence of values $y_0, y_1, y_2,...$ is a sequence $y_0, ay_0, a^2y_0...$

Ans.

Another Method: The given differential equation can be written as

$$\frac{y_{h+1}}{a^{h+1}} - \frac{y_h}{a^h} = 0 \Rightarrow \Delta\left(\frac{y_h}{a^h}\right) = 0$$

$$\therefore \frac{y_h}{a^h} = \text{constant}$$

$$\Rightarrow \quad y_h = \lambda a^h.$$ **Ans.**

Example 4:

Solve $y_h . y_{h+2} = y_{h+1}$ **(Meerut, B.Sc. Hons. 98)**

Solution:

The given differential equation can be written as

$$\frac{y_{h+2}}{y_{h+1}} = \frac{y_{h+1}}{y_h} = 0$$

$$\Rightarrow \quad \Delta\left(\frac{y_{h+1}}{y_h}\right) = 0 \quad \therefore \quad \frac{y_{h+1}}{y_h} = a \text{ (const.)}$$

$$\Rightarrow \quad y_{h+1} = ay_h$$

$$y_h - A.a^h$$ **Ans.**

Example 5:

Find the solutic n of

$$x_{h-1} = 2y_h - 1, \; h = 0, 1, 2, ...$$

with initial condition $y_0 = 5$. **(Meerut, M.Sc., 91)**

Solution:

Comparing the given equation with $y_{h+1} = Ay_h + C$ where have $A = 2$ and $C = -1$. Therefore the solution is

$$y_h = A^h y_0 + C \frac{1 - A^h}{1 - A}$$

$$= 2^h.5 - 1. \frac{1 - 2^h}{1 - 2} \qquad (\because \; y_0 = 5)$$

$$= 5.2^h + 1 - 2^h = 4.2^h + 1, \qquad h = 0, 1, 2, ...$$

From here we can write the consecutive values of y serially starting with $y_0 = 5$. If gives the increasing sequence of values

5, 9, 17, 32, 65, 129,... **Ans.**

Remark:

The result already obtained does not depend on the assumptions, made up to now. It is for convenience that the differential equation is defined over the set of h-values 0, 1, 2,...

We shall now be studying properties of linear differential equations.

Theorem:

If $y^{(1)}$ and $y^{(2)}$ are any two solutions of the linear homogeneous differential equation then $\lambda_1 y^{(1)} + \lambda_2 y^{(2)}$ is also a solution, where λ_1 and λ_2 are arbitrary constants.

Proof:

Let $y^{(1)}$ and $y^{(2)}$ be the solution of linear homogeneous differential equation

$$y_{h+n} + b_1\, y_{h+1+n-1} + \ldots + b_n y_h^{(1)} = 0 \qquad \ldots(1)$$

$$\therefore\; y^{(1)}_{h+n} + b_1 y^{(1)}_{h+n-1} + \ldots + b_n y_h^{(1)} = 0 \qquad \ldots(2)$$

$$y^{(2)}_{h+n} + b_1 y^{(2)}_{h+n-1} + \ldots + b_n y_h^{(2)} = 0 \qquad \ldots(3)$$

Multiplying (2) by λ_1 and (3) by λ_2 and adding, we have

$$[\lambda_1 y^{(1)}_{h+n} + \lambda_2 y^{(2)}_{h+n}] + b_1\, [\lambda_1 y^{(1)}_{h+n-1} + \lambda_2 y^{(2)}_{h+n-1}]$$

$$+ \ldots + b_n\, [\lambda_1 y_h^{(1)} + \lambda_2 y_h^{(2)}] = 0 \qquad \ldots(7)$$

which is true for $h = 0, 1, \ldots$

which shows that $y = \lambda_1 y^{(1)} + \lambda_2 y^{(2)}$ is also solution of (1).

Remark:

It is important to note that the proof in no way uses the fact that the coefficients are constants, so that this is indeed a property of linear equations, whether with constants coefficients or not.

13.8 Finite Linear Combination of Solution

Definition:

By a **finite linear combination of solution** we mean the sum of finite number of solutions, each multiplied by an arbitrary constant

i.e. $\lambda_1 y^{(1)} + \lambda_2 y^{(2)} + \ldots + \lambda_n\, y^{(n)}$

The reader can verify that any finite linear combination of solutions of the linear homogeneous differential equation is also a solution of this differential equation.

Theorem:

If Y is a solution of the linear homogeneous equation and y is a solution of the complete equation, then Y + y* is a solution of the complete equation.*

(Meerut, M.Sc. 88)

Proof:

We write the details of the proof for the linear differential equation of order 2, i.e., for $y_{h+2} + b_1 y_{h+1} + b_2 y_h = R(h)$.

The same method is applicable in the general case.

We are given

$$y_{h+2} + b_1 Y_{h+1} + b_2\, Y_h = 0$$

and $$y^*_{h+2} + b_1 y^*_{h+1} + b_2 y_h^* = R(h).$$

Now, adding these equations we obtain.

$$(Y_{h+2} + y^*_{h+2}) + b_1 (Y_{h+1} + y^*_{h+1}) + b_2 (Y_h + y^*_h) = R(h)$$

This is the desired conclusion.

This theorem points to a possible mode of approach in seeking to solve the complete differential equation. It appears that given a non-homogeneous differential equation it will be better to break the problem in two parts.

1. Non-homogeneous itself.
2. Corresponding to homogeneous part.

First we find any one solution of the complete equation, instead of finding all solutions call it as a particular solution.

Now, we look at this problem from a different angle, y* is already obtained. Can there be a choice for Y such that Y + y* includes all possible solutions of the complete equation? We feel inteutively that if Y is taken as the general solution of homogeneous equation then Y + y* may give all the solution of the non homogeneous equation.

As a matter of fact it can be proved that the general solution of the complete equation is obtainable as the sum of the general solution of the homogeneous equation and particular solution of the complete equation.

We shall illustrate it for first order differential equations.

Theorem:

Let us consider the linear first order differential equation with constant coefficients

$$y_{h-1} + b_l y_h = R(h)$$

(a) The function Y given by

$$Y_h = \lambda(-h_l)^h \qquad ...(2)$$

with λ and an arbitrary constant, is the general solution of the corresponding equation

$$y_{h-1} + b_l y_h = 0 \qquad ...(3)$$

(b) If y is any particular solution of the complete equation then Y + y* is the general solution of the complete equation. That is, if y is any solution of the complete differential equation, there is a value of the constant λ for which*

$$y_h = l\ (b_l)^h + y_h^*.$$

Proof:

It is easy to verify that (2) gives the solution of (3).

To prove the theorem, we shall take any solution y of the complete equation. Then we shall show that there exists a value of λ for which Y + y* and y are the same solutions.

In fact $\lambda(-b_1)^h + y_h^*$ is a solution of (1) for all possible value of λ. We shall show that for some λ, y become $\{\lambda(-b_1)^h + y_h^*\}$. We shall use **the Uniqueness Theorem** which says that two solutions of a linear first order differential equation which have the same value where h = 0 are identical.

Now choose λ such that

$$y_n - \lambda(-b_1)^0 + y_0^*$$

$$\Rightarrow \qquad \lambda = y_0 - y_0^*$$

Then $y_h \equiv (y_0 - y_0^*)(-b_1)^h + y_h^*$

This proves the theorem.

13.9 Fundamental set of Solution (or Linearly Independent Solutions)

(Meerut, B.Sc. 91)

Definition: *The solutions $y^{(1)}, y^{(2)}, \ldots y^{(n)}$ of nth order homogeneous differential equation*

$$y_{h+n} + b_1 y_{h+n-1} + b_2 y_{h+n-2} + \ldots + b_n y_h = 0 \qquad \ldots(1)$$

are said to form a **fundamental set** (or fundamental system) *of solutions of (1), if the nth order determinant*

$$\begin{vmatrix} y_0^{(1)} & y_0^{(2)} \ldots y_0^{(n)} \\ y_1^{(1)} & y_1^{(2)} \ldots y_1^{(n)} \\ \ldots & \ldots \quad \ldots \\ \ldots & \ldots \quad \ldots \\ y^{(1)}_{n-1} & y^{(2)}_{n-1} \ldots y^{(n)}_{n-1} \end{vmatrix} \text{ is different from zero.}$$

Theorem:

Let $y^{(1)}$ and $y^{(2)}$ be two solutions of the second order homo differential equation

$$y_{h+2} + b_1 y_{h+1} + b_2 y_h = 0 \qquad \ldots(1)$$

and let $Y = c_1 y^{(1)} + c_2 y^{(2)}$, where c_1 and c_2 are constants.

Then Y is the general solution of (1)

$$\text{if } \begin{vmatrix} y_0^{(1)} & y_0^{(2)} \\ y_1^{(1)} & y_1^{(2)} \end{vmatrix} = y_0^{(1)}.\, y_1^{(2)} - y_0^{(2)}.\, y_1^{(1)} \neq 0. \qquad \textbf{(Meerut, M.Sc. 99)}$$

PF. since $y^{(1)}$ and $y^{(2)}$ are two solutions of differential equation (1) therefore by $Y = c_1y^{(1)} + c_2y^{(2)}$ is also a solution of the differential equation (1).

Now, we need only to prove that if y is any solution of (1) then we can determine c_1 and c_2 such that Y and y are identical. By uniqueness theorem we know that two solutions of the linear differential equation of order 2 are identical if they coincide in value at two consecutive values of h say at h = 0 and h = 1. Therefore determine c_1 and c_2 so that for any choice of y_0 and y_1, $Y_0 = y_0$ and $Y_1 = y_1$.

i.e., $$Y_0 = c_1y_0^{(1)} + c_2y_0^{(2)} = y_0 \quad ...(2)$$

and $$Y_1 = c_1y_1^{(1)} + c_2y_1^{(2)} = y_1 \quad ...(3)$$

Now the above simultaneous linear equations (2) and (3) for unknowns c_1 and c_2 has unique solution if and only if

$$\begin{vmatrix} y_0^{(1)} & y_0^{(2)} \\ y_1^{(1)} & y_1^{(2)} \end{vmatrix} = y_0^{(1)}\, y_1^{(2)} - y_0^{(2)}\, y_1^{(1)} \neq 0.$$

i.e., for each choice of y_0 and y_1 we can find a unique pair of values of c_1 and c_2. Hence Y is a general solution of equation (1).

Example 1:

Solve, $y_{h-1} - 2y_h = h + 1.$ ...(1)

(Meerut M.Sc. 97)

Solution:

The corresponding homo. equation is

$$y_{h+1} - 2y_h = 0.$$

$\therefore$ Its solution is given by $y_h = \lambda.2^h$ ...(2)

To find the particular solution of (1), we apply the **Hit and Trial Method.** Let us try the constant function $y_h^* = A$.

Substituting in equation (1) we have

$$A - 2A = h + 1$$

$\Rightarrow$ $-A = h + 1$ which is not true since R.H.S. is not constant.

$\therefore$ Now, we try $y_h^* = Ah + B$.

Substituting in (1),

$$A(h + 1) + B - 2(Ah + B) = h + 1$$

$\Rightarrow$ $(-A - 1)h + (A - B - 1) = 0$

This will be an identity i.e. for every value of h the terms within brackets are zero.

$\therefore A = -1$ and $B = -2$, so that

$$y_h^* = -h - 2.$$

The general solution of equation (1) therefore is given by

$$y_h = \lambda.2^h - h - 2 \qquad ...(3)$$

Remark:

One can find any particular solution of equation *(1) satisfying any initial conditions.*

Let

$$y_0 = 4 \qquad ...(4)$$

(3) gives $4 = \lambda - 2 \Rightarrow l = 6$

$$y_h = 6.2^h - h - 2. \qquad \textbf{Ans.}$$

Example 2:

Solve completely the following diff. eq^n.

$2y_{h\ 1} - y_h = 2.$ **(Meerut 88)**

Solution:

The given differential equation can be written as

$$y_{h+1} - \frac{1}{2} y_h = 1 \qquad ...(1)$$

The corresponding homogeneous equation is

$$y_{h+1} - \frac{1}{2} y_h = 0.$$

On comparing it with standard equation

$$y_{h+1} = Ay_h + C \text{ we have, } A = \frac{1}{2}, C = 0$$

∴ The solution of homo. equation is given by

$$y_h = \lambda A^h + \frac{C(1 - A^h)}{1 - A} \Rightarrow y_h = \lambda \left(\frac{1}{2}\right)^h.$$

Particular Solution: We shall apply **Hit and Trial Method.** We first try see whether the constant function is a solution or not

Let $y_h^* = A.$

Substituting in (1), we have

$$A - \frac{1}{2} A = 1 \qquad \therefore A = 2$$

∴ The general solution of (1) is given by

$$y_h + \lambda\left(\frac{1}{2}\right)^h + 2$$ **Ans.**

Example 3:

Solve completely the following diff. equation

$$y_{h\ 1} + 5y_h = 2h.$$ **(Meerut, M.Sc. 92)**

Solution:

The corresponding homo. equation is

$$y_{h+1} + 5y_h = 0$$

Comparing with the standard equation -

$$y_{h+1} = Ay_h + C, \quad \text{we have } A = -5, C = 0$$

$\therefore$ The solution is given by

$$y_h = \lambda A^h + \frac{C(1-A)^h}{1-A} = \lambda(-5)^h$$

Particular Solution: To find particular solution we apply **Hit and Trial Method.** We consider the trial function $y_h^* = A$.

Substituting in the given equation, we have

$$A\ 5A = 2h$$

$$\Rightarrow \quad 6A = 2h$$

This is not true, since R.H.S. is not constant.

$\therefore$ We now try the function $y_h^* = Ah + B$.

Substituting in the given equation, we have

$$A(h+1) + B + 5(Ah + B) = 2h.$$

$$\Rightarrow \quad 6Ah + A + 6B = 2h.$$

Comparing the coefficients of like terms, we have

$$6A = 2 \text{ and } A + 6B = 0$$

$$\therefore \quad A = \frac{1}{3} \quad \text{and} \quad B = -\frac{1}{18}.$$

$$\therefore \quad y_h^* = \frac{1}{3}h - \frac{1}{18}.$$

$\therefore$ The general solution of the given equation is given by

$$y_h = \lambda(-5)^h + \frac{1}{3}h - \frac{1}{18}.$$ **Ans.**

Example 4:

Solve $y_{h\ 1} - 2y_h = 5.$ ...(1)

(Meeurt, M.Sc. 76, 79, 83)

Solution:

The corresponding homo. equation is

$$y_{h+1} - 2y_h = 0$$

$$\Rightarrow \quad y_{h+1} = 2y_h \quad ...(2)$$

On comparing it with standard equation

$y_{h+1} + Ay_h + C$ we have $A = 2,\ C = 0$

The solution is given by

$$y_h = \lambda A^h + \frac{C(1 - A^h)}{1 - A} = \lambda.2^h \quad ...(3)$$

To find particular solution of (1). We shall use **Hit and Trial Method** for the moment. We first try to see whether the constant function is a solution or not.

Let y_h A be solution of (1) then it must satisfy (1).

$$\Rightarrow \quad A - 2A = 5 \qquad \therefore \ A = -5$$

Hence, $y_n{}^* = = 5$ is a particular solution of equation (1).

Hence, the general solution of equation (1) is given by

$$y_h = \lambda.2^h - 5. \quad ...(4)$$

Remark:

From (4) it is always possible to find any solution satisfying given condition. Suppose we are interested in a solution for which

$$y_0 = 4 \quad ...(5)$$

Then (4) gives $\quad 4 = \lambda.2^0 - 5 \quad \Rightarrow \quad \lambda = 9$

$$\therefore \quad y_h = 9.2^h - 5. \qquad \textbf{Ans.}$$

This satisfies both the differential equation (1) and the initial condition (5).

Example 5:

Solve completely the following differential equation

$2y_{h+1} - 5y_h = 3h + 1$ **(Meerut, B.Sc. Hons. 96)**

Solution:

The given diffcrential equation can be written as

$$y_{h+1} - \frac{5}{8}y_h = \frac{1}{2}(3h+1) \qquad ...(1)$$

The corresponding homogeneous equation is

$$2y_{h+1} - 5y_h = 0$$

$$\Rightarrow \qquad y_{h+1} - \frac{5}{2}y_h = 0.$$

On comparing it with standard equation

$$y_{h+1} = Ay_h + C, \text{ we have } A = 5/2,\ C = 0.$$

The solution is given by

$$y_h = \lambda A^h = \frac{C(1 - A^{\lambda})}{1 - A} = \lambda.(5/2)^{\lambda}$$

Particular Solution: We shall apply **Hit and Trial Method.** We first try to see whether the constant function is a solution or not.

Let $y_h^* = A$ be a solution of (1).

Then $A - \frac{5}{2}A = \frac{1}{2}(3h+1) \quad \Rightarrow \quad \frac{3}{2}A = \frac{1}{2}(3h+1)$

This is not true since R.H.S. is not constant.

We now try $\quad y_h^* = Ah + B.$

$\therefore$ from (1), We have

$$A(h+1) + B \frac{5}{2}(Ah + B) = \frac{1}{2}(3h+1)$$

$$2Ah + 2A + 2B - 5Ah - 5B = 3h + 1$$

$$-3Ah + 2A - 3B = 3h + 1.$$

On comparing the coefficients we have

$$-3A = 3 \qquad ...(2)$$

$$2A - 3B = 1 \qquad ...(3)$$

from (2) $A = -1$; putting in (3)

$$-2 - 3B = 1 \quad \therefore \quad B = -1$$

$$\therefore \quad y_h^* = -h - 1.$$

The general solution of (1) is given by

$$y_h = \lambda \left(\frac{5}{2}\right)^h - (h+1). \qquad \textbf{Ans.}$$

Example 6:

Solve completely the following differential equation

$$y_{h+1} - y_h = 1. \qquad ...(1)$$

Solution:

The corresponding homogeneous equation is

$y_{h+1} - y_h = 0.$

On comparing it with standard equation

$y_{h+1} = Ay_h + C,$ we have $A = 1, C = 0$

The solution is given by

$$y_h = \lambda A^h + \frac{C(1 - A^h)}{1 - A)} = \lambda.(1)^h = \lambda.$$

Particular Solution: We shall apply **Hit and Trial Method.** We first try to see whether constant function is a solution or not.

Let $y_h^* = A$ be a solution of (1), then we have

$$A - A = 1 \quad \Rightarrow \cdot 0 = -1$$

This is not true since $0 \neq -1$.

We now try $y_h^* = Ah + B$ so that

$$y^*_{h+1} = A(h + 1) + B$$

$\therefore$ from (1) we have $A(h + 1) + B - (Ah + B) = 1$

$$Ah + A + B - Ah - B = 1 \quad \Rightarrow \quad A = 1$$

Hence, $y_h^* = h$ is a particular solution of (1).

The general solution of (1) is given by

$$y_h = \lambda + h.$$ **Ans.**

13.10 General Solution of the Homogeneous Equation of Order Two

The Homogeneous Equation of order 2, is given by

$$y_{h+2} + b_1 y_{h+1} + b_2 y_h = 0 \qquad ...(1)$$

where $b^2 \neq 0$

We suppose that the solution of equation (1) is of the type

$$y_h = m^h \qquad ...(2)$$

where m is a suitable chooses constant different from zero (we disallow $m = 0$ because it makes y identically zero, a solution which cannot be a fundamental set).

On substituting equations (2) in (1) we have

$$m^{h+2} + b_1 m^{h+1} + b_2 m^h = 0$$

On dividing each term by m^h we have

$$m^2 + b_1 m + b_2 = 0. \qquad ...(3)$$

The quadratic equation is called the auxiliary equation or characteristic equation of the differential equation (1). Therefore if m is a root of (3), then (2) will be a solution of the differential equation (1).

The auxiliary equation is a quadratic algebraic equation and therefore has two non-zero roots m_1 and m_2.

Remark:

Zero roots are excluded since $b_2 \neq 0$, a requirement which is needed for (1) to be of second order.

To these roots there corresponds the solution $y^{(1)}$ and $y^{(2)}$ given by

$$y_h^{(1)} = m_1^{\,h} \quad \text{and} \quad y_h^{(2)} = m_2^{\,h}. \qquad ...(4)$$

Now there are three cases.

(a) The roots m_1 and m_2 are real numbers and unequal.

(b) The two roots are real and equal, and

(c) The roots are complex numbers.

Since (3) is a quadratic equation, it is not possible to have one real and one complex root.

We now wish to study the above cases.

Case 1. Roots Real and Unequal. (i.e. $m_2 \neq m_1$).

In this case, the solution in (4) forms a fundamental set. To prove this we calculate the determinant.

$$\Delta = \begin{vmatrix} y_0^{(1)} & y_0^{(2)} \\ y_1^{(1)} & y_1^{(2)} \end{vmatrix} = \begin{vmatrix} 1 & 1 \\ m_1 & m_2 \end{vmatrix} = m_2 - m_1$$

which is different from zero, since $m_2 \neq m_1$.

Hence when the two roots m_1 and m_2 of A.E are real and unequal, the general solution of the differential equation (1) is given by

$$Y_h = C_1 m_1^{\,h} + C_2 m_1^{\,h}.$$

Case 2. Roots Real and Equal. (i.e. $m_2 = m_1$).

Now, the solutions $y^{(1)}$ and $y^{(2)}$ no longer form a fundamental set since the determinant Δ is zero.

We keep $y^{(1)}$, but find a function $y^{(2)}$ which together with $y^{(1)}$ from a fundamental set.

Now, our claim is that such a solution is given by

$$y_h^{(2)} = hm_1^h. \quad ...(5)$$

Proof:

Since m_1 and m_2 are the roots of (3), we have

$$m_1^2 + b_1m_1 + b_2 = 0 \quad ...(6)$$

and sum of roots $= m_1 + m_2 = -b_1$

$$\Rightarrow \quad 2m_1 + b_1 = 0 \quad (\because \ m_2 = m_1). \quad ...(7)$$

Now, to show that $y_h^{(1)} = hm_1^h$ is a solution of (1) we put it in (1) to yield

$$y^{(2)}_{h+2} + b_1y^{(2)}_{h+1} + b_2y_h^{(2)} = (h + 2)\, m_1^{h+2} + b_1\,(h + 1)\, m_1^{h+1}\, b_2.hm_1^h$$
$$= hm_1^h[m_1^2 + b_1m_1 + b_2] + m_1^{h+1}\,[2m_1 + b_1]$$
$$= 0.$$

In view of (6) and (7).

Hence, equation (5) indeed give a solution of (1).

To show that $y_h^{(1)} = m_1^h$, $y_h^{(2)} = h,\, m_1^h$

form a fundamental set

We have

$$\Delta = \begin{vmatrix} y_0^{(1)} & y_0^{(2)} \\ y_1^{(1)} & y_1^{(2)} \end{vmatrix} = \begin{vmatrix} 1 & 0 \\ m_1 & m_2 \end{vmatrix} = m_1$$

which is different from zero since, as we have already noted, no root of the auxiliary equation is zero.

Thus when the two roots m_1 and m_2 of A.E are equal, the general solution of (1) is given by

$$Y_h = C_1m_1^h + C_2hm_1^h$$

$$\Rightarrow \quad Y_h = m_1^h\,(C_1 + hC_2).$$

Case 3. Complex Roots. We know that complex roots of a quadratic equation occurring conjugate paris. Therefore if m_1 and m_2 are complex roots of auxiliary equation, then $m_1 \neq m_2$ and it can be checked that the two solutions

$$y_h^{(1)} = m_1^h \text{ and } y_h^{(2)} = m_2^h$$

form a fundamental set. The general solution is given by

$$y_h + C_1m_1^h \text{ and } y_h^{(2)} = m_2^h. \quad ...(8)$$

The only difficulty with this solution (8) is that it may be a complex number if m_1 and m_2 are themselves complex. But we require solutions which have real values for all h values for which they are defined. It is possible to show that if C_1 and C_2 are complex conjugates. Y_h is always a real number.

To prove this we write all complex numbers in polar form.

Let us write

$$\left.\begin{aligned} m_1 &= \rho\,(\cos\theta + i\sin\theta) \\ \text{and} \qquad m_2 &= \rho\,(\cos\theta + i\sin\theta) \end{aligned}\right\}. \qquad ...(9)$$

Further we are concerning that C_1 and C_2 are complex conjugates.

$$\left.\begin{aligned} \text{Hence } C_1 &= a\,(\cos B + i\sin B) \\ \text{and} \qquad C_2 &= a\,(\cos B + i\sin B) \end{aligned}\right\} \qquad ...(10)$$

$\therefore \qquad m_1{}^h = \rho^h\,(\cos h\theta + i\sin h\theta)$

and $\qquad C_1m_1{}^h = a\rho^h\,[\cos(h\theta + B) + i\sin(h\theta + B)]. \qquad ...(11)$

In the same way

$$C_2m_2{}^h = a\rho^h[\cos.\,(h\theta + B) - i\sin(h\theta + B)] \qquad ...(12)$$

$\therefore \quad Y_h = 2a\rho^h\{\cos(h\theta + B)].$

If we write $2b = A$.

Then $Y_h = A\rho^h\,[\cos(h\theta + B)] \qquad ...(13)$

and r is determined from (9).

Thus when the roots m_1, m_2 of A.E are complex conjugates with the polar form $\rho(\cos\theta \pm i\sin\theta)$, the general solution of (1) is given by

$$Y_h\ A\rho^h\cos(h\theta + B)$$

$$\Rightarrow \qquad \rho^h(A\cos h\theta + B\sin h\theta).$$

We summarize the above results as follows:

Let the homogeneous differential equation of order 2 be given by

$$y_{h+2} + b_1\,y_{h+1} + b_2\,y_h = 0 \qquad ...(1)$$

where b_1, b_2 are constants and $b_2 \neq 0$. If m_1 and m_2 are the roots of the auxiliary equation

$$m^2 + b_1m + b_2 = 0, \text{ than} \qquad ...(2)$$

Then general solution of the differential equation (1) is given (i) by

$$y_h = C_1m_1{}^h + C_2m_2{}^h.$$

If m_1 and m_2 are real and unequal,

(ii) by $\quad y_h = (C_1 + C_2h)\,m_1{}^h \quad$ if $\quad m_2 = m_1$

and (iii) by $\quad y_h = C_1r^h\cos(h\theta + C_2)$

$$\Rightarrow \qquad p^h\,(C_1\cos h\theta + C_2\sin h\theta).$$

If m_1 and m_2 are complex conjugates with polar form

p(cos θ ± i sin θ).

13.11 General Solution of the Homogeneous Differential Equation of Order n.

We generalize the results and summarize as follows:

If the nth order homogeneous differential equation is given by

$$y_{h+n} + b_1 y_{h+n-1} + b_2 y_{h+n-2} + \dots + b_n y_h = 0,\ b_n \neq 0 \qquad \dots(i)$$

then substituting $y_h = m^h$.

A.E. is $m^n + b_1 m^{n-1} + b_2 m^{n-2} + \dots + b_n = 0 \qquad \dots(ii)$

Solve the A.E. (i) and let $m_1, m_2, \dots, m_n$ be its n roots. The general solution of the homogeneous differential equation (i) depend on the nature of roots, $m_1, m_2, \dots, m_n$. The following cases arise.

(i) **Roots are real and distinct.**

In this case the general solution is given by

$$y_h = C_1 m_1^h + C_2 m_2^h + \dots + C_n m_n^h.$$

(ii) **Some of the roots are equal.**

If the root m_1 of the A.E. is repeated k times i.e. $m_1 = m_2 = \dots m_k$, then the term in general solution corresponding this root is $(C_1 + C_2 h + C_3 h^2 + \dots + C_k h^{k-1}),\ m_1^h$.

(iii) **Some of the roots are complex numbers.**

If the roots m_1 and m_2 of A.E. are complex conjugate with the polar form p (cos θ ± i sin θ), then the term is general solution corresponding to these roots is $C_1 p^h \cos(h\theta + C_2)$

$$\Rightarrow \qquad p^h (C_1 \cos h\theta + C_2 \sin h\theta)$$

Example 1:

Solve $3y_{h-2} - 6y_{h-1} + 4y_h = 0.$

(Meerut, M.Sc. 85; B.Sc. Hons. 89)

Solution:

Substituting $y_h = m^h$, A.E. is given by

$$3m^2 - 6m + 4 = 0$$

$$\therefore \qquad m = \frac{6 \pm \sqrt{(36-48)}}{6} = 1 \pm \frac{1}{\sqrt{3}}t$$

In polar form,

Let $\qquad m = 1 \pm \frac{i}{\sqrt{3}} = \rho(\cos\theta \pm i \sin\theta)$

On equating real and imaginary parts.

$$1 = \rho \cos\theta \text{ and } \frac{1}{\sqrt{3}} = \rho \sin\theta$$

solving, we have $\rho = \frac{2}{\sqrt{3}}$ and $\theta = \frac{\pi}{6}$.

$$m = 1 \pm \frac{t}{\sqrt{3}} = \frac{2}{\sqrt{3}}\left(\cos\frac{\pi}{6} \pm t \sin\frac{\pi}{6}\right)$$

The solution is given by

$$y_h = A\left(\frac{2}{\sqrt{3}}\right)^h \cos\left(\frac{h\pi}{6} + B\right).$$ **Ans.**

Example 2:

Solve $9y_{h-2} - 6y_{h-1} + y_h = 0.$

Also find the particular solution satisfying the initial conditions $y_0 = 0$ *and* $y_1 = 1.$

Solution:

Substituting $y_h = m^h$, A.E. is

$$9m^2 - 6m + 1 = 0$$

$\Rightarrow$ $$(3m - 1)^2 = 0$$ $\therefore \quad m = \frac{1}{2}, \frac{1}{3}$

$\therefore$ the general solution is given by

$$y_h = (C_1 + C_2 h) . (\frac{1}{3})^h.$$ **Ans.**

Putting h = 0 and h = 1 in the above solution, we have

$$y_0 = C_1 = 0$$

and $$y_1 = (C_1 + C_2) . \frac{1}{3} = 1$$

solving, $C_1 = 0$ and $C_2 = 3$.

$\therefore$ Particular solution satisfying the given initial conditions is

$$y_h = 3h . (\frac{1}{3})^h.$$ **Ans.**

Example 3:

Solve $y_{h-2} + y_h = 0$ *satisfying* $y_n = 0, y_1 = 1.$

Solution:

Substituting $y_h = m^h$, A.E. is

$$m^2 + 1 = 0$$

$$\Rightarrow \quad m = \pm i = \cos\left(\frac{\pi}{2} \pm i \sin \frac{\pi}{2}\right)$$

$$\therefore \quad \theta = \frac{\pi}{2}, \text{ and } \rho = 1.$$

The solution is given by

$$y_h = A \cos\left[h \frac{\pi}{2} + B\right]. \qquad \ldots(i)$$

We are interested in a solution for which

$$y_0 = 0, \; y_1 = 1.$$

Putting h = 0 and h = 1 in (i), we have

$$y_0 = A \cos B = 0$$

and
$$y_1 = A \cos\left(\frac{\pi}{2} + B\right) - 1.$$

Solving A = – 1 and B = π/2.

Hence the solution is given

$$y_h = -\cos \frac{\pi}{2}(h + 1) \quad \text{or} \quad y_h = \sin\left(\frac{h\pi}{2}\right).$$ **Ans.**

Example 4:

Solve $Y_{h+2} - 2Y_{h+1} + 2Y_h = 0.$ **(Meerut M.Sc. 85, 90)**

Solution:

Substituting $Y_h = m^h$, A.E. is

$$m^2 - 2m + 2 = 0$$

$$\therefore \quad m = \frac{2 \pm \sqrt{(4-8)}}{2} = 1 \pm 1.$$

Let $m = 1 \pm 1/\sqrt{3} = \rho(\cos\theta \pm i \sin\theta)$

Let $m = 1 \pm i = \rho(\cos\theta \pm i \sin\theta)$

On equating real and imaginary parts we have

$$1 = \rho \cos\theta \qquad \ldots(i)$$

$$1 = \rho \sin\theta. \qquad \ldots(ii)$$

Solving (i) and (ii), we have

$$\rho = \sqrt{2} \text{ and } \rho = \frac{\pi}{4}.$$

∴ The solution is given by

$$Y_h = A\,(\sqrt{2})^h \cos\left(\frac{h\pi}{4} + B\right).$$ **Ans.**

Example 5:

Solve $y_{h-3} + y_{h-2} - 8y_{h-1} - 12y_h = 0.$ **(Meerut B.Sc. Hons., 98)**

Solution:

Substituting $y_h = m^h$, A.E. is

$$m^3 + m^2 - 8m - 12 = 0$$

$\Rightarrow$ $(m-3)(m+2)^2 = 0 \quad \therefore \quad m = 3, -2, -2$

∴ The general solution is given by

$$y_h = C_1 \,.\, 3^h + (C_2 + C_3h) \,.\, (-2)^h.$$ **Ans.**

Example 6:

Solve $y_{h+4} - 4y_{h+2} + 6y_{h+2} - 4y_{h+1} + y_h = 0.$

(Meerut, B.Sc. Hons. 95)

Solution:

Substituting, $y_k = m^h$, A.E. is

$$m^4 - 4m^3 + 6m^2 - 4m + 1 = 0$$

$\Rightarrow$ $(m-1)(m^4 - 3m^2 + 3m - 1) = 0$

$\Rightarrow$ $(m-1)(m-1)(m^2 - 2m + 1) = 0$

$\Rightarrow$ $(m-1)^4 = 0 \quad \therefore \quad m = 1, 1, 1.$

The general solution is given by

$$y_0 = (C_1 + C_2h + C_3h^2 + C_4h^3) \,.\, 1^h$$
$$= C_1 + C_2h + C_3h^2 + C_3h^2.$$ **Ans.**

Example 7:

Solve $2h_{h+2} - 5y_{h+1} + 2y_h = 0.$

Also find the particular solution satisfying the initial conditions $y_o = 0$ *and* $y_1 = 1.$ **(Meerut M.Sc. 95)**

Solution:

Substituting $y_h = m^h$, A.E. is

$$2m^2 - 5m + 2 = 0$$

$\Rightarrow$ $(2m - 1)(m - 2) = 0 \quad \therefore \quad m = \frac{1}{2}, 2$

$\therefore$ the general solution is given by

$$y_h = C_1 \cdot (\frac{1}{2})^h + C_2 \cdot 2^h.$$ **Ans.**

Putting h = 0 and h = 1 in the above solution we have

$$y_0 = C_1 + C_2 = 0$$

$$y_1 = \frac{1}{2}C_1 + 2C_2 = 1.$$

Solving, we have $C_1 = -\frac{2}{3}$ and $C_2 = \frac{2}{3}$

$\therefore$ Particular solution satisfying the given initial conditions is

$$y_h = y_h = -\frac{2}{3}\left(\frac{1}{2}\right)^h + \frac{2}{3}.2^h.$$ **Ans.**

Example 8:

Solve $y_{h-2} + 6y_{h-1} + 25y_h = 0.$ **(Meerut, B.Sc. Hons. 97)**

Solution:

We have $y_{h+2} + 6y_{h+1} + 25y_h = 0$

Substituting $y_h = m^h$

A.E. is $m^2 + 6m + 25 = 0$

$$\therefore \quad m = \frac{-6 \pm \sqrt{(36 - 100)}}{2} = - \pm 4t.$$

In polar form.

Let $m = -3 \pm 4t = \rho(\cos\theta \pm t \sin\theta)$

$\therefore$ $\rho \cos\theta = -3$ ...(i)

$\rho \sin\theta = 4$ ...(ii)

Solving, $\rho = 5$ and $\theta = \tan^{-1}\left(-\frac{4}{3}\right)$

The solution is given by

$$y_h = \lambda(5)^h \cos(h\theta + D) \text{ where } \theta = \tan^{-1}\left(-\frac{4}{3}\right).$$ **Ans.**

13.12 Particular Solution of the Complete Differential Equation

We now turn to methods for finding the solution of the complete equation

$$y_{h+2} + b_1 y_{h+1} + b_2 h_h = R\ (h). \qquad ...(1)$$

We can find the general solution of the corresponding homogeneous equation. So if we are able to find the particular solution of (1) then we shall know the solution of (1) by adding the two solutions.

There are a number of special techniques for finding the particular solution of (1). Hence, we shall discuss the following two important methods for obtaining such solutions which will be found useful.

1. *Method of undetermined coefficients, and*
2. *Special operator methods.*

13.13 Method of Undetermined Coefficients to Find the Particular Solution

The method of undetermined coefficients is useful in finding the particular solutions of the complete equation when the R.H.S. R(h) consists of terms having certain special forms. Corresponding to each such term which is present in R(h) we consider a trial solution containing a number of unknown constant coefficients which are to be determined by substitution in the diff. equation. The trial solutions to be used in each case are shown in the following table where A, B, A_0, A_1 ... represent the unknown constant coefficients to be determined.

Table

R(h)	Trial function y_h*
b^h	Ab^h
Sin bh or cos bh	A sin bh + B cos bh
Polynomial P(h) of degree n	$A_0 + A_1 h + A_2 h^2 + ... + A_n h^n$
$a^h.P(h)$	$a^h (A_0 + A_1 h + ... + A_n h^n)$
a^h sin bh	a^h (A sin bh + B cos bh)
a^h cos bh	a^h (A sin bh + B cos bh).

Remark:

If R(h) is a linear combination of these functions then the trial solution to taken as the sum of corresponding trial functions with different unknown constant coefficients to be determined.

13.14 Special Operator Methods to Find the Particular Solution

The linear diff. equation of order n is given by

$$y_{h+n} + b_1 y_{h+n-1} + \ldots + b_n y_n = R(h),\ b_n \neq 0$$

which can be written as

$$(E^n + b_1 E^{n-1} + \ldots + b_n)\, y_n = R(h)$$

$$\Rightarrow \qquad \phi(E) \, . \, y_n = R(h)$$

Then the particular solution is given by $\frac{1}{\phi(E)} \, . \, R(h)$.

This method is very useful and easy in case R(h) takes on special forms as follows:

1. **If R (h) = b^h.**

   Then particular solution $= \frac{1}{\phi(E)} \, . \, b^h = \frac{b^h}{\phi(b)}$

   provided f (b) ≠ 0.

2. **If R (h) = sin bh or cos bh.**

   Then particular solution $= \frac{1}{\phi(E)}$ sin bh or $\frac{1}{\phi(E)}$ cos bh

   Write $\sin bh = \frac{e^{ibh} - e^{-ibh}}{2}$, $\cos bh = \frac{e^{ibh} - e^{-ibh}}{2}$

   and use method 1.

3. **If R (b) = P (h), Polynomial of degree n.**

   The particular solution $= \frac{1}{\phi(E)} P(h) = \frac{1}{\phi(1 + \Delta)} . P(h)$

   $$= (a_n + a_1\Delta + \ldots + a_m\Delta^m + \ldots)\, P(h)$$

   where the expansion need be carried out only as far as Δ^m.

   Since $\Delta^n P(h) = 0$ if $x \geq m + 1$.

4. **If R (h) = b^h . P (h).**

   Then particular solution $= \frac{1}{\phi(E)} \, . \, b^h P(h) = b^h \, . \, \frac{1}{\phi(E)} P(h)$

   Then use method 3.

Example 1:

Solve $y_{h+2} - 3y_{h+1} + 2y_h = b^h$ ...(1)

where b is some constant.

Solution:

Substituting $y_h = 3^h$, in the reduced homo. equation the A.E. is

$$m^2 - 3m + 2 = 0$$

$\Rightarrow$ $(m - 1)(m - 2) = 0 \quad \therefore \quad m = 1, 2$

$\therefore$ the general sol. of reduced homo. equation is $y_h = C_1 + C_2 \cdot 2^h$.

To find the particular solo. of the complete equation let the trial function be

$$y_h^* = A \cdot b^h. \qquad ...(2)$$

$\therefore$ Substituting in the given equation, we have

$$A(b^{h+2} - 3b^{h+1} + 2b^h) = b^h$$

$\Rightarrow$ $A(b^2 - 3b + 2) = 1$

$\Rightarrow$ $A(b - 2)(b - 1) = 1$ ...(3)

The equation (3) given A, provided b is not equal to 1 or 2 i.e. provided that b is not a root of the auxiliary equation. Thus, if b is not equal to 1 or 2.

$$y_h^* = \frac{1}{(b-1)(b-2)} \cdot b^h. \qquad ...(4)$$

Yields a particular solution of the given equation (1).

From this it follows that a separate treatment is needed when b = 1 or b = 2.

In these circumstances, we modify the trial function by multiplying by h and then to determine A so that the new function

$$y_h^* = Ahb^h \qquad ...(5)$$

becomes a solution of the given equation (1). Let us first do it for b = 2. Then (1) and (5) are

$$y_{h+2} - 3y_{h+1} + 2y_h = 2^h$$

$$y_h^* = Ah \cdot 2^h$$

$$A(h + 2)2^{h+2} - 3A \cdot (h + 1)2^{h+1} + 2Ah \cdot 2^h = 2^h$$

$\Rightarrow$ $A[4h + 8 - 6h - 6 + 2h] = 1 \quad \therefore \quad A = 1/2$

Hence, $y_h^* = \frac{1}{2} h \cdot 2^h$.

Hence, the general solution if b is not equal to 1 or 2 is

$$y_h = C_1 + C_2.2^h + \frac{1}{(b-1)(b-2)} b^h$$

and general sol. if b = 2 is

$$y_h = C_1 + C_2 \cdot 2^h + \frac{1}{2} h \cdot 2^h$$

$$= C_1 + C_2 \cdot 2^h + h \cdot 2^{h-1}$$

Similarly, the general sol. when b = 1 is

$$y_h = C_1 + C_2 - h.$$ **Ans.**

Example 2:

Solve $y_{h+2} - 4y_{h+1} + 4y_h = 3h + 2h$

(Meerut, M.Sc. 89)

Solution:

Substituting $y_h = m^h$ in the reduced homo. equation, the A.E. is

$$m^2 - 4m + 4 = 0 \quad \text{or} \quad (m-2)^2 = 0 \quad \therefore \quad m = 2, 2.$$

The general solution of the reduced equation is

$$y_h = (C_1 + C_2 h) 2^h \qquad ...(1)$$

To find the **particular solution** of the given equation let the trial solution be

$$y_h^* = A_0 + A_1 h + Bh^2 \cdot 2^h.$$

The question is why $Bh^2 \cdot 2^h$ has been chosen corresponding to 2^h and not $Bh \cdot 2^h$ or $B \cdot 2^h$. The reason is obvious. Since 2^h is a solution of the reduced equation, we do not take $B \cdot 2^h$ as our trial function. Therefore we multiply be h and try $Bh. 2^h$, but since this too is a solution of the reduced equation we must again multiply by h and so arrive at the trial solution $Bh^2 \cdot 2^h$.

Now, put y_h^* in (1), we have after simplifications

$$(A_0 - 2A_1) + A_1 h + 8B \cdot 2^h = 2^h + 3h.$$

On comparing the coefficients on both sides, we get

$$8B = 1 \quad \text{or} \quad B = \frac{1}{8}, A_1 = 3, A_0 - 2A_1 = 0 \quad \therefore \quad A_0 = 6$$

Hence, the general solution of the complete equation is given by

$$y_h = (C_1 + C_2 h) 2^h + 6 + 3h + \frac{1}{8} h^2 \cdot 2^h.$$ **Ans.**

Example 3:

Solve $y_{h+2} - 5y_{h+2} + 8y_{h+1} - 4y_h = h \cdot 2^h$.

(Meerut, M.Sc. 98, B.Sc. Hons. 96)

Solution:

Substituting $y_h = m^h$, in the reduced homo. equation, the A.E. is

$$m^3 - 5m^2 + 8m - 4 = 0$$

$$\Rightarrow \qquad (m-1)(m-2)^2 = 0 \quad \therefore \quad m = 1, 2, 2.$$

$\therefore$ the general solution of the reduced homo. equation is

$$y_h = C_1 + (C_2 + C_3h) \cdot 2^h.$$

To find the particular solution of the given equation, let the trial solution be

$$y_h^* = 2^h \cdot (A_0 + A_1h) \cdot h^2 = 2^h (A_0 + A_1h^2)$$

[we cannot take the trial solution $2^k (A_0 + A_1h)$ or $2^h (A_0 + A_1h)h$].

Substituting in the given equation, we have

$$2^{h+3} [A_0 (h+3)^2 + A_1 (h+3)^3] - 5 \cdot 2^{h+2} [A_0 (h+2)^2 + A_1 (h+2)^2]$$
$$+ 8 \cdot 2^{h+1} [A_0 (h+1)^2 + A_1 (h+1)^3]$$
$$- 4 \cdot 2^h (A_0h^2 + A_1h^2) = h \cdot 2^h$$

$$\Rightarrow \qquad 2^h [(8A_n + 72A_1) + 24 A_1h] = h \cdot 2^h$$

$\therefore 8A_0 + 72A_1 = 0$ and $24 A_1 = 1$

solving, $A_0 = -3/8$, $A_1 = 1/24$

$\therefore$ the particular solution is $y_h^* = 2^h \left(-\frac{3}{8}h^2 + \frac{1}{24}h^3\right)$.

Hence, the solution of the complete equation given by

$$y_n = C_1 + (C_2 + C_3h) \cdot 2^h + 2^h \left(-\frac{3}{8}h^2 + \frac{1}{24}h^3\right). \qquad \textbf{Ans.}$$

Example 4:

Solve $y_{h+2} + y_h = \sin \dfrac{h\pi}{2}$.

(Meerut, M.Sc. 90, 92, 93(P)

Solution:

Substituting $y_h = m^h$ in the reduced homo. equation, the A.E. is

$$m^7 + 1 = 0 \quad \text{or} \quad m = \pm i$$

$$\Rightarrow \qquad m = \cos \frac{\pi}{2} \pm i \sin \frac{\pi}{2}.$$

The general solution of the reduced equation is

$$y_h = A\cos\left(\frac{h\pi}{2} + B\right)$$

From the table for the particular solution, we have

$$y_h^* = A\sin\frac{h\pi}{2} + B\cos\frac{h\pi}{2}.$$

Substituting in the given equation, we have

$$A\sin(h+2)\frac{\pi}{2} + B\cos(h+2)\frac{\pi}{2} + A\sin\frac{h\pi}{2} + B\cos\frac{h\pi}{2} = \sin\frac{h\pi}{2}$$

$$\Rightarrow \quad -A\sin\frac{h\pi}{2} - B\cos\frac{h\pi}{2} + A\sin\frac{h\pi}{2} + B\cos\frac{h\pi}{2} = \sin h\frac{\pi}{2}$$

$$\Rightarrow \quad 0 = \sin h\frac{\pi}{2}$$

$$\Rightarrow \quad y_h^* = A\sin\frac{h\pi}{2} + B\cos\frac{h\pi}{2} \text{ fails.}$$

Now, we suppose

$$y_h^* = A\,h\sin\frac{h\pi}{2} + Bh\cos\frac{h\pi}{2}$$

Substituting in the given equation, we have

$$A(h+2)\sin(h+2)\frac{\pi}{2} + B(h+2)\cos(h+2)\frac{\pi}{2} + Ah\sin\frac{h\pi}{2} + Bh\cos\frac{h\pi}{2} = \sin\frac{h\pi}{2}$$

$$-A(h+2)\sin\frac{h\pi}{2} - B(h+2)\cos\frac{h\pi}{2} + Ah\sin\frac{h\pi}{2} + Bh\cos\frac{h\pi}{2} = \sin\frac{h\pi}{2}$$

$$\Rightarrow \quad \left(\sin\frac{h\pi}{2}\right)(-2A) + \left(\cos\frac{h\pi}{2}\right)(-2B) = \sin\frac{h\pi}{2}.$$

On comparing the coefficients of $\sin h\frac{\pi}{2}$ and $\cos h\frac{\pi}{2}$, we have

$$-2A = 1 \quad \text{and} \quad -2B = 0$$

$$\therefore \qquad A = -\frac{1}{2} \quad \text{and} \qquad B = 0$$

Hence, the general solution of the complete equation is given by

$$y_h = A \cos\left(\frac{h\pi}{2} + B\right) - \frac{1}{2} h \sin \frac{h\pi}{2}. \qquad \textbf{Ans.}$$

(the reader can verify that the trial solutions $A_5 \sin 2h\pi + A_6 \cos 2h\pi$ and $hA_5 \sin 2h\pi + hA_6 \cos 2h\pi$ fails).

Substituting the trial solution in the L.H.S., we have

$$y_{h+2} - 2y_{h+1} + y_h = (h+2)^2 [A_5 \sin 2(h+2)\pi + A_6 \cos 2(h+2)\pi]$$
$$- 2 (h+1)^2 [A_6 \sin 2 (h+1) \pi + A_6 \cos 2 (h+1) \pi]$$
$$+ h^2 [A_6 \sin 2h\pi + A_6 \cos 2h\pi]$$
$$= 2A_5 \sin 2h\pi + A_6 \cos 2h\pi.$$

Comparing with $\sin 2h\pi$, we have

$$2A_5 = 1 \qquad \text{and} \qquad 2A_6 = 0$$

$$\therefore \qquad A_5 = \frac{1}{2} \qquad \text{and} \qquad A_6 = 0.$$

$\therefore$ the particular solution corresponding to the term $\sin 2h\pi$ on R.H.S. of the diff. equation is $\frac{1}{2} \sin 2h\pi$.

Hence, the particular solution of the given differential equation is given by

$$y_h = 2^h (-5 + h) + \left(\frac{17}{6} h^2 + \frac{1}{6} h^3 + \frac{1}{12} h^4\right) + \frac{1}{2} \sin 2h\pi. \qquad \textbf{Ans.}$$

Example 5:

Solve the following equation

$$y_{h+2} - 3y_{h+1} + 2y_h = 1. \qquad \textbf{(Meerut, B.Sc. 96)}$$

Solution:

Substituting $y = m^h$ in the reduced homo. equation, the A.E. is

$$m^2 - 3m + 2 = 0 \quad \text{or} \quad (m-1)(m-2) = 0$$

$$\therefore \qquad m = 1, 2.$$

The general solution of the reduced homo. equation is

$$y_h = C_1 + C_2 \cdot 2^h$$

To find the **Particular Solution** of the complete equation let the trial function be

$$y_h^* = A.$$

Substituting in the given equation, we have

$$A - 3A + 2A = 1$$

which is not possible.

$\therefore$ Let the particular solution be $y_h^* = Ah$.

(Since A is the solution of reduced equation)

$$\therefore \quad y_h^* \neq A.$$

Substituting $y_h^* = Ah$ in the given equation, we have

$$A(h+2) - 3A(h+1) + 2Ah = 1$$

$$\Rightarrow \quad A = -1$$

$$\therefore \quad y_h^* = Ah = -h.$$

Hence, the general solution of the complete equation is given by

$$y_h = C_1 + C_2 \cdot 2^h - h.$$ **Ans.**

Example 6:

Solve the following

$y_{h+2} - y_{h+1} - 2y_h = h^2.$ **(Meerut, M.Sc. 93 P, 94, 90)**

Solution:

Substituting $y_h = m^h$ in the reduced homo. equation A.E. is

$$m^2 - m - 2 = 0$$

$$\Rightarrow \quad (m+1)(m-2) = 0 \quad \therefore \quad m = -1, 2$$

The general solution of the reduced homo. equation be

$$y_h = C_1 (-1)^h + C_2 \cdot 2^h$$

Let the particular solution of the given eq^n. be

$$y_h^* = A_0 + A_1 h + A_2 h^2$$

(Since R (h) = h^2 is a polynomial of degree 2).

Substituting in the given equation, we have

$$A_0 + A_1 (h+2) + A_2 (h+2)^2 - [A_0 + A_1 (h+1) + A_2 (h+1)^2] - 2(A_0 + A_1 h + A_2 h^2) = h^2$$

$$\Rightarrow \quad (-2A_0 + A_1 + 3A_2) + (-2A_1 + 2A_2)h - 2A_2 h^2 = h^2$$

On comparing the coefficients of like powers of h, we have

$$-2A_0 + A_1 + 3A_2 = 0 \quad \text{...(a)}$$

$$-2A_1 + 2A_2 = 0 \quad \text{...(b)}$$

$$-2A_2 = 1 \qquad \ldots(c)$$

From (c) $\qquad A_2 = -\frac{1}{2}$

We have from (2) $\qquad A_1 = A_2 = -\frac{1}{2}$

From (a) $-2A_0 - \frac{1}{2} - \frac{3}{2} = 0 \quad \therefore \quad A_0 = -1$

Therefore, particular solution of given eq^n. is

$$y_h^* = -1 - \frac{1}{2}h - \frac{1}{2}h^2.$$

Hence the general solution of the given equation is

$$y_h = C_1(-1)^h + C_2 \cdot 2^h - 1 - \frac{1}{2}h - \frac{1}{2}h^2. \qquad \textbf{Ans.}$$

Note: Here the particular solution can be found easily be special operator method as follows:

The given equation can be written as

$$(E^* - E - 2)\, y_h = h^2$$

Particular solution $= \dfrac{1}{E^2 - E - 2} \cdot h^2$

$$= \frac{1}{(1+\Delta)^2 - (1+\Delta) - 2} \cdot h^2$$

$$= \frac{1}{2 + \Delta + \Delta^2} \cdot h^2$$

$$= -\frac{1}{2}\left[1 - \frac{\Delta + \Delta^2}{2}\right]^{-1} \cdot h^2$$

$$= -\frac{1}{2}\left[1 + \frac{\Delta + \Delta^2}{2} + \left(\frac{\Delta + \Delta^2}{2}\right)^2 + \ldots\right] \cdot h^2$$

$$= -\frac{1}{2}\left[1 + \frac{1}{2}\Delta + \frac{3}{4}\Delta^2 + \ldots\right] \cdot h^2$$

$$= -\frac{1}{2}\left[h^2 + \frac{1}{2}\Delta h^2 + \frac{3}{4}\Delta^2 h^2\right]$$

$$= -\frac{1}{2}\left[h^2 + \frac{1}{2}(2h+1)\frac{3}{4}.2\right]$$

$$= -\frac{1}{2}h^2 - \frac{1}{2}h - 1.$$

Example 7:

Solve the following equation

$$y_{h+2} - 5y_{h+1} + 6y_h = 2.$$

Also find the solution satisfying the initial conditions

$y_0 = 1$ *and* $y_1 = -1.$ **(Meerut, 98)**

Solution:

Substituting $y_h = m^h$ in the reduced homo. equation, the A.E. is

$$m^2 - 5m + 6 = 0 \quad \text{or} \quad (m-3)(m-2) = 0$$

$$\therefore \quad m = 3, 2.$$

The general solution of the reduced homo. equation is

$$y_h = C_1 \,.\, 3^h + C_2 \,.\, 2^h.$$

To find **Particular solution** of the complete equation let the trial function be $y_h^* = A$.

Substituting in the given equation, we have

$$A - 5A + 6A = 2 \quad \text{or} \quad A = 1$$

$$\therefore \quad y_h^* = A = 1.$$

Hence the general solution of the complete equation is given by

$$y_h = C_1 \,.\, 3^h + C_2 \,.\, 2^h + 1.$$ **Ans.**

Putting h = 0 and 1 in the above solution,

we have, $y_0 = C_1 + C_2 + 1 = 1$

and $\quad y_1 = 3C_1 + 2C_2 + 1 = -1.$

Solving $C_1 = -2$ and $C_2 = 2$.

$\therefore$ the solution satisfying the initial conditions is

$$y_h = -2 \,.\, 3^h + 2^{h+1} + 1.$$ **Ans.**

Example 8:

Obtain the particular solution of the diff. equation by the method of undetermined coefficients.

$$y_{h+2} - 2y_{h+1} + y_h = 2^h (h - 1) + 5 + 3h + h^2 + \sin 2h\pi.$$

Solution:

Corresponding to the term $2^h (h - 1)$ on the R.H.S. of the diff. equation we assume the trial solution $y_h = 2^h (A_0 + A_1 h)$.

Substituting in L.H.S. we have

$$y_{h+2} - 2y_{h+1} + y_h = 2^{h+2} \{A_0 + A_1 (h + 2)\}$$
$$- 2.2^{h+1} \{A_0 + A_1 (h + 1)\} + 2^h (A_0 + A_1 h)\}$$
$$= 2^h \{hA_1 + (4A_1 + A_0)\}$$

Comparing with $2^h (h - 1)$, we have

$A_1 = 1$ and $4A_1 + A_0 = -1$ so that $A_1 = 1$ and $A_0 = -5$

$\therefore$ the particular solution corresponding to the term $2^h (h - 1)$ on R.H.S. of the diff. equation is $y_h = 2^h (-5 + h)$.

Now, corresponding to the term $5 + 2h + h^2$ on the R.H.S. of the diff. equation we assume the trial solution

$$y_h = h^2 [A_2 + A_3 h + A_4 h^2]$$

(We can not assume the trial solution $A_2 + A_3 h + A_4 h^2$ or $h (A_2 h + A_3 h + A_4 h^2)$ since $C_1 + C_2 h$ is the solution of the reduced homo. equation).

Substituting this trial solution in the L.H.S., we have

$$y_{h+2} + 6y_{h+1} + y_h = (h + 2)^2 [A_2 + A_3 (h + 2) + A_4 (h + 2)^2]$$
$$- 2 (h + 1)^2 [A_2 + A_3 (h + 1)$$
$$+ A_4 (h + 1)^2] + h^2 [A_2 + A_3 h + A_4 h^2]$$
$$= (2 A_2 + 6A_3 + 14A_4) + (6A_3 + 24A_4) h + 12A_4 h^2$$

Comparing with $5 + 3h + h^2$, we have

$2A_2 + 6A_3 + 14A_4 = 5$, $6A_3 + 24A_4 = 3$, and $12A_4 = 1$

solving, $A_2 = \frac{17}{6}, A_3 = \frac{1}{6}, A_4 = \frac{1}{12}$

$\therefore$ the particular solution corresponding to the term $5 + 3h + h^2$ on R.H.S. of the diff. equation is

$$h^2\left[\frac{17}{6} + \frac{1}{6}h + \frac{1}{12}h^2\right].$$

Again corresponding to the term sin 2hπ on the R.H.S. of the diff. equation we assume the trial solution

$$y_h = h^2 [A_5 \sin 2h\pi + A_6 \cos 2h\pi].$$

13.15 Solution of Simultaneous Differential Equations

If one or more differential equations are given with the same number of unknown functions we can solve such equations simultaneously by using a procedure which eliminates all but one of the unknowns.

For clear under standing of the method see the following example.

Example 1:

Solve the system.

$$x_{h+1} - x_h + 2y_{h+1} = 0 \qquad ...(1)$$

$$y_{h+1} - y_h - 2x_h = 2^h \qquad ...(2)$$

(Meerut, B.Sc. Hons. 98)

Solution:

The given system of equations, can be written as

$$(E - 1)\, x_h + 2Ey_h = 0 \qquad ...(3)$$

and

$$-2x_h + (E - 1)\, y_h = 2^h \qquad ...(4)$$

Operating on the equation (3) with E – 1 and equation (4) with 2E and then substituting, we have

$$\{(E - 1)^2 + 4E\}\, x_h = -2E2^h = -2 \,.\, 2^{h+1}.$$

$$\Rightarrow \qquad (E + 1)^2\, x_h = -4 \,.\, 2^h \qquad ...(5)$$

which is a linear diff. equation with constant coefficient.

A.E. is $(m + 1)^2 = 0$

$\therefore \quad m = -1, -1$

$\therefore$ general solution of reduced homo. equation is

$$x_h = (C_1 + C_2h)(-1)^h.$$

$$\text{Particular solution} = \frac{1}{(E+1)^2} \,.\, (-4 \,.\, 2^g)$$

$$= -4 \,.\, \frac{1}{(E+1)^2} \,.\, 2^h = -4\, \frac{2^h}{(2+1)^2} = -\frac{4}{9}.2^h.$$

$\therefore$ general solution of (3) is given by

$$x_h = (C_1 + C_2h)(-1)^h = \frac{4}{9} \,.\, 2^h. \qquad ...(6)$$

$\therefore$ from the equation (1)

$$2y_{h+1} = x_h - x_{h+1}$$

$$= (C_1 + C_2h)(-1)^h - \frac{4}{9} \cdot 2^h$$

$$- [\{(C_1 + C_2(h+1)\}(-1)^{h+1} - \frac{4}{9} \cdot 2^{h+1}]$$

$$= (2C_1 + C_2 + 2C_3h)(-1)^h + \frac{4}{9} \cdot 2^h$$

$$\Rightarrow \quad y_{h+1} = (C_1 + \frac{1}{2}C_2 + C_3h)(-1)^h + \frac{2}{9} \cdot 2^h$$

$$\therefore \quad y_h = \{C_1 + \frac{1}{2}C_2 + C_3(h-1)\}(-1)^{h-1} + \frac{2}{9} \cdot 2^{h-1}$$

$$\Rightarrow \quad y_h = (-C_1 + \frac{1}{2}C_2 - C_2 h)(-1)^h + \frac{1}{9} \cdot 2^h. \quad \ldots(7)$$

Equations (6) and (7) give the solution of the given system. **Ans.**

13.16 Matrix Method for Solving a System of Linear Differential Equation **(Meerut, M.Sc. 90, 92)**

Now, we shall develop a matrix method for solving a system of linear differential equations of order 1.

Consider the system of simultaneous diff. equations

$$\left.\begin{aligned} x_{t+1} &= \alpha x_t + \beta y_t \\ & \qquad t = 0, 1, 2, \ldots \\ y_{t+1} &= \alpha' x_t + \beta' y_t \end{aligned}\right\} \quad \ldots(1)$$

Now, we want to find expressions for x_t and y_t in terms of the given initial values x_0, y_0, constants α, β, α', β' and t.

The above system of two simultaneous diff. equations can be written as a single matrix differential equation as follows:

$$\begin{bmatrix} x_{t+1} \\ y_{t+1} \end{bmatrix} = \begin{bmatrix} \alpha & \beta \\ \alpha' & \beta' \end{bmatrix} \cdot \begin{bmatrix} x_t \\ y_t \end{bmatrix}$$

$$\Rightarrow \quad V_{t+1} = AV_t$$

where $V_t = \begin{bmatrix} x_t \\ y_t \end{bmatrix}$ and $A \begin{bmatrix} \alpha & \beta \\ \alpha' & \beta' \end{bmatrix}$ ...(2)

Since x_0 and y_0 are given $\therefore$ we know $V_0 = \begin{bmatrix} x_0 \\ y_0 \end{bmatrix}$

From (2), we have $V_1 = AV_0$,

$$V_2 = AV_1 = 4(AV_0) = A^2V_0$$
$$V_3 = AV_2 = A\ (A^2V_0) = A^3V_0.$$

Proceeding similarly, we have $V_t = A^tV_0$ ...(3)

From (3) it is obvious that we can find V_t, by multiplying V_0 by A^t. Thus, our problem is now reduced to the problem of finding t the power of matrix A.

Method of Finding A^t: This method depends on the Cayley-Hamilton theorem of matrices which states that every square matrix satisfies the characteristic equation. Here A is a square matrix of order 2, so its characteristic equation is a quadratic equation

$$f(\lambda) = |A - \lambda I| = a_0 + a_1\lambda + a_2\lambda^2 = 0$$

which is satisfied by matrix A i.e. f (A) = 0.

Now, we divide λ^t by f (λ) and let q (λ) be the quotient and r (λ) the remainder.

$$\therefore \quad \lambda^t = f(\lambda)\ .\ q(\lambda) + r(\lambda). \qquad ...(4)$$

Since f (λ) is a quadratic polynomial, the remainder r (λ) is at most of first degree.

$$\therefore \quad \text{Let } r(\lambda) = a + b\lambda \qquad ...(5)$$

where a and b are constants.

But (4) is an identity in λ, we may replace λ by matrix A

$$\therefore \quad A^t = f(A)\ .\ q(A) + r(A) = r(A) \qquad \because \quad f(A) = 0.$$

But from (5) $\quad r(A) = aI + bA$

$$\therefore \quad A^t = aI + bA \qquad ...(6)$$

If λ_1 and λ_2 are the characteristic roots of matrix A,

then $\quad f(\lambda_1) = 0$ and $f(\lambda_2) = 0$

$\therefore$ From (4) and (5), we have

$$\left.\begin{aligned} \lambda_1{}^t &= r(\lambda_1) = a + b\lambda_1 \\ \text{and} \quad \lambda_2{}^t &= r(\lambda_2) = a + b\lambda_2 \end{aligned}\right\} \qquad ...(7)$$

Now, we can determine a and b uniquely from two simultaneous liner equations (7), if

$$\begin{vmatrix} 1 & \lambda_1 \\ 1 & \lambda_2 \end{vmatrix} = \lambda_2 - \lambda_1 \neq 0$$

i.e. if $\quad \lambda_2 \neq \lambda_1$

$\therefore$ if the characteristic roots of the matrix A are distinct then we can find unique values of a and b from (7). Substituting these values of a and b in (6) the value of A' is calculated which enable us to get V_t from (3).

13.17 Working Method for Solving a Second Order Homogeneous Differential Equation with Constant Coefficients

(Meerut, M.Sc. 93, 95)

Let the second order homo. diff. equation be given by

$$x_{t+1} - b_1 x_t = b_2 x_{t-1} = 0, \quad t = 1, 2, \ldots \qquad \ldots(1)$$

with x_0 and x_1 as prescribed initial values.

First introduce the new function y defined for t = 1, 2 .. such that the second order diff. equation (1) can be written as a system of two first order diff. equations.

$$\left.\begin{aligned} x_{t+1} &= b_1 x_t + b_2 y_t \\ & \qquad\qquad t = 1, 2, \ldots \\ y_{t+1} &= x_t \end{aligned}\right\} \qquad \ldots(2)$$

Now, defining the column vector $v_t = \begin{bmatrix} x_t \\ y_t \end{bmatrix}$, t = 1, 2, ...

the system (2) may be written in matrix form as

$$V_{t+1} = A.V_t, \quad t = 1, 2, \ldots \qquad \ldots(3)$$

where $A = \begin{bmatrix} b_1 & b_2 \\ 1 & 0 \end{bmatrix}$

From (3),

$$V_2 = AV_1,\ V_3 = A \,.\, V_2 = A\,(AV_1) = A^2V_1.$$

Proceeding similarly $V_{t+1} = A^tV_1 = A^t \begin{bmatrix} x_t \\ y_t \end{bmatrix} = A^t \begin{bmatrix} x_1 \\ x_0 \end{bmatrix}$ ...(4)

We have $A^t = aI + bA$ ...(5)

where a and b are calculated from

$$\left.\begin{aligned} \lambda_1^t &= a + b\lambda_1 \\ \lambda_2^t &= a + b\lambda_2 \end{aligned}\right\} \qquad \ldots(6)$$

Where λ_1, λ_2 are the roots of characteristic equation matrix A.

Solving the characteristic equation

$$|\,A - \lambda I\,| = 0,$$

we get λ_1 and λ_2 and then from (6) a and b are calculated.

Substituting the values of a and b in (5), we calculate the value of A^t and then substituting the value of A^t in (4), we get V_{t+1}.

From V_{t+1}, we can find the value of x_{t+1} and hence the value of x_t, the solution of given diff. equation.

Example 1:

Apply matrix method to solve

$x_{t+1} = 3x_t + 2x_{t-1} = 0,\ t = 1, 2, 3, \ldots$

$x_0 = 0,\ x_1 = 1.$

(Meerut, B.Sc. Hons., 98, 90)

Solution:

Introducing the new function y_t (defined for t = 1, 2, ...) the given equation reduce to the following two first order diff. equations

$$\left.\begin{aligned} x_{t+1} &= 3x_t - 2y_t \\ & \qquad\qquad t = 1, 2, \ldots.. \\ y_{t+1} &= x_t \end{aligned}\right\} \qquad \ldots(1)$$

In matrix from the above equation can be written as

$$\begin{bmatrix} x_{t+1} \\ y_{t+1} \end{bmatrix} - \begin{bmatrix} 3 & -2 \\ 1 & 0 \end{bmatrix} \begin{bmatrix} x_t \\ y_t \end{bmatrix}$$

$\Rightarrow$ $$V_{t+1} = A \cdot V_t. \qquad \ldots(2)$$

where $$A = \begin{bmatrix} 3 & -2 \\ 1 & 0 \end{bmatrix} \text{ and } V_t = \begin{bmatrix} x_t \\ y_t \end{bmatrix} \qquad \ldots(3)$$

Now $$V_1 = \begin{bmatrix} x_1 \\ y_1 \end{bmatrix} = \begin{bmatrix} x_1 \\ x_0 \end{bmatrix} = \begin{bmatrix} 1 \\ 0 \end{bmatrix}$$

from (2), $V_2 = AV_1$, $V_3 = AV_2 = A \cdot (AV_1) = A^2V_1$

Proceeding similarly, we have

$$V_{t+1} = A^t \cdot V_1 = A^t \cdot \begin{bmatrix} 1 \\ 0 \end{bmatrix} \qquad \ldots(4)$$

Now, we have $A^t = aI + bA$ ...(5)

where a and b are calculated from

$$\left.\begin{aligned} \lambda_1^t &= a + b\lambda_1 \\ \lambda_2^t &= a + b\lambda_2 \end{aligned}\right\} \qquad \ldots(6)$$

where λ_1, λ_2 are characteristic roots of matrix A. Characteristic equation of matrix A is given by

$$|A - \lambda I| = \begin{vmatrix} 3-\lambda & -2 \\ 1 & 0-\lambda \end{vmatrix} = 0$$

$\Rightarrow$ $$\lambda^2 - 3\lambda + 2 = 0$$

$\Rightarrow$ $$(\lambda - 1)(\lambda - 2) = 0$$

$\therefore \quad \lambda_1 = 1, \lambda_2 = 2.$

Putting the values of λ_1 and λ_2 in (6) we have

$$1 = a + b$$

and $\quad 2^t = a + 2b$

solving, we have $a = 2 - 2^t$ and $b = 2^t - 1$,

$$\therefore \text{ from (5), } A^t = aI + bA = (2 - 2^t)\begin{bmatrix} 1 & 0 \\ 0 & 1 \end{bmatrix} + (2^t - 1)\begin{bmatrix} 3 & -2 \\ 1 & 0 \end{bmatrix}$$

$$= \begin{bmatrix} 2^{t+1} - 1 & -2^{t+1} + 2 \\ 2^t - 1 & 2 - 2^t \end{bmatrix}$$

$\therefore$ from (4), we have

$$V_{t+1} = \begin{bmatrix} 2^{t+1} - 1 & -2^{t+1} + 2 \\ 2^t - 1 & 2 - 2^t \end{bmatrix}\begin{bmatrix} 1 \\ 0 \end{bmatrix}$$

$$\Rightarrow \quad \begin{bmatrix} x_{t+1} \\ y_{t+1} \end{bmatrix} = \begin{bmatrix} 2^{t+1} - 1 \\ 2^t - 1 \end{bmatrix}$$

$\therefore \quad x_t = y_{t+1} = 2^t - 1, t = 0, 1, 2.$ **Ans.**

Example 2:

Reduce the second order system to two first order diff. equations and hence use matrix method to solve

$$x_{t-1} = x_{t-1} = 0, x_0 = 1, x_1 = 2.$$

(Meerut, B.Sc. Hons. 98, 89)

Solution:

Introducing the new function y (defined for t = 1, 2, ...) the given equation reduce to the following two first order diff. equations

$$\left.\begin{aligned} x_{t+1} &= y_t = 0, x_t + 1 \cdot y_t \\ y_{t+1} &= x_t = 1, x_t + 0 \cdot y_t \end{aligned}\right\} t = 1, 2, \ldots \qquad \ldots(1)$$

In matrix form the above equations can be written as

$$\begin{bmatrix} x_{t+1} \\ y_{t+1} \end{bmatrix} = \begin{bmatrix} 0 & 1 \\ 1 & 0 \end{bmatrix}\begin{bmatrix} x_t \\ y_t \end{bmatrix}$$

$$\Rightarrow \quad V_{t+1} = AV_t \qquad \ldots(2)$$

where $\quad A = \begin{bmatrix} 0 & 1 \\ 1 & 0 \end{bmatrix}$ and $V_t = \begin{bmatrix} x_t \\ y_t \end{bmatrix} \qquad \ldots(3)$

Now $\quad V_1 = \begin{bmatrix} x_1 \\ y_1 \end{bmatrix} = \begin{bmatrix} x_1 \\ x_0 \end{bmatrix} = \begin{bmatrix} 3 \\ 1 \end{bmatrix}$

from (3), $V_2 = AV_1$, $V_2 = AV_2 = A \,.\, (AV_1) = A^2V_1$.

Proceeding similarly, we have

$$V_{t+1} = A^t \,.\, V_1 = A^t \begin{bmatrix} 3 \\ 1 \end{bmatrix} \qquad ...(4)$$

Now, we have $A^t = aI + bA$ ...(5)

where a and b are calculated from

$$\text{and} \quad \left.\begin{aligned} \lambda_1^{\,t} &= a + b\lambda_1 \\ \lambda_2^{\,t} &= a + b\lambda_2 \end{aligned}\right\} \qquad ...(6)$$

$\therefore \ \lambda_1 = 1$ and $\lambda_2 = -1$

where λ_1, λ_2 are characteristic roots of matrix A. Characteristic equation of matrix A is given by

$$| A - \lambda I | = \left| \begin{bmatrix} 0 & 1 \\ 1 & 0 \end{bmatrix} - \lambda \begin{bmatrix} 1 & 0 \\ 0 & 1 \end{bmatrix} \right| = \begin{vmatrix} -\lambda & 1 \\ 1 & -\lambda \end{vmatrix} = 0$$

$\Rightarrow \quad \lambda^2 - 1 = 0 \quad$ or $\quad \lambda = \pm 1$.

Putting the values of λ_1 and λ_2 in (6), we have

$$1 = a + b$$

and $$(-1)^t = a - b$$

solving, we have $a = \dfrac{1}{2}\,[1 + (-1)^t]$, $b = \dfrac{1}{2}\,[1 - (-1)^t]$

$$\therefore \text{ from (5), } A^t = aI + bA = \frac{1}{2}\,\{1 + (-1)^t\} \begin{bmatrix} 1 & 0 \\ 0 & 1 \end{bmatrix} + \frac{1}{2}\,\{(1 - (-1)^t\} \begin{bmatrix} 0 & 1 \\ 1 & 0 \end{bmatrix}$$

$$= \frac{1}{2} \begin{bmatrix} 1 + (-1)^t & 1 - (-1)^t \\ 1 - (-t)^t & 1 + (-1)^t \end{bmatrix}$$

$\therefore$ from (4), we have

$$V_{t+1} = \frac{1}{2} \begin{bmatrix} 1 + (-1)^t & 1 - (-1)^t \\ 1 - (-1)^t & 1 + (-1)^t \end{bmatrix} \begin{bmatrix} 3 \\ 1 \end{bmatrix}$$

$$\Rightarrow \qquad \begin{bmatrix} x_{t+1} \\ y_{t+1} \end{bmatrix} = \frac{1}{2} \begin{bmatrix} 4 + 2(-1)^t \\ 4 - 2(-1)^t \end{bmatrix}$$

$$\therefore \qquad x_t = y_{t+1} = \frac{1}{2} \{4 - 2(-1)^t\} = 2 - (-1)^t$$

$t = 0, 1, 2, \ldots.$ **Ans.**

EXERCISES

1. Solve by matrix method

   (i) $y_{h+2} - 4y_{h+1} + 6y_h = h^2,\ y_0 = 1,\ y_1 = 2.$

   (Meerut, B.Sc. Hons. 98)

   Ans. $2^{h+3} - \frac{1}{2} \cdot 3^{h+2} + \frac{1}{2} h^2 + \frac{3}{2} /t - \frac{3}{2}.$

   (ii) $y_{h+1} = 2y_h + z_h$
   $z_{h+1} = y_h + 2z_h$ where $y_1 = 1,\ z_1 = 0.$

   (Meerut, B.Sc. Hons. 98)

   Ans. $y_h = \frac{1}{2} + \frac{1}{6} \cdot 3^h,\ z_h = -\frac{1}{2} + \frac{1}{6} \cdot 3^h.$

   (iii) $y_{h+1} = 2y_h + z_h + 1$
   $z_{h+1} = y_h + 2z_h + 2,\ h = 0, 1, 2, \ldots$
   $y_0 = 0,\ z_0 = 1.$ **(Meerut, M.Sc. 90)**
   [**Hint :** Take $Z_h + 1 = v_h$].

   Ans. $y_h = 3^{h-1} - 1,\ z_h = 3^{h-1}.$

2. Write an essay on the application of differential equation in social sciences. **(Meerut, M.Sc. 92)**

3. Define a fundamental set of solutions of a second order diff. equation.

4. If $y^{(1)}$ and $y^{(2)}$ constitute a fundamental set for

   $y_{h+2} + a_1 y_{h+1} + a_2 y_2 = 0,\ h = 0, 1, 2, \ldots$

   Prove that a necessary and sufficient condition for $\alpha_1 y^{(1)} + \beta_1 y^{(2)}$ and $\alpha_2 y^{(1)} + \beta_2 y^{(2)}$ to be a fundamental set is $\alpha_1 \beta_2 - \alpha_2 \beta_1 \neq 0.$

   (Meerut, B.Sc. Hons. 99)

5. If u (x) and v (x) are the solutions of the homo. diff. equation.
   $y(x + 2) + p(x)\, y(x + 1) + q(x)\, y(x) = 0,$

   then $C_1 u(x) + C_2 v(x)$ is also a solution of the above equation.

6. Describe the matrix method for solving a system of linear differential equations. **(Meerut, M.Sc. 92, 93 (P)**

7. What do you understand by linear differential equation of the nth order.

8. Solve the following homo. equations.

(i) $y_{h+2} - y_h = 0.$ **Ans.** $y_h = C_1 + C_2 (-1)^h.$

(ii) $y_{h+2} + 2y_{h+1} + y_h = 0$ **Ans.** $y_h = (C_1 + C_2h)(-1)^h.$

(iii) $y_{h+2} + 8y_{h+1} + 16y_h = 0$ **Ans.** $y_h = (C_1 + C_2h)(-4)^h.$

(iv) $y_{h+2} - 5y_{h+1} + 6y_h = 0, y_0 = 1, y_1 = 2.$

Ans. $y_h = 3^h - 2^h + 1.$

(v) $y_{h+2} + 3y_{h+1} + 2y_h = 0$ **(Meerut, M.Sc. 85, 90)**

Ans. $y_h = C_1 + C_2 . 2^h.$

(vi) $y_{h+2} + 3y_{h+1} + y_h = 0$

$$\textbf{Ans. } y_h = C_1 \left(\frac{-3+\sqrt{5}}{2}\right)^h + C_2 \left(\frac{-3-\sqrt{5}}{2}\right)^h$$

(vii) $2y_{h+3} - 7y_{h+2} + 5y_{h+1} + 2y_h = 0$

$$\textbf{Ans. } y_h = C_1 . 2^h + C_2 . \left(\frac{3+\sqrt{17}}{3}\right)^h + C_2 \left(\frac{3-\sqrt{17}}{2}\right)^h.$$

(viii) $y_{h+4} + y_h = 0$ **(Meerut, M.Sc. 96(P))**

$$\textbf{Ans. } y_h = C_1 \cos\left(\frac{3\pi}{4}h + C_2\right) + C_2 \cos\left(\frac{\pi h}{4} + C_4\right).$$

(ix) $y_{h+3} + y_{h+2} - 8y_{h+1} - 12y_h = 0$ **(Meerut, B.Sc. Hons. 98)**

Ans. $y_h = C_1 . 3^h + (C_2 + C_3h)(-2)^h.$

(x) $y_{h+2} - y_{h+1} + y_h = 0$ **(Meerut, M.Sc. 94(P))**

$$\textbf{Ans. } y_h = C_1 \cos\left(\frac{\pi h}{4} + C_2\right)$$

9. Solve the following diff. equations.

(i) $y_{h+2} - 2y_{h+1} + y_h = 5 + 3h.$ **(Meerut, M.Sc. 91)**

$$\textbf{Ans. } y_h = (C_1 + C_2/t) + h^2 + \frac{1}{2}h^3.$$

(ii) $y_{h+2} - 3y_{h+1} + 2y_h = 1, y_h = 1, y_1 = -1.$

Ans. $y_h = 2 - 2^h - 1$

(iii) $8y_{h+2} - 6y_{h+1} + y_h = 2^h$.

Ans. $y_h = C_1\left(\frac{1}{4}\right)^h + C_2\left(\frac{1}{2}\right)^h + \frac{1}{21} \cdot 2^h$.

(iv) $8y_{h+2} - 6y_{h+1} + y_h = 5 \sin \frac{h\pi}{2}$. **(Meerut, M.Sc. 91)**

Ans. $y_h = C_1\left(\frac{1}{4}\right)^h + C_2\left(\frac{1}{2}\right)^h + \frac{5}{3} \sin \frac{h\pi}{2}$.

(v) $y_{h+2} - 3y_{h+1} + 2y_h = 3^h$ **(Meerut, M.Sc. 98)**

Ans. $y_h = C_1 + C_2 \cdot 2^h + \frac{1}{2} \cdot 3^h$.

(vi) $y_{h+2} - y_{h+1} + \frac{1}{2} y_h = 1, y_0 = 2, y_1 = 3$

(Meerut, M.Sc. 96(P))

Ans. $-(2)^{h/2} \cdot \cos\left(\frac{\pi h}{4} + \frac{\pi}{2}\right) + 2$.

(vii) $(E^2 - 5E + 6)\, y(x) = x^2, y(0) = 1, y(1) = -1$.

(Meerut, M.Sc. 96(P))

Ans. $y(x) = 2^{x+2} - 3^{x+1}$.

(viii) $y_{h+2} - 5y_{h+1} + 6y_h = h^2 - 1$.

(Meerut, B.Sc. Hons. 94; M.Sc. 97)

Ans. $y_h = C_1 \cdot 2^h + C_2 \cdot 3^h + 2 + \frac{3}{2} h + \frac{1}{2} h^2$.

(ix) $y_{h+2} + y_{h+1} - 12y_h = 3^h + 10$. **(Meerut, B.Sc. Hons. 97)**

Ans. $y_h = C_1 \cdot 3^h + C_2 (-4)^h - 1 + \frac{1}{21} h \cdot 3^h$.

(x) $(E - 2)(E - 3)\, y_h = 4^h (h^2 - h + 5)$

(Meerut, B.Sc. Hons. 97)

Ans. $y_h = C_1 \cdot 2^h + C_2 \cdot 3^h + \frac{4^h}{2} (h^2 - 13h + 61)$

(xi) $y_{h+2} - 4y_h = 9h^2$. **(Meerut, M.Sc. 78, 90)**

Ans. $y_h = C_1 \cdot 2^h + C_2 \cdot (-2)^h - \left(3h^2 + 4h + \frac{20}{3}\right)$

(xii) $y_{h+1} \cdot y_h^2 = y^2_{h+1}$ if $y_1 = 1, y_2 = 2$.

[**Hint.** Taking logarithm, $\log y_{h+2} - 3 \log y_{h+1} + 2 \log y_h = 0$ let $v_h = \log y_h$. $\therefore\ v_{h+2} - 3v_{h+1} + 2v_h = 0$.

which is L.D.E. with constant coefficient]. **Ans.** $y_h = 2^{2^{h-1}-1}$.

10. Solve the following systems of simultaneous equations.

(i) $x_{h+1} = 2x_h + y_h + h$

$y_{h+1} = 2x_h + 3y_h - h$. **(Meerut, B.Sc. Hons. 97)**

Ans. $X_h = C_1 + C_2 \cdot 4^h + \frac{1}{2}(h^2 - h)$.

$y_h = -C_1 + 2C_2 - \frac{1}{2}(h^2 - h)$.

(ii) $2u_{n+1} - 3u_n + 5y_n = 2$

$2u_n + v_{n+1} - 2v_h = 7$.

Ans. $u_n = C_1 \cdot 4^n + C_2 (-\frac{1}{2})^n + \frac{37}{9}$,

$y_h = -C_1 \cdot 4^n + \frac{4}{9}(-\frac{1}{2})^n C_2 + \frac{11}{9}$.

(iii) $u_{n+1} - v_h = 2(n+1)$

$v_{n+1} + v_n = -2(n+1)$.

Ans. $u_n = C_1 + C_2(-1)^n + n$, $V_n = C_1 - C_2(-1)^n - (n+1)$

(iv) $2u_{n+1} + v_{n+1} = u_n + 3v_n$

$u_{n+1} + v_{n+1} = u_n + v_n$.

Ans. $U_n = C_1(-2)^n + C_2$, $V_n = C_1(-2)^n + \frac{1}{2}C_2$.

11. $y_{h+1} - y_h = 4, h = 0, 1, 2$ **(Meerut, M.Sc. 93 (P))**

Ans. $y_h = \lambda(-1)^h + 1$.

12. $2y_{h+1} + y_h = 2, h = 0, 1, 2$ **(Meerut, M.Sc. 83 (P))**

Ans. $y_h = \lambda(\frac{1}{2})^h + 4$.

13. $y_{h+1} + 3y_h = 8$. **Ans.** $y_h = \lambda(-3)^h + 2$.

14. $y_{h+1} - 2y_h = 6$. **(Meerut, M.Sc. 97)**

Ans. $y_h = \lambda \cdot 2^h - 6$.

15. $y_{h+1} + y_h = 1.$ **Ans.** $y_h = \lambda\,(-1)^h + \frac{1}{2}.$

16. $y^k_{h+4} - 4y^k_{h+3} + 6y^k_{h+2} - 4y^k_{h+1} + y^k_h = 0$

(Meerut, B.Sc. 97)

Ans. $y_h = C_1 + C_2 n + C_3 n^2 + C_4 n^3.$

17. $3y_{h+2} - 7y_{h+1} - 6y_h = 0,\ y_0 = 0,\ y_1 = 1.$

(Meerut, B.Sc. Hons. 98)

Ans. $y_h = C_1\left(\frac{3}{11}\right)^n + C_2\left(-\frac{3}{11}\right)^n.$

18. $9y_{h+2} + 42y_{h+1} + 49y_h = 0,\ y_0 = 0,\ y_1 = -7.$

(Meerut, B.Sc. Hons. 94; M.Sc. 97)

Ans. $3h\left(-\frac{7}{3}\right)^h.$

19. $y_{h+2} - 5y_{h+1} + 6y_h = 0$ **(Meerut, M.Sc. 92)**

Ans. $V_h = C_1 \,.\, 3^h + C_2 \,.\, 2^h.$

14

Cobweb Phenomenon and Generating Functions

14.1 Cobweb Phenomenon (or Cobweb Cycles)

(Meerut, M.Sc. 98, 99)

In this chapter, we shall discuss the very important phenomenon known as **Cobweb Phenomenon.** The so called Cobweb Phenomenon in economics efforts excellent examples in which first order difference equations may be used.

In this section we wish to study this cycle on a mathematical basis.

We study three functions defined for t = 0, 1, 2, 3,... as follows.

Let S_t denote the number of units supplied in time t.

D_t stands for the suits demenaded in time t.

P_t = price per unit in period t.

We Now make the following assumptions.

(a) A price demand relation is specified in which quantity demanded is deterined by thrice at e of purchase or suppose a functin g exists fo which we have

$$D_t = g\,(p_t) \qquad \text{...(1)}$$

(b) There exists a price supply curve that relates the supply in any period with the price one period before or in other words there exists a function h such that

$$S_{t+1} = h(p_t) \qquad ...(2)$$

(c) The market price is determined by the available supply, transactions occurring at the price at which the quantity demanded and the quantity supplied are equal

$$\Rightarrow \qquad S_t = D_t \qquad ...(3)$$

Now, suppose p_0 is persecribed if the function g and h are known, we may calculate S_t from (2) and D_1 from (3) and so obtain p_1 from the price demand curve in (1). The process may be repeated. After obtaining p_1 we obtain p_2 and so on. Our main purpose is to study the sequence of prices

$$p_0 \cdot p_1, p_2, ...$$

Here we are very particular to study the oscillatory be haviour.

If we wish to make a thorough study we have to make certain hypothesis to study the functions g and h.

We suppose that g and h are linear functins or equivalently the price demand and price supply curve are straight lines. The equation in (1) and (2) are there by simplified to

$$D_t = n_d p_t + c_d \quad x_d > 0, c_d > 0 \qquad ...(4)$$

$$S_{t+1} = n_s p_t + c_s \quad n_s > 0, c_s > 0 \qquad ...(5)$$

where $-n_d$, are the slope and intercept for the demand curve, with s_s and c_s the corresponding constants for the supply curve.

Here the slope of the demand curve is taken to be negative and that of supply curve positive. The resson is evident that an increase of one unit in price produces a decrease in demand and increase in supply.

By hypothesis equation (3) $\quad D_{t+1} = S_{t+1} \qquad ...(6)$

and hence $\quad n_s y_t + c_r = -n_d p_{t+1} + c_d \qquad ...(7)$

$$\Rightarrow \qquad p_{r+1} = Ap_t + B \qquad ...(8)$$

where $$A = \frac{-n_s}{n_d} \qquad ...(9)$$

and $$B = \frac{c_d - c_s}{n_d}$$

Remark:

We have already studied the first order difference equation (8) and the following conclusions are worth noting.

Since A is negative, the solution sequence $\{p_1\}$ is always oscillatory.

(i) being dameped or convergent with limit

$$p^* = \frac{B}{1 - A} \text{ if } -1 < A < 0$$

(ii) If $A = -1$, it undergoes fiite oscillations.

(iii) If $A < -1$ it has infinite oscillations.

Thus, the behaviour or proce sequence is depending upon A which is the ration of the supply and demand curves.

To have a more clear understanding let us have a graphical analysis.

We start with p_0 find S_1 by moving vertically to the supplier line, then move horizontally.

(Since $D_1 = \delta_1$) to find D_1, which determines the price p_1 on the price axis. The supply δ is found on the supply line directly above p_1 and again $D_2(=\delta)$ is found moving horizontally to the demand line etc.

The intersection of the suply and demands line determines the equilibrium price p*, (a) we find a cobweb moving towards this equilibrium point prices alternating above and below but converging to p*.

The Cobweb consists of one square endlessly repeated prices oscillatons finitely etween just o oscillations finitely between just two values.

The Cobweb moves from the intersection oint, prices oscillating infinitely about the equaillbium value.

Example 1:

Suppose the slopes and intercepts in the price demand and price supply equations $n_d = c_d = 2$ *and* $n_s = c_s = 1$.

Sketch a graph and analyse the behaviour of the price sequence. Find p and determine whether the equilibrium is stalbe.*

Solution:

We have here

$$p_{t+1} = Ap_t + B$$

where $$A = \frac{-n_s}{n_d} = -\frac{1}{2},$$

$$B = \frac{c_d - c_s}{n_d}$$

$$= \frac{2-1}{2} = \frac{1}{2}$$

$$\therefore \quad p_{t+1} = -\frac{1}{2} p_t + \frac{1}{2}.$$

Also $$p^* = \frac{B}{1-A} = \frac{1/2}{1+\frac{1}{2}} = \frac{1}{3} \text{ Stable.}$$ **Ans.**

Example 2:

Find p and determine whether the equailibirum is stable, when*

$$n_d = 1,\ c_d = 5,\ n_s = c_s = 1.$$

Solution:

We have here

$$p_{t+1} = Ap_t + B$$

where $$A = \frac{n_r}{d_d} = -\frac{1}{1} = -1$$

$$B = \frac{C_d - C_s}{n_d} = \frac{5-1}{1} = 4$$

$$\therefore \quad p_{t+1} = -p_t + 4$$

Also $$p^* = \frac{B}{1-A} = \frac{4}{1+1} = 2 \text{ unstable.}$$ **Ans.**

Example 3:

For the find p and determine whether the equaillbrilum is stable, when*

$$n_d = 1,\ c_d = 7,$$
$$n_s = 2,\ c_s = 1.$$

Solution:

We have here

$$p_{t+1} = Ap_t + B$$

where $$A = \frac{n_s}{n_d} = -2$$

$$B = \frac{c_d - c_s}{n_d} = \frac{7-1}{1} = 6$$

$$\therefore \quad p_{t+1} = -2p_t + 6$$

Also $p^* = \frac{B}{1 - A} = \frac{6}{1 + 2} = 2$

here $A < -1$. Unstable. **Ans.**

14.2 Generating Functions

The method of generating sequences is a powerful method to solve and study the solution of the difference equation.

Definition of Generating Function for a Sequence : *If y_0, y_1, y_2... is scquence of a real number, the function Y defined for some interval of real numbers containing zero whose value at t is given by the series.*

$$Y(t) = y_0 + y_1 t + y_2 t^2 + ... + y_h t^h + ... = \sum_{h=0}^{\infty} y_h . t^h \quad ...(1)$$

is the generating function of the sequence $\{y_t\}$.

14.3 Some Special Generating Functions

1. *The function given by*

$$Y(t) = \frac{1}{1 - t}$$

is the generating function of the constant sequence 1, 1, 1,...

P.F. We have write

$Y(t) = (1 - t)^{-1} = 1 + t + t^2 + ...$ by Binomial Theorem.

On comparing it with

We ahve $y_0 = 1$, $y_1 = 1$, $y_2 = 1$,...

$\therefore$ $Y_t = \frac{1}{1 - t}$ is the generating function of the constant sequence 1, 1, 1,...

2. *The function given by*

$$Y(t) = \frac{1}{(1 - t)^2}$$

is the generating function of the sequence 1, 2, 3, 4...

P.F. We have

$$Y(t) = \frac{1}{(1 - t)^2} = (1 - t)^{-2}$$

$$= 1 + 2t + 3t^2 + ... + (h + 1) t^h + ...$$

On comparing it with we have

$$y_0 = 1,\ y_1 = 2,\ y_2 = 3, \ldots$$

$$\therefore \quad Y_t = \frac{1}{(1-t)^2} \text{ is generating function of sequence } 1, 2, 3, 4 \ldots$$

3. *To find sequence generated by the function*

$$Y(t) = \frac{t}{(1-t)^2}.$$

Solutioin:

We have, $Y(t) = t\,(1 - t)^{-2}$

$$= t[1 + 2t + 3t^2 + \ldots] \qquad \text{(By Binomial Theorem)}$$

$$= 0 + 1.t + 2.t^2 + 3.t^3 + ..$$

Now on comparing with

We have $y_0 = 0,\ y_1 = 1,\ y_2 = 2, \ldots$

This given $\{y_k\} = \{k\}$.

4. *To find the generating junction of $\{B^h\}$.*

$$B \neq 0.$$

Solution:

We know that Y(t) must be given by

$$Y(t) = 1 + Bt + B^2t^2 + \ldots B^h t^h + \ldots \qquad \ldots(1)$$

Now, this generates $\{B^h\}$ but right hand side of (1) is an infinite Geometric Progression with common ratio Bt.

Now we so restrict t so that the common ratio

$$|t| < \frac{1}{|B|}$$

or this common ratio is < 1.

$$\because \quad \text{Sum of an infinte G.P.} = \frac{a}{1-r}$$

where a is the first term and r is the common ratio.

We can express equation (1) as

$$Y(t) = \frac{1}{1 - Bt}.$$

14.4 The Linearty Property of the Generating Function Transformation

Statement : *If $\{y_h^{(1)}\}$ has the generating function Y_1 and $\{y_h^{(2)}\}$ has the generating function Y_2 then $\{\lambda_1 h_h^{(1)} + \lambda_2 h_h^{(2)}\}$ has the generating function $\lambda_1 Y_1 + \lambda_2 Y_2$ for any constants λ_1 and λ_2.*

Proof:

We know that

$$Y_1(t) = y_0^{(1)} + y_1^{(1)} t + y_2^{(1)} t^2 + ... + y_h^{(1)} t^h + ... \quad ...(1)$$

and
$$Y_2(t) = y_0^{(2)} + y_1^{(2)} t + y_2^{(2)} t^2 + ... + y_h^{(2)} t^h + ... \quad ...(2)$$

On multiplying the first equation by λ_1 and second equation by l_2 and on adding we have

$$\lambda_1 Y_1 + \lambda_1 y_2 = \{\lambda_1 y_0^{(1)} + \lambda_2 y_0^{(2)}\} + \{\lambda_1 y_1^{(1)} + \lambda_2 y_1^{(2)}\} t$$
$$+ \{\lambda_1 y_2^{(1)} + \lambda_2 y_2^{(2)}\} t^2 + ... + \{\lambda_1 y_h^{(1)} + \lambda_2 y_h^{(2)}\} t^h + ... + ...$$

This shows that $\lambda_2 Y_1 + \lambda_2 Y_2$ is the generating function of the sequence with general term

$$\lambda_1 y_h^{(1)} + \lambda_2 y_h^{(2)}.$$

Before applying enerating function to the study of difference equation, let us examine one more result of importance. *This gives that if we know the generating function of rth sequence $\{y_h\}$, then we can easily find the generating function of the sequences*

$$\{y_{h+1}\}, \{h_{h+2}\}, ...$$

Suppose Y(t) is the generating function Y_h or,

$$Y(t) = y_0 + y_1 t + y_2 t^2 + ... + y_h t^h + ...$$

This give

$$\frac{Y(t) - y_0}{t} = y_1 + y_2 + ... + y_h t^{h-1} + y_{h+1} t^h + ...$$

This shows that, the function $\left\{\dfrac{Y(t) - y_0}{t}\right\}$ is the generating function for the sequence $\{y_{h+1}\}$.

Similarly, $\dfrac{Y(t) - y_0 - y_1 t}{t^2} = y_2 + y_2 t + y_4 t^2 + ...$ is the generating function for $\{y_{h+2}\}$.

We summarise the results of and the following Table.

S.N.	General Term of Sequence y_h	Generating Function Y(t)
1.	1	$\frac{1}{1-t}$
2.	h + 1	$\frac{1}{(1-t)^2}$
3.	h	$\frac{1}{(1-t)^2}$
4.	h(h + 1)	$\frac{2t}{(1-t)^2}$
5.	B^h	$\frac{1}{1-Bt}$
6.	${}^nC_h B^{n-h}.c^h$, h = 0, 1, 2,...n n > h.	$(B + ct)^n$
7.	y_h	Y(t)
8.	y_{h+1}	$\frac{Y(t) - y_0}{t}$
9.	y_{h+2}	$Y(t) - y_0 - y_1.t$
10.	y_{h+n}	$\frac{Y(t) - y_0 - t - ... - y_{n=1}t^{n-1}}{t^n}$
11.	$\lambda_1 y_h^{(1)} + \lambda_2 y_h^{(2)}$	$\lambda_1 y_1(t) + \lambda_2 y_2(t)$

Example 1:

Find the sequence $\{y_h\}$ having the generating function Y(t) given by

$$Y(t) = \frac{t}{(1-t)^2} + \frac{t}{1-t}.$$ **(Meerut, B.Sc. Hons. 1999)**

Solution:

$$\text{Here } Y(t) = \frac{t}{(1-t)^2} + \frac{(1-t)-1}{1-t} = \frac{t}{(1-t)^2} - 1 + \frac{1}{1-t}.$$

From the table if the generating function is Y(t) $\frac{t}{(1-t)^2}$, the sequence is $\{y_h\} = \{h\}$

and if the generating function is Y(t) = 1, the sequence is $\{y_h\} = 1$. Also if the generating function is

$$Y(1) = \frac{1}{1-t} \text{ the sequence is } \{y_h\} = \{1\}.$$

∴ The general term of the required sequence is

$$y_h = h - 1 + 1 = h$$

∴ Required sequence = $\{y_h\} = \{h\}$.

Aliter. $Y(t) = t(1-t)^{-2} + (1-t)^{-1} = t\{1 + 2t + 3t^2 + ...$

$$+ (h+1)t^h + ...\} + t(1 + t + t^2 + ... + t^h + ...)$$

$$= 0 + 2t + 3t^2 + ... + (h+1)t^h + ...$$

∴ Sequence is given by $y_0 = 0$, $y_h = (h + 1)$. **Ans.**

Example 2:

Find the generating function of the sequence

$$y_h = p^h q$$

Solution:

We have from the fourth result of the table taht the generating function of B is

$$\frac{1}{1 - Bt}$$

Similarly generating function of p is $1/(1 - pt)$.

∴ The generating function of $p^h q$ is $q/(1 - pt)$. **Ans.**

Example 3:

Find the generating function of the sequence y_h if

$$y_h = \frac{h(h-1)}{2}$$ **(Meerut, B.Sc. Hons. 96)**

Solution:

$$\text{Now, } \frac{h(h-1)}{2} = \frac{h}{2}[(h+1) - 2]$$

$$= \frac{h(h+1)}{2} - h.$$

Thus, the required generating function is equal to the difference of the generating functions of two sequences whose general terms are $\frac{h(h+1)}{2}$ and h.

From table generating function of the sequence whose general term is

$$\frac{h(h+1)}{2} \text{ is } \frac{2t}{(1-t)^3}$$

and that for the sequence whose general term is h is

$$\frac{t}{(1-t)^2}$$

$\therefore$ the required generating function

$$= \frac{1}{2} \cdot \frac{2t}{(1-t)^2} - \frac{t}{(1-t)^2}$$

$$= \frac{2t}{(1-t)^3} - \frac{t}{(1-t)^2}$$

$$= \frac{t - t(1-t)}{(1-t)^3} = \frac{t^2}{(1-t)^3}.$$

Ans.

Example 4:

Find the generating function of the sequence y_h if

$y_h = 2 + 3h$ **(Meerut, B.Sc. Hons. 1999)**

Solution:

From the table, the generating function of a sequence whose general term is 1 is $\frac{1}{1-t}$

If the general term is 2 then it is $\frac{2}{1-t}$

In case the general term is h then generating function is

$$\frac{t}{(1-t)^2}.$$

If the general term is 3h then generating function is $\frac{3t}{(1-t)^2}$.

Now, by Linearity property of generating function of required generating function is

$$\frac{2}{1-t} + \frac{3t}{(1-t)^2}.$$

Ans.

Example 5:

Find the sequence $\{y_h\}$ having the generating function given by

$$Y(t) = \frac{3}{1-t} + \frac{1}{1-2t}$$

From the table if the generating function

$$Y(t) = \frac{3}{1-t}, \text{ the sequence is } \{y_h\} = \{1\}$$

$\therefore$ if the generating function is

$$Y(t) = \frac{3}{1-t}, \text{ the sequence is } \{y_h\} = \{3\}$$

Also form the table if the generating function is

$$Y(t) = \frac{1}{1-At}, \text{ the sequence is } \{y_h\} = \{A^h\}$$

$\therefore$ if the generating function is

$$Y(t) = \frac{1}{1-2t}, \text{ the sequence is } \{y_h\} = \{2^h\}$$

Hence, the required sequence is $\{y_h\} = \{3 + 3^h\}$. **Ans.**

Aliter. We have

$$Y(t) = \frac{3}{1-t}, + \frac{1}{1-2t}$$

$$= 3\,(1-t)^{-1} + (1-2t)^{-1}$$

$$= 3\,(1 + t + t^2 + ... + t^h + ...)$$

$$+ (1 + 2t + 2^2t^2 + ... + 2^ht^h + ...)$$

$$= (3 + 1) + (3 + 2)\,t + (3 + 2^2)\,t^2 + ... + (3 + 2^h)\,t^h + ...$$

$$= \sum_{h=0}^{\infty} (3 + 2^h).\, t^h$$

$$= \sum_{h=0}^{\infty} y^h .\, t^h, \text{ where } y_h = 3 + 2^h$$

$\therefore$ The required sequence is $\{y_h\} = \{3 + 2^h\}$. **Ans.**

Example 6:

If $\{y_h\}$ has a generating function Y(t), whow that the generating function fo $\{\Delta y_h\}$ where $\Delta y_h = y_{h+1} - y_h$ is given by

$$\left(\frac{1}{t}\right) [(1-t)\, Y(t) - y_0]$$

Solution:

$\because \quad \Delta y_h = y_{h+1} - y_0.$

Now, from the table the generating function of the sequence whose general term is $y_h = Y(t)$ and that of the sequence whose general term is y_{h+1} is $\dfrac{Y(t) - y_0}{t}$

The required generating function is

$$\frac{Y(t) - y_0}{t} + Y(t)$$

$$= \frac{Y(t) - y_0 = t\ Y(t)}{t}$$

$$= \frac{Y(t)\ (1 - t) - y_0}{t}$$

$$= \left(\frac{1}{t}\right) \{(1 - t)\ Y(t) - y_0].$$ **Ans.**

Example 7:

Find the genrating function of the sequence $\{y_h\}$ *if* $y_0 + y_1 = 0,$ $y_h = 1$ *if* $h > 1.$ **(Meerut M.Sc. 88)**

Solution:

Let the generating function be given be

$$Y(t) = y_0 + y_1 t + y_2 t^2 + ... + y_h t^h + ...$$

Here $\quad y_0 = y_1 = 1$ and $y_h = 1$ if $h > 1$

i.e., $\quad y_2 = y_2 = y_4 = ... = 1$ (each)

$$\therefore \quad Y(t) = t^2 + t^3 + t^4 + ...$$

$$= t^2(1 + t + t^2 + ...) = t^2 (1 - t)^{-1}$$

$$= t^2/(1 - t).$$ **Ans.**

14.5 Generating Function Method of Solving a Linear Differential Equation

(Meerut M. Sc 97, 98)

Consider the differential equation

$$y_{h+1} = Ay_h + C,\ h = 0, 1, 2,... \qquad ...(1)$$

where A and C are constants.

We already know that, if y_0 if prescribed then the unique solution of (1) is given by

$$y_h = \begin{cases} y_0 . A^h + \dfrac{C}{1-A}(1 - A^h) \text{ if } A \neq 1 \\ y_0 + Ch \qquad\qquad \text{if } A = 1 \end{cases} \qquad ...(2)$$

Now, we shall find the solution of (1) with the method of generating functions.

Consider the generating function

$$Y(t) = \sum_{h=0}^{\infty} yh.\, t^h = y_0 + y_1 t + y_2 t^2 + ... \qquad ...(3)$$

Multiplying the diff. equation (1) by t^h and summing from $h = 0$ to ∞, we have

$$\sum_{h=0}^{\infty} yh + 1.t^h = A \sum_{h=0}^{\infty} yh.t^h + C \sum_{h=0}^{\infty} t^h$$

$$\Rightarrow \quad y_1 + y_2 t + {}_3t^2 + ... = A\,(y_0 + y_1 t + y_2 t^2 + ...) + C\,(1 + t + t^2 + ...)$$

$$\Rightarrow \quad \frac{Y(t) - y_0}{t} = A.\, Y\,(t) + C.\, \frac{1}{1-t}$$

$$\Rightarrow \quad Y(t) - At\; Y(t) = y_0 + C\, \frac{1}{1-t}$$

$$\therefore \quad Y(t) = \frac{y_0}{1 - At} + \frac{Ct}{(1-t)\,(1-At)}. \qquad ...(4)$$

$$\Rightarrow \quad Y(t) = \frac{y_0}{1-At} + \frac{C}{1-A}\left\{\frac{1}{(1-t)} - \frac{1}{(1-At)}\right\} \text{ if } A \neq 1$$

(Resolving into Partial Fraction)

$$= y_0(1 - At)^{-1} + \frac{C}{1-A}\,\{(1-t)^{-1} - (1-At)^{-1}\}$$

$$= y_0(1 + At + A^2t^2 + ...) + \frac{C}{1-A}\,\{(1 + t + t^2 + ...)$$

$$-(1 + At + A^2t^2 + ...)\}$$

$$\Rightarrow \quad \sum_{h=0}^{\infty} yh.t^h = y_0 \sum_{h=0}^{\infty} At^h + \frac{1}{1-A} \sum_{h=0}^{\infty} t^h \; \frac{C}{1-A} \sum_{h=0}^{\infty} A^h t^h$$

$$\Rightarrow \quad \sum_{h=0}^{\infty} yh.t^h = \sum_{h=0}^{\infty} \left\{y_0 Al^h + \frac{C}{1-A}(1 - A)\right\} t^h.$$

Equating the coefficients of t^h on the two sides, we have

$$y_h = y_0 A^h + \frac{C}{1-A}(1 - A^h) \text{ if } A \neq 1$$

If A = 1, then from (4), we have

$$Y(t) = \frac{y_0}{1-t} + \frac{Ct}{(1-t)^2}$$

$$= y_0(1 + t + t^2 + ...) + Ct(1 + 2t + 3t^2 + ...)$$

$$\Rightarrow \sum_{h=0}^{\infty} y_h t^h = y_0 \sum_{h=0}^{\infty} t^h + C. \sum_{h=0}^{\infty} ht^h$$

Equating the coefficients of t^h on two sides, we have

$$\therefore \quad y_h = y_0 + Ch \text{ if } A = 1.$$

Hence the solution of (1) is given by

$$y_h = \begin{cases} y_0 A^h + \dfrac{C}{1-A}(1 - A^h) & \text{if } A \neq 1 \\ y_0 + Ch & \text{if } A = 1 \end{cases}$$

Ans.

Example 1:

Apply the generating function technique to solve the intitial value problem

$$y_{h-2} - y_{h-1} + y_h = 0, y_0 = 1, y_1 = 1.$$ **(Meerut, 93, 98)**

Solution:

Consider the generating function

$$Y(t) = \sum_{h=0}^{\infty} y_h . t^h = y_0 + y_1 t + y_2 t^2 + y_3 t^3 + ... \quad ...(1)$$

Multiplying the given diff. equation by t^h and summing from h = 0 ¥, we have

$$\sum_{h=0}^{\infty} y_{h+2} . t^h - \sum_{h=0}^{\infty} y_{h12} t^h - \sum_{h=0}^{\infty} y_h t^h = 0$$

$$\Rightarrow \quad (y_2 + y_3 t + y_4 t^2 + ...) - (y_1 + y_2 t + y_3 t^2 + ...) - Y(t) = 0$$

$$\Rightarrow \quad \frac{Y(t) - y_0 - y_1 t}{r^2} - \frac{Y(t) - y_0}{t} \; Y(t) = 0.$$

$$\Rightarrow \quad Y(t) - y_0 \, ty_1 - t\,[Y(t) - y_0] - t^2 \, Y(t) = 0.$$

$$\Rightarrow \quad (1 - t - t^2) \, Y(t) = (1 - t) \, y_0 + ty_1$$

$$= t \qquad \because \; y_0 = 0 \text{ and } y_1 = 1$$

$$\therefore \quad Y(t) = \frac{t}{1-t-t^2}$$

$$= \frac{t}{\left(\frac{\sqrt{5}+1}{2}+t\right)\left(\frac{\sqrt{5}-1}{2}-t\right)}$$

$$= \frac{\frac{(\sqrt{5}+1)}{10}\sqrt{5}\;\frac{(\sqrt{5}-1)}{10}\sqrt{5}}{\left(\frac{\sqrt{5}+1}{2}+t\right)\left(\frac{\sqrt{5}-1}{2}-t\right)}$$

(Resolving in to P.F)

$$= -\frac{\sqrt{5}}{5}\left[1+\frac{2}{\sqrt{5}+1}.t\right]^{-1} + \frac{\sqrt{5}}{5}\left[1-\frac{2}{\sqrt{5}-1}.t\right]^{-1}$$

$$= -\frac{\sqrt{5}}{5}\left[1+\left(\frac{\sqrt{5}+1}{2}\right)t\right]^{-1} + \frac{\sqrt{5}}{5}\left[1-\left(\frac{\sqrt{5}-1}{2}\right).t\right]^{-1}$$

$$\sum_{h=0}^{\infty} y_h t^h = -\frac{\sqrt{5}}{5}\sum_{h=0}^{\infty}(-1)^h.\left(\frac{\sqrt{5}-1}{2}\right)^h t^h = \frac{\sqrt{5}}{5}\sum_{h=0}^{\infty}\left(\frac{\sqrt{5}+1}{2}\right)^h.t^h$$

Equationg the coefficients of t^h, on the two sides we have

$$y_h = -\frac{\sqrt{5}}{5}(-1)^h.\left(\frac{\sqrt{5}-1}{2}\right)^h + \frac{\sqrt{5}}{5}\left(\frac{\sqrt{5}+1}{2}\right)^h$$

$$\Rightarrow \qquad y_h = -\frac{\sqrt{5}}{5}\left(\frac{1-\sqrt{5}}{2}\right)^h + \frac{\sqrt{5}}{5}\left(\frac{1+\sqrt{5}}{2}\right)^h. \qquad \textbf{Ans.}$$

Example 2:

Apply the generating function method to solve the following L.D. equation.

$y_{h+2}..2y_{h-1} + y_h = 2^h$. *with* $y_0 = 2$ *and* $y_1 = 1$.

(Meerut, M.Sc. 97)

Solution:

Consider generating function

$$Y(t) = \sum_{h=0}^{\infty} y_h.t^h \quad y_0 + y_1 t + y_2 t^2 + y_2 t^3 + ... \qquad ...(1)$$

Multiplying the given diff. equation by t^h and summing from h = 0 to ∞, we have

$$\sum_{h=0}^{\infty} y_{h+2} t^h - 2\sum_{h=0}^{\infty} y_{h+1} t^h + \sum_{h=0}^{\infty} y_h t^h = \sum_{h=0}^{\infty} 2^h . t^h$$

$$\Rightarrow \quad (y_2 + y_3 t + y_4 t^2 + ...) - 2(y_1 + y_2 + y_3 t^2 ...) + Y(t) = 1 + 2t + (2t)^2 + ..$$

$$\Rightarrow \quad \frac{Y(t) - y_0 - y_1 t}{t^2} - 2\frac{Y(t) - y_0}{t} + Y(T) = (1 - 2t)^{-1}$$

$$\Rightarrow \quad (1 - 2t + t^2)\, Y(t) - (1 - 2t)\, y_0 - t y_1 = \frac{t^2}{(1 - 2t)}$$

$$\Rightarrow \quad (1 - t)^2\, Y(t) = 2(1 - 2t) + t + \frac{t^2}{(1 - 2t)}.$$

$\because y_0 = 2$ and $y_1 = 1$

$$\therefore \quad Y(t) = \frac{2 - 3t}{(1 - t)^2} + \frac{t^2}{(1 - t)^2 (1 - 2t)}$$

$$= \frac{3(1 - t) - 1}{(1 - t)^2} + \left[-\frac{1}{(1 - t)^2} + \frac{1}{1 - 2t}\right]$$

(Resolving in to PF)

$$= \frac{3}{1 - t} - \frac{1}{(1 - t)^2} - \frac{1}{(1 - t)^2} + \frac{1}{1 - 2t}$$

$$= \frac{3}{1 - t} - \frac{2}{(1 - t)^2} + \frac{1}{1 - 2t}$$

$$= 3(1 - t)^{-1} - 2(1 - t)^{-2} + (1 - 2t)^{-1}$$

$$= 3(1 + t + t^2 + ...) - 2(1 + 2t + 3^2 + ...)$$
$$+ (1 + 2t + 2^2 t^2 + ...)$$

$$\Rightarrow \quad \sum_{h=0}^{\infty} y_h . t^h = 3\sum_{h=0}^{\infty} t^h - 2.\sum_{h=0}^{\infty} (h + 1)\, t^h + \sum_{h=0}^{\infty} 2^h t^h.$$

Equating the coefficient of t^h, on the two sides, we have

$$y_h = 3 - 2(h + 1) + 2^h.$$

Example 3:

Solve $y_{n+1} = 2y_n + 3y_0 = 1$ *given that the generating function of*

(a) $f(n + 1)$ *is* $\frac{1}{n}[F(n) - F(0)]$

(b) $f(n)$ *is* $F(n)$

(c) *1 is* $\frac{1}{1-n}$ $|n| < 1$

(d) a^n is $\frac{1}{1-an}, |n| < \frac{1}{a}$. **(Meerut, B.Sc. Hons. 98)**

Solution:

With the help of the given generating function, from the given diff. equation, we have

$$\frac{1}{n}[Y(n) - Y(0)] = 2Y(n) + 3.\ \frac{1}{1-n} \quad ...(1)$$

where $Y(n) = y_0 + y_1.\ n + y_2.n^2 + ..., Y(0) = y_0 = 1$

$\therefore$ From (1), we have

$$Y(n) - 1 = 2n\ Y(n) + \frac{3n}{1-n}$$

$$\Rightarrow \qquad (1-2n)\ y(n) = 1 + \frac{3n}{1-n} = \frac{1+2n}{1-n}$$

$$\therefore \qquad Y(n) = \frac{1+2n}{(1-n)(1-2n)}$$

$$\Rightarrow \qquad Y(n) = -\frac{3}{1-n} + \frac{4}{1-2n}.$$

$\therefore$ From the given generating functions we have

$y_n = -3.1 + 4.2^n$. **Ans.**

EXERCISES

1. Find the general solution of the following difference equation and then find the particular solution for which $y_0 = 1, y_{h+1} - (h+1)\ y_h = 0$.

   Ans. $y_h = ch!, h!$

2. Find the sequence $\{y_h\}$ having the generating function Y given by

   (i) $\frac{t}{1-2t}$, (ii) $\frac{t^2}{1-t}$ (iii) $(3-t)^n$.

   Ans. (i) $y_0 = 0, y_h = 2h^{-1}$, if $h > 0$,

   (ii) $y_0 = y_1 = 0, y_h = 1$, if $h > 1$,

   (iii) $y_h = \left(\frac{n}{h}\right) (-1)^h, 3^{n-h}$

   if $h = 0, 1, 2, ...n$; yn 0 if $h > n$.

3. Apply the generating function technique to solve the following initial value problems.

(i) $y_{h+1} - 5y_{h+1} + 6y_h - 2$, if $y_0 = 1, y_1 = 2$.

(Meerut, M.Sc. 9; B.Sc Hons. 96)

(ii) $y_{h+1} - 3y_{h+1} + 2y_h = 0$, if $y_0 = 2, y_1 = 3$

(iii) $y_{h+2} - 6y_{h+1} + 8y_h = 0$, if $y_0 = 0, y_1 = 2$

(iv) $y_{h+2} + 4y_{h+1} + 4y_h = 0$, if $y_0 = 1, y_1 = 0$

(v) $y_{h+2} + 2y_{h+1} + y_h = 1 + h$.

Ans. (i) $y_h = 1 - 2^h + 3^h$,

(ii) $y_h = 1 + 2^h$

(iii) $y_h = 2^h (2^h - 1)$,

(iv) $y_h + 2^h (1 - h)$

(v) $y_h = (C_1 + C_2 h)(-1)^h + \frac{1}{6}(2h - 1)$.

4. Solve $y_{k+1} - 2y_{k+1} + y_k = 2^k$, by the method of generating function given that

General term of Sequence	Generating Function
y_k	$Y(s)$
y_{k+1}	$\frac{Y(s) - y_0}{s}$
y_{k+1}	$\frac{Y(s) - y_0 - y_1 s}{s^2}$

(Meerut, M.Sc. 1999)

Ans. $y_h = C_1 + C_2 h + 2^h$.